The Adrenal Gland

CLINICAL SURVEYS IN ENDOCRINOLOGY

C. R. Kannan, M.D.

Volume 1 THE PITUITARY GLAND
Volume 2 THE ADRENAL GLAND

A Continuation Order Plan is available for this series. A continuation order will bring delivery of each new volume immediately upon publication. Volumes are billed only upon actual shipment. For further information please contact the publisher.

The Adrenal Gland

C. R. Kannan, M.D.

Chairman
Division of Endocrinology and Metabolism
Department of Medicine
Cook County Hospital
and Associate Professor of Medicine
Rush–Presbyterian -St. Luke's Medical Center
Chicago, Illinois

PLENUM MEDICAL BOOK COMPANY
NEW YORK AND LONDON

Library of Congress Cataloging in Publication Data

Kannan, C. R. (Charkravarthy R.), 1943–
   The adrenal gland / C. R. Kannan.
      p.   cm. — (Clinical surveys in endocrinology; v. 2)
   Includes bibliographies and index.
   ISBN-13: 978-1-4612-8286-0   e-ISBN-13: 978-1-4613-1001-3
   DOI: 10.1007/978-1-4613-1001-3
   1. Adrenal gland — Diseases. I. Title. II. Series.
   [DNLM: 1. Adrenal Gland Diseases. 2. Adrenal Glands. W1 CL795UK v.
2 / WK 700 K16a]
RC659.K36   1988
616.4′5 — dc19
DNLM/DLC   88-19690
for Library of Congress   CIP

© 1988 Plenum Publishing Corporation
Softcover reprint of the hardcover 1st edition 1988
233 Spring Street, New York, N.Y. 10013

Plenum Medical Book Company is an imprint of Plenum Publishing Corporation

To S.V. I dedicate this tome — a small return for all I owe.

Preface

This volume, *The Adrenal Gland*, is the second in the Clinical Surveys in Endocrinology series. Like its predecessor on the pituitary gland, this work is written with one purpose in mind—to view the vast, relevant adrenal literature through a clinician's eyes. The intricate, and often complex, interrelationship between the clinical and research perspectives of "adrenology" poses a challenge. This is, in part, due to the commonly held belief that the milieux of steroid hormone research and clinical medicine are parallel phenomena, not destined to meet. But the twain do meet, and often with relative ease, when viewed as twin facets of the same gem. The view presented in this work is from the vantage point of the clinical endocrinologist who applies the research literature to understand adrenal diseases more clearly.

Adrenal pathology is arguably the most fascinating of all endocrinopathies. The images of patients suffering from adrenal diseases are of kaleidoscopic quality: the newborn child with ambiguous genitalia, in whom the very first ritual of assigning sex becomes shrouded with uncertainty; the revitalized patient with hitherto undiagnosed Addison's disease, who but for the cognitive powers of the endocrinologist would have ultimately succumbed, undiagnosed; the virilized female with adrenal tumor and its attendant onslaught on the body and mind; the febrile patient with pheochromocytoma masquerading as fever of undetermined origin for months. On a personal note, my decision to make endocrinology a career choice was due to the striking image of a moribund patient who would most certainly have died of undiagnosed Addisonian crisis, but for astute clinical observation and therapeutic intervention by an endocrinologist. These images form the focus of discussion in several chapters of this book.

Having learned and having taught at a large urban hospital for almost 20 years have indelibly colored my thinking and writing. Because of the richness of the clinical setting at Cook County Hospital, the reader will find that my focus is almost never shifted away from the patient. In addition to the pervading clinical theme, the historical and pathophysiological backgrounds have been explored. For instance, the life and times of Dr. Thomas Addison can be compelling and intriguing, as well as enlightening for the physician faced with the vicissitudes of the disease that Addison described. The pathophysiology of disease has been analyzed in depth, particularly in chapters dealing with primary hyper-

viii PREFACE

aldosteronism and Cushing's syndrome. Finally, the needs of the reader involved in research and teaching have been met by the approximately 2000
references provided.

The single authorship of this work notwithstanding, I am deeply indebted
to several people. As always, Plenum Publishing Corporation is an author's
dream, because of the care and attention to detail provided by their staff. I am
grateful to Mrs. Janice Stern, Senior Medical Editor at Plenum, for supporting
the concept of the Clinical Surveys series. My thanks to Ms. Gayla Blake, who
worked on the text with incomparable patience and care, and Mrs. Carolyn
Leach, who painstakingly typed the references. I am grateful to Dr. Gerald
Burke, Chairman of Medicine, for providing constant encouragement. And, of
course, I am grateful to my patients, from whom I have learned, and to my
family—Molly, Ashley, Alexander, and Margaret—for being patient with me.

C. R. KANNAN

Chicago, Illinois

Contents

The Adrenal Gland

The Glucocorticoid Hormones

Introduction

It has been said that not all hormones are created equal; some hormones are necessary for life, while other hormones make life worth living. Cortisol is a hormone essential for life. The characterization of the structure of various adrenal hormones and the synthesis of these hormones for therapeutic use represent major strides during the fourth and fifth decades of this century. The synthesis of glucocorticoids by the adrenal cortex and the regulatory mechanisms that govern the interplay between the adrenal cortex and the hypothalamic pituitary unit are fascinating examples of physiology applied to clinical medicine. This chapter focuses on the synthesis, transport, regulatory control, and physiological actions of glucocorticoids. The application of these principles to the diagnostic testing of patients with disordered adrenocortical function is also discussed.

Synthesis of Glucocorticoids

Cortisol is the biologically active glucocorticoid synthesized by the adrenal cortex. The steroidogenic pathway for cortisol synthesis is outlined in Figure 1. Five major steps are involved in the formation of cortisol from cholesterol:

1. Conversion of cholesterol to pregnenolone; this crucial initial step is mediated by an enzyme called cholesterol side-chain–cleaving enzyme.
2. The pregnenolone, thus formed, undergoes two major reactions: formation of 17-OH pregnenolone by 17-hydroxylation, or conversion to progesterone by the enzyme 3-β-hydroxysteroid dehydrogenase.
3. The next step in cortisol biosynthesis involves formation of 17-OH progesterone, which can be derived in one of two ways: either from progesterone (by 17-hydroxylation) or from 17-OH pregnenolone (by the action of 3-β-hydroxysteroid dehydrogenase.
4. 17-OH progesterone is further hydroxylated at C-21 position by the enzyme 21-hydroxylase to form 11-deoxycortisol (or compound S).
5. The final step in cortisol biosynthesis is the conversion of 11-deoxycortisol to cortisol by hydroxylation at C-11 position. This action is mediated by 11-β-hydroxylase.

The simplistic, easy-to-remember two-dimensional scheme in biosynthesis of steroid hormones in general, and the glucocorticoid hormone cortisol in particular, has undergone some conceptual reevaluations in the recent past.[1] Before examining these concepts, a general overview of the principles that underlie adrenal steroidogenesis—particularly in the zona fasciculata—are worthy of comment.

1. The zona fasciculata participates in two pathways of adrenal steroidogenesis—the 17-hydroxy pathway and the 17-deoxy pathway. The 17-hydroxy pathway involves the synthesis of cortisol from pregnenolone, 17-OH pregnenolone, 17-OH progesterone, and 11-deoxycortisol; the 17-deoxy pathway involves the production of corticosterone from pregnenolone, progesterone, and 11-dexoycorticosterone. While the zona fasciculata is the predominant, and the only, site for cortisol biosynthesis, 11-deoxycorticosterone synthesis takes place in both the zona glomerulosa and the zona fasciculata. However, the conversion of 11-deoxycorticosterone to aldosterone, the most potent mineralocorticoid, can take place only in the zona glomerulosa.
2. The five enzymes that are crucial for cortisol biosynthesis, in order of biosynthetic sequence, are the cholesterol side-chain–cleaving enzyme,

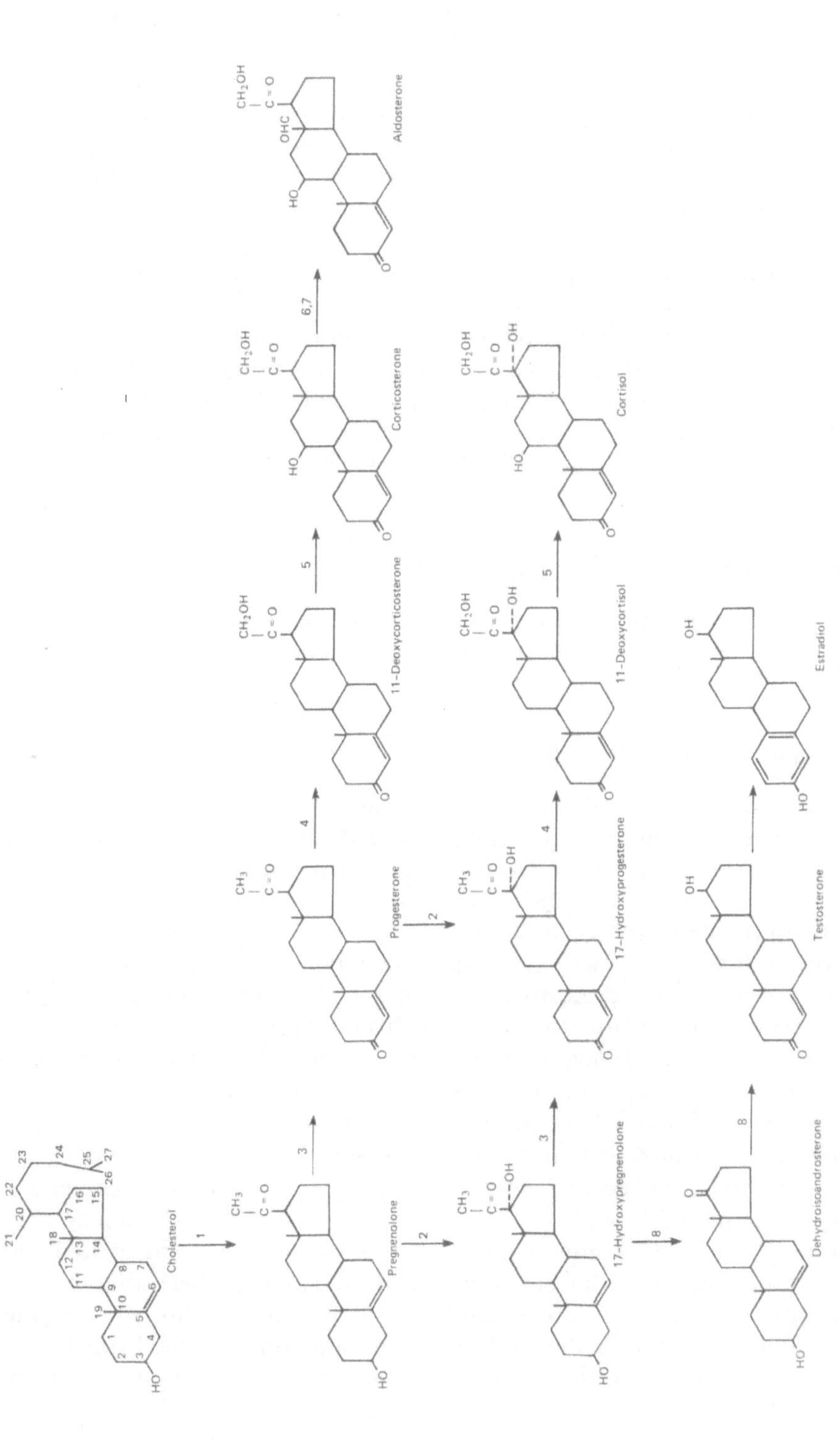

FIGURE 1. Pathways of stereoidogenesis. The enzymes involved in synthetic steps of the pathways are: 1. Cholesterol side-chain cleaving enzyme, 2. C-17-hydroxylase, 3. 3-β-hydroxysteroid dehydrogenase, 4. C-21-hydroxylase, 5. 11-β-hydroxylase, 6. C-18-hydroxylase, 7. 18-hydroxysteroid oxidase, and 8. C-17, C-20 lyase.

3-β-OH steroid dehydrogenase, 17-α-hydroxylase, 21-hydroxylase, and 11-β-hydroxylase. It is controversial whether the enzymes present in the zona fasciculata are identical to the enzymes present in the zona glomerulosa (the "one-enzyme" versus the "two-enzyme" theory).

3. The traditional view of the steroidogenic process holds that a single initial step, i.e., conversion of cholesterol to pregnenolone, regulates the entire chain of steroidogenic events. It is almost axiomatic that all adrenocorticotropic hormones (ACTH, angiotensin II, and rarely luteinizing hormone) exert their trophic effects by stimulating this rate-limiting step.

4. The traditional concept also assumes that steroidal intermediates are passed back and forth between the mitochondria (where certain enzymes are present) and the microsomes (where other enzymes are located).

5. The three steroidogenic enzymes located in the mitochondria are the cholesterol side-chain–cleaving enzyme, the C-11 hydroxylase, and the C-18 hydroxylase. The steroidogenic enzymes located in the microsomes are C-17 hydroxylase, C-21 hydroxylase, and the aromatase.

Enzymes Involved in Steroidogenesis

The steroidogenic processes in the adrenal cortex can be viewed from the perspective of the individual enzymes involved in each step.

Cholesterol Side-Chain–Cleaving Enzyme

The enzyme cholesterol side-chain–cleaving enzyme, or desmolase, is located in the inner membrane of the mitochondrial cristae. Sequential hydroxylation of cholesterol by this enzyme results in the formation of pregnenolone via a variety of intermediate compounds. These intermediate compounds include (22 R)-22-hydroxycholesterol, (20 S)-20-hydroxycholesterol, and (20 R, 22 R)-20, 22-dihydroxycholesterol as well as isocaproaldehyde.[2,3] The major intermediate metabolite seems to be (22 R)-hydrocholesterol. The side-chain–cleaving enzyme complex has the properties of a cytochrome P-450. This enzyme can be effectively blocked by aminoglutethimide.[4] Considerable controversy revolves around the role of these intermediate products in the formation of pregnenolone from cholesterol. Several studies have supported the notion that pregnenolone can be formed without the formation of stable cholesterol intermediates that are hydroxylated at C-20 and C-22 positions.[5-7] Using cholesterol analogs that cannot undergo hydroxylation, these workers[5-7] were able to demonstrate production of pregnenolone when these analogs were incubated with acetone powders of bovine adrenal mitochondria. The hypothesis that the true intermediates in cholesterol conversion to pregnenolone are transient, enzyme-bound complexes and that the hydroxylated compounds formed during the reaction are mere by-products of the reaction has not found complete acceptance.

3-β-Hydroxysteroid Dehydrogenase

This important enzyme is responsible for conversion of Δ^5 compounds to Δ^4 compounds. The major Δ^5 compounds are pregnenolone, 17-OH pregnenolone, DHEA, and Δ^5 androstenediol. The enzyme 3-β-hydroxysteroid dehydrogenase is necessary to effect conversion of these compounds into progesterone, 17-OH progesterone, androstenedione, and testosterone, respectively.

The action of 3-β-HSD is intimately linked with the presence of NAD^+ as cofactor. In the glucocorticoid pathway, 3-β-HSD with NAD^+ as cofactor converts 17-α-OH pregnenolone to the keto derivative of 17-α-OH pregnenolone called 17-hydroxypregn-5-ene-3,20-dione, which, by the action of Δ^5-Δ^4 isomerase, becomes 17-hydroxyprogesterone.

The importance of 3-β-HSD enzyme lies in the fact that such activity is crucial for the formation of progesterone and 17-OH progesterone (both Δ^4 compounds) from their Δ^5 precursors pregnenolone and 17-OH pregnenolone. 3-β-HSD activity is not limited to the adrenals or gonads exclusively. Such activity is found in extraadrenal and extragonadal locations, particularly the liver.[8]

17-α-Hydroxylase

In the conventional scheme for adrenal steroidogenesis, the pathways for both cortisol and the C-19 androgens is initiated by the 17-hydroxylation of pregnenolone or progesterone. 17-α-Hydroxylase catalyzes the formation of 17-α-OH pregnenolone and 17-α-OH progesterone. The intermediates formed by the reaction of 17-hydroxylase on its substrates are 17-hydroxy-20-ketosteroids. These intermediates subsequently are C-21 hydroxylated to form glucocorticoids or cleaved between C-17 and C-20 to form C-19 products. The 17-hydroxylase enzyme is a microsomal enzyme, its presence limited to the fasciculata–reticularis zones. This enzyme is also dependent on cytochrome P-450. The 17-hydroxylase enzyme is subject to trophic regulation and has been shown to be activated by ACTH.[9] There is accumulating evidence in the literature to suggest that 17-hydroxylation of progesterone to 17-OH progesterone and the cleavage of 17-OH progesterone to form androstenedione can be catalyzed by a single enzyme system containing a single cytochrome P-450. However, the concept that suggests the existence of isoenzymes of 17-hydroxylase, one for the formation of 17-hydroxylated corticosteroids and another for the formation of the 17-oxygenated C-19 androgen, is gaining popularity.

21-Hydroxylase

21-Hydroxylase enzyme converts 17-α-OH progesterone to 11-deoxycortisol and progesterone to 11-deoxycorticosterone. C-21 hydroxylation is catalyzed by a cytochrome P-450 located in the smooth endoplasmic reticulum. A large body of evidence has been accumulating in the literature that suggests the existence of more than one 21-hydroxylating system. This implies that different hormonal products—cortisol, aldosterone, and androgens—are produced through the mediation of different isoenzymes of 21-hydroxylase. This assumption has greatly facilitated explaining the heterogeneity in the expression of congenital deficiency of 21-hydroxylase enzyme. Thus, the "simple virilizing" form of congenital adrenal hyperplasia due to 21-hydroxylase deficiency reflects absence of the enzyme that converts 17-α-OH progesterone to 11-deoxycortisol in the zona fasciculata, with preservation of 21-hydroxylase activity in the zona glomerulosa. In contrast, patients with the salt-losing form suffer from deficiencies of both these hydroxylating enzymes in both the zona fasciculata as well as the zona glomerulosa, the latter precluding the formation of 11-deoxycorticosterone and the subsequent mineralocorticoid products of the zona glomerulosa. This clinical observation has been buttressed by ample laboratory studies that have attempted to isolate separate and different isoenzymes of 21-hydroxylase from the zona glomerulosa and the zona fasciculata.[10−13]

11-β-Hydroxylase

This final step in the glucocorticoid pathway results in conversion of 11-deoxycortisol (compound S) to cortisol (compound F). As with the 21-hydroxylase system, there appear to be separate isoenzymes for this reaction in the zona fasciculata and in the zona glomerulosa. The 11-hydroxylase enzyme is P-450 dependent and can be entirely blocked by the administration of metyrapone.

The central glucocorticoid pathway proceeds from pregnenolone by the formation of 17-α-hydroxypregnenolone. This reaction takes place in the smooth endoplasmic reticulum and requires 17-hydroxylase and cytochrome P-450. The conversion of 17-hydroxypregnenolone to 17-hydroxyprogesterone involves the concerted interplay between 3-β-HSD, NAD$^+$, and an isomerase. 17-OH-progesterone is converted to compound S by the action of cytochrome P-450-dependent 21-hydroxylase, and compound S is converted to cortisol by another cytochrome P-450-dependent 11-hydroxylase.

Recent Concepts in Steroidogenesis

Several concepts have recently been added to modify the conventional schema of adrenal steroidogenesis. Lieberman et al.[1] have garnered several lines of clinical and laboratory evidence to support the notion that adrenal steroidogenesis is not as simplistic or one-dimensional as it has been defined in the past. The authors have proposed that the processes of steroidogenesis are confined within biosynthetic functional units referred to as "hormonads." These workers have eloquently utilized the hormonad hypothesis to underscore several concepts— that steroidogenesis occurs through concerted processes that do not involve the release of stable intermediates; that isoenzymes are present for several hydroxylating enzyme systems; that steroidal end products may indeed have their own specific sterolic precursors; and that early as well as "late" steps of steroidogenesis are influenced by trophic hormones.

The role of the following phenomena in adrenal steroidogenesis continues to remain unclarified, elusive, or controversial:

1. Presence of sulfated precursors. Aside from the free form, steroids within the adrenal cortex exist in the form of steroid sulfate (the most prototypical of which is DHEA sulfate) and in the form of lipoidal derivatives. It is well known

that cholesterol sulfate can serve as a substrate for the formation of preg-nenolone sulfate. Fetal adrenal tissue, in particular, has the ability to form sul-fated steroids from cholesterol sulfate.[14] Thus, DHEA sulfate is a major secreto-ry product of fetal adrenals, quantitatively exceeding the amount of cortisol produced in this situation. Clearly, the metabolism of cholesterol sulfate is regu-lated differently in fetal and adult adrenal tissue. Although quantitatively less impressive, mature adrenal tissue can be shown to synthesize pregnenolone sulfate, 17-OH pregnenolone sulfate, 17-OH progesterone sulfate, and DHEA sulfate, both in vitro and in vivo.[15-18] The importance of cholesterol sulfate as a precursor for steroidogenesis has not been clearly elucidated. Similarly, the presence of extractable lipoidal derivatives of pregnenolone in bovine adrenal tissue has not found satisfactory explanation. Whether the sulfated or acyl esters of steroids serve as precursors for specific end products is also unclear.

2. Presence of binding proteins for steroids. Another intriguing phe-nomenon is the finding of binding proteins in the adrenal cortex of certain species. Strott and Lyons[19] have described a protein that binds pregnenolone sulfate in the adrenal cortex of the guinea pig. This protein is apparently differ-ent from the protein that binds pregnenolone. While the function of this protein is ill understood, it implies different mechanisms for the metabolism of preg-nenolone and pregnenolone sulfate.

3. Isoenzymes for cholesterol side-chain–cleaving enzyme. It has been shown that multiple forms of cholesterol side-chain–cleaving enzymes exist with-in the adrenal cortex. Considerable controversy exists regarding the meth-odology involved in demonstrating isoenzymes of the side-chain–cleaving en-zymes. Based on substrate affinity and kinetics, there appear to be at least two forms of this enzyme in bovine adrenal tissue.

4. Isoenzymes for the hydroxylases. Extraction studies performed on iso-lated adrenocortical tissue have suggested that the 18-hxdroxylase that catalyzes the 18-hydroxylation of DOC is different from that which catalyzes the 18-hydroxylation of corticosterone. It is of interest to note that the purified cyto-chrome P-450 that catalyzes the 11-β-hydroxylation of DOC is also involved in the in vitro 18-hydroxylation of DOC as well as the 19-hydroxylation of an-drostenedione, but not the 18-hydroxylation of corticosterone. This reaction is mediated by a different hydroxylase system probably located in the inner mem-brane of the mitochondria from the zona glomerulosa. Isoenzymes for 11-β-hydroxylase, 17-hydroxylase, and 21-hydroxylase probably exist, implying sepa-rate enzyme systems involved in the synthesis of separate sets of steroid hor-mones. This concept has gained momentum, based on clinical examples of vari-ous forms of congenital adrenal hyperplasia (Chap. 11).

5. The "two enzyme–two-gland theory." The theory that the zona fas-ciculata and the zona glomerulosa function as though they were two separate glands under separate trophic hormone control has been proposed by New and Seaman.[20] The adrenal zona glomerulosa and its synthetic products are pri-marily controlled by angiotensin II, while those of the zona fasciculata are con-trolled by ACTH. The catalytic activity for 11-β-hydroxylation, 18-hydroxyla-tion, and 18-hydrogenation resides in a single protein, encoded by a single gene,[21,22] notwithstanding the fact that mitochondria from bovine adrenal glomerulosa definitely demonstrate a greater activity for the conversion of cor-

ticosterone to aldosterone, when compared to the mitochondria from the adrenal fasciculata. It is generally agreed that any aldosterone extracted from the adrenal fasciculata in all probability represents contamination from the glomerulosa, since separation of the layers is technically difficult. The conversion of corticosterone to aldosterone is not expressed in the mitochondria of the fasciculata, while being enhanced in the glomerulosa. The adequate preservation of all steps involved in aldosterone synthesis in patients with the simple virilizing form of 21-hydroxylase deficiency implies selective involvement of the 21-hydroxylase system within the zona fasciculata with sparing of the 21-hydroxylase system within the zona glomerulosa. While such a theory admirably explains isolated enzyme deficiencies involving one or another part of the adrenal cortex, several recently reported findings have posed difficulties in reconciliation. For instance, the functional significance of 18-hydroxydeoxycorticosterone (18-OH-DOC) synthesized by the zona fasciculata has eluded explanation. The very fact that the zona fasciculata—a layer principally involved in the production of cortisol (and corticosterone)—should be involved in synthesizing 18-OH-DOC is intriguing.[23,24] The 18-OH-DOC synthesized by the fasciculata responds like cortisol and corticosterone to stimulation by ACTH and is unaffected by angiotensin.[25] Similarly, the functional significance of finding unusual steroids such as 18-hydroxycortisol,[26] 18-oxocortisol,[27] and 19-nordeoxycorticosterone in the zona glomerulosa has also defied explanation.

In summary, much remains unknown regarding the intricate details that involve adrenal steroidogenesis. The clarification of facts that are conflicting and confusing depends on evaluating newer approaches that analyze steroidogenesis in a different light—a perspective that focuses on the complexities, rather than the simplicity, of steroidogenesis.

Secretion, Release, and Transport

The major stimulus for glucocorticoid secretion is ACTH. Under the influence of this hormone, adrenal steroidogenesis takes place through all the necessary enzymatic steps. After secretion and release, cortisol becomes bound to a plasma globulin called transcortin. Transcortin has an extremely high affinity for cortisol (75% of all cortisol). Transcortin also binds to a variable extent with other steroids, such as progesterone, corticosterone, deoxycorticosterone, and deoxycortisol. The binding of cortisol to transcortin is reversible. Cortisol is also bound to albumin (15%). Approximately 10% of cortisol is free. It is the free fraction that moves into the cell to exert its action. The measurement of plasma cortisol (either by RIA or by fluorometric methods) measures only the cortisol bound to transcortin.[28,29] The plasma cortisol determination by radioimmunoassay has replaced the fluorometric methods for cortisol in plasma. Conditions that alter the transcortin would alter cortisol measurements. Thus, pregnancy, estrogen therapy, and obesity are associated with elevated cortisol levels due to elevated transcortin levels.

The glucocorticoids are mainly inactivated in the liver. By a series of reduction reactions, cortisol is converted to dihydro- and tetrahydrocortisol, which is then processed through the glucuronyl transferase system to become tetrahydrocortisol glucuronide. Similarly, dexoycortisol is excreted as tetrahydrodeoxy-cortisol. These two metabolites are measured as 17-hydroxycorticosteroids in the urine (Porter–Silber chromogens). Approximately 30% of cortisol secreted is metabolized to tetrahydrocortisol glucuronide. The collective measurement of 17–hydroxycorticosteroids in the urine, in general, reflects glucocorticoid activity, even though tetrahydrodeoxycortisol is not a biologically active compound (compound S). If one wishes to extract or fractionate the amount of tetrahydrocortisol from the tetrahydrodeoxycortisol, this can be done by extractions using carbon tetrachloride. In practice, however, this is not generally necessary, since the contribution of 11-deoxycortisol to the urinary 17-hydroxycorticoids is minimal. The 17-hydroxycorticoids can be affected by drugs such as spironolactone, dilantin, estrogens, and exogenous steroid administration.

In normal individuals, cortisol is secreted by the adrenal cortex in a series of bursts rather than in a steady, continuous fashion.[30,31] If 24-hr profiles of serum cortisols are obtained by measuring the cortisol level in the plasma every 20 or 30 min, 7–20 secretory spikes can be identified. This phenomenon of pulsatile or episodic secretion should be taken into account when interpreting a single sample of plasma cortisol drawn in an isolated period of time. In general, however, the amplitude of the spikes are greater in the early-morning period around midnight. The cortisol spikes are a reflection of ACTH spikes that precede them.

Regulation of Cortisol Secretion

Glucocorticoid secretion by the adrenal cortex is under the trophic control of ACTH. Regulation of cortisol secretion can be viewed from two perspectives: the trophic control exerted by ACTH on the adrenal cortex, and the negative feedback effect of glucocorticoids on the hypothalamic pituitary axis. The former is relatively straightforward, while the latter is characterized by considerable complexity.

Effect of ACTH on Cortisol Secretion

The predominant action of ACTH is stimulation of the growth and function of the adrenal cortex. Thus, both weight maintenance and steroidogenesis are dependent on ACTH. In the absence of ACTH, the adrenal glands undergo atrophy, a phenomenon that can be reversed by administering ACTH. In fact, the early bioassays for ACTH were based on the weight-maintaining activity of the sera administered to hypophysectomized rate. On the functional end, ACTH promotes the synthesis and release of all the adrenocortical hormones, particularly the glucocorticoids. While all steps of adrenal steroidogenesis are activated, the major step influenced by ACTH is conversion of cholesterol into pregnenolone. It is believed that cAMP is the intracellular mediator of the steroidogenic action of ACTH. There is controversy as to whether low- or high-affinity receptors in the adrenal cortex are linked to the activation of cAMP following ACTH stimulation.

While the secretion of glucorcorticoids and androgens is entirely dependent on the trophic action on ACTH, the secretion of mineralocorticoids by the zona glomerulosa is not. The predominant stimulus for mineralocorticoid secretion is the renin–angiotensin–aldosterone system. The zona glomerulosa has receptors for both angiotensin II and ACTH. Administration of pharmacological doses of ACTH is followed by a prompt increase in aldosterone levels.

The adrenal sensitivity to ACTH forms the basis for the ACTH stimulation tests used in clinical practice. Clearly, one of the most important factors that modify adrenal sensitivity to exogenous ACTH administration is the level of circulating endogenous corticotropin.[32] Acute ACTH deprivation leads to an impaired adrenocortical response to rapid ACTH administration within 8–12 days.[33] With chronically diminished endogenous ACTH secretion (either from ACTH deficiency or from ACTH suppression due to glucocorticoid therapy) the adrenal responsiveness to exogenous ACTH becomes attenuated, even abolished.[33] Thus, indirectly, the demonstration of a completely normal adrenal secretory response to rapid ACTH administration also indicates adequacy of endogenous corticotropin, except in the case of very early ACTH deficiency. The decreased responsiveness of the adrenal cortex in conditions characterized by diminished ACTH secretion assumes major importance in patients treated with glucocorticoids (Chap. 4).

Effect of Cortisol on the HP Axis

The inhibitory effects of glucocorticoids on ACTH secretion are more complex than simple negative feedback inhibition of ACTH [or corticotropin-releasing hormone (CRH)]. The issues that require elucidation are the locus at which glucocorticoids exert their negative feedback; the "time domains" of corticosteroid feedback; the effect of glucocorticoid feedback on nonpituitary, nonhypothalamic sites; the relationship between glucocorticoids and basal ACTH secretion; and the relationship between stress and corticosteroid feedback. Each of these facets represents highly complex and intricate phenomena.

Locus of Glucocorticoid Inhibition of HP Axis

There is evidence to suggest that glucocorticoids exert their negative feedback effects on both the pituitary and the hypothalamus.

Feedback on the Pituitary. Several lines of evidence strongly support the notion that glucocorticoids exert a dominant negative feedback effect on the corticotropes of the anterior pituitary.

1. In vitro and in vivo experiments have shown that injection of cortisol or dexamethasone directly into the pituitary gland decreases the corticosteroid responses to stress.[34–36]
2. Systemic administration of cortisone or dexamethasone to animals blunts the sensitivity of corticotropes to corticotropin-releasing factor (CRF) preparations.[37,38]
3. In humans receiving glucocorticoid therapy, the ACTH response to

ovine CRH is blunted on the day of treatment, while remaining mildly blunted on the day off treatment.[39]

4. In vitro work by Giguere et al.[40] has provided data supporting a predominantly pituitary locus for the negative feedback suppression of glucocorticoids. Using preparations of dispersed anterior pituitary cells maintained in primary culture, these workers showed that pretreatment with dexamethasone significantly inhibited the ACTH release induced by CRF as well as by a cAMP derivative. The failure of ACTH to respond to CRF in the presence of hypercortisolemia is strikingly reminiscent of the failure of thyroid-stimulating hormone to respond to thyrotropin-releasing hormone in the presence of hyperthyroidism.

5. The negative correlation between CRH-induced ACTH release and basal cortisol levels has been amply documented in the literature[41,42]; i.e., the lower the basal cortisol, the greater the response of ACTH to ovine CRH administration. The negative feedback effect of glucocorticoids on the pituitary forms the basis for the several dexamethasone suppression tests used for diagnostic purposes.

Feedback on the Hypothalamus. The evidence for the glucocorticoid inhibitory effects on the hypothalamus is derived from experimental studies in animals as well as measurements of CRH levels in the portal blood. In fact, some studies have suggested that the hypothalamus (or median eminence) may in fact demonstrate a greater sensitivity to glucocorticoid suppression than the pituitary itself. Local injections or implants of corticosterone or dexamethasone within the hypothalamus are quite effective in suppressing pituitary ACTH.[43−46] It has been experimentally shown that both basal and stress-induced CRH bioactivity are inhibited by the administration of dexamethasone or corticosterone.[47,48] The sensitivity of CRH release to falling glucocorticoid concentrations has been demonstrated by the observation that adrenalectomy causes a rapid increase in bioactive CRH content in the median eminence within 2.5 min after adrenalectomy in animals.[49] Taken together, clearly glucocorticoids exert a negative feedback effect on the hypothalamic CRF. The relative magnitude of this effect, in comparison to the negative feedback effects on the corticotropes, has not been completely eludicated.

The glucocorticoid inhibition of ACTH (and CRH) secretion is considerably influenced by glucocorticoid binding sites within the pituitary and the hypothalamus. Such sites have been demonstrated in the pituitary,[50,51] in the hypothalamus,[52] and elsewhere in the central nervous system.[53,54] These sites differ in their ability to bind different glucocorticoids. The differential suppressive effect of dexamethasone versus the other glucocorticoids may be due to differences in the binding potency, differences in the saturability, and differences in the number of sites in the various portions of the hypothalamic pituitary axis.

Time Domains of Corticosteroid Feedback

Much has been learned in the past two decades regarding the time domains of the negative feedback exerted by glucocorticoids on the hypothalamic–pituitary axis. Pioneering work in this field by Dallman and Yates[55] has been largely responsible for the unraveling of several facets of this complex phenomenon. It is believed that glucocorticoid feedback inhibition is a tricompartmentalized process, with three distinct time domains—a fast component, an intermediate component, and a slow component. The fast component occurs within seconds to minutes following glucocorticoid administration and is characterized by rate sensitivity; i.e., the magnitude of inhibition is dependent on the rate of rise of glucocorticoid concentration in the plasma. The intermediate component evolves after 2–10 hr following glucocorticoid administration, while the slow component reflects glucorticoid-induced ACTH suppression that persists over hours to days.

"Fast Feedback." This phase of suppression reflects the rapid inhibitory effect of glucocorticoids on ACTH that occurs while plasma concentrations of the glucocorticoids are rising. Dallman and Yates[55] first demonstrated the existence of this phase when they showed that injection of corticosteroids in rats abolished the corticosterone response to histamine if the injection preceded the histamine challenge by 15 sec or 5 min, but not when the corticosterone was injected 15 min before, or 2 min after the histamine challenge. The crucial aspect of this phase is that the suppressive effects last only as long as plasma concentrations of glucocorticoids are rising at a sufficient rate. Hence the term *rate-sensitive feedback* is applied to denote the fast component of feedback suppression. It has been proposed that the rate-sensitive fast feedback effect is a reflection of its dependency on the rate at which corticosteroids associate with its receptor, as well as the degree of saturability of the rate-sensitive feedback sites by glucocorticoids.[56–58] The rapidity with which this phase of suppression occurs is indicative of the fact that ACTH or CRH release, but not synthesis, is inhibited during this phase.

Intermediate and Slow Feedback. There seems to be a "silent" phase between the fast and delayed (intermediate and slow) phases of feedback inhibition. After a single injection of glucocorticoids, the fast, rate-sensitive phase fades off, and there is a period during which no ACTH inhibition occurs. During this phase, when the steroid concentrations are no longer continuing to rise at a certain rate, ACTH responses to stimuli are maintained. This silent phase ("silent" because it is characterized by lack of inhibition) is variable, dependent on whether a single injection or an infusion of steroid is used, and occurs at a time when the glucocorticoid concentration is not rising at a rapid rate, and before delayed feedback has had adequate time to develop. The time frame of intermediate negative feedback ranges between 2 and 20 hr following the injection of glucocorticoids, while the time frame of slow feedback suppression extends from hours to days. The intermediate phase of suppression blends temporally with the onset of the slow phase of suppression. These two phases of delayed suppression are dependent on the dose of glucocorticoid administered. The slow phase of suppression affects both secretion and synthesis of ACTH, with demonstrable changes in protein synthesis and RNA synthesis leading to decrease in synthesis of mRNA, Pro-opiomelanocortin (POMC), and ACTH.[59,60] In contrast to the slow feedback, the intermediate feedback does not appear to inhibit synthesis of ACTH. In an elegant study, Phillips and Tashjian[61] demonstrated that the cellular content, synthesis, and degradation of ACTH were not significantly affected by intermediate feedback.

The implications of these findings to the clinician are threefold; First, the stress-related increment in endogenous cortisol level would not exert slow feedback inhibition of ACTH or CRH due to the evanescent rise of cortisol levels; second, short periods (24 hr) of exposure to high levels of glucocorticoids may not be sufficient to cause decreased synthesis of ACTH[62]; and third, slow feedback inhibition assumes importance only in pathological states of cortisol excess or after pharmacological therapy with steroids for days.

Glucocorticoid Feedback on Nonpituitary, Nonhypothalamic Sites

While the hypothalamic–pituitary unit represents the major sites of negative feedback inhibition by glucocorticoids, local implantation of corticosteroids into serveral regions of the brain is attended by suppression of the pituitary adrenal axis. Experimentally, the sites of the brain which, when injected with cortisone or dexamethasone, can cause pituitary ACTH suppression include the mesencephalic reticular formation, hippocampus, anterior thalamus, and septum. The presence of corticosteroid binding sites in these regions has been experimentally documented. Since the distribution of CRH extends to several extrahypothalamic sites of the brain, it is possible that glucocorticoids suppress CRH at these sites, in a manner analogous to CRH suppression by steroids at the hypothalamic site or median eminence. Most studies that propose feedback suppression by glucocorticoids at extrahypothalamic sites of the brain involve animals. The role of such a phenomenon in humans is at best speculative.

Glucocorticoid Suppression of Basal ACTH

In general, basal ACTH release is less sensitive to the suppressive effect of glucocorticoids in comparison to stimulated ACTH release. Further, the time course of glucocorticoid-induced suppression of basal ACTH differs from that of suppression of stimulated ACTH release. For instance, even prolonged incubation of pituitary cultures with dexamethasone fails to inhibit basal ACTH secretion. Studies that examine in vitro suppression of basal ACTH have suffered from the inevitable consequence of ACTH leakage from the injured pituitary cells. In vivo studies support the notion that suppression of basal ACTH to glucocorticoids is less sensitive to feedback inhibition than hypoglycemia-stimulated ACTH release.[63]

Physiological Stress and Negative Feedback

The stress-mediated ACTH response is an adaptive process, mediated and integrated by the brain and other parts of the central nervous system. The two major participants of the stress response (general adaptation syndrome) are the hypothalamic–pituitary–adrenal axis and the sympathetic nervous system. Activation of the former results in secretion of cortisol, while activation of the latter leads to release of catecholamines. CRH probably plays an important role in mediating the ACTH response, because stalk section impairs activation of this system. In addition, the administration of CRH into the cerebral ventricles of laboratory animals elicits cardiovascular and metabolic responses identical to the stress syndrome.[64] The demonstration of CRH (and its receptors) in the limbic system and the hindbrain also provides suggestive evidence for CRH involvement in the stress response. The proximity of these structures to the central sympathetic system is also suggestive of possible catecholamine mediation of ACTH release, in a synergistic fashion. The afferent arc of the stress reflex is in the peripheral nervous system, because denervation abolishes the stress response secondary to limb injury. Various kinds of stress—burns, fever, trauma, anesthesia, hypoglycemia, surgery—can result in identical elevation of cortisol. The clinical importance of the stress-triggered CRH–ACTH–cortisol response is enormous, since patients with untreated primary or secondary adrenal insufficiency may do well in a basal state but may profoundly and critically decompensate under the stress of infection or surgery; unless urgent steroid replacement is instituted, these patients succumb to the stress.

The negative feedback effect of glucocorticoids can be overridden by stress, a situation characterized by continuous ACTH secretion despite hypercortisolemia. It is believed that the stress-induced CRH elevation is of a magnitude that far surpasses the negative feedback exerted on the corticotropes by glucocorticoids. Support for such a hypothesis is provided by the work of Kendall et al.,[65] which indicates that both the duration of prior corticosteroid administration and the intensity of the ACTH-releasing stimulus are important determinants of corticotrope activation. The stress-mediated release of ACTH is the most powerful stimulus for the corticotropes. The control of the hypothalamic–pituitary adrenal axis can be viewed as consisting of a "comparator unit" that balances the intensity of the afferent input (such as those mediated by CRH and other central nervous system commands) with the concentration of glucocorticoids in the plasma. It has been proposed by Yates et al.[66] that a change in the set point occurs with stress, resulting in signals that bring about the control variable (plasma glucocorticoid level) to a new higher level. Thus, during stress, the comparator element is reset, with more production of ACTH, since the ambient concentrations of glucocorticoids are perceived by the comparator as "low" in the reset milieu. Thus, increases in plasma glucocorticoids within the physiological range do not inhibit ACTH response to certain stresses. The role of glucocorticoids in this circumstance is best perceived as one that modulates ACTH secretion in response to multiple stresses, without compromising or exhausting the ability of the hypothalamic–pituitary–adrenal axis in adapting to life-threatening stresses. Indeed, it appears that the entire hypothalamic–pituitary–adrenal axis becomes sensitized to facilitate the stress response, the ultimate goal of which is to put out adequate amounts of cortisol to combat stress.

Testing Adrenocortical Function

The availability of highly specific radioimmunoassays for measurement of practically every steroid synthesized by the adrenal cortex has greatly enhanced and simplified evaluation of adrenocortical function. The tests involved in testing adrenal function can be classified into basal and dynamic tests for evaluation of adrenal integrity. The basal tests employ collection of a single sample of blood or urine for assay of adrenal hormones. The dynamic tests involve measurement of steroid responses to provocative or suppressive maneuvers.

Basal Tests to Evaluate Adrenocortical Function

As already indicated, the emergence of sophisticated radioimmunoassays has facilitated accurate measurement of practically every steroid secreted by the adrenal cortex. The convenience factor involved in measuring a single sample of blood or urine to establish the suspected clinical diagnosis is the singular advantage of basal tests. However, despite this advantage, basal tests have not eliminated the need for more elaborate testing. The two reasons for this are the pulsatile nature of secretion of these steroids and the great overlap in the ranges of basal hormones between normal and abnormal subjects, resulting in an unacceptable rate of false positives and negatives.

Table 1 illustrates the adrenal steroids in plasma and their products measured in the 24-hr urine. The three commonly used basal tests that involve plasma assays are cortisol, dehydroepiandrosterone (DHEA), and 17-hydroxyprogesterone. The three commonly used basal tests that involve urinary assays are 17-hydroxycorticosteroids, urine free cortisol, and 17-ketosteroids. A brief overview of each of these tests is necessary to help one understand the indications, use, and limitations involved.

Serum Cortisol (Compound F)

The availability of a radioimmunoassay for cortisol has eliminated the calorimetric and fluorometric methods for measurement of cortisol.[67-69] The disadvantages of these older methods were several, including interference from other drugs as well as from other steroids, particularly 11-deoxycortisol. The currently available radioimmunoassay for cortisol is exquisitely sensitive and specific for compound F, the only biologically effective glucocorticoid.[70]

The primary indication to obtain serum cortisol is when hypo- or hyperfunction of the adrenal cortex is suspected. Indeed, when the serum cortisol is extremely high or extremely low, the assay may adequately establish disordered function. Unfortunately, such instances in clinical practice are few. There are

TABLE 1.
Adrenal Steroids in Blood and Urine

Steroids	Blood	Urine
Mineralocorticoids	Aldosterone	Tetrahydroaldosterone
Glucocorticoids	Cortisol (compound F)	17-Hydroxycorticosteroids (S and F)
		Urinary free cortisol
Androgens	Dehydroepiandrosterone (DHEA)	17-Ketosteroids
Precursors	17-Hydroxyprogesterone Compound S	Pregnanetriol

two basic difficulties in interpreting a single sample of serum cortisol. First, although cortisol secretion does follow a circadian pattern, it is secreted in episodic bursts with variable peaks and nadirs. Therefore, depending on sampling time, a high or low value may be encountered, obliterating the clear difference between normal and abnormal ranges. Second, since the radioimmunoassay measures the cortisol in the circulation that is bound to the globulin transcortin, any factor that increases this globulin (such as obesity, pregnancy, alcohol, or estrogens) would spuriously elevate the cortisol level in the serum as measured by radioimmunoassay.

The normal range for cortisol in the serum is 5–22 μg/dl.

Dehydroepiandrosterone Sulfate (DHEA-S)

This steroid is the prototype of adrenal androgens and is an excellent indicator for the adequacy of androgen synthesis by the zona reticularis. The plasma level of DHEA-S is reasonably stable throughout the day but slightly higher in the morning. In females, the major source of DHEA-S is the adrenal cortex; in males, a small contribution (20%) is derived from testosterone secreted by the testes.

The assay of DHEA-S level in plasma has supplanted the cumbersome need for urine collections to estimate the 17-ketosteroids, the conventional indicator of adrenal androgen activity.

The primary indications to obtain DHEA-S level in plasma are in the evaluation of hirsutism or virilization in adult females, precocious puberty in boys, and premature adrenarche or virilization in girls. Since the adrenal androgens are "weak" androgens, the level of DHEA-S in plasma would be markedly elevated when virilization is secondary to adrenal androgen excess. The two important causes for increased synthesis of adrenal androgens by the zona reticularis are adrenogenital syndrome and tumors of the adrenal cortex, particularly carcinoma.

17-α-Hydroxyprogesterone

This substance is an important precursor in glucocorticoid synthesis. Adequate channeling of 17-α-hydroxygesterone into compound S (11-deoxycortisol) and compound F (cortisol) involves the integrity of two enzymes—21-hydroxylase and 11-hydroxylase, respectively. Hence, when a partial or complete block in either one of these two enzymes occurs, there is accumulation of 17-hydroxyprogesterone, the precursor proximal to the block. The measurement of serum level of 17-α-hydroxyprogesterone has supplanted the cumbersome need for urine collections to estimate pregnanetriol in urine, the conventional indicator of enzymatic blocks involving 21- or 11-hydroxylases.

The primary indication to obtain 17-hydroxyprogesterone level in plasma is in the evaluation of a patient suspected of having adrenogenital syndrome, i.e., ambiguous genitalia, hyponatremia, and hyperkalemia in the neonate, precocious puberty in the male child, virilization in the female child, hypertension and hypokalemia in young adults, and, most commonly, in the evaluation of hirsutism in adult females.

Urinary 17-Hydroxycorticosteroids

This measurement of the "Porter–Silber chromogens" in a 24-hr collection of urine estimates the metabolic products of glucocorticoid activity. Purportedly, this reflects the integrated secretory activity of cortisol in a 24-hr period as opposed to the serum cortisol, which merely reflects that activity at a given moment. Approximately one-third of the cortisol secreted during the day is excreted as 17-hydroxycorticosteroids (17-OHCS) in the urine. Since this measurement depends on intact hepatic and renal mechanisms of inactivation and clearance, the presence of liver or kidney damage would alter these levels.

As a screening test for hypo- or hyperfunction of the adrenal cortex, the urinary 17-OHCS are of help only when the values are extremely low or extremely high. The recent availability of 24-hr urine free cortisol determination is a much superior screening test for hypercortisolism than the 17-OHCS. The most significant value of the 17-OHCS is related to the fact that all the major dynamic tests for adrenal dysfunction have been standardized on the basis of the 17-OHCS in urine. Thus, the standard dexamethasone suppression test, the standard ACTH stimulation test, and the standard metyrapone test are all interpreted using 17-OHCS in urine as the measured parameter.

Urinary 17-Ketosteroids

These compounds, measured in the urine by the Zimmerman reaction, reflect the metabolic products of adrenal androgens such as dehydroepiandrosterone, androstenedione, and etiocholanolone. The 17-ketosteroid (17-KS) level in the urine is not a measure of testosterone activity, since the contribution of this potent adrogen to the urinary 17-KS is only minor. The advent of DHEA-S assays in plasma has decreased reliance on measuring urinary 17-KS to determine adrenal androgenicity.

Urinary Free Cortisol

In contrast to plasma cortisol, which is affected by changes in the level of transcortin, the urinary free cortisol parallels the true secretory rate of the glucocorticoids by the adrenal cortex. The measurement of "free" cortisol in the urine is a reflection of complete or near-complete saturation of transcortin by endogenously secreted cortisol. Thus, the only means of increasing free cortisol in the urine is by complete occupancy of transcortin by cortisol, resulting in a spillover of the "filtrable" cortisol, which is measured in urine as free cortisol. The urinary free cortisol is considered a sensitive screening test for hypercortisolism since it is independent of perturbations in transcortin and since it is nearly always elevated in that disorder. However, it should be recognized that endogenous depression and stress can also elevate urinary free cortisol. As a screening test for hypofunction, urinary free cortisol is unreliable, since the levels in normals and hypoadrenal patients overlap considerably.

Dynamic Tests to Evaluate Adrenocortical Function

These are tests that employ various maneuvers to test the physiology of the hypothalamopituitary–adrenal axis. As in other areas of endocrinology, stimulation tests are used when hypofunction is suspected, and suppression tests are used when hyperfunction is suspected. Table 2 summarizes the various dynamic tests employed in the evaluation of adrenocortical dysfunction. The stimulation test is first described, followed by an overview of the various suppression tests that have been devised. Finally, the metyrapone test and tests involving ACTH are mentioned.

The Rapid ACTH Stimulation Test

This is a simple screening test to detect adrenal insufficiency. In its simplest form, the test involves measuring the cortisol level before, and 30, 60, and 90 min following the administration of an intravenous bolus of 0.25 mg of synthetic ACTH. The usual preparation employed is cortrosyn (or cosyntropin, α-1-24-corticotropin), a synthetic water-soluble subunit of ACTH). The test has been widely used as a screening test for adrenocortical failure.[71–78] The "normal" response varies according to different workers, but one that is accepted by most, and applies the most rigid criteria for normalcy, is a cortisol response that doubles as well as increases by 10 μg over the baseline following ACTH. Additional criteria have been established to denote a normal response. These include a basal cortisol level greater than 5 μg/dl; a 30-min level that exceeds an absolute level of 18 μg/dl; an increment of 7 μg/dl at 30 min; or an increment of 11 μg/dl at 60 min. These diverse criteria underscore the difficulties in interpretation of the test when the results are borderline.

The simultaneous measurement of aldosterone levels in addition to cortisol levels following ACTH has added another diagnostic dimension to the rapid

TABLE 2.
Dynamic Tests in the Evaluation
of Adrenocortical Dysfunction

Test	Comment
Rapid ACTH stimulation test	Screening test for hypoadrenalism
Standard ACTH stimulation test	Definitive test to document hypoadrenalism as well as its origin, i.e., pituitary versus adrenal
Metyrapone test	A test to document ACTH deficiency as the cause for hypoadrenalism; also useful in differentiating the etiologies of hypercortisolism
Overnight dexamethasone suppression tests	Screening test for hypercortisolism
Standard dexamethasone suppression test	Assists in delineating the etiology of hypercortisolism

ACTH test. Dluhy et al.[79] have shown that subjects with intact function of the zona glomerulosa increase their aldosterone concentrations by an average of 14 ng/100 ml above the basal level (range 4–29 ng/100 ml). The preservation of a normal aldosterone response in the presence of a blunted or absent cortisol response is highly suggestive of secondary adrenal failure, whereas loss of both cortisol and aldosterone responses to ACTH is indicative of primary (Addisonian-type) adrenal failure.

Three statements need to be emphasized regarding the rapid ACTH stimulation test. First, the test is, at best, a screening procedure and separates the normal responder from the abnormal one; i.e., the demonstration of a clearly normal response precludes the need for further workup in the direction of adrenal failure. Second, a blunted or flat response can be encountered in both primary (Addison's) and secondary (pituitary) hypoadrenal states; in the former, the lack of response results from a decreased reserve of the zona fasciculata, whereas in the latter condition, this layer of the cortex is dormant because of chronic ACTH deprivation. Third, and most important, the response patterns of patients with partial adrenal insufficiency (the limited adrenal reserve syndrome) and patients under stress can be quite variable and can defy interpretation. For instance, some patients with partial disease may show a near-normal response of the rapid test, whereas normal subjects under stress, whose adrenals already are working at full capacity, may not demonstrate the expected peak following ACTH. It is for these reasons that the rapid ACTH test should be considered only as a screening test. No patient should be placed on lifelong steroid replacement solely on the basis of results of the rapid ACTH test performed in the simple manner described. For definitive documentation of adrenal failure, one resorts to the standard ACTH stimulation test.

The Standard ACTH Stimulation Test

This test evaluates the ability of the adrenal cortex to secrete glucocorticoids in response to protracted stimulation with ACTH. The parameter evaluated is the 24-hr urinary 17-OHCS (and in some protocols the 17-KS as well). The test, which requires hospitalization and the meticulous collection of 24-hr urines for 7 days at a stretch, is performed as follows. Baseline 24-hr urinary 17-OHCS (and 17-KS) are collected for 2 days. Plasma cortisol and the plasma level of circulating ACTH are also assayed at the basal state. Then, 40 IU of ACTH (crystalline, soluble ACTH) is infused in normal saline for an 8-hr period, usually from 8 A.M. to 4 P.M. The urinary 17-OHCS, 17-KS, and plasma cortisol are assayed daily and continuously during the 3 days of ACTH infusion as well as for 2 days thereafter. The total test period in this format is 7 days (Table 3).

In interpreting this cumbersome but "gold standard" of a test, three points are worth bearing in mind. First, the normal adrenal cortex puts out most of its glucocorticoids during the first day of ACTH administration, with the normal response defined as a two- to fourfold increase in the 17-OHCS following the infusion. Second, patients with Addison's disease demonstrate a characteristic flat response to the test, whereas patients with hypopituitarism (ACTH deficiency) demonstrate a "stepladder" pattern of increase, with responses after the second or third day of ACTH. Third, patients with "limited-reserve" syndrome

TABLE 3.
The Standard ACTH Stimulation Test

Day	ACTH	Cortisol	Urine 17-OHCS	17-KS
1 (Pretest)	Basal	Basal	Basal	Basal
2 (Pretest)	Basal	Basal	Basal	Basal
3 (40 IU ACTH)	—	√	√	√
4 (40 IU ACTH)	—	√	√	√
5 (40 IU ACTH)	—	√	√	√
6 (Posttest)	—	√	√	√

show their maximal response in the first day of ACTH, after which the magnitude of response diminishes. This pattern may overlap with the response of some normal subjects.

In general, the standard ACTH stimulation test permits the differentiation to be made among normals, primary hypoadrenalism, secondary hypoadrenalism, and the limited adrenal reserve syndrome. The limitations of the test are the time consumed and expense as well as the cumbersome need for collecting seven 24-hr urines.

The Suppression Tests

The suppression tests evaluate the ability of the hypothalamic–pituitary axis to suppress, in response to orally administered dexamethasone, a potent glucocorticoid. Obviously, these tests assume importance only when hyperfunction is suspected. The overnight dexamethasone test evaluates the plasma cortisol level following 1 mg dexamethasone given at midnight (before sleep). The dexamethasone given before sleep abolishes the nyctohemeral release of ACTH, and hence the cortisol level drawn at 8 A.M. will be below 5 μg/dl. Dexamethasone is potent at this dose but does not interfere with plasma measurements of cortisol. The value of the test lies in the fact that it is convenient and inexpensive, and if the patient suppresses below 5 μg, hypercortisolism is nearly completely excluded. Unfortunately, a wide variety of conditions are characterized by "nonsuppression"—obesity (30–45%), drug intake (estrogens, diphenylhydantoin), alcoholism, and endogenous depression. Thus, the only valuable information provided by the test is when the results are suppressible, i.e., a decline below 5 μg following oral dexamethasone.

The low-dose dexamethasone test involves administering 0.5 mg of dexamethasone four times a day orally for 2 days and evaluating the 24-hr urinary 17-OHCS before and after; normal suppression is defined as a decrease in 17-OHCS below 3.5 mg after dexamethasone. The only group of patients who suppress to the low dose in terms of those who showed nonsuppression to the overnight test are the obese patients and some patients on medication; since the urinary free cortisol is a superior index for hypercortisolism because it is not affected by obesity or medications, the low-dose test has nothing more to offer in these circumstances than the urinary free cortisol does. Therefore, in many centers, the low-dose test is circumvented in favor of the 24-hr urinary free cortisol.

The high-dose dexamethasone test evaluates the 17-OHCS before and after the administration of 2 mg of oral dexamethasone four times a day for 2 days. Suppression here is defined as a drop in the 17-OHCS by at least 50% of the baseline. This test has some discriminatory value in separating the various etiologies of hypercortisolism in the following manner:

1. The characteristic response of pituitary-dependent Cushing's syndrome is suppression to the high dose but nonsuppression to the low dose.
2. The characteristic response of adrenal tumors as well as those that ectopically secrete ACTH is nonsuppression to the high dose, since endogenous ACTH-secreting tumor is also nonsuppression to the high dose.

Despite these "classical" responses, there are several exceptions to the general rules:

1. Twenty to twenty-five percent of patients with pituitary-dependent Cushing's syndrome may not suppress to the high dose, requiring a "high-high" dose, since they are functioning at a phenomenally higher threshold.
2. The rare bronchial carcinoid, ectopically secreting ACTH, may also demonstrate preservation of suppression to the high-dose dexamethasone test.
3. Endogenous depression is often associated with abnormal steroid dynamics, the most frequent one being nonsuppression to the low dose but preservation of suppression to the high dose, mimicking pituitary disease. It is particularly important to keep this entity in mind, since the urinary free cortisol also can be elevated in depressed patients. The constellation of elevated urinary free cortisol and abnormal suppression data in such patients may result in an erroneous diagnosis of pituitary-dependent Cushing's syndrome.

The value of the dexamethasone suppression tests in the etiological diagnosis of Cushing's syndrome is outlined in Chapter 3.

The Metyrapone Test

Metyrapone is a drug that inhibits the conversion of 11-β-hydroxylase within the adrenal cortex. As a result, there is a decrease in compound F level, which stimulates the hypothalamic–pituitary axis to secrete more ACTH. The increased amounts of ACTH stimulate the adrenal cortex, with resultant activation of steroidogenesis. However, owing to the block in the final step, the steroidogenesis stops short of synthesizing increased amounts of compound S.

Therefore, the triple responses of a normal person, i.e., one with normal adrenals and an intact hypothalamic pituitary axis, to the oral administration of metyrapone are as follows: a decrease in cortisol (F), an increase in ACTH, and an increase in deoxycortisol (S), which is the precursor product proximal to the block created by metyrapone. This is reflected in the normal person as an increase in the urinary 17-OHCS, which measure compound S. (The 17-OHCS normally measure both compound F and compound S, but metyrapone precludes formation of compound F, and hence most of the 17-OHCS consists of the precursor, compound S.) (Also, see page 69.)

The proper interpretation of the metyrapone test depends on three prerequisites:

1. The completeness of the enzymatic block created by the drug. Unless the block is significant enough to lower the cortisol level, the subsequent phenomena will not take place. Adequacy of the block should always be confirmed by demonstrating a significant lowering of cortisol level in plasma.

2. The integrity of the hypothalamic–pituitary axis, since the ACTH response to declining cortisol levels is the key phenomenon.

3. The integrity of the adrenal cortex to respond to the endogenous ACTH drive. The metyrapone test cannot be intrepreted without knowing whether the adrenal glands are viable. For instance, a failure to increase urinary 17-OHCS following metyrapone can be indicative of either hypopituitarism (ACTH lack) or Addison's disease (primary adrenal disease). However, if the presence of adrenal responsiveness has been established prior to performance of the metyrapone test, the failure to increase the 17-OHCS following metyrapone can only mean hypopituitarism. It is therefore essential to perform an adrenal stimulation test by the administration of exogenous ACTH before doing the metyrapone tests. If the adrenal glands are shown to respond to exogenous ACTH, one can assume that a similar response can be expected with increases in the endogenous ACTH; thus, attempts to stimulate the endogenous ACTH reserve by metyrapone are valid. If, on the other hand, the adrenals fail to respond to exogenous ACTH, the metyrapone test is not indicated, since lack of response to metyrapone can no longer be interpreted.

The test is performed by measuring basal levels of urinary 17-OHCS, serum cortisol, and ACTH before and following the oral administration of 750 mg metyrapone every 4 hr for six doses. A failure to increase the 17-OHCS in the urine collection the day after metyrapone is indicative of either inadequate block, ACTH lack, or primary adrenal insufficiency. Adequacy of block can be ensured by demonstrating a significant lowering of cortisol in the serum, while adequacy of adrenal function can be ensured by prior assessment of the adrenal response to exogenous ACTH. With these two prerequisites satisfied, a failure to increase the 17-OHCS following metyrapone is diagnostic of ACTH deficiency.

Patients receiving anticonvulsant therapy and those who are hypothyroid or depressed may also demonstrate blunted responses to metyrapone.

While metyrapone challenge is used primarily to diagnose ACTH deficiency, it can also be employed in the differential diagnosis of hypercortisolism (Cushing's disease). Patients with pituitary-dependent Cushing's disease increase their 17-OHCS following metyrapone, while those with suppressed ACTH secondary to an autonomous adrenal tumor demonstrate no response.

The Actions of Glucocorticoids

The actions of glucocorticoids encompass numerous organ systems. The term *glucocorticoids* is derived form one of the most primal effects of these steroids, namely, the ability to promote gluconeogenesis. The actions of these steriods can be viewed in terms of their effects on carbohydrate, protein, and lipid metabolism, as well as their effects on water, electrolyte, and mineral metabolism. In addition, glucocorticoids have unique anti-inflammatory potency, an effect that has made these steroids extremely useful in the treatment of clinical disorders caused by autoimmune processes. It is remarkable that several effects of glucocorticoids that are inconspicuous in physiological states are brought out with alarming clarity during supraphysiological or pharmacological therapy with these steroids. Appropriately, several facets of glucocorticoid action are discussed in Chapter 4, which deals with steroid therapy. This section merely deals with the physiological actions of glucocorticoids.

Carbohydrate Metabolism

Glucocorticoids possess the inherent property of promoting gluconeogenesis. This is reflected in the five phenomena seen in association with glucocorticoid use—increased hepatic glucose output, tendency toward hyperglycemia, glycosuria, insulin antagonism, and increased hepatic glycogen deposition. The major source for the steroid-induced hepatic gluconeogenesis are the amino acids derived from muscle catabolism. Hepatic glucose production has been shown to increase by as much as seven times compared to the basal state following glucocorticoid administration.[80] In addition, glucocorticoid-induced secretion of glucagon may also play an important role in the increased hepatic glucose production. In addition to utilizing amino acid substrates, as well as glucose precursors such as pyruvate, for gluconeogenesis, glucocorticoids inhibit the conversion of pyruvate to acetyl CoA and CO_2. This effect permits increased availability of pyruvate to serve as glucose precursors.

Protein Metabolism

The trinity of effects on protein metabolism are inhibition of protein synthesis, enhancement of protein catabolism, and induction of negative nitrogen balance.[81] The impact of these effects assumes full significance in patients with Cushing's syndrome in whom the decreased muscle mass and myopathy contrast sharply with the truncal deposition of fat. The increased aminoaciduria seen with glucocorticoid administration is a direct result of the catabolic effects of glucocorticoids. For the same reason, varying degrees of creatinuria are associated with glucocorticoid use. The myopathy that ensues from chronic glucocorticoid excess is discussed in chapter 4 in the section "Clinical Considerations."

Lipid Metabolism

Although increased fat deposition is a hallmark of glucocorticoid excess, the effect of glucocorticoids on adipocytes remains enigmatic. The information in the literature regarding this aspect of glucocorticoid action has yielded conflicting conclusions. Thus, glucocorticoids have been reported to decrease,[82,83] as well as increase,[84] free fatty-acid (FFA) synthesis by the liver. A similar controversy exists regarding the effect of glucocorticoids on FFA release by the adipose tissue. The effect of glucocorticoids on mobilization of lipids is also unsettled.

Indeed the only action of glucocorticoids on lipid metabolism that is universally accepted is its synergistic role in potentiating the action of other lipolytic hormones, particularly catecholamines.[85] Further, the effects of glucocorticoids on circulating concentrations of cholesterol, triglyceride, and lipoproteins have been too inconsistent to permit generalization.

Water and Electrolyte Metabolism

Glucocorticoids are well-recognized agents that promote water diuresis.[86] This effect is probably mediated by inhibiting release of vasopressin. Early experiments in the 1960s had established that cortisol not only modified the osmotic threshold for vasopressin release in normal subjects, but also inhibited the release of vasopressin from the neurohypophysis.[87,88] In addition to the action of promoting water diuresis, glucocorticoids can increase free water clearance by decreasing proximal tubular reabsorption of solute and water.[86] The combination of these two effects is conducive to the occasional development of hypernatremia and hyperosmolarity with glucocorticoid use.

In a different direction, large doses of glucocorticoids can lead to water retention, with increased plasma volume. It is probable that the renal effects of glucocorticoids on water and sodium absorption coupled with shifts of fluids from the intracellular to the extracellular compartments may account for this phenomenon.[89,90]

The effects of glucocorticoids on sodium or potassium handling by the renal tubules depend on the dose and the duration of administration. The acute effect, which is usually transient, is natriuresis. This can result from an increase in the glomerular filtration rate or from direct effects on the renal tubules.[91,92] However, with continued administration of glucocorticoids, especially in large doses, sodium retention develops. It should be noted that such an effect is also transient, lasting for several days, after which the tubules "escape" from the salt-retaining effects of these steroids.[93] Liddle et al.[94] have shown that the sodium-retaining effects of glucocorticoids can be reduced by concomitant administration of potassium.

The primary effect of glucocorticoids on potassium handling is promoting kaliuresis.[95] In the early stages of glucocorticoid therapy, the potassium excretion by the kidney reflects the increased load delivered to the kidney from egress of potassium from the body cells. With continued administration of glucocorticoids, direct tubular losses associated with sodium reabsorption account for the kaliuresis. At this stage hydrogen and ammonium ions are also lost along with potassium ions. With continued glucocorticoid treatment the renal tubules "escape" from the kaliuresis induced by the glucocorticoids. Sodium restriction will diminish the renal losses of potassium and ammonium ions. However, the early phase of kaliuresis occasioned by primary cellular loss of potassium will not be inhibited by sodium restriction. The effects of large doses of glucocorticoids on potassium handling by the tubules are conducive to the development of hypokalemic alkalosis.

Mineral Metabolism

The effects of glucocorticoids on calcium metabolism play a crucial role in the development of osteopenia—the worst complication of chronic steroid therapy. The triple deleterious effects of steroids on calcium metabolism are decreased intestinal absorption with increased fecal excretion, increased renal wasting with hypercalciuria, and antagonism to the effects of vitamin D. The net result is the development of negative calcium balance. The relationship between glucocorticoid use and osteopenia is discussed in Chapter 4.

Anti-inflammatory Effects

This unique effect of glucocorticoids has rendered this group of drugs invaluable in disorders characterized by autoimmune destruction of tissues. The effects of glucocorticoids on neutrophils, lymphocytes, and cell-mediated immunity are outlined in Chapter 4. This unique property of glucocorticoids that provides enormous therapeutic benefit to patients also, unfortunately, predisposes them to the development of numerous opportunistic infections.

References

1. Lieberman S, Greenfield NJ, Wolfson A: A heuristic proposal for understanding steroidogenic processes. *Endocrine Rev* **5:**128, 1984.
2. Roberts KD, Bandy L, Lieberman S: The occurrence and metabolism of 20α-hydroxycholesterol in bovine adrenal preparations. *Biochemistry* **8:**1259, 1969.
3. Dixon R, Furutachi T, Lieberman S: The isolation of crystalline 22R-hydroxycholesterol and 20α, 22R-dihydroxycholesterol from bovine adrenals. *Biochem Biophys Res Commun* **40:**161, 1970.
4. Kahn FW, Neher R: Adrenal steroid biosynthesis in vitro. 3. Selective inhibition of adrenal cortical function. *Helv Chim Acta* 49:725, 1966.
5. Luttrell B, Hochberg RB, Dixon WR, et al: Studies on the biosynthetic conversion of cholesterol into pregnenolone: Side chain cleavage of a *t*-butyl analog of 20-hydroxycholesterol, (20R)-20-t-butyl-5-pregnene-3β,20α diol, a compound completely substituted at C-22. *J Biol Chem* **247:**1462, 1972.
6. Hochberg RB, McDonald PD, Landany S, et al: Transient intermediates in steroidogenesis. *J Steroid Biochem* **6:**323, 1975.
7. Hoyte RM, Hochberg RB: Enzymatic side chain cleavage of C-20 alkyl and aryl analogs of (20-S)-20-hydroxycholesterol. Implications for the biosynthesis of pregnenolone. *J Biol Chem* **254:**2278, 1978.
8. Cara JF, Moshang T Jr, Bongiovanni AM: Elevated 17-hxdroxyprogesterone and testosterone in a newborn with 3-beta-hydroxysteroid dehydrogenase deficiency. *N Engl J Med* **313:**618, 1985.
9. Fevold HR, Wilson PL, Slanina SM: ACTH-stimulated rabbit adrenal 17α-hydroxylase. Kinetic properties and a comparison with those of 3β-hydroxysteroid dehydrogenase. *J Steroid Biochem* **9:**1033, 1978.
10. Mackler B, Haynes B, Tattoni DS, et al: Studies of adrenal steroid hydroxylation. 1. Purification of the microsomal 21-hydroxylase system. *Arch Biochem Biophys* **145:**194, 1971.
11. Kaufmann SHE, Sinterhauf K, Lommer D: 21-Hydroxylation of pregnenolone by microsomal preparations of rat and human adrenals. *J Steroid Biochem* **13:**101, 1980.
12. Kominami S, Mori S, Takemori S: Purification and optical studies of cytochrome P-450 from bovine adrenocortical microsomes. *FEBS Lett* **89:**215, 1978.
13. Kominami S, Oshi O, Kobayashi Y, et al: Studies on the steroid hydroxylation system in adrenal cortex microsomes. Purification and characterization of cytochrome P-450 specific for steroid C-21 hydroxylation. *J Biol Chem* **255:**3386, 1980.
14. Mason JI, Hemsell PG: Cholesterol sulfate metabolism in human fetal adrenal mitochondria. *Endocrinology* **111:**208, 1982.
15. Young DG, Hall PF: The side-chain cleavage of cholesterol and cholesterol sulfate by enzymes from bovine adrenal mitochondria. *Biochemistry* **8:**2987, 1969.
16. Hochberg RB, Landany S, Welch M, et al: Cholesterol and cholesterol sulfate as substrates for the adrenal side chain cleavage enzyme. *Biochemistry* **13:**1938, 1974.

17. Calvin HI, Lieberman S: Evidence that steroid sulfates serve as biosynthetic intermediates. II. In vitro conversion of pregnenolone-^{3}H sulfate-^{35}S to 17-hydroxypregnenolone-^{3}H sulfate-^{35}S. *Biochemistry* **3**:259, 1964.

18. Wallace EZ, Lieberman S: Biosynthesis of dehydroisoandrosterone sulfate by human adrenocortical tissue. *J Clin Endocrinol Metab* 23:90, 1963.

19. Strott CA, Lyons CD: Pregnenolone sulfate binding in the guinea pig adrenal cortex: Comparisons with pregnenolone binding. *Biochemistry* **17**:4557, 1978.

20. New MI, Seaman MP: Secretion rates of cortisol and aldosterone precursors in various forms of congenital adrenal hyperplasia. *J Clin Endocrinol Metab* 30:361, 1970.

21. Yanagibashi K, Haniu M, Shively JE, et al: The synthesis of aldosterone by the adrenal cortex: Two zones (fasciculata and glomerulosa) possess one enzyme for 11β-, 18-hydroxylation, and aldehyde synthesis. *J Biol Chem* **261**:3556, 1986.

22. Chua SC, John M, White PC: Cloning of cDNA encoding a human cytochrome P-450 for 11β-hydroxylase. *Pediatr Res* **20**:262A/650 1986 (Abstract).

23. Lucis OJ, Dyrenfurth I, Venning EH: Effect of various preparations of pituitary and diencephalon on the in vitro secretion of aldosterone and corticosterone by the rat adrenal gland. *Can J Biochem* 39:901, 1961.

24. Haning R, Tait SAS, Tait JF: In vitro effects of ACTH, angiotensins, serotonin and potassium on steroid output and conversion of corticosterone to aldosteorne by isolated adrenal cells. *Endocrinology* 87:1147, 1970.

25. Biglieri FG, Wajchenberg BL, Malerbi DA, et al: The zonal origins of the mineralocorticoid hormones in the 21-hydroxylation deficiency of congenital adrenal hyperplasia. *J Clin Endocrinol Metab* 53:964, 1981.

26. Chu MD, Ulick S: Isolation and identification of 18-hydroxycortisol from the urine of patients with primary aldosteronism. *J Biol Chem* 257:2218, 1982.

27. Chu MD, Ulick S: Biosynthesis of 18-oxocortisol by aldosterone producing adrenal tissue. *J Biol Chem* **258**:5498, 1983.

28. Gold EM: The Cushing syndromes: Changing views of diagnosis and treatment. *Ann Intern Med* **90**:829, 1979.

29. Daughaday WH, Mariz IK: Corticosteroid-binding globulin: Its properties and quantitation. *Metabolism* **10**:936, 1961.

30. Krieger DT, Allen W, Rizzo F, et al: Characterization of the normal temporal pattern of plasma corticosteroid levels. *J Clin Endocrinol Metab* **32**:266, 1971.

31. Weitzman ED, Fukushima D, Nogeire C, et al: Twenty-four hour pattern of the episodic secretion of cortisol in normal subjects. *J Clin Endocrinol Metab* **33**:14, 1971.

32. Lindholm J, Kehlet H, Blichert-Toft M, et al: Reliability of the 30-minute ACTH test in assessing hypothalamic–pituitary–adrenal function. *J Clin Endocrinol Metab* **47**:272, 1978.

33. Hjortrup A, Kehlet H, Lindholm J, et al: Value of the 30 minute adrenocorticotropin (ACTH) test in demonstrating hypothalamic–pituitary–adrenocortical insufficiency after acute ACTH deprivation. *J Clin Endocrinol Metab* **57**:668, 1983.

34. Dunn J, Critchlow V: Feedback suppression of pituitary adrenal function in rats with pituitary islands. *Life Sci* **8**:9, 1969.

35. Rose S, Nelson J: Hydrocortisone and ACTH release. *Aust J Biol Sci* 34:77, 1956.

36. Russell SM, Dhariwal APS, McCann SM, et al: Inhibition by dexamethasone of the in vitro pituitary response to corticotropin-releasing factor (CRF). *Endocrinology* **85**:512, 1969.

37. Gonzalez-Luque A, L'Age M, Dhariwal APS: Stimulation of corticotropin release by corticotropin-releasing factor (CRF) or by vasopressin following intrapituitary infusions in unanesthetized dogs: inhibition of the responses by dexamethasone. *Endocrinology* **86**:1134, 1970.

38. Arimura A, Bowers CY, Schally AV, et al: Effect of corticotrophin-releasing factor, dexamethasone, and actinomycin D on the release of ACTH from rat pituitaries in vivo and in vitro. *Endocrinology* **85**:300, 1969.

39. Schurmeyer TH, Tsokos GC, Avgerinos PC, et al: Pituitary–adrenal responsiveness to corticotropin-releasing hormone in patients receiving chronic, alternate day glucocorticoid therapy. *J Clin Endocrinol Metab* **61**:22, 1985.

40. Giguere V, Labrie F, Cote J, et al: Stimulation of cyclic AMP accumulation and corticotropin release by synthetic ovine corticotropin-releasing factor in rat anterior pituitary cells: Site of glucocorticoid action. *Proc Natl Acad Sci USA* **79**:3466, 1982.

41. Hermus A, Pieters G, Smals A, et al: Plasma adrenocorticotropin, cortisol and aldosterone

responses to corticotropin-releasing factor: Modulatory effect of basal cortisol levels. *J Clin Endocrinol Metab* **58**:187, 1984.

42. Lytras N, Grossman A, Perry L, et al: Corticotrophin releasing factor: Responses in normal subjects and patients with disorders of the hypothalamus and pituitary. *Clin Endocrinol (Oxford)* **20**:71, 1984.

43. Smelik PG, Sawyer CH: Effects of implantation of cortisol into the brain stem or pituitary gland on the adrenal response to stress in the rabbit. *Acta Endocrinol* **41**:561, 1962.

44. Chowers I, Feldman S, Davidson JM: Effects of intrahypothalamic crystalline steroids on acute ACTH secretion. *Am J Physiol* **205**:671, 1963.

45. Bohus B, Strashmirov D: Localization and specificity of corticosteroid "feedback receptors" at the hypothalamo–hypophyseal level; comparative effects of various steroids implanted in the median eminence or anterior pituitary of the rat. *Neuroendocrinology* **6**:197, 1970.

46. Stark E, Gyevai A, Acs Z, et al: The site of the blocking action of dexamethasone on ACTH secretion: In vivo and in vitro studies. *Neuroendocrinology* **3**:275, 1968.

47. Buckingham JC, Hodges J: The use of corticotrophin production by adenohypophyseal tissue in vitro for the detection and estimation of potential corticotrophin releasing factors. *J Endocrinol* **72**:187, 1977.

48. Takebe K, Kunita H, Sakamura M, et al: Suppressive effect of dexamethasone on the rise of CRF activity in the median eminence induced by stress. *Endocrinology* **89**:1014, 1971.

49. Keller-Wood ME, Dallman MF: Corticosteroid inhibition of ACTH secretion. *Endocrine Rev* **5**:1, 1984.

50. Koch B, Lutz-Bucher B, Briaud B: Relationship between ACTH secretion and corticoid binding to specific receptors in perifused adenohypophyses. *Neuroendocrinology* **28**:169, 1979.

51. Warembourg M: Radioautographic study of the rat brain and pituitary after injection of ^{3}H dexamethasone. *Cell Tissue Res* **161**:183, 1975.

52. Eik-Nes KB, Brizee KR: Concentration of tritium in brain tissue of dogs given [^{1,2,3}H$_2$] cortisol intravenously. *Biochem Biophys Acta* **97**:320, 1965.

53. Gerlach JL, McEwen BS: Rat brain binds adrenal steroid hormone: Radio-autography of hippocampus with corticosterone. *Science* **175**:1133, 1972.

54. McEwen BS, Weiss JM, Schwartz LS: Selective retention of corticosterone by limbic structures in rat brain. *Nature* **220**:911, 1968.

55. Dallman MF, Yates FE: Dynamic asymeteries in the corticosteroid feedback path and distribution-metabolism-binding elements of the adrenocortical system. *Ann NY Acad Sci* **156**:696, 1969.

56. Jones MT, Brush FR, Neame RLB: Characteristics of fast feedback control of corticotrophin release by corticosteroids. *J Endocrinol* **55**:489, 1972.

57. Jones MT, Tiptaft EM, Brush FR, et al: Evidence for dual corticosteroid-receptor mechanisms in the control of adrenocorticotrophin secretion. *J Endocrinol* **60**:223, 1974.

58. Kaneko M, Hiroshige T: Fast, rate-sensitive corticosteroid negative feedback during stress. *Am J Physiol* **234**:R39, 1978.

59. Roberts JL, Johnson LK, Baxter JD, et al: Effect of glucocorticoids on the synthesis and processing of the common precursor to adrenocorticotropin and endorphin in mouse pituitary tumor cells. In Sato GH, Ross R (eds): *Hormones and Cell Culture*, Book B. Cold Spring Harbor Laboratory, Cold Spring Harbor, NY, 1979, p 827.

60. Schacter BS, Johnson LK, Baxter JD, et al: Differential regulation by glucocorticoids of pro-opiomelanocortin mRNA levels in the anterior and intermediate lobes of the rat pituitary. *Endocrinology* **110**:1442, 1982.

61. Phillips M, Tashjian AH Jr: Characteristics of an early inhibitory effect of glucocorticoids on stimulated adrenocorticotropin and endorphin release from a clonal strain of mouse pituitary cells. *Endocrinology* **110**:892, 1982.

62. Stark E, Gyevai A, Acs Z, et al: The site of the blocking action of dexamethasone on ACTH secretion: In vivo and in vitro studies. *Neuroendocrinology* **3**:275, 1968.

63. Keller-Wood ME, Shinsako J, Dallman MF: Feedback inhibition of adrenocorticotrophic hormone by physiological increases in plasma corticosteroids in conscious dogs. *J Clin Invest* **71**:859, 1983.

64. Sutton RE, Koob GF, Le Moal M, et al: Corticotropin releasing factor produces behavioral activation in rats. *Nature (London)* **297**:331, 1982.

65. Kendall JW, Egans ML, Stott AK, et al: The importance of stimulus intensity and duration of steroid administration in suppression of stress induced ACTH secretion. *Endocrinology* **90**:525, 1972.

66. Yates FE, Leeman SE, Glenister DW, et al: Interaction between plasma corticosterone concentration and adrenocorticotropin-releasing stimuli in the rat: Evidence for the reset of an endocrine feedback control. *Endocrinology* **69**:67, 1961.
67. Bowman RE, DeLuna RF: Assessment of a protein-binding method for cortisol determination. *Anal Biochem* **26**:465, 1969.
68. Kraicer J, Conrad RG, Bicknese MB: Abnormally high plasma corticosterone values using the acid fluorescence method. *Clin Chim Acta* **23**:512, 1969.
69. Kendall JW, Egans ML, Stott AK: Fluorometric determination of corticosteroids: An interfering substance in impure dichloromethane which fluoresces with benzyl alcohol preservative in heparin. *J Clin Endocr* **28**:1373, 1968.
70. Nugent CA, Mayes DM: Plasma corticosteroids determined by use of corticosteroid-binding globulin and dextran-coated charcoal. *J Clin Endocr* **26**:1116, 1966.
71. Wood JB, Frankland AW, James VHT, et al: A rapid test of adrenocortical function. *Lancet* **1**:243, 1965.
72. Greig WR, Browning MCK, Boyle JA, et al: Effect of the synthetic polypeptide β1-24 (Synacthen) on adrenocortical function. *J Endocr* **34**:411, 1966.
73. Hicklin JA, Wills MR: Plasma "cortisol" response to Synacthen in patients on long-term small-dose prednisone therapy. Ann Rheum Dis **27**:33, 1968.
74. Musa BU, Dowling J: Rapid intravenous administration of corticotropin as a test of adrenocortical insufficiency. *JAMA* **201**:633, 1967.
75. Speckart PF, Nicoloff JT, Bethune JE: Screening for adrenocortical insufficiency with cosyntropin (synthetic ACTH). *Arch Intern Med* **128**:761, 1971.
76. Maynard DE, Folk RL, Riley TR, et al: A rapid test for adrenocortical insufficiency. *Ann Intern Med* **64**:552, 1966.
77. Arner B, Hedner P, Karlefors T, et al: One hour subcutaneous ACTH test with determination of plasma corticosteroids. *Acta Med Scand* **173**:91, 1963.
78. McGill PE, Greig WR, Browning MCK, et al: Plasma cortisol response to synacthen (β1-24 Ciba) at different times of the day in patients with rheumatic disease. *Ann Rheum Dis* **26**:123, 1967.
79. Dluhy RG, Himathongkam T, Greenfield M: Rapid ACTH test with plasma aldosterone levels: Improved diagnostic discrimination. *Ann Intern Med* **30**:693, 1974.
80. Eisenstein AB: Current concepts of gluconeogenesis. *Am J Clin Nutr* **20**:282, 1967.
81. Silber RH, Porter CC: Nitrogen balance, liver protein repletion and body composition of cortisone treated rats. *Endocrinology* **52**:518, 1953.
82. Brady RO, Lukins FWD, Gurin S: Synthesis of radioactive fatty acids in vitro and its hormonal control. *J Biol Chem* **193**:459, 1951.
83. Timiras PS, Koch P: Morphological and chemical changes elicited in liver of rabbits by cortisone and desoxycortisone acetate. *Anat Rec* **113**:349, 1952.
84. Hill RB, Drake WA: Production of fatty liver in the rat by cortisone. *Proc Soc Exp Biol NY* **114**:766, 1963.
85. Ramey ER, Goldstein MS: The adrenal cortex and the sympathetic nervous system. *Physiol Rev* **37**:155, 1957.
86. Raisz LG, McNeely WF, Saxon L, et al: The effects of cortisone and hydrocortisone on water diuresis and renal function in man. *J Clin Invest* **36**:767, 1957.
87. Aubry RH, Nankin HR, Moses AM, et al: Measurement of the osmotic threshold for vasopressin release in human subjects and its modification by cortisol. *J Clin Endocrinol* **25**:1481, 1965.
88. Dingman JF, Despointes RH: Adrenal steroid inhibition of vasopressin release from the neurohypophysis of normal subjects and patients with Addison's disease. *J Clin Invest* **39**:1851, 1960.
89. Dingman JF: Adrenal steroids and water metabolism. In Mills LC, Moyer JH (eds): *Inflammation and Diseases of Connective Tissue.* Saunders, Philadelphia, 1961.
90. Schayer RW, A unified theory of glucocorticoid action-II on a circulatory basis for the metabolic effects of glucocorticoids. *Perspect Biol Med* **10**:409, 1967.
91. Garrod O, Davies SA, Cahill G Jr: The action of cortisone and desoxycorticosterone acetate on glomerular filtration rate and sodium and water exchange in adrenalectomized dog. *J Clin Invest* **34**:761, 1955.
92. Dingman JF, Finkenstaedt JT, Laidlow JC, et al: Influence of intravenously administered adrenal steroids on sodium and water excretion in normal and Addisonian subjects. *Metabolism* **7**:608, 1958.

93. Relman AS, Schwartz WB: The metabolic effects of compound F acetate in man. *J Clin Invest* **31:**656, 1952.
94. Liddle GW, Bennett LL, Forsham PH: The prevention of ACTH-induced sodium retention by the use of potassium salts: A quantitative study. *J Clin Invest* **32:**1197, 1953.
95. Ross EJ: Modification of the effects of aldosterone on electrolyte excretion in man by simultaneous administration of corticosterone and hydrocortisone. Relevance to Conn's syndrome. *J Clin Endocrin* **20:**229, 1960.

2

Addison's Disease

Historical Perspectives

In 1855 Thomas Addison of Guy's Hospital, London, published a monograph entitled "On the Constitutional and Local Effects of Disease of the Suprarenal Capsules."[1] This 39-page monograph, regarded by medical historians as one of the classics of medical literature, contained the first perfect description of an endocrine disease. In painstaking detail, Addison described the clinical features of the disease that now bears his name and speculated on the pathogenesis of this uniformly fatal disorder that had no cure. The events that preceded and followed the publication of Addison's monograph make interesting reading and have all the markings of grand drama. It is noteworthy that the last decade of his life was punctuated by the excitement of his discovery of a new disease, the doubts and controversies generated by his peers, the misjudgments of the publishing hierarchy, and, above all, the tragedy that led to his untimely death within 5 years following the publication of his now celebrated, (but then ignored) monograph. Indeed, as for many geniuses, fame came to Addison, but posthumously.

    Addison literally "stumbled," as he said, on the discovery of this disease. He was pursuing the cause of pernicious anemia. On March 15, 1849, he read a paper to the South London Medical Society that was entitled "On Anemia: Disease of the Suprarenal Capsules." He described the symptoms and signs of some of his patients with a peculiar anemia, along with the autopsy findings in three. In two of them, the only abnormal lesions were limited to the suprarenal glands. He postulated that these glands were in some way. directly or indirectly, related to the anemia, pigmentation, and wasting seen in his patients. He called this disorder "melasma suprarenale." The paper generated little interest, and the significance of these pioneering observations were lost at that time—a time when the functional importance of the adrenals was not even known. In the following 4 years, he added more patients to his collection and was repeatedly able to demonstrate diseased adrenals in patients dying from a wasting illness. He was convinced that "the disease was peculiar, uniform in character and primary in nature." He submitted his papers to the Medico-Chirurgical Society of London, for consideration for publication, and was ironically turned down not once, but three times.[2] Vexed and despondent by such obstinate and persistent refusal by this august body, he continued to gather more data, driven by the courage of his conviction. At that time, very few intellectuals believed in him; one such person, who was later to play a major role in establishing the authenticity of his findings, was Samuel Wilks (1824–1911). Wilks, not only fervently believed in Addison's discovery, but helped him investigate additional cases. Having just qualified as a Fellow of the Royal College of Physicians, he took Addison's modified paper to the Society, which, despite the academic authenticity of Samuel Wilks, once again was refused publication. Wilks, with unflinching fervor, persuaded Addison to publish his findings on his own. Thus, 6 years later, Addison had his monograph published on his own, without the support or mediation of the Society.

The 11 patients described by Addison in his monograph had an assortment of pathological lesions at necropsy. Five had bilateral tuerculosis, one had unilateral tuberculosis (or a possible tumor), three had metastatic disease of the adrenal, one had apparent atrophy and fibrotic changes, while one had no obvious findings. Thus, even during that first description, Addison had identified the three major causes of primary adrenal failure—tuberculous, autoimmune, and idiopathic factors.

Addison's monograph, because of its literary mastery, received wide attention, particularly on the Continent. His monograph became the source of controversy in England, while becoming an object of celebration in France. In fact, the polemics surrounding the monograph reached a point where the Society had to set up a special committee to investigate the validity of Addison's findings.[3] Sporadic anecdotal cases soon began to surface in England, in support of Addison's observations. One of the committee members, George Harley (1829–1896), basking in the fame of his own new-found disease, paroxysmal nocturnal hemoglobinuria, vigorously argued against the functional importance of the adrenals. He went as far as adrenalectomizing white rats, which survived for months and did not change color. He presented these data to the society and proposed that the association drawn by Addison between pigmentation and diseased adrenals was faulty and mythical. (The fact that white rats possess accessory adrenocortical tissue between the testes and epididymes was not known then.) Despite the disbelief generated by many, Wilks continued to promulagate the existence of the disease described by his friend, Addison. In France, Armand Trousseau (1801–1867), a highly respected pioneer in endocrine pathology, not only acknowledged Addison's findings, but was largely responsible for renaming it "Addison's disease."[4] He described one case of tuberculosis of adrenals with premortem clinical findings identical to those of the patients described by Addison.

The controversy continued, with periodic opposition to the recognition of the new disease entity. The task of completely dispelling the doubts surrounding the reality of its existence was undertaken by E. H. Greenhow (1814–1888). He collected and analyzed a staggering 196 cases in the literature and proclaimed that 129 represented geniune cases of adrenal pathology. He published his book *Addison's Disease* in 1866. Two years later, Wilks and Daldy[5] edited the collected papers of Addison and remarked at the slow recognition of a well-documented disease entity. These papers, as well as the book by Greenhow, were widely read throughout the Western world. The description by Addison touched the fancy of nonmedical writers as well. The term *bronzed skin* found common usage in medical and nonmedical literature. The concept of being struck by a malady that gradually darkened the skin, with cachexia and slow but certain death, sparked the imagination of writers. One example in particular stands out; Oliver Wendell Holmes used this effectively in his book *The Poet at the Breakfast Table*, a widely read piece of the day.

Two decades after his original description the disease described by Addison had finally found acceptance. Tragically, Addison was not to know this. He was of melancholic temperament, given to profound attacks of depression. One cannot predict the impact that the controversy over acceptance of his discovery had on the tragic events to come. A glimpse of that dejection can be sensed in a discussion he gave at the Royal Medical and Chirurgical Society of London in 1858, 2 years before his death, and 3 years following the publication of his monograph. He said, "Who can tell what influence the contact of these diseased organs might have on these great nerve centers [the solar plexus] and what share the secondary effect might have on the general health and in the production of the symptoms presented."[2] These words, spoken in front of a Society that had repeatedly shown no faith in his works, do indicate a sense of faltering in the beliefs that he had once held so strongly. In 1860, he was forced to retire from his practice due to a "mental disorder." On June 29, 1860, he traveled to Brighton and committed suicide by throwing himself out of a window.

Etiology

The most common cause of primary adrenal failure is autoimmune disease of the adrenal cortex. Autoimmune adrenalitis occupies the center of the stage in syndromes characterized by polyglandular endocrine failure. Next to autoimmunity, infectious diseases constitute an important etiology. These include tuberculosis and fungal infections such as histoplasmosis and coccidioidomycosis. An emerging cause for adrenal failure is the setting of acquired immune deficiency syndrome (AIDS). The third category of causes for adrenocortical failure is destruction of the adrenal cortex by metastatic spread from diverse malignant diseases. The fourth cause of adrenal failure is adrenal hemorrhage. The term *Waterhouse–Friderichsen Syndrome* is used to denote acute adrenal hemorrhage that results from septicemic shock. While fulminant meningococcal infection was originally linked with the syndrome of bilateral adrenal hemorrhage, a growing list of organisms, such as *Hemophilus influenzae, Escherichia Coli*, pneumococci, and the newly identified agent the DF-2 bacillus, has been associated with this phenomenon. And, of course, bilateral adrenal hemorrhage is a well-recognized, but rare complication of anticoagulation. Finally, adrenocortical insufficiency can result from use of drugs that interfere with steroidogenesis. In addition to the five categories mentioned (autoimmune, infectious, metastatic, hemorrhagic, and drug related), rare causes of adrenal insufficiency include sarcoidosis and amyloidosis (Table 4). At times, none of the causes mentioned can be identified, and these represent the truly "idiopathic" variety of the disorder.

TABLE 4.
Etiology of Addison's Disease

I. Autoimmune adrenalitis
II. Infectious organisms
    Tuberculous
    Fungal infections
III. Metastatic disease
    Lung
    Breasts
    Stomach
    Non-Hodgkin's lymphoma
IV. Adrenal hemorrhage
    "Waterhouse–Fredrichsen syndrome"
      Meningococcus
      Pneumococcus
      E. coli
      Hemophilus
      DF-2 bacillus
    Anticoagulation therapy
    Bilateral adrenal vein catheterization
V. Drug-induced or related causes
    Withdrawal from steroid therapy
    Adrenolytic therapy
      o p' DDD
      Aminoglutethimide
      Trilostane
    Other agents
      Ketoconazole
      Etomidate
      Rifampin
      Cyproterone acetate
      Anticoagulation (coumadin, heparin)
VI. Rare causes
    Acquired immune deficiency syndrome
    Sarcoidosis
    Amyloidosis
VII. Neonatal adrenal insufficiency
    Enzymatic blocks in cortisol synthesis
    Maternal Cushing's syndrome
    Adrenal hypoplasia
    Adrenal leukodystrophy

Autoimmune Adrenal Failure

Autoimmune adrenalitis is the most common cause of adrenal failure. In a series of 108 patients with Addison's disease reported by Nerup,[6] 66% had autoimmune adrenal disease. This strikes a contrast with the declining incidence of tuberculous disease of the adrenal, which once dominated the etiological spectrum of Addison's disease. For instance, a 1930 review by Guttman[7] confers a striking 70% incidence to tuberculosis and only 17% to atrophic adrenal failure. These figures have reversed over the years, with tuberculous disease currently accounting for approximately 17% of Addison's disease.

Autoimmune adrenal failure results from progressive destruction of both adrenals by an autoimmune process. Antibodies to the adrenal cortex are found in the sera of approximately 60% of patients with "idiopathic" Addison's disease. The autoimmune nature of the nontuberculous variety of Addison's disease was suggested as far back as 1957, when Anderson et al.[8] demonstrated adrenal antibodies in the serum of two patients with Addison's disease. Since then, a plethora of papers has appeared, supporting the autoimmune mediation in development of adrenal failure.[9–16] The methodology for measuring these antibodies had generated some confusion in interpreting the data in earlier literature. The indirect immunofluorescence technique using unfixed human adrenal tissue has become established as a widely used, sensitive method for detecting complement-fixing adrenal antibodies. The ability to accurately measure adrenal antibodies has had a major impact in adrenal endocrinology in three ways: first, it has simplified the etinological approach to adrenal failure; second, it has permitted evaluation of the presence and prevalence of immune markers for the adrenals in patients with other autoimmune disorders; and third, the technique has led to an understanding of the natural history of euadrenal asymptomatic patients with positive antiadrenal antibody titers. Each of these aspects deserves brief mention.

Diagnostic Uility

In general, it is believed that adrenal antibodies are demonstrable in approximately two-thirds of patients with idiopathic adrenal failure. Nerup[17] measured adrenal antibodies in 106 patients with Addison's disease and was able to demonstrate organ-specific adrenal antibodies in 70% of patients with "idiopathic" disease. The specificity of these measurements was notable in that less than 0.1% of controls and less than 1.5% of other disease groups demonstrated such immunopositivity. In the same study, Nerup[17] showed a greater prevalence of thyroid, gastric, and gonadal antibodies in the sera of patients who were adrenal antibody positive. Adrenal antibodies are generally of the IgG class, often fix complement, and react with other steroid-producing cells,[18,19] usually cells of the gonads and placenta. Based on the differential staining pattern, it appears that these antibodies react with a range of antigens in steroid-producing cells. Despite this cross-reactivity, the demonstration of these antiadrenal antibodies in the circulation has great diagnostic value owing to the extremely low prevalence of these markers in normal subjects.

Prevalence of Adrenal Antibodies in Other Diseases

Autoantibodies to the adrenals have been demonstrated in euadrenal patients suffering from other autoimmune diseases. In one of the largest studies of its kind, Ketchum et al.,[20] using an indirect immunofluorescence method, sought the presence of adrenal antibodies in 1675 patients with insulin-dependent diabetes mellitus (IDDM), 2032 relatives of patients with IDDM, and 2543 normal subjects. These authors noted that the frequency of adrenal antibody detection was significantly greater in patients with IDDM and their relatives than in normal subjects. Again, they corroborated that those who were adrenal anti-body positive had a greater incidence of thryoid microsomal and gastric parietal cell antibodies than age-, sex-, and race-matched controls. Autoadrenal anti-bodies are also likely to be found in patients with hypoparathyroidism and ovarian failure.[9,21,22] The relationship between the presence of steroid cell anti-bodies and the presence of ovarian failure is strong.[23,24] Owing to the shared nature of antigens between the steroid-producing cells of the adrenal cortex, patients with autoimmune ovarian failure often develop autoimmune adrena-litis. Indeed, the age at diagnosis of primary autoimmune ovarian failure and the age at diagnosis of autoimmune Addison's disease in the same patient show significant correlation. Autoimmune Addison's disease plays a central role in both type I and type II varieties of pluriglandular failure. The frequency of finding other organ-specific antibodies in a patient with positive adrenal autoan-tibodies is greater than of finding adrenal antibodies in euadrenal patients suf-fering from other autoimmune disorders. For instance, Irvine and Barnes[16] reported that more than 25% of 102 patients with idiopathic Addison's disease had evidence of premature gonadal failure with increased luteinizing hormone (LH) and follicle-stimulating hormone (FSH) levels. In a smaller series, Turk-ington and Lebovitz[25] demonstrated that 23% of Addisonian patients exhibited this combination. While the combined occurrence of multiple-target glandular deficiencies should raise the suspicion of the primary autoimmune, pluriglan-dular end-organ failure, use of this "law of parsimony" may lead to missing the diagnosis of primary pituitary trophic hormone deficiency that mimics plu-riglandular failure.[26,27] Of course, measurement of pituitary hormones will clearly obviate this error.

Significance of Adrenal Antibodies Detected in Asymptomatic Patients

It is now becoming apparent that the presence of adrenocortical antibodies in subjects not diagnosed as having Addison's disease carries an extremely high risk for subsequently developing impaired adrenal function.[28] Scherbaum and Berg[29] identified 30 patients who were adrenal antibody positive and clinically euadrenal by screening 1036 sera. Prospective testing of adrenal function dem-onstrated a high incidence of development of impaired adrenal reserve in this group of patients. A similar number of asymptomatic, adrenal autoantibody–positive patients was studied by Ketchum et al.,[20] who showed that the mean plasma ACTH level in this group was significantly higher than in matched antibody-negative subjects. Despite the normal cortisol levels, this finding may indicate compensated adrenal function (analogous to the normal T_4–high thy-roid-stimulating hormone combination in compensated euthyroidism). This un-derscores the need for close observation to detect the subsequent occurrence of overt or subclinical decompensation, a phenomenon that could be precipitated by stress. Betterle et al.[30] prospectively evaluated nine patients with positive adrenal antibodies in an attempt to predict the subsequent development of Addison's disease. During a 42-month follow-up of these initially asymptomatic adrenal autoantibody–positive subjects, four developed Addison's disease within 1–31 months, and a fifth patient showed reduced adrenal reserve. The authors concluded that the demonstration of complement-fixing adrenal antibodies may be regarded as a valuable marker for individuals in whom clinical adrenal deficiency is highly likely to develop.

A large body of evidence points to cell-mediated immunity as the mechanism of adrenal insufficiency in these patients. The HLA typing of patients with autoimmune Addison's disease has yielded conflicting results. Thomsen et al.[31] reported a significant overrepresentation of HL-A8 and of the LD-8α determinant in 32 unrelated patients with autoimmune Addison's disease. Irvine[28] has noted that the prevalence of B-8 in autoimmune Addison's disease is essentially the same, regardless of whether or not associated with other autoimmune organ-specific disorders. Similarly, other workers[32,33] have failed to find a correlation between autoimmune Addison's disease and prevalence of HLA B-8. Extensive D or DR typing has not been performed in patients with Addison's disease. However, Nerup et al.[34] have indicated the existence of a close correlation between the demonstration of adrenal antibodies in the serum of patients with autoimmune Addison's disease and the presence of DRW 3 in the HLA typing. The consensus of opinion, based on the excellent correlation between the presence of autoantibody and development of the disease, is that genetic influences do play a major role in the development of autoimmune Addison's disease. This was pointed out as early as in 1963, when Dunlop[35] observed that patients with Addison's disease who had relatives suffering from the same disease invariably had the "idiopathic" form of the disease. More recently, Valdemarsson et al.[36] described two siblings with autoimmune Addison's disease whose HLA antigens were determined and compared with the HLA phenotypes of their parents. These studies were compatible with the concept that an Addisonian trait may segregate with the HLA complex in the familial form of Addison's disease. The familial nature of Addison's disease has been well established. This familial aggregation is most striking when Addison's disease occurs as part of the pluriglandular failure of the syndrome.[37] However, even the isolated variety of Addison's disease has been noted to be familial, with well-documented cases occurring in twins or siblings.[38–41] The mode of inheritance is far from understood, presumably owing to the limited number of studies conducted in these settings.

In summary, autoimmune destruction is the most common cause of Addison's disease. The salient clinical aspects are the following:

1. Autoimmune adrenalitis, in the United States, affects Caucasian patients more than Blacks.
2. This may occur as an isolated entity or as part of autoimmune alteration in the function of several endocrine glands, notably the ovaries, parathyroids, the thyroid, and the β cells of the pancreas.
3. Even in the absence of overt involvement of other endocrine glands, autoimmune adrenalitis is associated with a high prevalence of circulating antibodies against other endocrine glands.
4. Females are more commonly affected by autoimmune Addison's disease than males.
5. The condition can be familial. However, a clear pattern of inheritance or an HLA connection has not been convincingly established.

6. The evolution of adrenal failure is slow in autoimmune adrenalitis, the mean duration of the process being approximately 3 years.
7. Patients with autoimmune adrenal failure often go through a phase of compensated adrenal reserve followed by a limited adrenal reserve before eventually decompensating.
8. The autoimmune destruction is confined to the cortex and does not involve the medulla (in contrast to tuberculous adrenal disease).
9. More than 70% of patients with autoimmune adrenal failure demonstrate autoadrenal antibodies by indirect immunofluorescence.
10. The CT appearance, while of diverse nature, fails to show evidence of calcification, a frequent finding in tuberculous and other diseases that cause adrenal destruction.

Infections

Several infectious process can lead to Addison's disease by causing progressive destruction of the adrenal glands. Tuberculosis and fungal diseases are the two broad categories of chronic infections that merit consideration.

Tuberculosis

Tuberculosis of the adrenal gland was once considered the most common cause of adrenocortical failure. Although this incidence has dramatically declined at the present time,[42] tuberculous adrenalitis is still encountered in endemic areas of tuberculosis, which in the United States is synonymous with endemic areas of underprivilege and poverty. Tuberculosis in general, and of the adrenal glands in particular, is still a frequent cause of Addison's disease among American Indians, as well as the underprivileged of any race. This cause must always be considered in patients who have immigrated to the United States from developing nations.

As indicated earlier, a review in the early thirties by Guttman[7] conferred an all-important role for tuberculosis in the causation of Addison's disease. Much of the knowledge regarding tuberculous adrenalitis was gained by experience from the literature of the thirties and forties. Tuberculosis of the adrenals may appear as primary or secondary disease. The adrenal involvement can occur in conjunction with evidence of other tuberculous involvement or as an isolated manifestation with no clinical evidence of extraadrenal tuberculosis. This, however, is less common. For instance, Guttman[7] reported that of 243 patients with tuberculous disease of the adrenal only 7 (3%) had disease exclusively limited to the adrenal gland. A recent review on tuberculosis in the Boston area also emphasized the presence of extraadrenal tuberculosis in patients with tuberculosis of the adrenals.[43] Tuberculous Addison's disease can occur in patient's with *active* tuberculosis elsewhere or can occur years after the initial manifestation of tuberculosis at the primary focus. Thus, it is important to recognize that the absence of overt tuberculosis does not exclude the possibility of adrenal involvement by this disease. Among the usual locations affected by tuberculosis, pulmonary and genitourinary infections predominate. In the large series reviewed by Guttman,[7] there was a 15% incidence of genitourinary tract involvement. The development of tuberculous Addison's disease in association with extensive tuberculosis of the fallopian tubes, uterus, and gastrointestinal tract in the absence of pulmonary involvement has recently been emphasized.[44]

Histopathologically, tuberculosis of the adrenals usually involves both glands, but one adrenal is often more severely affected than the other. Histopathologically, the affected glands may show one of four types of lesions—gross enlargement of the glands by granulomatous inflammatory tissue with destruction of functional adrenals; caseating or noncaseating granulomas; development of a cold abscess in the gland, resembling a mass lesion by ultrasonography or computed tomography; and the eventual development of atrophied adrenal glands, which are reduced to thin streaks of fibrotic remnants. In all cases, varying degrees of calcification are invariably present, a sign that is sensitive but not particularly specific.

Clinically, tuberculous adrenal insufficiency evolves more rapidly than the autoimmune variety of the disease. The mean duration of history in tuberculosis of the adrenal gland is approximately 1 year.[7] Also, the destruction by tuberculosis is not limited to the adrenal cortex, since adrenal medullary involvement is often seen at autopsy. Perhaps for this reason, patients with tuberculous adrenalitis suffer more severe symptoms of orthostasis, which may reflect combined loss of mineralocorticoids and catecholamines. Indeed, tests of adrenomedullary reseve are often impaired in tuberculosis of the adrenals.

Computed tomography (CT) of the adrenals involved by tuberculosis may reveal enlargement with calcification or atrophy. The difference in morphology by adrenal CT may be a reflection of the duration of disease. When adrenal failure from tuberculosis occurs in persons with a history of recent disease (less than 10 years), adrenal enlargement is the likely sequel, while patients with tuberculous infection 20–30 years prior to performance of the tomographic study are likely to demonstrate atrophy.[45]

When evaluating patients suspected of having a tuberculous etiology for the adrenal failure, it is necessary for the clinician to evaluate all lines of evidence for tuberculous disease elsewhere. Several clues, when combined, can weigh in favor of tuberculous adrenal failure. Thus, a positive tuberculin test, old granulomatous disease or fibrotic changes in the lungs, pleural effusion, ascites, hyperglobulinemia, an elevated alkaline phosphatase (the latter two suggesting a systemic granulomatous process), and a pelvic mass in females should all be sought. The importance of documenting tuberculosis as the etiology of Addison's disease has important therapeutic connotations, since it is essential to initiate antituberculous treatment along with glucocorticorticoid replacement therapy. The possibility that antituberculous therapy may preserve and restore function of the adrenals is theoretical and often anecdotal, since such a phenomenon has not been convincingly documented in the literature.

In summary, the salient features of tuberculosis of the adrenals are

1. Tuberculosis of the adrenal continues to remain as important as an etiology for adrenal failure today as it did in the days of Thomas Addison.
2. A past history of tuberculosis or the presence of contemperaneous active disease elsewhere is usually evident; pulmonary, genitourinary, and even gastrointestinal tuberculosis may be present. The implied importance of a careful search for disease at these sites is obvious.

3. The evolution of adrenal failure is more rapid in comparison to autoimmune adrenal failure.
4. Both sexes are equally affected.
5. The destruction by tuberculosis does not spare the medulla, resulting in impairment of adrenomedullary reserve.
6. Autoantibodies to adrenals are characteristically absent in 98–99% of patients with tuberculosis of the adrenals.
7. The presence of adrenal calcification is a hallmark of tuberculous involvement of the adrenals and can be seen roentgenologically in more than 50% of patients. The figure may be higher with CT, a highly sensitive method to detect adrenal calification.
8. Enlargement of the adrenals is the most frequent finding seen by CT. In long-standing disease atrophy can be demonstrated.

Fungal Diseases

Adrenal involvement by histoplasmosis, coccidioidomycosis, or blastomycosis can result in the development of Addison's disease. Of these, histoplasmosis is the more frequently encountered fungal disease in the United States. Recognition of this group of etiological agents has a threefold importance: First, the administration of antifungal chemotherapy with ketoconazole can cause further deterioration in adrenal function already compromised by the fungal disease. The precipitation of adrenal crisis during antifungal therapy is well recognized. Second, early treatment with antifungal therapy, may restore adrenal reserve. Third, fungal disease can mimic a unilateral mass lesion and resemble an adrenal tumor. Correct preoperative diagnosis in such instances may obviate the need for surgery.

Histoplasmosis[46–48]

Histoplasma capsulatum is a fungus endemic to the midwestern region of the United States. Customarily, the manifestations of histoplasmosis are categorized into primary, pulmonic cavitary, and progressive disseminated forms. Adrenal involvement by histoplasmosis is generally encountered in the progressive disseminated form of that disease. In a prospective study of 26 patients with disseminated histoplasmosis, Smith and Utz[49] demonstrated overt Addison's disease in one, while limitation of adrenal reserve was seen in two. In a large multicenter trial reported from the Centers of Disease Control by Sarosi et al.,[50] the frequency of adrenal insufficiency in disseminated histoplasmosis was remarkably higher. In this study of 54 patients with disseminated histoplasmosis, adrenal insufficiency developed in half the patients regardless of treatment and was the most common cause of death. The presence of dissemination of histoplasmosis can be diagnosed by demonstration of systemic (hepatic, hematological, renal, gastrointestinal, or cutaneous) involvement by the fungus, or by culturing the fungus from blood, bone marrow, or tissues. The symptoms of disseminated histoplasmosis reflect systemic involvement and are characterized by fever, weight loss, night sweats, malaise, constitutional symptoms, and cough. Dissemination of histoplasmosis is more common in males, those with underlying illnesses, and those with occupational exposure. It is believed that all cases of primary histoplasmosis probably have hematogenous dissemination of the fungus, but in some the organisms continue to progressively disseminate.

Adrenal involvement may be seen during the active phase of dissemination or may evolve years later when the disease has become "inactive." There is little correlation between the presence of active or cavitary lung disease and the presence of dissemination. However, most patients with adrenal failure secondary to histoplasmosis demonstrate other evidence of systemic involvement by the fungus. Sarosi et al.[50] were unable to find significant differences in age, sex, duration of follow-up, or dose of amphotericin during initial therapy between the groups with and without adrenocortical insufficiency. Of the 27 patients who died, 21 showed bilateral adrenal replacement by granulomas containing *H. capsulatum*. The extraordinarily high incidence of adrenal failure in disseminated histoplasmosis mandates the careful and repeated evaluation of adrenocortical function in patients with progressive disseminated histoplasmosis.

The range of adrenal involvement by histoplasmosis is also impressive.[51-53]

1. The mildest form of adrenal involvement characterized by isolated cortical foci of parasitized macrophages.
2. Extensive caseation necrosis with swelling of both adrenals.
3. Extensive infarction.
4. Granulomatous replacement of the adrenals.
5. Calcified mass lesions that can mimic tuberculous or metastatic lesions; Gibb et al.[54] reported a 74-year-old woman who presented with adrenal failure with bilateral adrenal masses visualized by CT, resembling malignant disease. At necropsy both adrenals were replaced by necrotic masses teeming with *H. capsulatum*, which was also isolated from the hilar nodes and the kidneys. The patient had contracted the disease 20 years earlier. The infection lay dormant for two decades before surfacing as adrenocortical insufficiency. This case was illustrative of the mimicry of histoplasmosis to resemble malignant disease of the adrenals and underscores the need to consider fungal disease as the etiology of adrenal failure in all patients with a history of histoplasmosis, no matter how remote in their past.

Blastomycosis

South American blastomycosis, caused by the fungus *Paracoccidioides brasiliensis*, is known to frequently involve the adrenals. This involvement, which has been noted more frequently postmortem than antemortem, varies between 20 and 80%, with an average of 50%.[55-57] The clinical development of adrenocortical failure in patients with South American blastomycosis has been well documented.[58-60] Like *H. capsulatum* infection, the adrenal involvement by South American blastomycosis is often associated with presence of disease elsewhere, usually from an ulcerated lesion. Adrenal calcification is relatively uncommon. Blastomycosis of the adrenals results in a granulomatous form of Addison's disease, where appropriate antimicrobial therapy has been known to restore adrenal reserve. Osa et al.[61] described the case of an Ecuadorian man suffering from disseminated paracoccidioidomycosis with documented Addison's disease treated with amphotericin B and steroids. Following administration of 4000 mg of amphotericin B, the patient discontinued steroid replacement therapy on his own, but remained asymptomatic. Repeat evaluation demonstrated improvement in the basal steroid production as well as in the glucocorticoid response to exogenous ACTH administration. This is the first documented case in which adrenal reserve recovered following treatment of granulomatous Addison's disease with specific antimicrobial therapy.

Adrenal involvement by North American blastomycosis is also well known.[62-66] Blastomyces dermatitidis is a chronic granulomatous infection indigenous to the southern, southwestern, and midwestern portions of the United States. While clinical involvement of the adrenals by North American blastomycosisis is rare, autopsy series have revealed adrenal involvement in approximately 10% of cases.[67,68] The increasing use of CT in the evaluation of patients with Addison's disease, coupled with the CT-guided thin-needle aspiration, might increase the premortem diagnosis of adrenal blastomycosis.[69]

Metastatic Disease

Autopsy studies indicate that the adrenal glands are a frequent site of metastases. In a group of 1000 postmortem examinations comprised of diverse primary neoplasms, Abrams et al.[70] demonstrated a 27% incidence of metastases to the adrenals. This figure is consistent with the analysis of Glomsett,[71] who found metastatic adrenal involvement in 445 autopsy cases of patients with diverse malignant disorders. This contrasts with earlier reports that conferred an incidence of 9% for adrenal involvement by metastases at autopsy.[72] Antemortem demonstration of massively enlarged adrenals, presumably infiltrated by metastatic disease, is being recognized with increasing frequency as an incidental finding in patients with cancer undergoing CT examinations.[73-75] Despite this striking incidence of bilateral involvement of the adrenals, noted both during autopsy and by antemortem CT evaluations, it has traditionally been held that adrenal insufficiency only infrequently results from metastatic destruction. In a review by Gutman,[7] statistical analysis of 566 cases revealed that metastatic carcinoma accounted for less than 1% of the reported cases of Addison's disease. A review in 1965 by Hill and Wheeler,[76] identified only 22 reported cases of Addison's disease due to metastatic carcinoma. A more recent report and review of the literature by Black et al.[77] found fewer than 50 cases of adrenal insufficiency secondary to metastatic disease of adrenals, most of which had inadequate laboratory documentation. Yet, sporadic, well-documented cases began to appear in the literature recognizing adrenal metastases as an important cause of Addison's disease,[78-80] an observation that dates back to the original report by Addison himself in 1855.

The relative infrequency of clinical recognition of adrenal failure, despite the high frequency of finding metastatic involvement of the adrenals at autopsy, may be due to several factors.

First, it is commonly held that more than 90% of the adrenal tissue needs to be replaced before adrenal insufficiency results.[52,81] This concept, of course, implies the development of complete adrenocortical insufficiency and fails to address the development of limited adrenal reserve due to metastatic destruction (a "pre-Addisonian" state).

Second, the symptoms experienced by patients suffering from disseminated malignancy are strikingly similar to those caused by adrenal insufficiency. Thus, weight loss, weakness, fatigue, lassitude, anorexia, vomiting, and orthostatic hypotension are shared by both groups. Consequently, these symptoms are likely to be attributed to malignant disease, resulting in missing or underdiagnosing adrenocortical insufficiency. Further, the increased pigmentation and gastrointestinal symptoms such as nausea and vomiting may be mistakenly attributed to the chemotherapy that most such patients receive, when in fact these symptoms may have been caused by adrenocortical insufficiency.

The problem can be further compounded by the fact that hyponatremia, a reflection of adrenocortical failure, is often mistakenly attributed to the syndrome of inappropriate ADH secretion (SIADH), admittedly a far more frequent accompaniment of malignant disease than adrenocortical failure.

Finally, the steroid therapy given to many patients with malignant diseases, as part of the chemotherapy protocols, can lead to masking of partial or complete adrenal insufficiency.

For all these reasons, it is probably true that adrenal insufficiency in malignant disease is probably underdiagnosed, and underrepresented in the literature. Yet, recognition of adrenal insufficiency assumes importance, since specific therapy can add an immense measure of palliation to the quality of life of patients suffering from disseminated malignancy. The adrenal glands can be involved by metastastic spread from any primary focus, but the four most common primaries are from the lung, breasts, melanoma, and lymphoma. Thus, autopsy studies have demonstrated adrenal metastases in 42% of lung tumors,[72] in 58% of breast cancers,[71] and in 50% of malignant melanomas.[82] Sahagian-Edwards and Holland[83] found evidence of adrenal infiltration in 28.7% of bronchogenic tumors and in 34% of breast cancers. Autopsy data from patients with non-Hodgkin's lymphoma suggest that the adrenal glands may be involved by lymphoma in 25% of instances.[84,85] The adrenal involvement by non-Hodgkin's lymphoma is usually seen when the lymphoma is widespread and simultaneously present in the retroperitenal area or the kidney.[86–88] Very rarely, non-Hodgkin's lymphoma may be limited only to the adrenal gland.[89] Also rarely, primary adrenal failure may be the initial manifestation of malignant lymphoma. Osei et al.[90] reported a case of a 55-year-old man presenting with clinical and biochemical evidence of primary adrenal insufficiency, with subsequent development of malignant lymphoma. At autopsy the architecture of both adrenal glands was completely effaced by malignant plasmacytoid cells. Thus, adrenal involvement by lymphoma can antedate the generalized manifestations of lymphoma. These reports challenge the notion that adrenal involvement by metastatic disease occurs only in the advanced stages of malignancy.

In addition to primary adrenal failure that occurs due to neoplastic spread, occasionally the chemotherapy used to treat leukemia or lymphoma may cause secondary adrenal insufficiency. Tobin et al.[91] described two patients with chronic myelogenous leukemia who developed a form of ACTH deficiency following treatment with 6-mercaptopurine. Similarly, the use of busulfan may be associated with the development of a syndrome consisting of hyperpigmentation, severe weakness, fatigue, anorexia, nausea, and loss of weight.[92] The mechanisms for the development of such a phenomenon are unclear.

The adrenal insufficiency that develops in patients with metastatic disease can be partial or complete. The existence of limited adrenal reserve in patients with metastatic adrenal disease is illustrated by the case reported by Shea et al.[89] of a patient with symptoms of adrenal insufficiency who showed normal basal cortisol levels but failed to mount a response to ACTH challenge. Since adrenal insufficiency may develop abruptly during intercurrent stress, patients with malignancy, especially those with enlarged adrenals demonstrated by CT, need to be screened for adrenal insufficiency. In a series of 21 patients with metastatic cancer and enlarged adrenals who were studied by Seidenwurm et al.,[93] 19% developed symptomatic adrenal insufficiency. The authors went as far as to suggest initiating prophylactic maintenance glucocorticoid therapy as soon as the diagnosis of adrenal metastases is made. In addition to causing adrenal destruction by infiltration, the metastatic adrenal glands are also prone to hemorrhage.

The diagnosis of adrenal metastases can be readily made by the use of CT. Using equipment with a fast scan time, and narrow (<1 cm) interval sections, it is possible to identify both adrenals in 97–99% of patients.[94,95] Adrenal metastases are generally bilateral and reveal an enlarged appearance on CT. Tumors as small as 5 mm have been detected by this imaging technique. The excellent correlation between abnormal adrenal tomography in patients with advanced malignancy and demonstration of disease by surgery or at autopsy has been pointed out by Cedermark and Ohlsen.[75] If histological confirmation is desired, CT guided per cutaneous fine-needle biopsy can be performed with safety and simplicity[96,97] where expertise is available.

In summary, the following observations are pertinent as they relate to metastatic etiology of adrenal failure.

1. The increasing survival rate of patients with cancer and the increasing use of CT are resulting in an increased recognition of adrenal metastastes antemortem.
2. Adrenal insufficiency in patients with malignant disease is probably underdiagnosed, perhaps owing to the mistaken attribution of symptoms to the underlying neoplasm.
3. Since CT of the adrenals is a simple, noninvasive procedure, it should be recommended as a part of the workup for patients with disseminated malignancy.
4. The demonstration of enlarged adrenal glands by CT is reason enough to recommend adrenal function testing even in the absence of symptoms of adrenal insufficiency.
5. Compromised adrenal function in metastatic adrenal disease can be partial or complete. The diagnosis of limited adrenal reserve in such a setting would necessitate prophylactic glucocorticoid therapy.
6. While the routine prophylactic administration of maintenance glucocorticoid therapy to all patients with malignancy and enlarged adrenals by CT has been recommended, this practice is not universally accepted.
7. The diagnosis of adrenal insufficiency and the resultant institution of therapy can considerably palliate the suffering of these patients.

Adrenal Hemorrhage

The first description of hemorrhage into adrenals probably dates back to the days of Griselius of Vienna in 1670. Although rare, bilateral adrenal hemorrhage is an important cause of sudden death from adrenal crisis.

Histopathologically two anatomical types of adrenal bleeding can be recognized at necropsy. The term *adrenal apoplexy* is used by British workers to describe a form of adrenal hemorrhage characterized by large central, clotted hemorrhage that distorts the gland and stretches the capsule. The other form of adrenal hemorrhage is characterized by multiple coalescing hemorrhages throughout the cortex and medulla, leaving the gland swollen.

Etiologically, the three classic settings associated with bilateral adrenal hemorrhage are fulminant infections, anticoagulant therapy, and trauma to the abdomen or thorax.

Waterhouse–Friderichsen Syndrome

The adrenal hemorrhage associated with fulminant septicemia is termed the Waterhouse–Friderichsen syndrome (WF syndrome). This syndrome, first recognized by Rupert Waterhouse and Carl Friderichsen, was observed in infants and children with overwhelming septicemia caused by *Neisseria meningitidis*. In 1936, Aegerter[98] reviewed the literature that then consisted of 56 reported cases of the syndrome. The notable feature described in most cases was the enormous size of the adrenal glands, which exhibited a hemorrhagic purplish hue. The shock occurring during the course of endotoxemia seems to be the cause of adrenal hemorrhage, and not vice versa. Levin and Cluff[99] in 1965 showed that an adrenal gland actively engaged in manufacturing steroids is sensitive to endotoxic damage. A Schwartzman-like phenomenon affecting the blood vessels may result from the injurious effects of the endotoxins. The WF syndrome bears a close resemblance to the experimental findings seen in connection with the Schwartzman phenomenon.[100,101] The role of the meningococcus is believed to be important in causing endothelial damage and subsequent thrombosis in addition to the endotoxin release. The development of WF syndrome in septicemic shock is multifactorial and unresolved. The relative roles of endotoxemia, vasoconstriction, the Schwartzman reaction, and the "stressed adrenal" are not clear. Nevertheless, the presentation is characterized by overwhelming sepsis, hypotension progressing to shock, a hemorrhagic diathesis, and often a purpuric rash. The bleeding is, at least in some cases, related to the rapidly developing disseminated intravascular coagulation caused by endotoxins.[102,103] Death results from the combination of adrenocortical failure and irreversible septic shock.

As more cases of WF syndrome were recognized, organisms other than *N. meningitidis* emerged as causative organisms associated with the syndrome. Grant et al.[104] reported the development of WF syndrome in a patient with pneumococcic shock. The relationship with pneumoccocal-mediated WF syndrome and splenectomy is intriguing. Splenectomy, splenic agenesis, and splenic disease have frequently been recognized in patients who develop adrenal hemorrhage as a consequence of fulminant pneumococcal bacteremia.[105–107] Rarely, overwhelming sepsis caused by *Hemophilus influenzae* type b may be associated with the WF syndrome.[108] More recently, a new species of organism called DF-2 bacillus (dysgonic fermenter bacillus) has been associated with the development of WF syndrome.[109] DF-2 bacteremia usually follows a dog bite and, as with pneumococcal bacteremia, appears to affect splenectomized individuals. Since its original description by Butler et al.[110] as a "new disease of man," several cases have been reported.[111–114] The clinical features consist of a febrile illness, symmetrical peripheral gangrene, ecchymosis, coagulopathy, adrenal hemorrhage with shock, and sometimes endocarditis. A generalized Schwartzman-like reaction is believed to exist in patients with DF-2 bacteremia. WF syndrome caused by septicemia due to bacterial causes carries a high mortality rate. In patients who survive, the adrenal failure is generally permanent. Reversible adrenocortical insufficiency has been reported rarely following survival of WF syndrome.[115]

Anticoagulation

The sudden development of adrenal insufficiency due to adrenal hemorrhage is a well-recognized complication of chronic anticoagulant therapy. Several isolated case reports[116–123] and reviews[124,125] have highlighted this occurrence. The sudden development of abdominal or back pain, anorexia, nausea, vomiting, hypotension, and altered sensorium in a patient on anticoagulation therapy should immediately raise the possibility of adrenal hemorrhage. While this complication can occur in any person anticoagulated with both coumadin and heparin, its occurrence is more common in older patients with increased capillary fragility and in those overanticoagulated. The observation by Albert et al.[126] that acute bilateral hemorrhage into the adrenals can be demonstrated by CT even in the absnece of abnormal serum electrolytes is important. It underscores the need for performing adrenal reserve testing in all anticoagulated patients who develop acute abdominal pain and hypotension regardless of the electrolytes or coagulation profile. Prompt clinical recognition of acute adrenal hemorrhage in anticoagulated patients can be quite difficult owing to its similarity to other disorders. A high index of suspicion coupled with immediate adrenal reserve testing and CT would readily permit accurate diagnosis. This is especially relevant since early institution of therapy is lifesaving. The value of CT of the adrenals in establishing the diagnosis of acute adrenal hemorrhage has been well established.[126,127]

Trauma

Adrenal hemorrhage can occur secondary to thoracic or abdominal trauma.[128,129] Crushing of one or both adrenals, particularly the right, against the vertebral columns can result in rupture of the central vessels. As a result of the bleeding, the adrenal glands are distended by the blood, with destruction of the cortex and medulla. Thrombosis of the adrenal veins with hemorrhagic infarction can also result in loss of adrenal function. Fox[130] reviewed 78 cases of adrenal hemorrhage and necrosis and found adrenal vein thrombosis in 32. It is unclear whether this finding is primary or secondary.

In addition to external trauma, adrenal hemorrhage can result from investigational trauma caused by invasive procedures, particularly bilateral adrenal venography.[131–133] The occurrence of this complication should be suspected when the pain associated with injection of dye persists for more than a few minutes. The development of adrenocortical insufficiency following venography has been well documented. In fact, recognition of this complication has in large part led to the obsolescence of this procedure.

In addition to overwhelming septicemia, anticoagulation therapy, and trauma, several other causes for adrenal hemorrhage have been reported in the literature. Table 5 outlines the common and rare causes of adrenal hemorrhage.

Drug-Induced Adrenocortical Insufficiency

Drug-induced adrenocortical insufficiency constitutes a rare, but important cause of hypoadrenalism. Several classes of drugs can contribute to partial or complete adrenal failure.

1. Of course, the classic example of adrenal failure occurs after withdrawal of steroid therapy. Chronic suppression of the hypothalamic–pituitary–adrenocortical axis by exogenous steroid therapy underlies the hypoadrenalism of this category. The clinical and biochemical facets of the steroid withdrawal syndrome are discussed in Chapter 4.
2. The adrenal insufficiency that results from the use of antiadrenal agents is often deliberate and intentional to treat hypercortisolism. The use of adrenolytic therapy with o p′ DDD, aminoglutethimide and the use of agents such as metyrapone or trilostane to effect enzyme inhibition belong in this category.
3. The adrenal insufficiency that results from anticoagulation therapy with heparin and coumadin occurs as a consequence of adrenal hemorrhage. In addition to this mechanism, use of heparin can result in hypo-

TABLE 5.
Causes of Adrenal Hemorrhage

I. Overwhelming septicemia (Waterhouse–Friderichsen syndrome)
    Neisseria meningitidis
    Diplococcus pneumoniae
    Hemophilus influenzae type b
    DF-2 bacillus
II. Anticoagulant therapy
III. Trauma
    Abdominal and thoracic surgery
    Postadrenal venography
IV. Rare causes
    Hemorrhagic diathesis
    Neonatal adrenal hemorrhage
    Burns
    Pregnancy
    Hematological diseases (leukemia)
    Pancreatitis
V. Metastatic disease

aldosteronism even in the absence of adrenal hemorrhage. The mechanism of heparin-induced selective hypoaldosteronism are outlined in chapter 8.

4. Finally, drugs administered for other purposes may inadvertently result in the development of adrenal insufficiency. Four such drugs deserve mention—ketoconazole, rifampin, etomidate, and cyproterone acetate. The mechanisms underlying the adrenal insufficiency associated with each of these agents will be outlined next.

Ketoconazole

Ketoconazole, an imidazole derivative, is a broad-spectrum antifungal drug, widely used in treatment of infections caused by the *Coccidioides, Histoplasma, Candida, Blastomyces,* and *Cryptococcus* species of fungi. The ease of oral administration (usually as a single dose of 200–400 mg/day) coupled with an extremely low incidence of adverse reactions has made ketoconazole a popular antifungal drug. The observation that gynecomastia developed in some males taking ketoconazole[134,135] led to further evaluation of its effects on gonadal steroidogenesis. Since the basic mechanism of ketoconazole is inhibition of 14-demethylation of lanosterol to ergosterol in fungi,[136] it seemed probable the ketoconazole may also inhibit important enzymatic pathways in gonadal steroid synthesis. Several reports have established that ketoconazole blocks testosterone synthesis by the testes,[137–139] probably by creating a block in the enzyme 17-hydroxyprogesterone aldolase.[140] Once development of gynecomastia and hypogonadism became recognized as adverse side effects of ketoconazole therapy, these observations were naturally extended in studying the effects of this drug on adrenal stereoidogenesis. Pont et al.[141] evaluated the cortisol response to ACTH in healthy volunteers following oral ketoconazole and demonstrated significant blunting of cortisol response to ACTH 4 hr after the drug, persisting up to 8 hr. Further in vitro studies showed that the drug virtually eliminated corticosterone production by isolated adrenal cells of rats. These in vitro data are similar to its potent inhibitory effects on testosterone synthesis by isolated Leydig cells from rats.[142]

The mechanism of ketoconazole-induced adrenal suppression is believed to be mediated by inhibition of the cytochrome P-450–dependent enzymes. Loose et al.[143] used radiolabeled substrates and high-performance liquid chromatography to elucidate the site of inhibition in the adrenal gland. They were able to demonstrate two important blocks caused by ketoconazole: inhibition of 11-β-hydroxylation and inhibition of a more proximal step in the side-chain cleavage that converts cholesterol to pregnenolone. Both 11-β-hydroxylase and the side-chain–cleaving enzyme are mitochondrial enzymes that are cytochrome P-450 dependent. Ketoconazole did not inhibit conversion of progesterone to DOC or the other steps in adrenal steroid synthesis mediated by non-P-450–dependent enzymes. Thus, it appears that ketoconazole has a predilection to inhibit mitochondrial enzymes, especially those that are dependent on P-450.[144] In addition, ketoconazole binds to the glucocorticoid receptor and exerts an antagonist activity at the target level.[145]

Although the potential for adrenal suppression by ketoconazole is well recognized,[146] the clinical development of hypoadrenalism with the use of ketoconazole has been an extremely rare event. Several explanations have been cited to account for the relative rarity of hypoadrenalism, despite the demonstration of striking and potent in vitro suppression of adrenal steroidogenesis. First, ketoconazole is given usually as a single daily dose. Since the drug-induced inhibition of adrenal steroidogenesis is only short lived (8 hr), recovery from the block results in normal synthetic activity for the most part of the day. Second, there may be adaptations to overcome these blocks. For example, increased cholesterol uptake by the adrenals, augmented by ACTH, may override the effects of drug-induced inhibition. Third, limited adrenal reserve developing on ketoconazole therapy may go undetected since these patients are asymptomatic. Despite the reported rarity of hypoadrenalism with the drug, the use of high doses of ketoconazole may precipitate adrenal failure. Reversible adrenal insufficiency has been described by Tucker et al.[147] in a patient receiving experimentally high doses of ketoconazole for pulmonary blastomycosis. Individual susceptibility probably plays an important role in expression of ketoconazole-induced adrenal suppression. Suffice to say that the monitoring of adrenal (and gonadal) functions is mandatory in patients receiving ketoconazole therapy.

Etomidate

Etomidate is an intravenous sedative–hypnotic widely used in Europe as a "safe" anesthetic.[148,149] Ledingham et al.,[150] in 1983, reported low plasma cortisol levels in patients receiving etomidate. Since etomidate is an imidazole derivative, structurally analogous to ketoconazole, considerable attention has been focused on the effect of this anesthetic on adrenal steroid synthesis. Wagner et al.[151] evaluated the cortisol and aldosterone responses to ACTH in five patients receiving etomidate. Marked adrenal suppression was found in all five, persisting in one case as long as 4 days after discontinuation of the drug. In vitro studies corroborated this finding and showed that etomidate produced a concentration-dependent blockade of two mitochondrial enzymes, the cholesterol side-chain–cleaving enzyme and 11-β-hydroxylase, both dependent on cytochrome P-450. Thus, the antisteroid effects of etomidate bear a striking resemblance to those of ketoconazole, without the peripheral effects of the latter that cause glucocorticoid antagonism at the end-organ level.

Although, overt adrenal insufficiency has not been reported with the use of etomidate, the blunted cortisol and aldosterone reserve may assume critical importance during stress. Thus, the development of postoperative complications in etomidate-treated patients may require treatment with glucocorticoids. Two reports[152,153] have substantiated the development of a clinical picture consistent with adrenal insufficiency in patients who had blunted cortisol responses to ACTH while receiving etomidate.

Rifampin

Rifampin, a macrocyclic antibiotic produced by *Streptomyces mediterrranei*, is an invaluable component in the chemotherapy of tuberculosis. Edwards et al.[154] were among the first to recognize that rifampin increased cortisol catabolism by accentuating hepatic microsomal induction. While administration of rifampin per se does not induce adrenocortical failure, the use of rifampin in patients with borderline adrenal disease may result in precipitation of adrenal crisis. Two cases of tuberculosis patients in whom use of rifampin resulted in precipitation of acute adrenal crisis were reported by Elansary and Earis.[155] Rifampin increases the oxidation of cortisol to 6-β-hydroxycortisol.[156] While this represents only a minor pathway of cortisol catabolism, in the patient with limited adrenal function any reduction in the amount of metabolically active glucocorticoids has a tremendous impact. In most patients in whom rifampin precipitated acute adrenal crisis, the event occurred within 2 weeks of institution of therapy. Therefore, in patients with tuberculosis placed on rifampin the importance of considering adrenal insufficiency in the first 2 weeks of rifampin therapy is emphasized.

In addition, Addisonian patients receiving treatment with glucocorticoids may require larger doses of cortisone when rifampin therapy is instituted. The patient reported by Edwards et al.[154] required increased corticosteroid dosage while receiving rifampin. The pharmacological half-life of cortisol, which was reduced during rifampin treatment, reverted to normal when the rifampin was stopped. This important effect of rifampin must be kept in mind whenever rifampin is administered to Addisonian patients on replacement glucocorticoid therapy.

Tuberculosis continues to represent an important cause for Addison's disease in some parts of the United States. Since rifampin is almost always included in the first line of therapy for tuberculosis, and since many patients with tuberculous Addison's disease may have near-normal basal plasma cortisol levels, our recommendation is to perform a rapid ACTH stimulation test in all patients with tuberculosis prior to placement on rifampin. In patients with evidence of limited adrenal reserve, we strongly consider glucocorticoid supplementation at least during the first 2 weeks of rifampin therapy, since such a practice clearly averts the development of rifampin-induced adrenal crisis in patients with limited adrenal reserve.

Cyproterone Acetate

Cyproterone acetate possesses antiandrogenic and progestational properties; it has found therapeutic application in Europe in treatment of such diverse disorders as virilization in women, hypersexuality and "deviationism" in males, and precocious puberty in children of both sexes. Cyproterone is known to cause secondary adrenal insufficiency.[157] In addition, Savage and Swift[158] reported the development of primary adrenal suppression in children treated with cyproterone acetate for precocious puberty. An editorial in *Lancet*[159] advocated steroid cover for intercurrent illness and surgery during treatment with cyproterone and 12 months thereafter. The experience in the United States with cyproterone is too limited to confirm these findings.

Rarer Causes of Primary Adrenal Failure

Rare causes of a rare disease need to be considered only rarely. Sarcoidosis can affect the adrenal glands. Most reports of sarcoid involvement of the adrenal glands are based on autopsy findings.[160] Adrenal failure may occur during the granulomatous phase or during the phase of healing when fibrosis supervenes. In general, adrenal failure due to sarcoidosis occurs in the presence of diffuse sarcoid involvement of several organs, such as lung or liver. The ability of sarcoidosis to involve multiple endocrine glands can result in the development of multiple-target-organ failure mimicking the autoimmune pluriglandular failure syndromes. The combination of sarcoid-induced Addison's disease and sarcoid-induced hypothyroidism has been recognized.[161] In clinical practice, the diagnosis of sarcoid-related adrenal failure rests on excluding other disease, coupled with evidence for widespread sarcoidosis often with elevated serum level of angiotensin-converting senzyme.

Adrenocortical failure secondary to disseminated cytomegalovirus (CMV) infection used to be considered a rare event.[162,163] This fact, however, may have to be reconsidered in light of the increase in the incidence of AIDS. CMV inclusions have been demonstrated at autopsy in the adrenals of patients with AIDS.[164]

Patients with AIDS may develop primary adrenal failure. Disseminated infection involving the adrenal glands has occasionally been described in AIDS.[165,166] Several infections seen in patients with AIDS can involve the adrenals. Thus, fungi, mycobacteria, or CMV can invade the adrenals and cause adrenal insufficiency. The situation can be compounded by the administration of drugs such as ketoconazole which may further compromise adrenal function. Guenthner et al.[167] reported a case of documented Addison's disease in a patient who developed AIDS 4 months later.

Amyloidosis, by virtue of its diffuse infiltration of almost every organ, can also involve the adrenals. Hepatomegaly is invariably present, and the diagnosis of amyloidosis can be established by demonstrating amyloid deposits in the rectal mucosa, liver, kidney, or peripheral nerve.

Hemochromatosis can infiltrate the adrenals and result in adrenal insufficiency. Secondary hemochromatosis is more likely to cause iron deposition in the adrenals than the primary form. Primary hemochromatosis involves the pituitary more frequently than it involves the adrenals.[168] Adrenal involvement by secondary hemochromatosis (iron overload) often shows a characteristic hyperdense appearance on the CT scan,[169] while the size of the adrenals may be normal or reduced. In a large series of patients with secondary hemochromatosis Long et al.[170] found poor correlation between the increased glandular density by CT and adrenal hypofunction.

Adrenal failure in the newborn can be due to hypoplasia of the adrenals. Karenyi[171] described two types of adrenal hypoplasia in infancy—the anencephalic (or secondary) type, where the adrenals were hypoplastic, and the cytomegalic type, where the adrenal cortical cells were large with abnormal nuclear and cytoplasmic staining. Both forms are believed to represent developmental failure and are an imprtant etiology for death during infancy.[172] Neonatal adrenal failure can also result from congenital adrenogenital syndrome with complete blocks in 21-hydroxylation or in the conversion of cholesterol to pregnenolone. Another rare cause of neonatal adrenal insufficiency is maternal Cushing's syndrome.[173] Finally, adrenal hemorrhage, probably during birth trauma, represents an important cause of neonatal adrenal failure and death.

Adrenocortical failure associated with degenerative neurological changes (adrenomyelopathy) is discussed elsewhere in this chapter in the section dealing with the syndromes associated with adrenal failure.

Clinical Features

Addison's disease can be the proverbial masquerader. The symptoms can be so nonspecific as to be passed off as insignificant. The presentations of the disease can be so diverse and the recognition so difficult that the patient may have seen several specialists in widely different disciplines before the diagnosis is made. Thus, it is not uncommon for the chronically ill patient to have seen the gastroenterologist (for diarrhea), neurologist (for muscle weakness), hematologist (for anemia), cardiologist (for "dizzy spells"), psychiatrist (for depression), and dermatologist (for skin problems) before the entire picture is put together by the astute generalist and referred to the endocrinologist.

Addison's disease, in its early stages, is insidious in onset and evolves gradually and imperceptibly. The symptoms of weakness and lassitude my persist on such a chronic basis that many patients hopelessly attribute their state of ill being as a basal state. Many such patients, after having sought several medical opinions that disclosed "nondisease," are placed on vitamins. Some turn to bizarre remedies rather than face multiple physicians and a possible diagnosis of hypochondriasis. For instance, Cotterill and Cunliffe[174] described a patient who turned to excessive ingestion of licorice, which with its mineralocorticoidlike properties warded off symptoms of mineralocorticoid failure. Others take to excessive salt intake to avoid symptoms of volume depletion. Most patients give the typical history that their energy level is fair in the morning, but experience rapid tiring as the day progresses. The classic picture described by Addison consisting of "typical pigmentary changes and wasting in a patient with a feeble pulse" should not be allowed to develop if the diagnosis is suspected early.

The three major symptoms of Addison's disease are weight loss, fatigue, and pigmentary changes. In the series of 108 patients with Addison's disease reported by Nerup[6] weight loss and fatigability (weakness) were present in 100% of cases, while 92% showed hyperpigmentation. Since fatigability and weight loss may occur in the absence of other stigmata of Addison's disease, it is reasonable to screen all patients with unexplained weight loss and fatigue for the presence of adrenocortical failure.

Addison's disease can involve almost all the systems. Therefore, the symptoms and signs are best viewed in terms of individual systems.

Constitutional

These are the most frequent symptoms experienced by patients with adrenocortical failure. Malaise, loss of energy, lassitude, and a general feeling of ill health are important symptoms of adrenal failure, which are, unfortunately, highly nonspecific. These symptoms are mistakenly attibuted to overwork, stress, or psychological factors. If the question "Could these symptoms reflect Addison's disease?" were asked in every patient with these constitutional symptoms, the disease would be missed less often. It has been the experience of most endocrinologists that patients with Addison's disease have made several visits with persistent complaints, yet unfortunately the diagnosis is established often only when acute decompensation occurs. The importance of listening to these symptoms cannot be overemphasized. As anorexia and weight loss manifest, once again Addison's disease tends to be missed because the focus has often shifted to a malignancy workup. Once again, too many Addisonian patients too often have undergone gastrointestinal investigations and total body imaging before a simple and careful history is put together and an adrenal reserve testing is done. The dictum that all patients with weight loss must be screened for thyroid disease and adrenal insufficiency has served us well at our institution in identifying several patients each year with atypical hyperthyroidism as well as adrenal failure. Weight loss, weakness, and fatigability were also universally noted in the patients with Addison's disease reported by Dunlop[175] and the 26 patients reported by Ask-Upmark and Hall.[176]

Dermatological

Increased pigmentation of skin and mucous membrane, a reflection of an increase in circulating β-lipotropin levels, was encountered in 92% of the 108 patients reported by Nerup.[6] The increased pigmentation is seen in whites as well as in blacks. At our institution, where the majority of diagnosed Addisonians are black, the increased skin tone is sensed by patients readily. A "slatelike," shiny-dark hyperpigmentation can be quite readily recognized. The characteristic features of hyperpigmentary changes described in Addison's disease are most applicable to Caucasian patients. Patients often are intrigued by the ease with which their skin tans, and by the lingering of suntan for longer periods than usual. The degree of pigmentation can be highly variable. In some light-skinned subjects, the only sign of hyperpigmentation may be the development of "black freckles." At the other extreme, the hyperpigmentation can be intense enough to cast doubt on the patient's racial origin. This was illustrated in one of our Hispanic patients with Addison's disease, who was refused his welfare check by the clerk, who mistook him for being black while the computer insisted that he was not! The three characteristics of the hyperpigmentation of Addison's disease are:

1. Striking involvement of the extensor surfaces such as the dorsum of the hands, (especially over the joints), elbows, knees, and palmar creases.
2. Mucosal pigmentation involving the lips, buccal mucosa, dental gingival margin, and the tongue.
3. Pigmentation of the scars, especially those acquired after the onset of Addison's disease.

It is generally considered that hyperpigmentation in Addison's disease indicates chronicity. Guttman,[7] in his 1930 review of 566 cases, noted that when pigmentation was the initial presenting feature, the average duration of the disease was much longer than when weakness was the predominant symptom at onset. It was also generally observed that when hyperpigmentation is present in Addison's disease, the other manifestations of the disease are quite apparent.

Exceptionally, pigmentation can be an early and sole symptom of Addison's disease. Strakosch and Gordon[177] described a healthy 20-year-old woman with pigmentation of skin and buccal mucous membrane as the only symptom of Addison's disease, which was documented by endocrine testing. Rarely, Addisonian type of pigmentation may involve the nails, resulting in longitudinal banded pigmentation of nails.[178]

In addition to hyperpigmentation, vitiligo may be evident in patients with autoimmune adrenal failure. The incidence of vitiligo has been variably reported from 4%[6] to 20%.[16] The vitiligo often occurs in the hyperpigmented areas, striking a curious contrast.

Gastrointestinal

Gastrointestinal manifestations are present in more than 50% of patients with Addison's disease. Lower abdominal cramps, anorexia, and nausea are seen often in these patients. In some patients with Addison's disease, the gastrointestinal symptoms can be severe enough to dominate the clinical picture. The stress of an intense gastrointestinal workup, including panendoscopies, can be conducive to the precipitation of decompensation. In Addisonian crises, the gastrointestinal manifestations become severe and are invariably present. Adrenal crisis is often ushered in by nausea, vomiting, lower abdominal cramps, and diarrhea. Indeed, the diarrhea can be severe enough that the initial clinical suspicion may be gastroenteritis. The diarrhea and vomiting seen in Addisonian crisis contribute significantly to the dehydration and volume depletion seen with this catastrophe.

The reasons for the presence of gastrointestinal symptoms in Addison's disease and adrenal crisis are unclear. These symptoms are undoubtedly related to glucocorticoid deficiency, since dramatic resolution occurs with administration of hydrocortisone. Regardless of the mechanism, patients with unexplained gastrointestinal symptoms, particularly chronic diarrhea, should be tested for adrenal insufficiency.

Cardiovascular

Hypotension is present in 80–90% of patients with Addison's disease. Moderate to occasionally severe orthostasis is encountered in nearly all patients with hypotension due to Addison's disease. Infrequently, the supine blood pressure may be entirely normal, but may demonstrate striking changes when the patient assumes erect posture. A history of chronic hypertension that ameliorated spontaneously despite discontinuation of antihypertensive medications is an excellent clue to the presence of Addison's disease. While orthostatic hypotension is a frequently demonstrable sign, less than 20% of patients complain of postural dizziness. In fact, unlike patients with other forms of postural hypotension, Addisonian patients feel well in the morning when they arise from a long period of recumbency. In contrast to autonomic neuropathic situations that cause orthostatic hypotension, Addisonian patients demonstrate a well-preserved rise in the pulse rate upon standing.

The reasons for othostatic hypotension are twofold—volume and sodium depletion resulting from mineralocorticoid deficiency. In the acute setting, vigorous repletion of intravascular volume by administration of normal saline rapidly corrects the orthostatic hypotension. Replacement therapy with chronic glucocorticoid administration corrects the hypotension exceedingly well. The hypotension in Addison's disease is an excellent illustration of the relative importance of angiotensin II and sodium balance in maintenance of normal blood pressure. The serum concentration of angiotensin II is extremely high in patients with Addison's disease; despite this elevation, the pressor effects of angiotensin are not expressed, owing to the extreme negative sodium balance. Replacement of mineralocorticoid rapidly corrects the negative sodium balance and restores the blood pressure to normal. It is a clinical observation that the effect of pressor substances on vascular smooth muscle is blunted in the presence of volume and sodium depletion. The role of the sodium content in the cells of the arterial vasculature may arguably contribute to such blunting. Regardless, the administration of mineralocorticoids rapidly restores normotension. Indeed, patients with Addison's disease may, during the initial phase of replacement, demonstrate a heightened sensitivity to mineralocorticoid administration.

Adrenal insufficiency should always be excluded in the evaluation of patients with orthostatic hypotension. In spite of the fact that adrenal insufficiency is a relatively infrequent cause of orthostatic hypotension, it is one of the few causes amenable to cure of the orthostatic hypotension. Several clues may be helpful in distinguishing the orthostatic hypotension caused by adrenal failure. First, the postural drop in blood pressure is mild to moderate. Severe orthostasis is uncommon, since the autonomic cybernatic responses are intact. Second, the response of the heart rate that is normally seen after standing is usually preserved in Addisonian orthostatic hypotension. This contrasts with the orthostatic hypotension associated with primary and secondary autonomic neuropathy, where marked attenuation in the heart-rate response accompanies the drop in the blood pressure. Third, the symptoms of postural dizziness are only mild to moderate in patients with Addison's disease. Severe symptoms and sequelae such as falling down are uncommon with the orthostatic hypotension of adrenal failure. Absence of other clinical features of Addison's disease would be rather unusual. In summary, the magnitude and sequelae of the orthostatic phenomenon of Addison's disease are less pronounced than, for instance, in the autonomic dysfunction of Shy–Drager syndrome. Yet, since it is totally curable, Addison's disease must be excluded in all patients with orthostatic hypotension, regardless of the severity.

It is essential to underscore the fact that Addisonian patients on replacement are not exempt from subsequently developing hypertension and cardiac failure due to other causes. Knowlton and Baer[179] have observed the development of cardiac failure in 7 of 22 patients with primary adrenal insufficiency when followed prospectively. Revisions in the management of such patients require decrease in dietary sodium intake as well as elimination of mineralocorticoid supplementation. Similarly, hypertension and edema have been reported to complicate pregnancy in an Addisonian patient.[180] These reports highlight the fact that Addisonian patients, once they are treated are not immune from the vascular and cardiac factors that precipitate and perpetuate cardiac failure and hypertension.

Musculoskeletal

Muscle weakness is a subjective symptom that is present in most patients with adrenal failure. The lack of anabolic androgen steroids probably underlies the muscle weakness of adrenal failure. Although the subjective symptoms may be severe, muscle power is often reasonably well preserved. Rarely, flexion contractures have been reported in association with Addison's disease, gradually improving with replacement therapy.[181,182] In some cases, the muscle contractures can be associated with moderate to severe pain.[183] It has been postulated that the underlying mechanism for the muscle contractures involves an internal shift in water and electrolytes in muscle tissue, occurring as a consequence of glucocorticoid deficiency. This is supported by the observations of Wisenbaugh and Heller,[184] who described an Addisonian patient with disabling flexion contractures of the knee, which reversed with only glucorticoid but not mineralocorticoid replacement. Recent reviews on the musculoskeletal manifestations of Addison's disease[185,186] have emphasized the muscle aching and stiffness that accompany the disorder. An extreme example of the stiffness is seen in the "stiff-man syndrome" that has been described in association with hypocortisolism secondary to ACTH deficiency.[187] The hallmarks of this disorder are progressive muscular rigidity that impairs volitional movement coupled with paroxysms of painful muscle spasms often precipitated by emotional stimuli. The cause of the stiff-man syndrome is not known. Although the syndrome has been reported in association with thyrotoxicosis,[188] diabetes mellitus,[189] and ACTH deficiency,[187] it has not been reported in primary adrenal failure.

Rheumatic symptomatology in Addison's disease also includes the frequently described "pain in the loin" often with downward transmission. In fact, literature of the early thirties emphasizes pain in the costovertebral angle as a consistent feature of Addison's disease. First described by Rogoff,[190] this finding has come to be known as "Rogoff's sign." Recent reviews, however, have not mentioned this physical finding. The possibility that Rugoff's sign may in fact represent a finding in tuberculous adrenalitis might explain the relative infrequency of this finding in an era when tuberculosis is declining as a cause of Addison's disease.

Renal

Several abnormalities in renal function can be encountered in patients with Addison's disease.

1. A decrease in the glomerular filtration rate and in the renal blood flow is a frequent accompaniment of Addison's disease.[191]
2. An elevated blood urea nitrogen (BUN) and/or creatinine levels are often seen as a consequence of the decreased glomerular filtration rate (GFR) and the prerenal azotemia.
3. Decreased secretion of ammonia and hydrogen ion results in mild metabolic acidosis. Administration of an acid load is less readily cleared by the kidneys in patients with adrenal insufficiency. The acidifying defect is due to the absence of mineralocorticoids. The problems with urinary

acidification in aldosterone deficiency are discussed at greater length in Chapter 7.

4. In addition to the problem with hydrogen ion excretion, patients with Addison's disease demonstrate an inability to normally excrete a water load. It is believed that the abnormalities in urinary dilution seen in Addison's disease are a result of the decreased GFR caused by salt and water depletion. This results in increased reabsorption of water in the proximal tubule (vasopressin independent), reducing the delivery of filtrate to the diluting segments of the nephron. Consequently, the urine is less than maximally dilute. The natriuresis from the mineralocorticoid deficiency further contributes to urinary concentration. Several lines of evidence support this theory.

a. Ufferman and Schrier[192] studied the diluting capacity of the nephrons of adrenalectomized dogs. Their observations indicated that the free-water clearance was normal up to 4 days without any steroids as long as the animals were fed abundant sodium chloride. If salt and water depletion was allowed to occur, the animals rapidly became hyponatremic, with a decrease in free-water clearance and urine that was less than maximally dilute.

b. In patients with Addison's disease, restoration of water diuresis can be demonstrated simply by expanding the extracellular fluid volume by normal saline.[193]

c. A direct effect of glucocorticoids has also been suggested because of the striking improvement in free-water clearance following hydrocortisone therapy in patients with adrenal insufficiency.[194]

d. Finally, an impressive line of evidence that the impaired diluting ability seen in adrenocortical failure is non-ADH dependent comes from experiments with Brattleboro rats. These rats are a special strain and are predisposed to Diabetes Insipidus (DI) (lack of ADH). When these rats with DI were adrenalectomized, they became unable to maximally dilute urine, implicating adrenal insufficiency as the cause of the phenomenon and ruling out ADH mediation.[195]

Psychiatric

The psychiatric symptom seen most often in association with adrenal failure is depression. The lassitude seen in many Addisonian patients may reflect a mild form of endogenous depression. Addison, in his excellent monograph,[1] described a patient with primary adrenal failure who had developed severe depression 8 years before his death from the disease. Rajathurai et al.[196] reported an interesting case of a 14-year-old girl with self-mutilation and Addison's disease. The psychiatric profile in this withdrawn, silent child was highlighted by a 3-year history of her repeatedly gouging out pieces of skin from her arms and legs every night before falling asleep. Her condition continued despite intense counseling. Addison's disease was diagnosed when she developed acute adrenal insufficiency. The remarkable aspect of the case was the dramatic disappearance of self-mutilation and a change in her demeanor 1 week after steroid therapy was begun. Rare as these events are, they underscore the variegated manifestation of Addison's disease.

Gonadal

Oligomenorrhea, even amenorrhea, and impotence are often present in patients with Addison's disease. While these symptoms may occur as a consequence of the chronic illness and debilitation seen in Addison's disease, primary autoimmune gonadal failure may be the underlying mechanism in some cases. The development of the amenorrhea–galactorrhea syndrome in a patient with Addison's disease has been reported by Refetoff et al.[197] Replacement with conventional doses of cortisone acetate resulted in complete normalization of the previously elevated prolactin levels, along with spontaneous resumption of regular menses and disappearance of galactorrhea. At the other extreme, a rare association of primary adrenal failure is the development of precocious sexual development. Marilus et al.[198] described an 11-year-old boy with Addison's disease and advanced pubertal development, with gonadotropin responses to luteinizing-hormone releasing hormone (LHRH) that were appropriate for his clinical stage of puberty. The extraordinarily elevated plasma ACTH levels in this patient led to the proposal by the authors that the concomitant appearance of Addison's disease and true precocious puberty might be due to a "drift" phenomenon of LHRH and/or gonadotropins in response to the prolonged and profound increase in ACTH secretion. The phenomenon is analogous to the association of precocious puberty with the high thyroid-stimulating hormone (TSH) of juvenile hypothyroidism.[199,200]

Bronchopulmonary

The possibility that adrenocortical insufficiency may play a role in the natural history of bronchial asthma has been a matter of controversy. It is questionable if deteriorating adrenal function unmasks a latent asthmatic tendency. The general consensus favors the notion that the development of bronchial asthma in Addisonian patients is a purely fortuitous association. Green and Lim[201] evaluated two patients who presented with recent-onset bronchial asthma and weight loss and uncovered the presence of underlying Addison's disease in both. A similar case was reported by Hadley.[202] The frequency of bronchial asthma and the rarity of its association with Addison's disease diminish the impact of these reports. The role of glucocorticoid hormones in preventing the development of asthma remains heavily contested; in most reviews of Addison's disease, bronchial asthma has not been mentioned as an association.

In summary, the clinical features of Addison's disease can be striking enough to afford detection at first glance or can be subtle enough to be missed unless a high index of suspicion is entertained. Screening for Addison's disease is justifiable under the following circumstances:

1. Unexplained weight loss
2. Weakness and asthenia

3. Orthostatic hypotension
4. Hyperpigmentation
5. Presence of nonspecific symptoms in patients with pluriglandular failure or vitiligo
6. Electrolyte abnormalities (hyponatremia, hyperkalemia)
7. Dehydration and shock, especially in the presence of hypoglycemia

Associations of Addison's Disease

Addison's disease, especially the autoimmune variety, can be associated with several syndromes. The most important association of Addison's disease is with the polyglandular autoimmune syndrome. Less commonly, Addison's disease can be part of a neurological syndrome characterized by progressive demyelination of the white matter of the brain. The syndromes of adrenoleukodystrophy and adrenomyeloneuropathy belong in this category. Even rarer are the associations of adrenal failure with the beguiling POEMS syndrome, achalasia cardia, and renal microangiopathy.

Polyglandular Deficiency Syndromes

Autoimmune adrenal failure is often associated with deficiency syndromes that involve other endocrine glands. Primary hypoadrenalism, primary hypothyroidism, primary hypogonadism, and diabetes mellitus constitute an important tetrad, the common factor being autoimmunity.

The association of idopathic Addison's disease and primary hypothyroidism dates back to the original description by Schmidt.[203] In the early 1960s, Gastineau and Arnold[204] reviewed the Mayo Clinic experience with 500 patients with Addison's disease and noted that 11 patients had overt myxedema and 7% had abnormal palpatory findings of the thyroid. Carpenter et al.[205] found evidence of thyroid disease in 27 of 174 patients with Addison's disease. Primary hypoadrenalism and primary hypothyroidism are the two most frequent components of the type II pluriglandular syndrome, being present in 100% and 69% of patients, respectively.

Gonadal failure is especially prevalent in females with autoimmune adrenal disease. Irvine and Barnes[206] evaluated 157 females and 79 males with Addison's disease and found that amenorrhea or oligomenorrhea were present in 37% of postpubertal females. Antisteroid cell antibodies were demonstrable in more than 50% of patients with Addison's disease and secondary amenorrhea, whereas very few males had circulating anti–Leydig cell antibodies. Thus, gonadal failure is more often encountered in females with autoimmune Addison's disease than in males.

The association between diabetes mellitus and "idiopathic" adrenal failure has been emphasized by several reports.[207–216] This association probably occurs more frequently than recognized. Since the manifestations of diabetes are more familiar than those of adrenal insufficiency, it is easier to recognize diabetes in the Addisonian patient than the reverse situation. The temporal relation between the two diseases when they occur in the same patient can be impressively variable. In the series reported from Joslin Clinic by Yoo and Kozak[210] diabetes usually preceded the development of Addison's disease. Gittler et al.,[217] however, reported that in two of their three cases, Addison's disease preceded the development of diabetes by 9–15 years. In the collected series of 113 cases of concomitant diabetes and adrenal failure reviewed by Solomon et al.,[209] diabetes appeared first in 63% of cases; Addison's disease appeared first in 23%, while a simultaneous onset was found in 10% and the sequence was unspecified in 4%. The temporal separation in the evolution of these two diseases in the same patient can be as short as 1 year, or as long as 15, with an average of 6 years, when diabetes mellitus occurs as part of the polyglandular syndrome, especially type II. In this setting, the diabetes is more likely to be insulin dependent, often with circulating antibodies against the β cells.

In addition to its association with hypothyroidism, hypogonadism, and diabetes mellitus, autoimmune adrenal failure has been associated with hypoparathyroidism,[218–227] vitilgo,[228,229] and pernicious anemia.

The concept of polyglandular failure has evolved in the past decade and has become a well-defined entity. Neufeld et al.[230,231] have classified the polyglandular autoimmune syndromes into three types. This classification is based on age of onset, the association of certain endocrine disorders, and human leukocyte antigen (HLA) association. Autoimmune Addison's disease is an important constituent of types I and II.

Type I polyglandular autoimmune disease has its onset in childhood and bears no HLA association. The most prevalent endocrine dysfunction in type I is hypoparathyroidism (82%), followed by adrenal failure (67%). Less frequent endocrinopathies noted in the type I syndrome include gonadal failure (12–17%), autoimmune thyroid disease (10%), and diabetes mellitus (2–4%). Type I polyglandular autoimmune syndrome is highlighted by two other facets; mucocutaneous candidiasis is present in 73–78% of patients with this syndrome.[232–236] Trence et al.[237] have emphasized the nature of the mucocutaneous candidiasis seen in association with type I polyglandular autoimmune syndrome. The fungal infection is often chronic, can be quite resistant to conventional antifungal therapy, and can become quite extensive with diffuse involvement. The other interesting facet of the syndrome is the association with autoimmune disorders of the gastrointestinal tract such as chronic active hepatitis and malabsorption syndromes. The dermatological components include vitiligo, and alopecia (Table 6). The only hyperfunctional state that can be associated with the pluriglandular failure syndrome is Graves' hyperthyroidism, a condition characterized by circulating antibodies against the TSH receptor.

The type II polyglandular autoimmune syndrome is characterized by the universal presence of autoimmune adrenal failure. This is followed by autoimmune thryoid disease, usually chronic lymphocytic thyroiditis,[238,239] seen in approximately 69% of patients, and diabetes mellitus, usually insulin dependent, in 52% of patients. Less common associations include gonadal failure, vitiligo, pernicious anemia, and celiac disease (Table 7). One characteristic of the diabetes that develops in type II polyglandular autoimmune syndrome is that insulin-dependent diabetes (IDDM) occurs at a much older age (median 36 years) when compared to age of onset of IDDM in the general population.

The type III polyglandular autoimmunity syndrome is characterized by

TABLE 6.
Type I Polyglandular Autoimmune
Syndrome[230,231,237]

A. Components
  1. Endocrine
      Hypoparathyroidism (82%)
      Addison's disease (67%)
      Hypogonadism (12–17%)
      Autoimmune thyroid disease (10%)
      Diabetes mellitus (2–4%)
  2. Nonendocrine
      Chronic mucocutaneous candidiasis
        (73–78%)
      Pernicious anemia (13–15%)
      Chronic active hepatitis (12%)
      Malabsorption syndromes (22–24%)
      Vitiligo (8%)
      Alopecia (26–32%)
B. Age
    Childhood onset
C. HLA association
    None

absence of Addison's disease, but the presence of IDDM, pernicious anemia, vitiligo, alopecia, or other organ-specific autoimmune disorders. It is a "catchall" category to include autoimmune disorders other than Addison's disease. Should Addison's disease eventually evolve in this category of patients, this would necessitate reclassification into type I or type II. The list of organ-specific antibodies that can be demonstrated in patients with all three types of polyglandular autoimmune syndromes is bewildering.[240] These antibodies are listed in Table 8.

The importance of recognizing the coexistence of multiple endocrinopathies in patients with Addison's disease has several clinical implications. First,

TABLE 7.
Type II Polyglandular Autoimmune
Syndrome[230,231,237]

A. Components
  1. Endocrine
      Addison's disease (100%)
      Autoimmune thyroid disease (69%)
      Diabetes (usually IDDM) (52%)
      Hypogonadism (3.5%)
  2. Nonendocrine
      Pernicious anemia (less than 1%)
      Vitiligo (4.5%)
      Celiac disease
B. Age
    Adult onset
C. HLA type
    Higher incidence of B8-Dw3-DR3

TABLE 8.
Organ-Specific Antibodies in Polyglandular Autoimmune Syndrome

1. Thyroid	Antithyroglobulin; antimicrosomal; antibodies against TSH receptors and thyroid cell surface antigen; antibodies against retroorbital tissues
2. Pancreas	Antibodies against pancreatic β cells, α cells, and δ cells; antibodies against insulin receptors
3. Parathyroid	Anti–chief cell antibodies; anti–oxyphil cell antibodies
4. Gonads	Antibodies against steroid-producing cells; antiova and antisperm antibodies; anti–Leydig cell antibodies
5. Pituitary	Antibodies against lactotropes and somatotropes
6. Other organ-specific antibodies	Antiparietal cell gastric antibodies; anti–intrinsic factor antibodies; antimelanocyte antibodies; anti–smooth muscle antibodies; anti–acetylcholine receptor antibodies
7. Nonorgan-specific antibodies	Antimitochondrial antibodies; anti–DNA and RNA antibodies

the presence of hypogonadism is common to polyglandular syndromes as well as hypopituitarism. Thus, the clinical mimicry of polyglandular autoimmune syndrome to panhypopituitarism has been well recognized in the literature[224,241–245]; of course, the measurement of pituitary trophic hormones would clearly differentiate between the two entities. Second, the likelihood of patients with autoimmune Addison's disease developing a second or even third endocrinopathy is as high as 25%. This, obviously, underscores the need for careful search for the presence of other endocrine dysfunction not only at the time at which Addison's disease manifested, but prospectively for the rest of the patient's life. Third, and from a nonendocrine perspective, the patient with autoimmune adrenal failure needs to be screened for other systemic autoimmune disorders, particularly pernicious anemia. Fourth, the role of HLA antigens in predisposing patients to the polyglandular autoimmune syndrome is undergoing intense investigation.[246–250] The association of the HLA B8-DW-DR3 with the development of type II polyglandular autoimmune syndrome has been emphasized in the literature. Therefore, the screening of family members in this context assumes relevance.

Adrenoleukodystrophy and Adrenomyeloneuropathy

Adrenoleukodystrophy and adrenomyeloneuropathy are progressive degenerative disorders of the brain and spinal cord occurring in association with primary adrenal failure. The classic adrenoleukodystrophy (ALD) is characterized by progressive development of dementia, blindness, quadriparesis, and adrenal insufficiency.[251] Progressive degeneration of the white matter of brain ultimately results in death. Histopathologically, linear cycloplasmic inclusion bodies are found in brain macrophages and adrenocortical cells. The condition is believed to be inherited as an X-linked disorder.

Adrenoleukodystrophy resembles other demyelinating disorders of the brain in its tendency to relentlessly progress. The clinical lines that separate adrenoleukodystrophy, Schilder's disease, metachromatic leukodystrophy, and subacute sclerosing panencephalitis are extremely nebulous. The combination of a demyelinating disease and adrenocortical deficiency was first called "bronzed Schilder's disease."[252] Schaumburg et al.[253] have identified 50 cases in the literature, to which they have added the clinical and pathological findings in 17 of their cases. The disorder is not as rare as once considered, and awareness has led to discovering more cases. In addition to the systematic description of the clinical features of the syndrome, Schaumburg and colleagues have provided valuable insight into the histological and biochemical basis for the syndrome.[253–258]

Adrenoleukodystrophy typically affects children during the first decade of life. Behavioral problems, school failure, visual disturbances, and motor deficits with caudal–rostral progression is the usual presentation. Inexorable decline, progressing to death within 6 years of developing neurological symtoms, is the usual course of this disease. The adrenal insufficiency can evolve before, simultaneously, or after the development of the neurological syndrome. In addition to this childhood form, ALD can also afflict adults. Visual deterioration, mental confusion, behavioral changes, and spastic paraplegia eventually supervene. The neurological syndrome can evolve as late as 20 years after the development of Addison's disease.

The tissues affected in ALD demonstrate characteristic changes. The adrenocortical cells demonstrate crystal aggregates, a feature also seen in the Schwann cells, peripheral nerves, testes, and macrophages of the central nervous system. A defect in cholesterol metabolism involving unusually long-chain fatty acids is believed to underlie the pathophysiology of ALD.[259,260]

Adrenomyeloneuropathy (AMN) is a clinical variant of ALD. In addition to the association of primary adrenal failure and neurological deterioration, this entity is characterized by variable degrees of hypogonadism and distal polyneuropathy.[261,262] The same pathological process that involves the white matter of brain also involves the Leydig cells of the testes. In addition to primary hypogonadism, the spectrum of adrenomyeloneuropathy has extended to involve the hypothalamus and the pituitary. Peckham et al.[263] described a 32-year-old man with contractures, peripheral neuropathy, primary adrenal insufficiency, and abnormal pituitary function. The abnormalities in pituitary function included low gonadotropin levels despite hypotestosteronemia, minimal gonadotropin responses to LHRH, impaired growth hormone reserve, and mild hyperprolactinemia. Another interesting case, report by Fettes et al.,[264] described a 36-year-old man who developed hypogonadotropic hypogonadism 12 years prior to the onset of neurological dysfunction and primary adrenal failure.

The temporal relationships between adrenal failure and the development of neurological symptoms of AMN have been outlined well by Schaumburg and colleagues.[253,256] The adrenal insufficiency in this variant of ALD usually has its onset during childhood. The other components of the syndrome usually evolve in the third decade and include the development of spastic paraparesis, sphincter disturbances, peripheral neuropathy, and impotence.

Both adrenoleukodystrophy and adrenomyeloneuropathy are inherited as X-linked recessive traits. Davis et al.[265] have documented four cases of ALD and one case of AMN in one kindred over three generations, suggesting that AMN may be a clinical variant of ALD. The observation that the adrenal insufficiency in these patients can be partial stresses the need for careful evaluation of adrenal function in patients with ALD and AMN.

Rare Associations

Rarely, Addison's disease can be part of the POEMS syndrome, may be associated with achalasia cardia with alacrimation, or can occur with renal microangiopathy.

The term *POEMS syndrome* refers to the acronym coined by Bardwick et al.[266] to describe a constellation characterized by polyneuropathy, organomegaly, endocrinopathy, M proteins, and skin changes (POEMS). The mechanism for the development of this multisystem disorder is not known, but since its original description in 1968 by Shimpo,[267] at least 40 patients with this syndrome have been indentified.[268–275] The disease is predominantly seen in middle-aged men, particularly in Japanese, and is characterized by the universal occurrence of polyneuropathy, which is both sensory and motor, and is usually progressive. Biopsy of the nerves reveals varying degrees of axonal degeneration. Hepatomegaly, splenomegaly, and lymphadenopathy are present in 68%, 39%, and 65% of patients, respectively. The endocrinopathies encountered in the syndrome include gynecomastia (69%), impotence (68%), amenorrhea (100%), glucose intolerance (50%), primary hypothyroidism (12%), and Addison's disease (9%). One of the inherent problems in evaluating the data in this rare disorder is that endocrine evaluation of patients reported to have the POEMS syndrome has been incomplete; moreover, a large number of these patients had received chemotherapy and/or radiation to the lumbosacral spine and the pelvic bones, procedures that might account for the high incidence of impotence and gynecomastia. The monoclonal gammopathy of the syndrome is predominantly of the IgG lambda chain type, often with increased plasma cells in the bone marrow. Thus, the basic disorder is a plasma cell dyscrasia, characterized by the unique occurrence of sclerotic bone lesions. The cause of POEMS syndrome is not known.

Achalasia, Alachrymia Syndrome

The rare syndrome of isolated adrenal glucocorticoid failure with normal aldosterone production associated with achalasia cardia and defective tear production (alacrimation) has recently been discribed.[276–279] There has been no satisfactory explanation for this rare syndrome. Postulated mechanisms include ACTH insensitivity, degenerative processes that affect the adrenal glands and the autonomic nerve structure, and a loss of parasympathetic input to the adrenal gland.

Finally, Sachdev et al.[280] have described a rare syndrome of Addison's disease associated with renal microangiopathy and renal failure. The renal lesions were characterized by glomerular damage and thrombomicroangiopathic changes in the afferent arterioles and the intralobular arteries. Both patients described in the report continued to deteriorate despite glucocorticoid replacement and died as a result of their renal disease.

Laboratory Diagnosis

The laboratory diagnosis of Addison's disease can be viewed in terms of several perspectives. First, the nonspecific clues derived from the routine tests will be outlined. Second, the screening tests for adrenocortical insufficiency will be discussed. Third, the confirmatory tests to establish the diagnosis will be discussed. Fourth, the steps taken to identify the cause of adrenal insufficiency will be outlined. Finally, the extraadrenal workup to detect other concomitant hormonal abnormalities will be reviewed.

Nonspecific Routine Tests

Nonspecific, but important, information from routine tests might often raise the first suspicion of adrenal insufficiency. Particularly relevant are the changes in serum electrolytes, glucose, BUN/creatinine, and calcium.

The typical electrolyte abnormality in Addison's disease is the combination of hyponatremia and hyperkalemia. The two reasons for these changes are the loss of mineralocorticoid effect on the renal handling of these ions and the loss of glucocorticoid action on the sodium–potassium pump that maintains a normal gradient between the intracellular and extracellular sodium and potassium. The fact that hyponatremia occurs more frequently in Addisonian patients than hyperkalemia is supported by the observation of Nerup,[6] who noted that hyponatremia was present in 88% of 106 patients while hyperkalemia was seen in only 64%. Hyperkalemia can be masked in Addisonian patients with potassium loss from vomiting, diarrhea, or diuretic abuse. Hyperkalemia is nearly always present during adrenal crises. Treatment with glucocorticoids alone can lower the serum potassium level, ascribing an important role to the effect of glucocorticoids on the sodium–potassium pump. In the absence of cortisol, there is loss of the normal $Na^+–K^+$ gradient across the cell, with resultant movement of potassium out of the cell, and sodium into it. With glucocorticoid administration, the gradient is reestablished, and the hyperkalemia promptly resolves.

The BUN and creatinine are mildly elevated in patients with adrenococortical failure due to a decrease in the GFR and renal flow. In patients with acute adrenal crises, the presence of dehydration magnifies these effects, often resulting in impressive prerenal azotemia.

Hypoglycemia in Addison's disease is due to inadequate hepatic gluconeogenesis. The decreased food intake due to anorexia usually compounds the problems. Patients with adrenocortical failure poorly tolerate fasting. Profound, symptomatic hypoglycemia may develop when Addisonian patients are fasted for diagnostic procedures. Also, hypoglycemia with hypotension may be seen in patients with Addisonian crisis. The counterregulatory mechanisms to combat hypoglycemia are extremely impaired in these patients owing to loss of glucocorticoids and often catecholamines. The presence of hypoglycemia in a patient with "shock" must immediately bring Addisonian crisis to mind, since in most other situations characterized by shock, the blood sugar is not low, because the "stress hormones" are present in adequate amounts.

Hypercalcemia is a rare feature of Addison's disease.[281–284] The degree of hypercalcemia is generally mild to modest, although occasionally hypercalcemic crisis has been reported.[285] The mechanism of hypercalcemia is unclear, but seems to be related to cortisol deficiency, since it completely normalizes after cortisol replacement. Glucocorticoids increase urinary excretion of calcium, probably as a consequence of expansion of extracellular fluid volume which suppresses reabsorption of sodium and calcium by the proximal nephron. It is, therefore, possible that mineralocorticoid deficiency with resultant contraction of extracellular fluid volume could lead to enhanced proximal tubular reabsorption of calcium.

The development of hypercalcemia has been reported to occur in patients with idiopathic hypoparathyoidism, when Addison's disease supervenes.[286] This interesting observation is particularly significant since these two diseases may occur in the same patient as part of the polyglandular autoimmune syndrome. Thus, the development of hypercalcemia in hypoparathyroid patients may not only be secondary to overzealous treatment, but may also herald the development of autoimmune Addison's disease. Reductions in the filtered load of calcium, coupled with its enhanced tubular reabsorption by the kidneys, are conducive for the hypercalcemia in this setting. Concomitant adrenal failure may mask the characteristic "renal tubular leak" of calcium that is usually associated with hypoparathyroidism; following cortisone therapy the serum calcium promptly normalizes and will continue to drop unless vitamin D and calcium therapy are instituted.

The hemogram in patients with Addison's disease often shows mild anemia, moderate eosinophilia, and relative lymphocytosis. These findings revert to normal following institution of glucocorticoid therapy.

Screening Tests

The three screening tests utilized for detecting adrenal insufficiency are measurement of basal steroid levels in plasma; evaluation of the cortisol and aldosterone response to an intravenous bolus of cosyntropin; and evaluation of plasma 11-deoxycortisol level following an overnight metyrapone test. Of these, the cosyntropin test is the most sensitive.

Basal Steroid Measurement

Measurement of 8 A.M. (or 4 P.M.) cortisol levels in the plasma is a procedure with low sensitivty for screening. When the 8 A.M. plasma cortisol level is below 5 μg/dl, the diagnosis of adrenal insufficiency is strongly suspect, while a level in excess of 25 μg/dl excludes it. Plasma cortisol levels in the range of 10–20 μg/dl can still be associated with adrenal insufficiency, when the disease is partial and the adrenals are functioning at maximal capacity. Thus, in such instances, screening with only basal cortisol levels in plasma carries the dangerous risk of missing the diagnosis of a potentially fatal disease. When the disease is suspected, the presence of normal basal cortisol levels should not deter one from pursuing further diagnostic tests.

For the same reasons, measurement of basal 24-hr urinary free cortisol and 17-hydroxycorticosteroids fails to achieve a desirable sensitivity in screening for Addison's disease. The inconvenience of urine collection, and the low sensitivity of these tests have been major reasons for discarding these determinations for screening purposes. These tests, however, are invaluable as parameters that reflect adrenal responsiveness to exogenous ACTH administration.

The Rapid ACTH Stimulation Test

This dynamic procedure evaluates the immediate response of the adrenal cortex to the exogenous administration of synthetic ACTH ("cosyntropin"). Cosyntropin contains the first 24 amino acids of ACTH and has all the biological activity of the intact hormone. This is a simple test to detect adrenal insufficiency. The test involves, in its simplest form, measuring the serum cortisol at the basal state and 30, 60, and 90 min following the intravenous administration of 2.5 μg of cosyntropin. The "normal" response varies according to different workers, but a widely accepted criterion is a peak level that doubles *and* increases by at least 10 μg/dl above the baseline, following ACTH. An absolute peak level of 25 μg/dl of cortisol has also been considered as a normal response (chapter 1, in the section "Testing Adrenal Cortical Function").

Three important statements need to be emphasized regarding the rapid ACTH test.

First, the test is, at best, only a screening procedure and separates the normal responder from the nonresponder. The demonstration of a clearly normal response precludes the possibility of primary adrenal failure.

Second, a blunted or flat cortisol response to rapid ACTH administration can be seen in both primary and secondary (pituitary) hyponadrenalism.

The blunted adrenal response in pituitary disease is due to the development of adrenocortical dormancy from chronic lack of ACTH. The atrophied zona fasciculata, deprived of its trophic stimulus, fails to show an immediate response to a single bolus of ACTH. However, the simultaneous measurement of aldosterone in conjunction with cortisol may help differentiate between primary and pituitary hypoadrenalism on the basis of the rapid ACTH stimulation test. In patients with primary adrenal failure, both the cortisol and aldosterone responses to ACTH are abolished owing to intrinsic disease of the adrenal cortex. However, in patients with pituitary hypoadrenalism, although the zona fasciculata may remain dormant without responding to intravenous administration of ACTH, the zona glomerulosa responds to a pharmacological bolus of ACTH with a normal rise in serum aldosterone concentrations following ACTH. Thus, the demonstration of a rise in aldosterone following ACTH in the hypoadrenal patient who showed an impaired cortisol response to the same trophic stimulus favors the diagnosis of secondary hypoadrenalism. Cunningham et al.[287] compared the rapid ACTH test with the overnight metyrapone test in five patients who had undergone hypophysectomy and 27 who had recently been treated with glucocorticoids. They found that 11 patients who had normal adrenal responses to consyntropin failed to show adequate responses to metyrapone. They also noted that no patient who responded normally to metyrapone failed to show a response to cosyntropin. Thus, a normal cortisol response to cosyntropin may be encountered in some patients with secondary adrenal failure. A clearly normal response, however, does indeed exclude primary adrenal failure (Addison's type).

Finally, and most important, the response patterns of patients with partial adrenal insufficiency (the limited adrenal reserve syndrome) and those under stress can be quite variable and can often defy interpretation. For instance, some patients with partial adrenal insufficiency may show a near-normal response on the rapid challenge, while decompensating under continuous stimulation. In contrast, normal subjects under stress may show no appreciable rise following ACTH since they are functioning at full capacity.

In patients with acute adrenal crisis, where withholding of therapy is not advisable, the rapid ACTH test can be still performed in the following manner. After the basal sample is drawn, 16–20 mg of methylprednisolone and 10 mg of desoxycorticosterone are administered to provide immediate glucocorticoid and mineralocorticoid support. Simultaneously, the cosyntropin is administered intravenously and blood is sampled at 30, 60, and 90 min following ACTH for cortisol and aldosterone level in the plasma. The rapid ACTH test is merely a screening test, and confirmation of the diagnosis by other standard tests is crucial prior to committing the patient to lifelong replacement therapy.

The Metyrapone Test

Metyrapone is an inhibitor of the 11-β-hydroxylase enzyme that converts 11-deoxycortisol (compound S) to cortisol. The "short" version of the test is performed by administering 30 mg/kg of metyrapone orally at midnight and measuring the levels of cortisol, compound S, and ACTH at 8 A.M. the next morning.[288,289] The danger of precipitating acute adrenal insufficiency by further lowering the cortisol due to administration of metyrapone is not a significant problem with the overnight test. Normal subjects respond to metyrapone by demonstrating an increase in the levels of ACTH and compound S, following overnight administration of metyrapone. Of course, hypocortisolemia should be demonstrated to ensure adequacy of the metyrapone-induced block in 11-β-hydroxylation. At our institution, the normal rise in compound S following overnight metyrapone is greater than 60 ng/ml. The demonstration of a subnormal rise in compound S levels despite hypocortisolemia and adequate ACTH release is strong evidence for primary adrenocortical failure.

At our institution, Cook County Hospital, the screening for adrenal insufficiency in a patient receiving no steroid coverage consists of the following steps:

1. The basal plasma cortisol at 8 A.M. is first measured. If the 8 A.M. cortisol is greater than 25–30 μg/dl, the diagnosis of adrenal insufficiency is no longer considered, and no further investigations are pursued in that direction.

2. If the cortisol level is below normal, a rapid ACTH test, using cosyntropin, is performed. A basal sample for plasma ACTH is collected and saved for future assay if necessary. Similarly, the samples for aldosterone are also saved for future assay, if deemed appropriate. Thus, for cost-effectiveness, only the cortisol determinations are performed in the samples before and after ACTH. If the cortisol response is unequivocally normal, the diagnosis of primary adrenal insufficiency has been excluded with reasonable certainty. The other stored samples are discarded.

3. If the cortisol response to ACTH is subnormal, or absent, the diagnosis of hypoadrenalism is entertained. Now, the plasma aldosterone levels are determined in the sample that was saved. An absent or impaired aldosterone response, in the presence of an absent or impaired cortisol response to ACTH, would suggest primary adrenal disease. If the aldosterone response is normal, pituitary disease needs to be excluded. The plasma ACTH level in the stored sample can now be assayed to provide additional help.

4. If the cortisol response to cosyntropin is borderline (i.e., neither completely normal nor completely abnormal), an overnight metyrapone test is performed. This test has the advantage of screening the entire hypothalamic–pituitary–adrenal axis. The demonstration of a subnormal rise in compound S levels in the plasma following metyrapone, despite the demonstration of an adequate block as well as an adequate rise in plasma ACTH, has almost established the diagnosis of primary adrenal insufficiency.

Confirmatory Tests

The confirmation of primary adrenal failure rests on demonstrating impaired or absent adrenal responsiveness to protracted exogenous ACTH administration, coupled with elevated basal ACTH levels.

The diagnostic application of the rapid as well as the standard ACTH stimulation tests is discussed in detail in Chapter 1.

The use of other probes for adrenocortical function has not found wide application. The concept that theophylline may be used to evaluate adrenocortical function is interesting. Theophylline presumably stimulates intracellular steroidogenesis by circumvention of ACTH–receptor interaction. Geffner et al.[290] evaluated the adrenal response to theophylline infusion in five patients with Addison's disease. Interestingly, it was demonstrated that some patients with Addison's disease who failed to respond to ACTH demonstrated cortisol responsiveness to theophylline infusion. It was postulated that preservation of theophylline responsiveness in spite of loss of ACTH responsiveness might be related to the duration of adrenal insufficiency or to the etiological mechanism. With the limited data available, theophylline is still regarded as an investigational tool in the elevation of adrenocortical reserve.

Etiological Diagnosis of Addison's Disease

Once the diagnosis of Addison's disease has been established, it is important to find the etiology of the primary adrenal failure for several reasons. First, if autoimmune disease of the adrenal, the most common cause today, is established, the physician should screen for endocrine dysfunction involving other endocrine glands. Second, if an infectious cause is identified [tuberculosis (TB), fungal disease], appropriate therapy for the offending pathogen needs to be instituted. Rarely, the institution of such therapy can result in "reversing" adrenal insufficiency. Third, if metastatic disease is established as the cause of adrenal disease, appropriate therapy for the primary neoplasm may become necessary. Rarely, the adrenal manifestation may indeed be the first clue to the presence of neoplastic disease.

With this reasoning in mind, several lines of approach can be utilized for the etiological evaluation of primary adrenocortical failure. These include the use of CT, measurement of antibodies against the adrenal cortex, and reviewing routine biochemical data to detect evidence of infectious disease.

Computerized Tomography of the Adrenals

CT of the adrenals can provide extremely valuable information regarding the etiology of primary adrenal failure. The size, the density, the contours, and the presence of calcification can provide clues to the underlying cause.

Size of the Adrenals. The adrenal size can be variable even in normal subjects. The normal left adrenal gland is 21.5 ± 4.6 mm in greatest diameter (length), and 6.7 ± 1.7 mm in thickness; the normal right adrenal measures 22.8 ± 6.3 mm in greatest diameter, and 5.1 ± 1.1 mm in thickness.[291] Demonstration of enlarged adrenals in a patient with Addison's disease is compatible with tuberculosis, metastatic disease, and fungal infections. Decreased size of the adrenals, on the other hand, is indicative of autoimmune disease, idiopathic atrophy, and late sequelae of TB. Thus, demonstration of enlarged adrenals by CT almost excludes autoimmune disease as a cause, while decreased size or atrophy is consistent with both autoimmune disease and end-stage tuberculosis of the adrenals.

Density. Acute tuberculous involvement of the adrenals may cause multiple lucent areas seen by CT, representing necrosis. Bilaterally enlarged, nonhomogenous adrenals can also be seen with histoplasmosis. Secondary hemochromatosis presents as bilaterally hyperdense glands with normal or reduced size, the hyperdensity being caused by iron deposition within the adrenals. Malignant disease generally appears as bilaterally enlarged glands of homogenous density, except in the case of metastases from malignant melanoma, which tend to bleed and necrose. Adrenal hemorrhage also causes irregular enlargement with nonhomogenous densities due to the blood within.

Contours. Metastatic disease typically causes round or oval intraadrenal masses, altering the normal adrenal contours. Inflammatory causes generally are associated with preservation of normal adrenal contours.

Calcification (Fig. 2). Historically, radiographic demonstration of intraadrenal calcification has been regarded as a classic sign for tuberculous origin of adrenal failure. The incidence of detecting calcification by plain films of the abdomen in tuberculosis of the adrenals ranges from 30 to 50%. Computed tomography of the adrenals is the most sensitive method for detecting calcification. Adrenal calcification is extremely unusual when autoimmune atrophy underlies the etiology of Addison's disease. While tuberculosis is associated with calcification in more than half the cases,[292] rarely other diseases processes can cause calcification; thus, fungal infections,[69,293] metastatic disease,[294] and adrenal hemor-

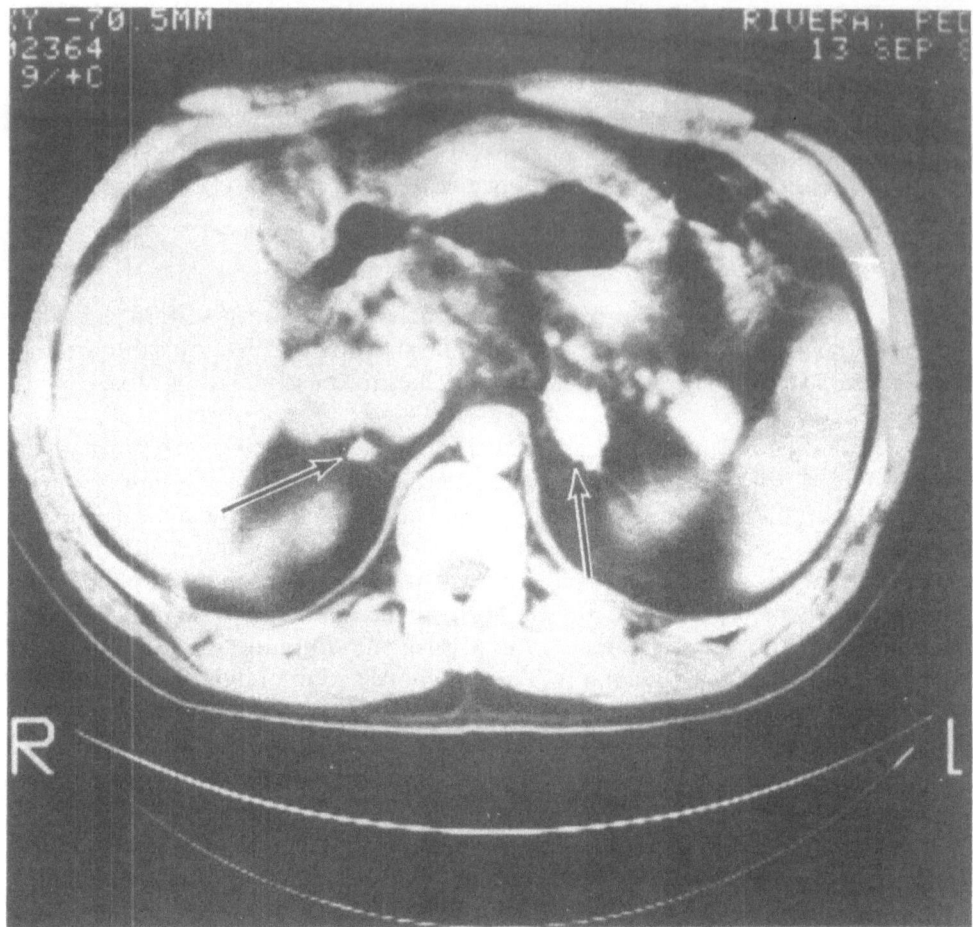

FIGURE 2. CT scan of patient with tuberculous adrenalitis demonstrating bilateral calcification of the adrenals.

rhage[295] may produce calcific changes detectable by CT. In summary, CT has been suggested as an important tool to determine the etiology of Addison's disease.[296–299] The size of the adrenals and the presence of calcification are two important and helpful radiographic signs. Yet the limitations of the procedure are illustrated in patients with tuberculous Addison's disease with small-sized, noncalcified adrenals, resembling the "classic" picture of autoimmune Addison's disease. To compound matters even more, pulmonary tuberculosis can coexist with idiopathic atrophy. In a series of 13 patients with idiopathic adrenal atrophy reported by Vita et al.,[292] pulmonary and other extraadrenal tuberculosis was present in six patients. The fact that the size of the adrenals affected by tuberculosis can be enlarged or atrophied depending on the duration of the disease can cause extreme confusion in diagnosing tuberculous adrenal disease. This differential morphology of the adrenals associated with the timing of tuberculous infection may have important therapeutic implications as well.

Measurement of Adrenal Antibodies

When available, this measurement can be extremely helpful in documenting the presence of autoimmune adrenalitis. The fact that the titers of antiadrenal antibodies are elevated in more than 80% of patients with autoimmune disease while being undetectable in tuberculous adrenalitis is of enormous help in distinguishing these two forms of Addison's disease.

Tests for Infectious Disease

The presence of TB elsewhere, the background of endemicity for fungal disease, positive skin tests or serology (ELISA test for TB or fungal serology), and nonspecific biochemical clues, such as elevated globulins or alkaline phosphatase, would help in marshalling evidence for the presence of underlying infections.

CT-guided biopsy of the adrenal masses can establish the diagnosis of metastasis or fungal disease.[69,96,97] However, in clinical practice such a procedure is not often employed.

The etiological separation between tuberculous and autoimmune adrenal disease can present difficulties. The four most important clues to assist in making the separation are duration of symptoms, size of adrenal glands, presence of calcification, and presence of extraadrenal tuberculosis or other autoimmune diseases (Table 9). Of course, none of these factors are of universal precision. The fact that medullary function is spared in autoimmune adrenal failure may assist in separating the two, but this has not found wide application in the literature.

Extraadrenal Endocrine Dysfunction in Addison's Disease

In patients diagnosed as having autoimmune adrenal failure, it is mandatory to evaluate other endocrine glandular function. Thus, screening for primary

TABLE 9.
Tuberculous versus Autoimmune Etiology

Features	TB	Autoimmmune
Incidence	Decreasing (17–30%)	60%
History of TB (past or present)	Obtainable in 70–80%	
Duration of symptoms	Short (6–9 mo)	Longer (6 mo–2 yr)
Calcification of adrenals	30–50%	Rarely may be seen
Associated autoimmunopathy		40–70%
Medullary function (catecholamine response to posture)	Decreased	Intact
Genitourinary TB	30–40%	

hypothyroidism, hypoparathyroidism, diabetes mellitus, and gonadal failure (particularly in females) constitutes part of the workup. The other part includes evaluation of pituitary hormones, since several of these demonstrate perturbations which are reversible with glucocorticoid treatment.

Screening for Polyglandular Failure Syndrome

Evaluation of gonadal function, hypoparathyroidism, and diabetes mellitus in patients with Addison's disease is relatively simple and straightforward. Measurement of testosterone, 17-β-estradiol, LH, and FSH and the determination of antibodies that react with steroid-secreting cells would disclose the presence of autoimmune gonadal failure. The simple measurement of serum calcium level with simultaneous assay of parathyroid hormone would disclose the presence of parathyroid failure. Hyperglycemia or an abnormal glucose tolerance test, often in conjunction with the presence of circulating, anti–islet cell antibodies, would establish the autoimmune diabetes syndrome. The only endocrine component that may pose some difficulty in interpretation is thyroid failure.

Primary hypothyroidism clearly occurs with greater frequency in primary autoimmune adrenal failure than can be attributed to chance alone.[6,205,300,301] Measurement of basal TSH level in plasma is considered one of the hallmarks of diminished thyroid reserve.[302] However, in patients with Addison's disease, regardless of etiology, an elevated basal TSH need not always imply the presence of underlying thyroid failure. It is well recognized that glucocorticoid deficiency per se can cause an elevation in basal TSH levels, which is reversible upon restoration of eucortisolism. Several workers have reported "reversible hypothyroidism" in patients with Addison's disease,[303–305] underscoring the need for caution in committing patients with such a syndrome to lifelong levothyroxine therapy. The phenomenon of high TSH in Addison's disease is related to the effects of glucocorticoids on TSH and thyroxine dynamics. Pharmacological doses of glucocorticoids suppress basal TSH secretion[306–308] as well as the response of TSH to TRH stimulation.[309–311] Peripherally, glucocorticoids inhibit conversion of thyroxine to tri-iodothyronine.[312] A decrease in circulating glucocorticoid levels would be expected to result in elimination of the restraining effect of glucocorticoids on the secretion and release of TSH by the thyrotropes. Some of the patients reported in the literature with elevated TSH levels in Addison's disease have also shown lowering of the thyroidal hormones,[304,305] while other patients have demonstrated elevated TSH with mild elevation of serum thyroid hormone levels.[313] Thus, the former profile (elevated TSH with subnormal T_4, T_3) mimics subclinical hypothyroidism, while the latter (elevated TSH with mild elevation in T_4, T_3) mimics the syndrome of inappropriate TSH secretion. The reversal of these abnormalities with glucocorticoid administration[304,305,313] underscores the need for following these parameters before and after glucocorticoid therapy while withholding thyroxine replacement until the irreversible nature of these findings has been established. Of course, in patients with marked lowering of T_4, T_3, with clinical evidence of overt hypothyroidism, or significantly elevated antimicrosomal antibody titers, thyroxine replacement should be initiated, with periodic reevaluation of the thyroid profile after glucocorticoid replacement. The mistake of attributing an elevated TSH in Addisonian disease to permanent hypothyroidism can be avoided if the relationship between TSH and glucocorticoids is kept in mind.

Evaluation of Pituitary Hormones

While an elevated ACTH is the hallmark of primary adrenal failure, β lipotropin, TSH, and prolactin can undergo significant alterations. The reversible elevation in TSH secondary to glucocorticoid lack has already been alluded to; the perturbations in ACTH, β-lipotropin, and prolactin in patients with Addison's disease deserve brief mention.

ACTH

Basal ACTH levels are variably elevated in almost all patients with Addison's disease, ranging between 200 and 1000 pg. In general, the degree of elevation bears little correlation to the severity of adrenal failure or the duration of the disease. The spontaneous fluctuation in plasma ACTH level can cause difficulties in interpretation. The availability of sensitive and standardized assays for ACTH has permitted evaluation of plasma levels of ACTH in patients with Addison's disease who are on long-term, conventional glucocorticoid replacement. While most patients demonstrate suppression of the raised ACTH following glucocorticoid therapy, some Addisonian patients demonstrate only partial suppression. Clayton et al.[314] described two patients with Addison's disease in whom hydrocortisone produced only a partial suppression of their elevated ACTH levels. It was postulated that secondary pituitary hyperplasia might exist in Addison's disease, resulting in some degree of autonomy. The relatively short half-life of the cortisone used for replacement might be conducive to the inability to maintain normal levels throughout the day, a fact that may contribute to hyperplasia of the pituitary corticotropes.

Since chronically untreated or undertreated target gland failure can result in secondary enlargement of the pituitary gland, it is interesting to note that five individual reports in the 1970s have described such an occurrence in untreated as well as conventionally treated Addisonian patients.[315–319] However, the diagnosis of ACTH-secreting adenoma was not unequivocally established in these cases, since morphological and immunocytochemical studies had not been performed. Krautli et al.[320] described two patients with Addison's disease on chronic conventional replacement treatment who developed hyperpigmentation and abnormal sellar radiography. One patient had visual loss and bilateral field cuts with marked dilatation and destruction of the sella turcica. The clinical presentation, the radiological features, and the aggressive course resembled those of Nelson's syndrome; which has a proclivity "to erupt, explode and compress parasellar structures."[321] The plasma level of ACTH was markedly elevated (>1000 pg) in both cases. Surgical exploration of the pituitary gland demonstrated ACTH-producing tumors in both by specific immunocytochemical stains. The rarity of this phenomenon contrasts with the "secondary" pituitary tumors associated with, for instance, primary hypogonadism[322] or primary hypothyroidism.[323–325] This rarity, coupled with the observation that these tumors developed in patients on adequate conventional glucocorticoid therapy, raises another possibility—that an ACTH-secreting adenoma had fortuitously developed in a patient with preexisting Addison's disease. Such a rare combination would represent the pathophysiological equivalent of Nelson's disease.

The clinical correlation of these observations is that ACTH measurements be obtained in all patients with Addison's disease, and if ACTH is found to be extraordinarily elevated, a CT scan of the pituitary is indicated. Also, the reappearance or intensification of pigmentation in Addisonian patients on replacement therapy would warrant a similar approach.

β-Lipotropin

β-Lipotropin and ACTH share "similar dynamics but different kinetics." Thus, β-lipotropin is more stable in the plasma and has a longer half-life. The correlation between the magnitude of elevation in plasma levels of ACTH and β-lipotropin is an assumed one and has not been demonstrated in any large series of Addisonian patients. Similarly, the correlation between the degree of hyperpigmentation and the magnitude of elevation of these pituitary hormones has not been conclusively established. Unless measurement of β-lipotropins in the plasma is routinely performed in all Addisonians before and during treatment, questions regarding these correlations cannot be resolved.

TSH

The elevated TSH levels seen in conjunction with untreated adrenal failure might lead to mistakenly conferring a diagnosis of concomitant primary hypothyroidism. The nature of the relationship between glucocorticoids and TSH has already been alluded to in this chapter in the section, "Extraadrenal Endocrine Dysfunction in Addison's Disease."

Prolactin

Prolactin dynamics can also undergo significant alteration as a consequence of primary adrenal failure. These changes parallel the changes in TSH. For instance, glucocorticoid administration suppresses prolactin secretion in humans, while adrenalectomy can result in elevations of serum prolactin.[310,326,327] Adrenalectomy has also been shown to increase the prolactin response to ether stress in rats.[328,329] Development of the amenorrhea galactorrhea syndrome in a patient with primary adrenal failure, which reversed on glucocorticoid replacement, has been reported.[197] More recently, Stryker and Molitch[313] described an Addisonian patient with hyperprolactinemia (along with hyperthyrotropinemia) in whom correction of these abnormalities occurred following glucocorticoid replacement. It has been postulated that deficiency of glucocorticoids results in loss of their normal inhibitory influence on prolactin (and TSH) secretion. Lever and McKerron[330] have described four patients with hyperprolactinemia associated with autoimmune adrenal insufficiency. The hyperprolactinemia completely resolved with glucocorticoid replacement. Thus, primary adrenal insufficiency should be included among the diverse causes of hyperprolactinemia.

Differential Diagnosis

Adrenal insufficiency is a slowly evolving disorder, often with nonspecific and vague symptoms, with few obvious signs. It is during this stage that the diagnosis is often missed. When the disease has evolved into its chronic, full-blown form, the weight loss and asthenia are striking enough to prompt screening for adrenal failure. Malignant disease, tuberculosis, apathetic hyperthyroidism, inflammatory bowel disease (particularly Crohn's disease), anorexia nervosa, and organic depression can mimic several aspects of adrenocortical insufficiency. In some of these situations, such as tuberculosis and malignant disease, adrenal failure can result from these primary disorders and contribute to the morbidity and even mortality.

Tuberculosis

The weight loss, asthenia, and systemic symptomatology of active and chronic tuberculosis can simulate adrenal failure. The orthostatic hypotension seen in many tuberculous patients and the hyponatremia occasioned by inappropriate ADH secretion can carry the similarity even further. Since the symptoms of adrenal failure can be completely overridden by and attributed to TB, the adrenal disease is often detected only when acute adrenal crisis supervenes. The effect of rifampin on cortisol clearance has an important impact on precipitating adrenal crisis in the marginally performing pre-Addisonian patient. As indicated earlier, all patients with tuberculosis could benefit from the simple screening test to determine adrenal reserve prior to placement on rifampin therapy. The inexpensiveness of this procedure weighted against the potentially lethal consequences well warrants employment of the test in all tuberculous patients.

Malignant Disease

Carcinomatosis, with its accompanying triad of anorexia, asthenia, and weight loss, closely resembles full-blown adrenal failure. As with tuberculosis, the tendency to attribute many symptoms of adrenal failure (as well as the hyponatremia) to neoplastic disease undoubtedly contributes to underrecognition of adrenal involvement by malignant disease. It is not uncommon to find patients with Addison's disease in whom the wasting has been severe enough to have subjected the patient to an intensive cancer-screening workup.

Apathetic Hyperthyroidism

This unusual variant of hyperthyroidism is particularly common in elderly subjects with thyroidal hyperfunction. Weight loss, weakness, and depression highlight the presentation. The presence of gastrointestinal symptoms and lack of thyromegaly may lead the physician away from the thyroid gland. The diagnosis, once considered, can be readily established by routine thyroid function tests.

Inflammatory Bowel Disease

Rarely, the systemic manifestations of Crohn's disease can supercede the gastrointestinal symptoms associated with this inflammatory bowel disorder. Thus, unexplained weight loss, asthenia, and mild abdominal cramps with some diarrhea can be the only features of this disorder. Unless granulomatous lesions are discovered and biopsied, the diagnosis of this form of Crohn's disease can be quite elusive.

Anorexia Nervosa and Organic Depression

These psychiatric disorders are highlighted by weight loss. A careful history is often required to trace the underlying psychiatric problem. The steroid dynamics in these disorders is clearly different from that of adrenal failure. Hypercortisolemia with varying degrees of resistance to dexamethasone suppression characterizes anorexia nervosa and organic depression.

SIADH

Adrenal insufficiency and SIADH share several similarities:

1. Both may occur in the background of TB and malignant disease.
2. Moderate to even severe hyponatremia is common to both.
3. Natriuresis is encountered in both diseases.
4. Several symptoms of adrenal failure, such as anorexia and muscle weakness, can be caused by the hyponatremia of SIADH.
5. Compensated Addisonian patients may clinically appear "euvolemic" even though mildly volume depleted. Patients with SIADH often appear clinically euvolemic.
6. Both conditions are characterized by the inability to completely excrete a water load.

The presence of hyperkalemia and the impaired cortisol response to exogenous ACTH permit clear distinction of Addison's disease from the syndrome of inappropriate ADH secretion.

Secondary Adrenal Failure

The distinction of Addison's disease from secondary (pituitary) adrenal failure can readily be made by the following means:

1. The plasma aldosterone response to exogenous ACTH administration is usually preserved in patients with secondary hypoadrenalism whereas it is lost in Addison's disease.
2. The basal plasma ACTH concentration in primary adrenal failure is generally elevated while being subnormal in patients with secondary hypoadrenalism. Of course, in a small group of patients with both primary and secondary hypoadrenalism the basal plasma ACTH may not be clearly diagnostic. In such instances the use of exogenous corticotropin-releasing hormone (CRH) may permit better delineation. This is analo-

gous to the use of TRH in the separation of primary from secondary hypothyroidism. Patients with Addison's disease characteristically demonstrate an exaggerated ACTH response to exogenous administration of CRH, while patients with hypopituitarism demonstrate an impaired or absent response.

3. The response of urinary 17-OHCS or urine free cortisol to the "standard ACTH stimulation" test permits clear separation of Addisonian patients (who show no response to continued administration of ACTH) from hypopituitary patients (who show a gradual "stepladder" type of increment in urinary steroid levels with continued administration of ACTH).

4. Finally, the demonstration of other pituitary trophic hormone deficiencies favors secondary hypoadrenalism. Of course, this would not be applicable to patients with the rare "isolated ACTH deficiency syndrome."

Complications (Acute Addisonian Crisis)

The most important complication of hypoadrenalism is the development of Addisonian crisis that is potentially fatal. Adrenal crisis develops when the marginally adequate adrenal reserve becomes suddenly inadequate to combat the rising needs for cortisol during stress. Addisonian crisis is almost always precipitated by some form of stress. Physical, emotional, anesthetic, or traumatic stress can all be responsible for triggering a crisis. But the most notable factor that precipitates adrenal crisis is infection. Even a seemingly simple viral cold can trigger adrenal crisis. At times adrenal crisis can be triggered by the use of medications. Initiating thyroxine therapy in a patient with marginal adrenal insufficiency is a classic setting where the sudden increase in metabolic demands cannot be met by the compromised adrenals. Similarly, the use of rifampin, which accelerates the turnover of cortisol, can be conducive for precipitating decompensation in a marginally compromised situation. Of course, the sudden withdrawal of steroid therapy is another classic setting where the chronically suppressed hypothalamic pituitary adrenal axis cannot muster a response to stress.

The clinical presentation of acute adrenal crisis is easy to recognize if the background of adrenocortical failure is known. For instance, the development of crisis in the Addisonian patient on steroid replacement is often detected even at the very early stages of its evolution. In contrast, the diagnosis is often missed when the adrenal crisis is the presenting manifestation of Addison's disease. The syndrome is ushered in by anorexia, nausea, vomiting, and diarrhea with vague abdominal cramps. Fever, chills, and diaphoresis may accompany the gastrointestinal manifestations of acute adrenal crisis. As the untreated syndrome evolves, patients develop volume depletion and dehydration, often severe enough to culminate in shock. Hyponatremia and hyperkalemia are usually apparent, often associated with mild to moderate metabolic acidosis, hypoglycemia, and occasionally hypercalcemia. Unfortunately, when the hyperkalemia is masked by severe vomiting and diarrhea, the diagnosis of adrenal crisis can be missed by even experienced clinicians. The list of initial diagnoses that are often made in patients who were subsequently documented as being in

Addisonian crisis includes acute gastroenteritis, salmonellosis, ruptured ectopic pregnancy, toxic shock syndrome, septic shock, acute myocardial failure with shock, and others.

Acute adrenal crisis should be a diagnostic concern under the following conditions.

1. Dehydration and electrolyte abnormalities in patients with a chronic history of ill health characterized by weight loss and weakness.
2. Shock in the background of tuberculosis, or carcinomatosis.
3. Diarrhea and dehydration in the presence of normal or elevated potassium and low sodium concentrations in the plasma.
4. The unusual combination of hypoglycemia in association with hypotension; most conditions characterized by hypotension are associated with a brisk release of several counterinsulin hormones that usually increase the blood glucose levels.
5. The presence of hyperpigmentation or absence of axillary hair in females in the setting of hypotension should strongly point to underlying adrenal disease as the cause of illness.

The recognition of Addisonian crisis is crucial since early administration of glucocorticoids can make the difference between life and death. The beneficial effects of therapy are often dramatically evident within 24 hr of steroid therapy. Little time should be wasted on performing tests when acute adrenal crisis is considered. A pretreatment cortisol, often coupled with a rapid ACTH stimulation test, is all that is required. When the patient's condition is deemed too dangerous to allow even a 1-hr delay in therapy, the rapid ACTH test can be still performed while covering the patient with a bolus of 20 mg of methylprednisolone and 10 mg of desoxycorticosterone acetate. Vigorous normal saline coupled with stress doses of hydrocortisone (100 mg intravenously three times a day) rapidly corrects the dehydration, electrolyte disturbances, and cortisol deficiency of acute adrenal crisis. Of course, treatment toward any precipitating factors, when identified, constitutes an important part of therapy.

Treatment

Addison's disease is indeed one of the most gratifying diseases to treat. The transformation of a listless, withdrawn, helpless person into a normal being, capturing the lost vigor and vitality with a vengeance, is an unforgettable experience. The replacement therapy for Addison's disease consists of glucocorticoid and mineralocorticoid administration, which, in physiological doses, is remarkably devoid of adverse effects. The treatment of Addison's disease can be viewed from four perspectives—glucocorticoid replacement, mineralocorticoid replacement, treatment of acute adrenal crisis, and therapy for the underlying disease that caused adrenal failure.

Glucocorticoid Therapy

Replacement glucocorticoid therapy consists of administering cortisone acetate or hydrocortisone hemisuccinate. To simulate the physiological diurnal rhythm, cortisone acetate is administered orally as 25 mg in the morning and 12.5 mg in the evening, while hydrocortisone hemisuccinate is given in the dosage of 20 mg in the morning and 10 mg in the evening. Administration of these two glucocorticoids, although well tolerated, entails some inherent problems. First, since these steroids are short acting, it is inevitable that such therapy be associated with intermittent periods of unsuppressed ACTH levels in the plasma.[331,332] Feek et al.[333] evaluated the patterns of plasma cortisol and ACTH concentrations in patients with Addison's disease treated with conventional corticosteroid replacement and showed that some patients failed to show adequate suppression of ACTH with the use of conventional doses of either drug. Second, therapy with cortisone acetate may be hampered by inadequate absorption of steroid acetate by the gastrointestinal tract and impaired in vivo conversion of cortisone to cortisol. Obviously, these phenomenona may be conducive to reducing the bioavailability of the drug. Third, unlike replacement therapy of primary hypothyroidism, where the dosage of thyroxine can be finely adjusted, there are no scientific tools to provide fine titration of glucocorticoid therapy. Thus, adequacy of replacement is often judged on the basis of clinical response, and on the subjective symptoms of the patient.

Khalid et al.[334] studied the efficacy of four different glucocorticoids (cortisone, hydrocortisone, dexamethasone, and prednisolone) as replacement therapy for Addison's disease. Their data suggest that dexamethasone was strikingly potent in suppressing ACTH levels at the lowest total glucocorticoid dose. However, the same potency of dexamethasone may also play an important role in predisposing patients to adverse effects of long-term glucocorticoid activity. Besides, dexamethasone possesses little mineralocorticoid activity. These two factors limit the usefulness of dexamethasone in treatment of Addison's disease, despite its longer-acting effects and its ability to suppress ACTH more effectively than cortisone or hydrocortisone. In addition, it is unclear whether complete suppression of ACTH is necessary in treatment of Addison's disease, in contrast to adrenogenital syndrome where total suppression of ACTH has great therapeutic benefit.

The parameters to follow in Addisonian patients on glucocorticoid replacement are mostly clinical.[335] Titration of the maintenance dose is largely empirical. Thus, if fatigue, malaise, weight loss, and hyperpigmentation persist, the dose of cortisone or hydrocortisone is increased, while the presence of signs indicative of steroid excess necessitates reduction of glucocorticoid dosage. In general, measurement of serum cortisol is not a good parameter for dose titration in patients receiving cortisone or hydrocortisone.[336] However, measurement of serum cortisol might prove valuable in studying the absorption characteristics of these medications when a question of bioavailability arises.[337,338] Of course, serum cortisol measurements are of absolutely no value in patients receiving dexamethasone. As for measurement of serum ACTH level to titrate the dose of glucocorticoid, this marker has not assumed the status or sophistication of measurement of TSH for adjustment of thyroxine dosage in patients with primary hypothyroidism. The relatively short half-life of cortisone (or hydrocortisone), the pulsatile nture of ACTH secretion, the heterogeneity of ACTH levels in treated Addisonians, and the limited availability of ACTH determinations have all limited the use of serum ACTH level as a parameter for titrating the glucocorticoid dosage. Burch[339] has suggested that measurement of urine free cortisol might be a useful tool for detecting undertreatment or excess dose of the orally administered cortisone or hydrocortisone. Notably, the 24-hr urine free cortisol showed little day-to-day variation, and in the nine patients with Addison's disease on oral replacement regimens with cortisone or hydrocortisone, the urine free cortisol correlated well with clinical euadrenalism. Again, this measurement is meaningless in patients receiving dexamethasone or prednisolone, steroids that do not get converted into hydrocortisone in the body.

Individual variations in the absorption of the orally administered cortisone acetate or hydrocortisone hemisuccinate are the most frequent reasons for undertreatment despite administration of a conventional replacement dose. The development of diarrhea, vomiting, and other side effects would mandate an increment (usually doubling) in the dose of oral cortisone. If the problem is severe, as with gastroenteritis, parenteral hydrocortisone becomes necessary. Patients with Addison's disease should be repeatedly reinforced as to the need for increasing their glucocorticoid dose during periods of stress. Also, it is mandatory that such patients wear on their person some form of identification as to their disease, with instructions to use intravenous hydrocortisone in the event the patient is found unconscious.

Mineralocorticoid Replacement

Mineralocorticoid replacement for patients with Addison's disease is provided by oral administration of 9-α-fludrocortisone, 0.05–0.1 mg/day. Initially, nearly all patients with Addison's disease require both gluco- and mineralocorticoids. However, after several months to years, many patients continue to do well despite exclusion of mineralocorticoids.

Adjustment of the dose of mineralocorticoid is also empirical and is based on clinical and biochemical data. Thus, the persistence of orthostatic hypotension, dizziness, hyperkalemia, or hyponatremia is an indication to start, or increase the preexisting dose of, mineralocorticoid. Similarly, the development of hypertension, edema, or cardiac failure is an indication to decrease the dose or discontinue the mineralocorticoid altogether. During the initial stages of therapy with florinef, some patients with Addison's disease may demonstrate exquisite mineralocorticoid sensitivity and may rapidly develop edema. In such cases withdrawal of the drug with very careful reinstitution becomes necessary. While clinical judgment and electrolyte monitoring continue to remain the parameters to evaluate dosage of mineralocorticoid replacement, measurement of plasma renin activity (PRA) has been advocated as a sensitive tool to assess the need and dose of mineralocorticoid replacement.[340,341] A raised PRA indicates the presence of sodium depletion and hence the need for more mineralocorticoid, even when the serum potassium concentrations are still within the normal range. However, PRA determinations cannot clearly distinguish between adequate dosage and overdosage.

Patients with Addison's disease generally learn to include a high salt intake in their diet. This, coupled with adequate glucocorticoid replacement, may allow elimination of all mineralocorticoid. However, a careful assessment of blood pressure, pulse, cardiac status, electrolytes, and perhaps plasma renin is indicated before oral mineralocorticoid replacement is discontinued. Occasionally, patients may resort to self-medication with licorice, which because of its deoxy-cortone-mimetic action can provide mineralocorticoid converge.[174] This, however, is not to be recommended since sooner or later these effects are lost and acute mineralocorticoid deficiency is likely to manifest.

Adrenal Crisis

This represents an emergency situation where early therapeutic intervention can be lifesaving. The four-pronged approach to treatment of Addisonian crisis is administration of hydrocortisone, 100 mg three times a day, intravenously; use of mineralocorticoid, 10 mg DOC intramuscularly once or twice a day, or 9-α fludrocortisone, 0.1 mg orally; liberal administration of normal saline until the patient has been adequately hydrated and volume-repleted; and treatment of precipitating causes, if identified. Glucose-containing solutions should be incorporated in the fluid regimen to provide adequate caloric support.

With timely intervention, patients with Addisonian crisis improve with dramatic and startling rapidity. The key here is early intervention. Once the patient has developed hypoperfusion and shock, the prognosis is gloomy. The electrolyte abnormalities and metabolic acidosis rapidly improve with hydration and steroid therapy. Within 24–48 hr, the patient can be switched to oral replacement therapy.

Treatment of Underlying Cause

This aspect of therapy assumes importance when infectious etiologies underlie the adrenal failure. Two in particular merit attention—TB and fungal disease.

Although the scourge of TB has declined over the past two decades, it still continues to be an important cause of Addison's disease. Several important statements are relevant in terms of specific therapy.

1. The mere presence of a reactive tuberculin test should not deter the physician from searching for another cause for Addison's disease. The coexistence of idiopathic adrenal atrophy in patients exposed to tuberculosis has been emphasized.[292] In such cases, treatment for "active TB" with 18–24 months of intensive antituberculous therapy is hardly justified.
2. The presence of "active" tuberculosis elsewhere (e.g., lung, genitourinary tract) in a patient with Addison's disease clearly indicates the need for conventional antituberculous therapy with three drugs—isoniazid, rifampin, and ethambutal—along with replacement therapy.
3. When active tuberculosis cannot be demonstrated in extraadrenal locations, the combination of a reactive tuberculin test, exclusion of other causes of Addison's disease, and demonstration of bilaterally enlarged adrenal glands by CT constitutes an adequate basis for the diagnosis of tuberculous Addison's disease. In these cases, institution of a full course of antituberculous treatment is justified, along with, of course, full hormonal replacement.

4. When active tuberculosis cannot be demonstrated in extraadrenal locations, but the patient demonstrates a reactive tuberculin test with bilaterally atrophied adrenals by CT and without evidence of other causes for Addison's disease, the diagnosis of "end-stage tuberculous adrenalitis" can be presumed. In such cases, institution of a full course of antituberculous therapy is of dubious value. When adrenal atrophy occurs secondary to chronic tuberculosis, there is no current evidence that indicates need for treatment beyond hormonal replacement.

5. When antituberculous therapy is administered to patients with Addison's disease on replacement, the effect of rifampin on the catabolism of cortisol should be borne in mind. An increase in the glucocorticoid dosage should be anticipated, since rifampin increases the hepatic catabolism of glucocorticoids. Failure to realize this fact may result in the development of acute adrenal crisis due to inadequate bioavailability of the glucocorticoid administered.

6. Finally, there are reports of improvement in adrenal function after treatment of tuberculous adrenalitis.[342-344] However, these cases were reported in the early 1960s, without complete documentation by present standards. A more recent study from France[345] reported the findings in eight patients with tuberculous Addison's disease following a full course of antituberculous therapy. While the urinary 17-hydroxycorticosteroids had normalized in all eight patients, the basal plasma ACTH continued to remain elevated in all, and only two showed a rise in urinary steroids in response to exogenous ACTH administration. Thus, complete recovery of adrenal function in Addisonian patients following a full course of antituberculous therapy must be a rare event.

Fungal infections, when documented as a cause of adrenal failure, demand antifungal chemotherapy with either ketoconazole or amphotericin B. Recovery of adrenal reserve following amphotericin B treatment for disseminated South American blastomycosis has been convincingly documented in a single case.[61]

References

1. Addison T: *On the Constitutional and Local Effects of Disease of the Suprarenal Capsules.* S. Highley, London, 1855.
2. Rolleston HD: *The Endocrine Organs in Health and Disease with an Historical Review.* Oxford University Press, London, 1936, p. 340.
3. Medvei VC: *A History of Endocrinology.* MTP Press Limited, Lancaster/Boston/The Hague/Dordrecht, 1982, p. 230.
4. Trousseau A: Bronze Addison's disease. *Arch Gen Med* **8:**478, 1856.
5. Wilks B, Daldy P (eds): *Addison's Collected Papers.* New Sydenham Society, London, 1868.
6. Nerup J: Addison's disease—Clinical studies. A report of 108 cases. *Acta Endocr (Kbh)* **76:**127, 1974.
7. Guttman PH: Addison's disease. A statistical analysis of 566 cases and a study of the pathology. *Arch Pathol* **10:**742, 1930.
8. Anderson JR, Goudie RB, Gray KG, et al.: Auto-antibodies in Addison's disease. *Lancet* **1:**1123, 1957.
9. Blizzard RM, Kyle M: Studies of the adrenal antigens and antibodies in Addison's disease. *J Clin Invest* **42:**1653, 1963.

10. Goudie RB, Anderson JR, Gray KK, et al.: Autoantibodies in Addison's disease. *Lancet* **1**:1173, 1966.

11. Blizzard RM, Chee D, Davies D: The incidence of adrenal and other antibodies in sera of patients with idiopathic adrenal insufficiency (Addison's disease). *Clin Exp Immunol* **2**:19, 1967.

12. Nerup J, Halberg P, Soborg M, et al: Organ specific antibodies in Addison's disease. *Acta Med Scand* (Suppl 445): 383, 1966.

13. Irvine WJ, Stewart AG, Scarth L: A clinical and immunological study of adrenocortical insufficiency (Addison's disease). *Clin Exp Immunol* **2**:31, 1967.

14. Pousset G, Monier JC, Thivolet J: Anticorps antisurrenaliens et maladie d'Addison. *Ann Endocr (Paris)* **31**:995, 1970.

15. Irvine WJ: Clinical and immunological associations in adrenal disorders. *Proc Roy Soc Med* **61**:271, 1968.

16. Irvine WJ, Barnes EW: Adrenocortical insufficiency. *Clin Endocr Metab* **1**:549, 1972.

17. Nerup J: Addison's disease—Serological studies. *Acta Endocrinol* **76**:142, 1974.

18. Irvine WJ, Chan MMW, Scarth L: The further characterization of autoantibodies reactive with extra-adrenal steroid-producing cells in patients with adrenal disorders. *Clin Exp Immunol* **4**:489, 1969.

19. Sotsiou F, Bottazzo GF, Doniach D: Immunofluorescence studies on autoantibodies to steroid-producing cells, and to germline cells in endocrine disease and infertility. *Clin Exp Immunol* **39**:97, 1980.

20. Ketchum CH, Riley WJ, MacLaren NK: Adrenal dysfunction in asymptomatic patients with adrenocortical autoantibodies. *J Clin Endocrinol Metab* **58**:1166, 1984.

21. Blizzard RM, Chee D, Davis W: The incidence of parathyroid and other antibodies in the sera of patients with idiopathic hypoparathyroidism. *Clin Exp Immunol* **1**:119, 1966.

22. Irvine WJ, Barnes EW: Addison's disease, ovarian failure and hypoparathyroidism. *Clin Endocrinol Metab* 4:379, 1975.

23. Irvine WJ: Immunological aspects of endocrine disease. *Proc R Soc Med* **67**:548, 1974.

24. Irvine WJ, Chan MMY, Scarth L, et al: Immunological aspects of premature ovarian failure associated with idiopathic Addison's disease. *Lancet* **2**:883, 1968.

25. Turkington RW, Lebovitz ME: Extra-adrenal endocrine deficiencies in Addison's disease. *Am J Med* **43**:499, 1967.

26. Rupp JJ, Paschkis KE: Pluriglandular insufficiency simulating panhypopituitarism: Report of 2 cases. *Am J Med* **18**:507, 1955.

27. Nicholls MG, Espiner EA, Donald RA: Schmidt's syndrome presenting as hypopituitarism. *Ann Intern Med* **80**:505, 1974.

28. Irvine WJ: Autoimmunity in endocrine disease. *Rec Prog Horm Res* **36**:509, 1980.

29. Scherbaum WA, Berg PA: Development of adrenocortical failure in non-Addisonian patients with antibodies to adrenal cortex. *Clin Endocrinol* **16**:345, 1982.

30. Betterle C, Zanchetta R, Trevisan A, et al: Complement-fixing adrenal autoantibodies as a marker for predicting onset of idiopathic Addison's disease. *Lancet* **1**:1238, 1983.

31. Thomsen M, Platz P, Ortved Andersen O, et al: MLC typing in juvenile diabetes mellitus and idiopathic Addison's disease. *Transplant Rev* **22**:125, 1975.

32. Ludwig H, Mayr WR, Pacher M, et al: HL-A antigens in idiopathic Addison's disease. *Z Immun Forsch* **149**:423, 1975.

33. Fairchild RS, Schimke RN, Abdou NI: Immunoregulation abnormalities in familial Addison's disease. *J Clin Endocrinol Metab* **51**:1074, 1980.

34. Nerup J, Christy M, Kroman H, et al: In Irvine, WJ (ed): *Immunology of Diabetes* Teviot Sci Publ, Edinburgh, 1980, Chapter 2, p. 55.

35. Dunlop D: Eighty-six cases of Addison's disease. *Br Med J* **2**:887, 1963.

36. Valdemarsson S, Hedner P, Low B: HLA antigens in two siblings with autoimmune Addison's disease. *Acta Med Scand* **210**:517, 1981.

37. Spinner MW, Blizzard RM, Childs B: Clinical and genetical heterogeneity in idiopathic Addison's disease and hypoparathyroidism. *J Clin Endocrinol Metab* **28**:795, 1968.

38. Heggarty H: Addison's disease in identical twins. *Br Med J* **1**:559, 1968.

39. Simmonds JP, Lister J: Auto-immune Addison's disease in identical twins. *Postgrad Med J* **54**:552, 1978.

40. Smith ME, Gough J, Galpin OP: Addison's disease in identical twins. *Br Med J* **2**:1316, 1963.

41. Frey HMM, Vogt JH, Nerup J: Familial poly-endocrinology. *Acta Endocrinol (Copenh)* **72:**401, 1973.

42. Stuart-Mason A, Meade TW, Lee JAH, et al: Epidemiological and clinical picture of Addison's disease. *Lancet* **2:**744, 1968.

43. Alvarez S, McCabe WR: Extra-pulmonary tuberculosis revisited: A review of experience of experience at Boston City and other hospitals. *Medicine (Baltimore)* **63:**25, 1984.

44. Ludmerer KM, Kissane JM: Wasting illness in a 33-year-old woman. *Am J Med* **76:**302, 1984.

45. McMurry, JF Jr, Long D, McClure R, et al: Addison's disease with adrenal enlargement on computed tomographic scanning: Report of two cases of tuberculosis and review of the literature. *Am J Med* **77:**365, 1964.

46. Rubin H, Furcolow ML, Yates JL, et al: The course and pregnosis of histoplasmosis. *Am J Med* **27:**277, 1959.

47. Parsons RJ, Zarafonetis CD: Histoplasmosis in man-report of seven cases and a review of seventy-one cases. *Arch Intern Med* **75:**1, 1945.

48. Reddy P, Gorelick DF, Brasher CA, et al: Progressive disseminated histoplasmosis as seen in adults. *Am J Med* **48:**629, 1970.

49. Smith JW, Utz JP: Progressive disseminated histoplasmosis: A prospective study of 26 patients. *Ann Intern Med* **76:**557, 1972.

50. Sarosi GA, Voth DW, Dahl BA, et al: Disseminated histoplasmosis: Results of long-term follow-up. A center for disease control cooperative mycoses study. *Ann Intern Med* **75:**511, 1971.

51. Johnston AW, Postlethwaite R, Ewen SWR, et al: Disseminated histoplasmosis. *J Infect* **9:**79, 1984.

52. Crispell KR, Parson W, Hamlin J, et al: Addisons disease associated with histoplasmosis. *Am J Med* **20:**23, 1956.

53. Wilson DA, Muchmore HG, Tisdal RG, et al: Histoplasmosis of the adrenal glands studied by CT. *Radiology* **150:**779, 1984.

54. Gibb WRG, Ramsay AD, McNeil NI, et al: Bilateral adrenal masses. *Br Med J* **291:**203, 1985.

55. Pena CE: Deep mycotic infections in Columbia: a clinicopathologic study of 162 cases. *Am J Clin Pathol* **47:**505, 1967.

56. Brass K: Observaciones sobre la anatomia patologica, pathgenesis y evolucion de la paracoccidioidomicosis. *Mycopathologica* **27:**119, 1969.

57. Salfelder K, Doehnart G, Doehnert HR: Paracoccidioidomycosis: anatomic study with complete autopsies. *Virchows Arch (Pathol Anat)* **248:**51, 1969.

58. Delnegro G, Wajchenberg BL, Pereira VG, et al: Addison's disease associated with South American blastomycosis. *Ann Intern Med* **54:**189, 1961.

59. Marsiglia I, Pinto J: Adrenal cortical insufficiency associated with paracoccidioidomycosis (South American blastomycosis). Report of four patients. *J Clin Endocrinol* **26:**1109, 1966.

60. Murray HW, Littman ML, Roberts RB: Disseminated paracoccidioidomycosis (South American blastomycosis) in the United States. *Am J Med* **56:**209, 1974.

61. Osa SR, Peterson RE, Roberts RB: Recovery of adrenal reserve following treatment of disseminated South American blastomycosis. *Am J Med* **71:**298, 1981.

62. Kunkel WM Jr, Weed LA, McDonald JR, et al: Collective review, North American blastomycosis-Gilchrist's disease: Clinicopathologic study of 90 cases. *Int Abstr Surg* **99:**1, 1954.

63. Fish RG, Takaro T, Lovell M: Coexistent Addison's disease and North American blastomycosis. *Am J Med* **28:**152, 1960.

64. Abernathy RS, Melby JC: Addison's disease in North American blastomycosis. *N Engl J Med* **266:**552, 1962.

65. Chandler PT: Addison's disease secondary to North American blastomycosis. *South Med J* **70:**863, 1977.

66. Eberle DE, Evans RB, Johnson RH: Disseminated North American blastomycosis. Occurrence with clinical manifestations of adrenal insufficiency. *JAMA* **238:**2629, 1977.

67. Witorsch P, Utz JP: North American blastomycosis: A study of 40 patients. *Medicine (Baltimore)* **47:**169, 1968.

68. Schwarz J, Baum GL: Blastomycosis. *Am J Clin Pathol* **21:**999, 1951.

69. Halvorsen RA Jr, Heaston DK, Johnston WW, et al: CT guided thin needle aspiration of adrenal blastomycosis. *J Comp Assist Tomogr* **6**(2):389, 1982.

70. Abrams, HL, Spiro R, Golstein N: Metastases in carcinoma. *Cancer* **3:**74, 1950.

71. Glomsett DA: The incidence of metastasis of malignant tumors of the adrenal. *Am J Cancer* **32:**57, 1938.
72. Willis RA: *Pathology of Tumors.* Mosby, St. Louis, 1953, 178 pp.
73. Karstaedt N, Sagel SS, Stanley RJ, et al: Computed tomography of the adrenal gland. *Radiology* **129:**723, 1978.
74. Korobkin M, White EA, Kressel HY, et al: Computed tomography in the diagnosis of adrenal disease. *Am J Radiol* **132:**231, 1979.
75. Cedermark BJ, Ohlsen H: Computed tomography in the diagnosis of metastases of the adrenal glands. *Surg Gynecol Obstet* **152:**13, 1981.
76. Hill GJ II, Wheeler HB: Adrenal insufficiency due to metastatic carcinoma of the lung: Case report and review of Addison's disease caused by adrenal metastases. *Cancer* **18:**1467, 1965.
77. Black RM, Daniels GH, Coggins CH, et al: Adrenal insufficiency from metastatic colon carcinoma masquerading as isolated aldosterone deficiency. *Acta Endocrinol (Cophenh)* **98:**586, 1981.
78. Vieweg WVR, Reitz RE, Weinstein RL: Addison's disease secondary to metastatic carcinoma: An example of adrenocortical and adrenomedullary insufficiency. *Cancer* **31:**1240, 1973.
79. Alpers DH, Engelman K, Foley FD: Addison's disease secondary to carcinoma of the breast. *Ann Intern Med* **57:**464, 1962.
80. Galloway JA, Perloff WH: Addison's disease secondary to adrenocortical destruction by metastatic cancer of the breast. *Am J Med* **28:**156, 1960.
81. Barker NW: The pathologic anatomy in twenty-eight cases of Addison's disease. *Arch Pathol* **8:**432, 1929.
82. Gupta DT, Brasfield R: Metastatic melanoma. *Cancer* **17:**1323, 1964.
83. Sahagian-Edwards A, Holland JF: Metastatic carcinoma to the adrenal glands with cortical hypofunction. *Cancer* **7:**1242, 1954.
84. Rosenberg SA, Diamond HD, Jaslowitz B, et al: Lymphosarcoma: A review of 1269 cases. *Medicine (Baltimore)* **40:**31, 1961.
85. Richmond J, Sherman RS, Diamond HD, Craver LF: Renal lesions associated with malignant lymphomas. *Am J Med* **32:**184, 1962.
86. Paling MR, Williamson BRJ: Adrenal involvement in non-Hodgkin's lymphoma. *Am J Roentgenol* **141:**303, 1983.
87. Jafri SZH, Francis IR, Glazer GM, et al: CT detection of adrenal lymphoma. *J. Comput Assist Tomogr* **7:**254, 1983.
88. Foster SC, Gauvin E: A man with a large suparenal mass. *Urol Radiol* **5:**58, 1983.
89. Shea TC, Spark R, Kane B, et al: Non-Hodgkin's lymphoma limited to the adrenal gland with adrenal insufficiency. *Am J Med* **78:**711, 1985.
90. Osei K, Falko J, Pacht E, et al: Primary adrenal insufficiency manifesting as malignant lymphoma. *Arch Intern Med* **143:**1791, 1983.
91. Tobin MS, Kyung-Suk K, Kossowsky WA: Adrenocorticotrophic-hormone deficiency in chronic myelogenous leukemia after treatment. *N Engl J Med* **282:**187, 1970.
92. Harrold BP: Syndrome resembling Addison's disease following prolonged treatment with busulphan. *Br Med J* **1:**463, 1966.
93. Seidenwurm DJ, Elmer EB, Kaplan KM, et al: Metastases to the adrenal glands and the development of Addison's disease. *Cancer* **54:**552, 1984.
94. Reynes CJ, Churchill R, Moncada R, et al: Computed tomography of adrenal glands. *Radiol Clin North Am* **27:**91, 1979.
95. Wilms G, Baert A, Marchal G, et al: Computed tomography of the normal adrenal glands: correlative study with autopsy specimens. *J Comput Assist Tomogr* **3:**467, 1979.
96. Heaston DK, Handel DB, Ashton PR, et al: Narrow gauge needle aspiration of solid adrenal masses. *Am J Roentgenol* **138:**1143, 1982.
97. Meyer JE, Halperin EC, Levene SR, et al: Adrenal insufficiency secondary to metastatic lung carcinoma: CT aided diagnosis. *J Comp Assist Tomogr* **7(6):**1107, 1983.
98. Aegerter EE: Waterhouse–Friderichsen syndrome: Review of literature and report of two cases. *JAMA* **106:**1715, 1936.
99. Levin J, Cluff LE: Endotoxemia and adrenal hemorrhage: Mechanism for Water–Friderichsen syndrome. *J Exp Med* **121:**247, 1965.

100. Margaretten W, McAdams AJ: An appraisal of fulminant meningococcemia with reference to the Shwartzman phenomenon. *Am J Med* **25**:868, 1958.
101. Hjort PF, Rapaport SI: The Shwartzman reaction. *Annu Rev Med* **16**:135, 1965.
102. Ratnoff OD, Nebehay WC: Multiple coagulative defects in a patient with the Waterhouse–Friderichsen syndrome. *Ann Intern Med* **56**:627, 1962.
103. Hardaway RM: Disseminated intravascular coagulation in experimental and clinical shock. *Am J Cardiol* **20**:161, 1967.
104. Grant MD, Horowitz HI, Lorian V, et al: Waterhouse–Friderichsen syndrome induced by pneumococcemic shock. *JAMA* **212**:1373, 1970.
105. Maldacea F: Waterhouse–Friderichsen syndrome with pneumococcal origin in a splenectomized microcythemic patient. *Policlinico (Prat)* **76**:206, 1969.
106. Myerson RM, Koelle WA: Congenital absence of the spleen in an adult: Report of a case associated with recurrent Waterhouse–Friderichsen syndrome. *N Engl J Med* **254**:1131, 1956.
107. Parr LJA, Shipton EA, Holland EH: Fatal case of Still's disease associated with Waterhouse–Friderichsen syndrome due to pneumococcal septicemia. *Med J Aust* **1**:300, 1953.
108. Beach RC, Glayden GS, Eykyn SJ, et al: Waterhouse–Friderichsen syndrome caused by haemophilus influenzae type b. *Br Med J* **3**:1111, 1979.
109. Chaudhuri AK, Hartley RB, Maddocks AC: Waterhouse–Friderichsen syndrome caused by a DF-2 bacterium in a splenectomised patient. *J Clin Pathol* **34**:172, 1981.
110. Butler T, Weaver RE, Ramani TKV, et al: Unidentified gram-negative rod infection: A new disease of man. *Ann Intern Med* **86**:1, 1977.
111. Findling JW, Pohlmann GP, Rose HD: Fulminant gram negative bacillemia (DF-2) following a dog bite in an asplenic woman. *Am J Med* **68**:154, 1980.
112. Shankar PS, Scott JH, Anderson CL: Atypical endocarditis due to gram-negative bacillus transmitted by dog bite. *South Med J* **73**:1640, 1980.
113. Martone WJ, Zuehl RW, Minson GE, et al: Postsplenectomy sepsis with DF-2: Report of a case with isolation of the organism from the patient's dog. *Ann Intern Med* **93**:457, 1980.
114. Schlossberg D: Septicemia caused by DF-2. *J Clin Microbiol* **9**:297, 1979.
115. Bosworth DC: Reversible adrenocorticol insufficiency in fulminant meningococcemia. *Arch Intern Med* **139**:823, 1979.
116. Harper JR, Ginn WM Jr, Taylor WJ: Bilateral adrenal hemorrhage: A complication of anticoagulant therapy. *Am J Med* **32**:984, 1962.
117. Lansing PF: Adrenal hemorrhage associated with dicumarol anticoagulation. *J Maine Med Assoc* **52**:207, 1961.
118. Fragge RG, Bernstein LL, Bell J: Fatal "Waterhouse–Friderichsen syndrome" due to dicumarol. *Ann Intern Med* **52**:923, 1960.
119. Chokas WV: Bilateral adrenal hemorrhage complicating dicoumarol therapy for myocardial infarction. *Am J Med* **24**:454, 1958.
120. Knight LL, Valentine EH: Spontaneous bilateral adrenal hemorrhage: Report of a case occurring during heparin therapy. *JAMA* **182**:1312, 1962.
121. Muller RE, Ceballos R: Anticoagulant therapy as a cause of bilateral adrenal necrosis. *Alabama J Med Sci* **1**:404, 1964.
122. Botteri A, Orell SR: Adrenal hemorrhage and necrosis in the adult: A clinicopathological study of 23 cases. *Acta Med Scand* **175**:409, 1964.
123. McDonald FD, Myers AR, Pardo R: Adrenal hemorrhage during anticoagulant therapy. *JAMA* **198**:126, 1966.
124. Portnay GI, Vagenakis AG, Braverman LE, et al: Anticoagulant therapy and acute adrenal insufficiency. *Ann Intern Med* **81**:115, 1974.
125. O'Connell TX, Aston SJ: Acute adrenal hemorrhage complicating anticoagulant therapy. *Surg Gynecol Obstet* **139**:355, 1974.
126. Albert SG, Wolverson MK, Johnson FE: Bilateral adrenal hemorrhage in an adult: Demonstration by computed tomography. *JAMA* **247**:1737, 1982.
127. Liu L, Haskin ME, Rose LI, et al: Diagnosis of bilateral adrenocortical hemorrhage by computed tomography. *Ann Intern Med* **97**:720, 1982.
128. Sevitt SJ: Post-traumatic adrenal apoplexy. *Clin Pathol* **8**:185, 1955.
129. Editorial: Adrenal haemorrhage, apoplexy, and infarction. *Lancet* **2**:295, 1976.
130. Fox B: Venous infarction of the adrenal glands. *J Pathol* **119**:65, 1976.

131. Eagan RT, Page MI: Adrenal insufficiency following bilateral adrenal venography. *JAMA* **215:**115, 1971.

132. Bookstein JJ, Conn J, Reuter SR: Intra-adrenal hemorrhage as a complication of adrenal venography in primary aldosteronism. *Radiology* **90:**778, 1968.

133. Reuter SR, Blair AJ, Schteingart DE, et al: Adrenal venography. *Radiology* **89:**805, 1967.

134. DeFelice R, Johnson DG, Galgiani JN: Gynecomastia with ketoconazole. *Antimicrob Agents Chemother* **19:**1073, 1981.

135. Stevens DA, Stiller RL, Williams PL, et al: Experience with ketoconazole in three major presentations of progressive coccidioidomycosis. *Am J Med* **74:**58, 1983.

136. Van Den Bossche H, Willemsens G, Cools W, et al: In vitro and in vivo effects of the antimycotic drug ketoconazole on sterol synthesis. *Antimicrob Agents Chemother* **17:**922, 1980.

137. Pont A, Graybill JR, Craven PC, et al: High-dose ketoconazole therapy and adrenal and testicular function in humans. *Arch Intern Med* **144:**2150, 1984.

138. Pont A, Williams PL, Azhar S, et al: Ketoconazole blocks testosterone synthesis. *Arch Intern Med* **142:**2137, 1982.

139. Schurmeyer T, Nieschlag E: Ketoconazole induced drop in serum and saliva testosterone. *Lancet* **2:**1098, 1982.

140. Santen RJ, Brugmans J, Symoens J, et al: Ketocoanzole inhibits androgen production by blocking the 17-hydroxyprogesterone aldolase (C17-20) lyase enzyme [Abstract]. *Clin Res* **31:**473A, 1983.

141. Pont A, Williams PL, Loose DS, et al: Ketoconazole blocks adrenal steroid synthesis. *Ann Intern Med* **97:**370, 1982.

142. Pont A, Williams PL, Azhar S, et al: Ketoconazole blocks testosterone synthesis [Abstract]. *Clin Res* **30:**520A, 1982.

143. Loose DS, Kan PB, Hirst MA, et al: Ketoconazole blocks adrenal steroidogenesis by inhibiting cytochrome P450-dependent enzymes. *J Clin Invest* **71:**1495, 1983.

144. Kowal J: The effects of ketoconazole on steroidogenesis in cultured mouse adrenal cortex tumor cells. *Endocrinology* **112:**1541, 1983.

145. Loose DS, Stover EP, Feldman D: Ketoconazole binds to glucocorticoid receptors and exhibits glucocorticoid antagonist activity in cultured cells. *J Clin Invest* **72:**404, 1983.

146. Graybill JR: Summary: Potential and problems with ketoconazole. *Am J Med* **74:**86, 1983.

147. Tucker WS Jr, Snell BB, Island DP, et al: Reversible adrenal insufficiency induced by ketoconazole. *JAMA* **253:**2413, 1985.

148. Criado A, Maseda J, Navarro E, et al: Induction of anesthesia with etomidate: Hemodynamic study of 36 patients. *Br J Anaesth* **52:**803, 1980.

149. Gooding JM, Weng J-T, Smith RA, et al: Cardiovascular and pulmonary responses following etomidate induction of anesthesia in patients with demonstrated cardiac disease. *Anesth Analg* **58:**40, 1979.

150. Ledingham LM, Finlay WEI, Watt I, et al: Etomidate and adrenocortical function. *Lancet* **1:**1434, 1983.

151. Wagner RL, White PF, Kan PB, et al: Inhibition of adrenal steroidogenesis by the anesthetic etomidate. *N Engl J Med* **310:**1415, 1984.

152. Fellows JW, Bastow MD, Byrne AJ, et al: Adrenocorticol suppression in multiply injured patients: A complication of etomidate treatment. *Br Med J* **287:**1835, 1983.

153. Chee HD, Bronsveld W, Lips PTAM, et al: Adrenocorticol suppression in multiply-injured patients: A complication of etomidate treatment. *Br Med J* **288:**485, 1984.

154. Edwards OM, Galley JM, Courtenay-Evans RJ, et al: Changes in cortisol metabolism following rifampicin therapy. *Lancet* **2:**549, 1974.

155. Elansary EH, Earis JE: Rifampicin and adrenal crisis. *Br Med J* **286:**1861, 1983.

156. Ohnhaus EE, Park BK: Measurement of urinary 6β-hydroxycortisol excretion as an in vivo parameter in the clinical assessment of the microsomal enzyme-inducing capacity of antipyrine, phenobarbitone and rifampicin. *Eur J Clin Pharmacol* **15:**139, 1979.

157. Girard J, Baumann JB: Secondary adrenal insufficiency due to cyproterone acetate. *Pediatr Res* **9:**669, 1975.

158. Savage DCL, Swift PGF: Effect of cyproterone acetate on adrenocortical function in children with prococious puberty. *Arch Dis Child* **56:**218, 1981.

159. Editorial: Adrenal suppression by cyproterone acetate. *Lancet* **2:**290, 1981.

160. Mayock RL, Bertrand P, Morrison CE, et al: Manifestations of sarcodosis: Analysis of 145 patients, with a review of nine series selected from the literature. *Am J Med* **35**:67, 1963.
161. Karlish AJ, MacGregor GA: Sarcoidosis, thyroiditis, and Addison's disease. *Lancet* **2**:330, 1970.
162. Freinkel JK: Pathogenesis of infections of the adrenal glands leading to Addison's disease in man: The role of corticoids in adrenal and generalized infection. *Ann NY Acad Sci* **84**:393, 1960.
163. Irvine WJ, Toft AD, Feek CM: Addison's disease. In James VHT, (ed): *The Adrenal Gland.* Raven Press, New York, 1979, pp. 131–164.
164. Reichert CM, O'Leary TJ, Levens DL, et al: Autopsy pathology in the acquired immune deficiency syndrome. *Am J Pathol* **112**:357, 1983.
165. Macher AM, Reichert CM, Straus SE, et al: Death in the AIDS patient: role of cytomegalovirus [Letter]. *N Engl J Med* **309**:1454, 1983.
166. Tapper ML, Rotterdam HZ, Lerner CW, et al: Adrenal necrosis in the acquired immunodeficiency syndrome. *Ann Intern Med* **100**:239, 1984.
167. Guenthner EE, Rabinowe SL, Van Niel A, et al: Primary Addison's disease in a patient with the acquired immunodeficiency syndrome. *Ann Intern Med* **100**:847, 1984.
168. Neinhuis AW, Peterson DT, Henry W: Evaluation of endocrine and cardiac function in patients with iron overload on chelation therapy. In Zaino EC, Roberts RH (eds): *Chelation Therapy in Chronic Iron Overload.* Symposia Specialists, Miami, 1977, pp. 1–15.
169. Doppman JL, Gill JR Jr, Nienhuis AW, et al: CT Findings in Addison's disease. *J Comput Assist Tomogr* **6**:757, 1982.
170. Long JA Jr, Doppman JL, Nienhuis AW, et al: Computed tomographic analysis of beta-thalassemic syndromes with hemochromatosis: Pathologic findings with clinical and laboratory correlations. *J Comput Assist Tomogr* **4**:159, 1980.
171. Karenyi A: Congenital adrenal hypoplasia. *Arch Pathol* **71**:336, 1961.
172. Baker WdeC, Wise G, Mezger ML: Cytomegalic adrenal hypoplasia in a 4½-year-old boy. *Am J Dis Child* **114**:180, 1967.
173. Kreines K, DeVaux WD: Neonatal adrenal insufficiency associated with maternal Cushing's syndrome. *Pediatrics* **47**:516, 1971.
174. Gotterill JA, Cunliffe WJ: Self medication with liquorice in a patient with Addison's disease. *Lancet* **1**:294, 1973.
175. Dunlop D: Eighty-six cases of Addison's disease. *Br Med J* **2**:887, 1963.
176. Ask-Upmark E, Hull R: Addison's disease in a university medical department during 20 years. *Acta Med Scand* **192**:445, 1972.
177. Strakosch CR, Gordon RD: Early diagnosis of Addison's disease; pigmentation as sole symptom. *Aust NZ J Med* **8**:189, 1978.
178. Bisell GW, Surakomol K, Greenslit F: Longitudinal banded pigmentation of nails in primary adrenal insufficiency. *JAMA* **215**:1665, 1971.
179. Knowlton AI, Baer L: Cardiac failure in Addison's disease. *Am J Med* **74**:829, 1983.
180. Normington EAM, Davies D: Hypertension and oedema complicating pregnancy in Addison's disease. *Br Med J* **2**:148, 1972.
181. Thorn GW: *The Diagnosis and Treatment of Adrenal Insufficiency,* 2nd ed. Charles C. Thomas, Springfield, IL, 1951, p. 150.
182. Adams RD, Denny-Brown D, Pearson CM: *Diseases of Muscle.* Hoeber, New York, 1953, pp. 477–478.
183. Aubertin PJ: Painful muscular contractions during Addison's disease successfully treated with cortisone. *Ann Endocrinol* **12**:888, 1951.
184. Wisenbaugh H, Heller J: Flexion contractures in Addison's disease. *J Clin Endocrinol Metab* **20**:792, 1960.
185. Calabrese LH, White CS: Musculoskeletal manifestations of Addison's disease. *Arthritis Rheum* **22**:558, 1979.
186. Almog C, Menachem S: Flexion contractures in Addison's disease. *Confin Neurol* **32**:33, 1970.
187. George TM, Burke JM, Sobotka PA, et al: Resolution of stiff-man syndrome with cortisol replacement in a patient with deficiencies of ACTH, growth hormone, and prolactin. *N Engl J Med* **310**:1511, 1984.
188. Werk EE Jr, Sholiton LJ, Mamell RT: The "stiff-man" syndrome and hyperthyroidism. *Am J Med* **31**:647, 1961.
189. Howard RM Jr: A new and effective drug in the treatment of the stiff-man syndrome: Preliminary report. *Proc. Mayo Clin* **38**:203, 1963.

190. Rogoff JM: Diagnosis and treatment of Addison's disease. *Can Med Assoc J* **24**:43, 1931.

191. Gerrod O, Davis SA, Cahill G Jr: The action of cortisone and desoxycorticosterone acetate on glomerular filtration rate and sodium-water exchange in the adrenalectomized dog. *J. Clin Invest* **34**:761, 1955.

192. Ufferman RC, Schrier RW: Importance of sodium intake and mineralocorticoid hormone in the impaired water excretion in adrenal insufficiency. *J Clin Invest* **51**:1639, 1972.

193. Gill JR, Gann DS, Bartter F, et al: Restoration of water diuresis in Addisonian patients by the expansion of the volume of extracellular fluid. *J Clin Invest* **41**:1078, 1962.

194. Agus ZS, Goldberg M: Role of antidiuretic hormone in the abnormal water diuresis of the anterior hypopituitarism in man. *J Clin Invest* **50**:1478, 1971.

195. Green HH, Harrington AR, Valtin, H, et al: On the role of antidiuretic hormone in the inhibition of acute water diuresis in adrenal insufficiency and the effects of gluco- and mineralocorticoids in reversing the inhibition. *J Clin Invest* **49**:1724, 1970.

196. Rajathurai A, Chazan BI, Jeans JE: Self mutilation as a feature of Addison's disease. *Br Med J* **287**:1027, 1983.

197. Refetoff S, Block MB, Ehrlich EN, et al: Chiari–Frommel syndrome in a patient with primary adrenocortical insufficiency. Cure by glucocorticoid replacement. *N Engl J Med* **287**:1326, 1972.

198. Marilus R, Dickerman Z, Kaufman H, et al: Addison's disease associated with precocious sexual development in a boy. *Acta Paediatr Scand* **70**:587, 1981.

199. Laron Z, Karp M, Dolberg L: Juvenile hypothyroidism with testicular enlargement. *Acta Paediatr Scand* **59**:317, 1970.

200. Van Wyk JJ, Grumbach MM: Syndrome of precocious menstruation and galactorrhea in juvenile hypothyroidism: An example of hormonal overlap in pituitary feedback. *J Pediatr* **57**:416, 1960.

201. Green M, Lim KH: Bronchial asthma Addison's disease. *Lancet* **1**:1159, 1971.

202. Hadley RA: Bronchial asthma with adreno-cortical hypofunction *Ann Allerg* **27**:121, 1969.

203. Schmidt MB: Eine Biglandulare Erkrankung (Nebennieren und Schiddruse) bei morbus Addisonii. *Verh Deutsch Pathol* **21**:212, 1926.

204. Gastineau CF, Arnold JW: Thyroid disorders in Addison's disease. *Proc Staff Meet Mayo Clin* **38**:323, 1963.

205. Carpenter CCJ, Solomon N, Silverberg SG: Schmidt's syndrome. A review of the literature and a report of 15 new cases including 10 instances of coexistent diabetes. *Medicine (Baltimore)* **43**:153, 1964.

206. Irvine WJ, Barnes EW: Addison's disease, ovarian failure and hypoparathyroidism. *Clin Endocrinol Metab* **4**:379, 1975.

207. Beaven DW, Nelson DM, Renold AE, et al: Diabetes mellitus and Addison's disease. *N Engl J Med* **261**:443, 1959.

208. Wehrmacher WH: Addison's disease with diabetes mellitus. *Arch Intern Med* **108**:114, 1961.

209. Solomon N, Carpenter CCJ, Bennett IL, et al: Schmidt's syndrome (thyroid and adrenal insufficiency) and coexistent diabetes mellitus. *Diabetes* **14**:300, 1966.

210. Yoo J, Kozak GP: Diabetes and Addison's disease. *Postgrad Med* **55**:62, 1974.

211. Adler DK: Atypical Addison's disease associated with diabetes mellitus: Report of case. *N Engl J Med* **237**:805, 1947.

212. Baird IM, Munro DS: Addison's disease with diabetes mellitus: Case treated with cortisone. *Lancet* **1**:962, 1954.

213. Bartels EC, Fields ML, Murphy R: Addison's disease complicated by development of diabetes mellitus. *Lahey Clin Bull* **10**:234, 1958.

214. Breslaw L, Lashof J, Klein C: Diabetes mellitus, hypothyroidism and Addison's disease in one patient. *Ann Intern Med* **38**:338, 1953.

215. Coulshed N, Jones EW: Diabetes mellitus and Addison's disease: Discussion with report of two new cases. *Postgrad Med J (London)* **33**:60, 1957.

216. Faber V, Gronbaek P: Diabetes mellitus and Addison's disease. *Acta Endocrinol* **22**:145, 1956.

217. Gittler RD, Fajans S, Conn J, et al: Coexistence of Addison's disease and diabetes mellitus: Report of three cases with a discussion of metabolic interrelationships. *J Clin Endocrinol Metab* **19**:797, 1959.

218. Kenny FM, Holliday MA: Hypoparathyroidism, moniliasis, Addison's and Hashimoto's diseases. *N Engl J Med* **271**:708, 1964.

219. Leifer E, Hollander W Jr: Idiopathic hypoparathyroidism, and chornic adrenal insufficiency: A case report. *J Clin Endocrinol Metab* **13**:1264, 1953.

220. Leonard MF: Chronic idiopathic hypoparathyroidism with superimposed Addison's disease in a child. *J Clin Endocrinol Metab* **6**:493, 1946.

221. Perlmutter M, Ellison RR, Norsa L, et al: Idiopathic hypoparathyroidism and Addison's disease. *Am J Med* **21**:634, 1956.

222. Spinner MW, Blizzard RM, Childs B: Clinical and genetic heterogeneity in idiopathic Addison's disease and hypoparathyroidism. *J Clin Endocrinol Metab* **28**:795, 1968.

223. McMahon FG, Cookson DV, Inhorn SL: Idiopathic hypoparathyroidism and idiopathic adrenal cortical insufficiency occurring with cystic fibrosis of the pancreas. *Ann Intern Med* **51**:371, 1959.

224. Carter AC, Kaplan SA, DeMayo AP, et al: An unusual case of idiopathic hypoparathyroidism, adrenal insufficiency, hypothyroidism and metastatic calcification. *J Clin Endocrinol Metab* **19**:1633, 1959.

225. Morse WI, Cochrane WA, Landrigan PL: Familial hypoparathyroidism with pernicious anemia steatorrhea and adrenocortical insufficiency. *N Engl J Med* **264**:1021, 1961.

226. Szczepanski H, Sapiecha J: Addison's disease with hypofunction of the parathyroid glands. *Arch Dis Child* **34**:498, 1959.

227. Whitaker, J, Lamding BH, Esselborn VM, et al: The syndrome of familial juvenile hypoadrenocorticism, hypoparathyroidism, and superficial moniliasis. *J Clin Endocrinol* **16**:1374, 1956.

228. Forcier RJ, McIntyre OR, Frey WG, et al: Autoimmunity and multiple endocrine abnormalities. *Arch Intern Med* **129**:638, 1972.

229. McGregor BC, Katz HI, Doe RP: Vitiligo and multiple glandular insufficiency. *JAMA* **219**:724, 1972.

230. Neufeld M, Maclaren N, Blizzard R: Autoimmune polyglandular syndromes. *Pediatr Ann* **9**:154, 1980.

231. Neufeld M, Maclaren NK, Blizzard RM: Two types of autoimmune Addison's disease associated with different polyglandular autoimmune (PGA) syndromes. *Medicine (Baltimore)* **60**:355, 1981.

232. Hermans PE, Ulrich JA, Markowtiz H: Chronic mucocutaneous candidiasis as surface expression of deep-scaled abnormalities: A report of a syndrome of superficial candidiasis, absence of delayed hypersensitivity and aminoaciduria. *Am J Med* **47**:503, 1969.

233. Lehner T: Classification and clinopathological features of Candida infections in the mouth. In Winner HI, Herly R (eds): *Symposium on Candida infections*. E. and S. Livingstone, Edinburgh, 1966, pp. 119–136.

234. Kenny FM, Holliday MA: Hypoparathyroidism, monillasis, Addison's disease and Hashimoto's diseases: Hypercalcemia treated with administered sodium sulfate. *N Engl J Med* **271**:708, 1964.

235. Blizzard RM, Gibbs JH: Candidiasis: Studies pertaining to its association with endocrinopathies and pernicious anemia. *Pediatrics* **42**:231, 1968.

236. Wells RS, Higgs JM, MacDonald A, et al: Familial chronic mucocutaneous candidiasis. *J Med Genet* **9**:302, 1972.

237. Trence DL, Morley JE, Handwerger ES: Polyglandular autoimmune syndromes. *Am J Med* **77**:107, 1984.

238. McKenzie JM, Zakarija M, Sato A: Humoral immunity in Graves' disease. *Clin Endocrinol Metab* **7**:31, 1978.

239. Burke G: The cell membrane: A common site of action of thyrotrophin (TSH) and long-acting thyroid stimulator (LATS). *Metabolism* **18**:720, 1969.

240. Donaich D, Bottazzo GF: Polyendocrine autoimmunity. In Franklin EC (ed): *Clinical Immunology, Update*. Elsevier/North Holland, New York, 1981, pp. 96–109.

241. Bernstein DE: Diabetes mellitus followed by Addison's disease and hypothyroidism, simulating panhypopituitarism. *J Clin Endocrinol Metab* **8**:687, 1948.

242. Christy NP, Holub DA, Tomasi TB: Primary ovarian, thyroidal and adrenocortical deficiencies simulating pituitary insufficiency, associated with diabetes mellitus. *J Clin Endocrinol Metab* **22**:155, 1962.

243. Crispell KR, Parson W: The administration of purified growth hormone to a female with

hypoadrenalism, hypothyroidism, diabetes mellitus, and secondary amenorrhea, simulating panhypopituitarism. *J Clin Endocrinol Metab* **22:**881, 1962.

244. Lucky AW, Rebar RW, Blizzard RM, et al: Pubertal progression in the presence of elevated gonadotropins in girls with multiple endocrine deficiencies. *J Clin Endocrinol Metab* **45:**673, 1977.

245. Appl GB, Holub DA: The syndrome of multiple endocrine gland insufficiency. *Am J Med* **61:**129, 1976.

246. Farid NR, Bear JC: The human major histocompatibility complex and endocrine disease. *Endocr Rev* **2:**50, 1981.

247. Farid NR, Larsen B, Payne R, et al: Polyglandular autoimmune disease and HLA. *Tissue Antigens* **16:**23, 1980.

248. Van Thiel DH, Smith WI Jr, Rabin BS, et al: A syndrome of immunoglobulin—A deficiency, diabetes mellitus, malabsorption and a common HLA haplotype—Immunologic and genetic studies of forty-three family members. *Ann Intern Med* **86:**10, 1977.

249. Eisenbarth G, Wilson P, Ward F, et al: HLA type and occurrence of disease in familial polyglandular failure. *N Engl J Med* **298:**92, 1978.

250. Eisenbarth GS, Wilson PW, Ward F, et al: The polyglandular failure syndrome: Disease inheritance, HLA type and immune function. *Ann Intern Med* **91:**528, 1979.

251. Schaumburg HH, Powers JM, Raine CS, et al: Adrenoleukodystrophy: A clinical and pathological study of 17 cases. *Arch Neurol* **32:**577, 1975.

252. Case records of the Massachusetts General Hospital (Case 6—1962). *N Engl J Med* **266:**191, 1962.

253. Schaumburg HH, Richardson EP, Johnson PC, et al: Schilder's disease: Sex-linked recessive transmission with specific adrenal changes. *Arch Neurol* **27:**458, 1972.

254. Powers JM, Schaumburg HH: Adreno-leukodystrophy: Similar ultrastructural changes in adrenal cortical cells and Schwann cells. *Arch Neurol* **30:**406, 1974.

255. Powers JM, Schaumburg HH: Adreno-leukodystrophy (sex-linked Schilder's disease): A pathogenic hypothesis based on ultrastructural lesions in adrenal cortex, peripheral nerve and testis. *Am J Pathol* **76:**481, 1974.

256. Schaumburg HH, Powers JM, Suzuki K, et al: Adreno-leukodystrophy (sex-linked Schilder's disease): Ultrastructural demonstration of specific cytoplasmic inclusions in the central nervous system. *Arch Neurol* **31:**210, 1974.

257. Powell H, Tindall R, Schultz P, et al: Adrenoleukodystrophy: Electron microscopic findings. *Arch Neurol* **32:**250, 1975.

258. Ogino T, Schaumburg HH, Suzuki K: Cholesterol ester metabolism in adrenoleukodystrophy (ALD). *Trans Am Soc Neurochem* **9:**209, 1978.

259. Burton BK, Nadler HL: Schilder's disease: Abnormal cholesterol retention and accumulation in cultivated fibroblasts. *Pediatr Res* **8:**170, 1974.

260. Igarashi M, Schaumburg HH, Powers J, et al: Fatty acid abnormality in adrenoleukodystrophy. *J Neurochem* **26:**851, 1976.

261. Griffin JW, Goren E, Schaumburg H, et al: Adrenomyeloneuropathy: A probable variant of adrenoleukodystrophy. I. Clinical and endocrinologic aspects. *Neurology* **27:**1107, 1977.

262. Schaumburg HH, Powrs JM, Raine CS, et al: Adrenomyeloneuropathy: A probable variant of adrenoleukodystrophy. II. General pathologic, neuropathologic, and biochemical aspects. *Neurology* **27:**1114, 1977.

263. Peckham RS, Marshall MC Jr, Rosman PM, et al: A variant of adrenomyeloneuropathy with hypothalamic–pituitary dysfunction and neurologic remission after glucocorticoid replacement therapy. *Am J Med* **72:**173, 1982.

264. Fettes I, Killinger D, Volpe R: Adrenoleukodystrophy: Report of a familial case. *Clin Endocrinol* **11:**151, 1979.

265. Davis LE, Snyder RD, Orth DN, et al: Adrenoleukodystrophy and adrenomyeloneuropathy associated with partial adrenal insufficiency in three generations of a kindred. *Am J Med* **66:**342, 1979.

266. Bardwick PA, Zvaifler NJ, Gill GN, et al: Plasma cell dyscrasia with polyneuropathy, organomegaly, endocrinopathy, M protein, and skin changes: The POEMS syndrome. Report on two cases and a review of the literature. *Medicine* **59:**311, 1980.

267. Shimpo, S: [Solitary myeloma causing polyneuritis and endocrine disorders.] *Jpn J Clin Med* **26:**2444, 1968 [Jap].

268. Crow RS: Peripheral neuritis in myelomatosis. *Br Med J* **2**:802, 1956.
269. Imawari M, Akatsuka N, Ishibashi M, et al: Syndrome of plasma cell dyscrasia, polyneuropathy, and endocrine disturbances. Report of a case. *Ann Intern Med* **81**:490, 1974.
270. Iwashita H, Ohnishi A, Asada M, et al: Polyneuropathy, skin hyperpigmentation, edema, and hypertrichosis in localized osteosclerotic myeloma. *Neurology* **27**:675, 1977.
271. Meshkinpour H, Myung CG, Kramer LS: A unique multisystemic syndrome of unknown origin. *Arch Intern Med* **137**:1719, 1977.
272. Morley JB, Schwieger AC: The relation between chronic polyneuropathy and osteosclerotic myeloma. *J Neurol Neurosurg Psychiatry* **30**:432, 1967.
273. Saihan EM, Burton JL, Heaton KW: A new syndrome with pigmentation, sclerodema, gynaecomastia, Raynaud's phenomenon and peripheral neuropathy. *Br J Dermatol* **99**:437, 1978.
274. Takatsuki K, Uchiyama T, Sagawa K, et al: Plasma cell dyscrasia with polyneuropathy and endocrine disorder: Review of 32 patients. In Seno S, Takaku F, Irino S (eds): *Topics in Hematology.* Proceedings of the 16th International Congress of Hematology, Kyoto, Sept. 5–11, 1976. Excerpta Medica, Amsterdam, 1977, pp. 454–457.
275. Resnick D, Greenway GD, Bardwick PA, et al: Plasma-cell dyscrasia with polyneuropathy, organomegaly, endocrinopathy, M-protein, and skin changes: The POEMS syndrome. *Radiology* **140**:17, 1981.
276. Allgrove J, Clayden GS, Grant DB, et al: Familial glucocorticoid deficiency with achalasia of the cardia and deficient tear production. *Lancet* **1**:1284, 1978.
277. Geffner ME, Lippe BM, Kaplan SA, et al: Selective ACTH insensitivity, achalasia, and alacrima: A multisystem disorder presenting childhood. *Pediatr Res* **17**:532, 1983.
278. Lanes R, Plotnick LP, Bynum TE, et al: Glucocorticoid and partial mineralocorticoid deficiency associated with achalasia. *J Clin Endocrinol Metab* **50**:268, 1980.
279. Pombo M, Devesa J, Taborda A, et al: Glucocorticoid deficiency with achalasia of the cardia and lack of lacrimation. *Clin Endocrinol* **23**:237, 1985.
280. Sachdev Y, Morley AR, Wilkinson R, et al: Addison's disease with renal microangiopahy and renal failure (a new syndrome). *Q J Med* **46**:151, 1977.
281. Paterson CR: *Metabolic Disorders of Bone.* Blackwell Scientific, Oxford, 1975.
282. Siegler DIM: Idiopathic Addison's disease presenting with hypercalcemia. *Br Med J* **2**:522, 1970.
283. Walser M, Robinson BHB, Duckett JW Jr: The hypercalcemia of adrenal insufficiency. *J Clin Invest* **42**:456, 1963.
284. Jorgensen H: Hypercalcemia in adrenocortical insufficiency. *Acta Med Scand* **193**:175, 1973.
285. Downie WW, Gunn A, Paterson CR, et al: Hypercalcaemic crisis as presentation of Addison's disease. *Br Med J* **1**:145, 1977.
286. Walker DA, Davies M: Addison's disease presenting as a hypercalcaemic crisis in a patient with idiopathic hypoparathyroidism. *Clin Endocrinol* **14**:419, 1981.
287. Cunningham SK, Moore A, McKenna TJ: Normal cortisol response to corticotropin in patients with secondary adrenal failure. *Arch Intern Med* **143**:2276, 1983.
288. Jubiz W, Meikle W, West CE, et al: Single-dose metyrapone test. *Arch Intern Med* **125**:482, 1970.
289. Spark RF: Simplified assessment of pituitary–adrenal reserve. *Ann Intern Med* **75**:717, 1971.
290. Geffner ME, Lippe BM, Kaplan SA, et al: The use of theophylline as an in vivo probe of adrenocortical function. *J Clin Endocrinol Metab* **55**:56, 1982.
291. Karstaedt N, Sagel SS, Stanley RJ, et al: Computed tomography of the adrenal gland. *Radiology* **129**:723, 1978.
292. Vita JA, Silverberg SJ, Goland RS, et al: Clinical clues to the cause of Addison's disease. *Am J Med* **78**:461, 1985.
293. Fishman EK, Siegelman SS: Computed body tomography. *Contemporary Issues in Computed Tomography* **3**:46–48.
294. Twersky J, Levin DC: Metastatic melanoma of the adrenal—An unusual cause of adrenal calcification. *Radiology* **116**:627, 1975.
295. Wolverson MK, Kannegiesser H: CT of bilateral adrenal hemorrhage with acute adrenal insufficiency in the adult. *Am J Roentgenol* **142**:311, 1984.
296. Wilms GE, Baert AL, Kint EJ, et al: Computed tomographic findings in bilateral adrenal tuberculosis. *Radiology* **146**:729, 1983.

297. Doppman JL, Gill JR, Nienhuis AW, et al: CT findings in Addison's disease. *J Comput Assist Tomogr* **6:**757, 1982.

298. Huebener KH, Treugut H: Adrenal cortex dysfunction: CT findings. *Radiology* **150:**195, 1984.

299. Hauser H, Battikha JB, Wettstein P: Pathology of the adrenal glands: Common and uncommon findings in computed tomography. *Eur J Radiol* **1:**215, 1981.

300. Volpe R: The role of autoimmunity in hypoendocrine and hyperendocrine function with special emphasis on autoimmune thyroid disease. *Ann Intern Med* **87:**86, 1977.

301. McHardy-Young S, Lessof MH, Maisey MN: Serum TSH and thyroid antibody studies in Addison's disease. *Clin Endocrinol (Oxf)* **1:**45, 1972.

302. Evered DC, Ormston BJ, Smith PA, et al: Grades of hypothyrodism. *Br Med J* **1:**657, 1973.

303. Gharib H, Hodgson SF, Gastineau CF, et al: Reversible hypothyroidism in Addison's disease. *Lancet* **2:**734, 1972.

304. Topliss PJ, White EJ, Stockigt JR: Significance of thyrotropin excess in untreated primary adrenal insufficiency. *J Clin Endocrinol Metab* **50:**52, 1980.

305. Barnett AM, Donald RA, Espiner EA: High concentrations of thyroid-stimulating hormone in untreated glucocorticoid deficiency: Indication of primary hypothyroidism. *Br Med J* **3:**172, 1982.

306. Wilbur JF, Utiger RD: The effect of glucocorticoids on thyrotropin secretion. *J Clin Invest* **48:**2096, 1969.

307. Re RN, Kourides IA, Ridgway ED, et al: The effect of glucocorticoid administration on human pituitary secretion of thyrotropin and prolactin. *J Clin Endocrinol Metab* **43:**338, 1976.

308. Nicoloff JT, Fisher DA, Appleman MD Jr: The role of glucocorticoids in the regulaiton of thyroid function in man. *J Clin Invest* **49:**1922, 1970.

309. Sowers JR, Carlson HE, Brautbar N: Effect of dexamethasone on prolactin and TSH responses to TRH and metoclopramide in man. *J Clin Endocrinol Metab* **44:**237, 1977.

310. Dussault JH: The effect of dexamethasone on TSH and prolactin secretion after TRH stimulation. *Can Med Assoc J* **111:**1195, 1974.

311. Otsuki M, Dakoda M, Baba S: Influence of glucocorticoids on TRF-induced TSH response in man. *J Clin Endocrinol Metab* **36:**95, 1973.

312. DeGroot LJ, Hoye K: Dexamethasone suppression of serum T_3 and T_4. *J Clin Endocrinol Metab* **42:**976, 1976.

313. Stryker TD, Molitch ME: Reversible hyperthyrotropinemia, hyperthyroxinemia, and hyperprolactinemia due to adrenal insufficiency. *Am J Med* **79:**271, 1985.

314. Clayton R, Burden AC, Shrieber V, et al: Secondary pituitary hyperplasia in Addison's disease. *Lancet* **2:**954, 1977.

315. Dexter RN, Orth DN, Abe K, et al: Cushing's disease without hypercortisolism. *J Clin Endocrinol Metab* **30:**573, 1970.

316. Dluhy RG, Moore TJ, Williams GH: Sella turcica enlargement and primary adrenal insufficiency. *Ann Intern Med* **89:**513, 1978.

317. Jara-Albarran A, Bayort J, Caballero A, et al: Probable pituitary adenoma with adrenocorticotropin hypersecretion (corticotropinoma) secondary to Addison's disease. *J Clin Endocrinol Metab* **49:**236, 1979.

318. Himsworth RL, Lewis JG, Rees LH: A possible ACTH secreting tumour of the pituitary developing in a conventionally treated case of Addison's disease. *Clin Endocrinol (Oxf)* **9:**131, 1978.

319. Aanderud S, Bassøe HH: A pituitary tumour with possible ACTH and TSH hypersecretion in a patient with Addison's disease and primary hypothyroidism. *Acta Endocrinol (Kbh)* **95:**181, 1980.

320. Krautli B, Muller J, Landolt AM, et al: ACTH-producing pituitary adenomas in Addison's disease: Two cases treated by transphenoidal microsurgery. *Acta Endocrinol* **99:**357, 1982.

321. Conomy JP: Nelson's syndrome: Chicken, eggs and sellar explosions [Editorial]. *Arch Intern Med* **138:**691, 1978.

322. Samaan NA, Stepanas AV, Danziger J, et al: Reactive pituitary abnormalities in patients with Klinefelter's and Turner's syndromes. *Arch Intern Med* **139:**198, 1979.

323. Leiba S, Landau B, Ber A: Target gland insufficiency and pituitary tumours. *Acta Endocrinol (Kbh)* **60:**112, 1969.

324. Vagenakis AG, Dole K, Braverman V: Pituitary enlargement, pituitary failure, and primary hypothyroidism. *Ann Intern Med* **85:**195, 1976.

325. Samaan NA, Osborne BM, Mackay B, et al: Endocrine and morphologic studies of pituitary adenomas secondary to primary hypothyroidism. *J Clin Endocrinol Metab* **45**:903, 1977.

326. Copinschi G, L'Hermite M, Leclerq R, et al: Effects of glucocorticoids on pituitary hormonal response to hypoglycemia. Inhibition of prolactin release. *J Clin Endocrinol Metab* **40**:442, 1975.

327. Bratusch-Marrain P, Vierhapper H, Waldhausl W, et al: Acute suppressive effect of ACTH-induced cortisol secretion on serum prolactin levels in healthy man. *Acta Endocrinol* **99**:352, 1982.

328. Leung FC, Chen HT, Verkaik SJ, et al: Mechanism(s) by which adrenalectomy and corticosterone influence prolactin release in the rat. *J Endocrinol* **87**:131, 1980.

329. Harms PG, Langlier P, McCann SM: Modification of stress-induced prolactin release by dexamethasone or adrenalectomy. *Endocrinology* **96**:475, 1975.

330. Lever EG, McKerron CG: Auto-immune Addison's disease associated with hyperprolactinemia. *Clin Endocrinol* **21**:451, 1984.

331. Scott RS, Donald RA, Espiner EA: Plasma ACTH and cortisol profiles in Addisonian patients receiving conventional substitution therapy. *Clin Endocrinol (Oxf)* **9**:571, 1978.

332. Smith R, Donald RA, Espiner EA, et al: The effect of different treatment regimens on hormonal profiles in congenital adrenal hyperplasia. *J Clin Endocrinol Metab* **51**:230, 1980.

333. Feek CM, Ratcliffe JG, Seth J, et al: Patterns of plasma cortisol and ACTH concentrations in patients with Addison's disease treated with conventional corticosteroid replacement. *Clin Endocrinol* **14**:451, 1981.

334. Khalid BAK, Burke CW, Hurley DM, et al: Steroid replacement in Addison's disease and in subjects adrenalectomized for Cushing's disease: Comparison of various glucocorticoids. *J Clin Endocrinol Metab* **55**:551, 1982.

335. Nelson DH: Diagnosis and treatment of Addison's disease. In DeGroot LJ (ed): *Endocrinology*, vol. 2. Grune & Stratton, New York, 1979, p. 1199.

336. Levin J, Zumoff B, Kream J, et al: Cortisol measurements in patients receiving oral corticosteroid replacement treatment. *J Clin Pharmacol* **21**:52, 1981.

337. Barbato AL, Landau RL: Serum cortisol appearance–disappearance in adrenal insufficiency after oral cortisone acetate. *Acta Endocrinol* **84**:600, 1977.

338. Kehlet H, Binder C, Blichert-Toft M: Glucocorticoid maintenance therapy following adrenalectomy: Assessment of dosage and preparation. *Clin Endocrinol* **5**:37, 1976.

339. Burch WM: Urine free-cortisol determination: A useful tool in the management of chronic hypoadrenal states. *JAMA* **247**:2002, 1982.

340. Thompson DG, Mason AS, Goodwin FJ: Mineralocorticoid replacement in Addison's disease. *Clin Endocrinol* **10**:499, 1979.

341. Smith SJ, Markandu ND, Banks RA, et al: Evidence that patients with Addison's disease are undertreated with fludrocortisone. *Lancet* **1**:11, 1984.

342. Nordin BEC: Addison's disease with partial recovery. *Proc R Soc Med* **48**:1024, 1955.

343. Annear TD, Baker GP: Tuberculous Addison's disease: A case apparently cured by chemotherapy. *Lancet* **2**:577, 1961.

344. Coleman EN, Arneil GC: Acute tuberculous adrenocortical failure with clinical recovery. *Lancet* **1**:886, 1962.

345. Vague P, Combes R, Altomare E, et al: La maladie d'Addison "blanchie": Huit cas apparement gueris. *Nouv Presse Med* **7**:1621, 1978.

3

Cushing's Syndrome

Introduction

The term Cushing's syndrome refers to a constellation of clinical findings that evolve from sustained hypersecretion of the glucocorticoid cortisol. The effects of chronic hypercortisolism can be devastating, with far-reaching adverse effects on several organ systems, resulting in considerable morbidity and even mortality. The most common form of hypercortisolism, besides the iatrogenic variety, is caused by ACTH hypersecretion and is referred to as "Cushing's disease" (CD) or pituitary-dependent, ACTH-mediated hypercortisolism. This accounts for 70–80% of all cases of hypercortisolism. In addition, Cushing's syndrome can result from autonomous secretion of cortisol by tumors of the adrenal cortex (adenoma or carcinoma) or from a heterogenous group of adrenal disorders loosely termed "nodular adrenocortical hyperplasia." Finally, Cushing's syndrome can also result from adrenocortical stimulation secondary to extraadrenal neoplasms that secrete ACTH and, less commonly, corticotropin-releasing hormone (CRH). Regardless of the etiology, the ultimate expression is hypercortisolism, with its attendant spectrum of pathophysiological effects. Identification of the correct etiology of hypercortisolism is crucial, owing to its impact on specific therapy. Such identification can, at times, be frought with inordinate difficulties, causing considerable frustration to patient and physician alike. The advances in our understanding of steroid physiology, the impact of the technical marvel of computerized tomography, the excitement surrounding the characterization of synthetic CRH with its use as a diagnostic tool, and the emergence of selective inferior petrosal sinus sampling to locate the source of trouble have all contributed to better and early diagnosis of the etiology of hypercortisolism. Yet, despite these advances, the correct diagnosis of the etiology of hypercortisolism can be quite challenging and at times may represent an unresolved diagnostic conundrum. Study of the nature and expressions of this fascinating disease carries with it the same excitement that it must have generated when first described by Harvey Cushing in 1932.[1] While many facets of Cushing's syndrome have been illuminated by current knowledge, there are several aspects of this syndrome that still remain tantalizingly unclear.

Etiology and Pathogenesis

The etiologies of hypercortisolism are outlined in Table 10. A brief description on the nature of each of these etiologies would provide a perspective that

TABLE 10.
Etiology of Hypercortisolism

I. Endogenous
   1. Pituitary-dependent, ACTH-mediated (Cushing's Disease)
      a. Microadenoma of the anterior lobe
      b. Macroadenoma of the anterior lobe
      c. Hyperplasia of corticotropes
      d. Tumors of the intermediate lobe
   2. Tumors of the adrenal cortex
      a. Adenoma
         Single
         Multiple
      b. Carcinoma
   3. Nodular hyperplasia
      a. Micronodular disease
      b. Macronodular disease
      c. Primary adrenocortical nodular dysplasia (PAND)
   4. Ectopic ACTH syndrome
      a. Ectopic secretion of ACTH
      b. Ectopic secretion of CRH
II. Exogenous
   1. Iatrogenic
   2. Factitious

facilitates the understanding of the clinical and hormonal expressions of Cushing's syndrome.

Pituitary-Dependent Hypercortisolism

Pituitary tumors are the most important and frequent anatomic etiology for CD. Thus, microadenomas (tumors under 10 mm), noninvasive macroadenomas (tumors larger than 10 mm, but confined to the sella), and invasive macroadenomas are all known to cause CD. Rare cases of CD secondary to hyperplasia of the corticotropes probably represent the early stages in the evolution of the disease. Microadenomas represent the most common etiology, accounting for nearly 80% of CD.

While the anatomical etiology of CD is relatively clear, the pathophysiological basis for the development of the disease is far from explicit. It is believed that an abnormality in the hypothalamopituitary–adrenal (HPA) axis sets the stage for the development of this disorder. The characteristic hallmark of a normal HPA axis is prompt recognition by and brisk suppression of ACTH to even minor increases in circulating cortisol levels. This unique sensitivity to suppress in response to negative feedback is blunted in pituitary-dependent CD. Consequently, the hypothalamic pituitary threshold for suppression is raised and the ACTH secretory rate is consistently inappropriate, relative to the circulating cortisol levels at any given time. The raised threshold to negative feedback is evidenced by the response of patients with CD to the standard dexamethasone suppression tests (i.e., lack of physiological suppression to low-dose dexamethasone, but preservation of suppression to "higher" doses). The degree of abnormality is variable from patient to patient, some requiring a higher dose than that used in the conventional test. Nevertheless, they do suppress, denoting preservation of physiological cues, albeit operating at a higher threshold. This response pattern is classic for pituitary-dependent CD.

The aforementioned simplistic concept of "reset" hypothalamic pituitary axis has come under rigorous scrutiny in the past decade. The controversy revolves around a very basic issue—is CD a "primary derangement of the pituitary gland" as characterized by Harvey Cushing, or is it a result of an overactive hypothalamic driving mechanism? The proponents of both theories have drawn on histological, biochemical, hormonal, and pharmacological facts to support their respective premise. The ongoing polemic has great significance in terms of its impact on choosing the options for therapy; if the disease is a result of a primary pituitary tumor, a properly performed transsphenoidal microadenectomy should be curative, while if a hypothalamic overdrive was causing the problem, then medical therapy to correct the overdrive would be the logical initial choice. The conflicting data in the literature can be sorted, if viewed from three perspectives—evidence that favors a primary *hypothalamic* defect, evidence that favors a primary *pituitary* defect, and evidence that points to the existence of anatomical *subtypes* encountered in seemingly identical corticotrope adenomas.

Primary Hypothalamic Etiology

There are several lines of evidence to suggest that CD is a result of corticotrope overactivity secondary to an accentuated hypothalamic drive. *Anatomically* this is supported by the observation that a small, but significant number of patients with CD fail to show any tumor in the pituitary gland despite the most scrutinizing exploration. Indeed, remission of pituitary-dependent CD after removal of nonneoplastic pituitary gland has been reported.[2] There is a large body of data derived from dynamic studies. Two earlier reports, one by Liddle[3] and another by Orth and Liddle,[4] documented that some patients with pituitary-dependent CD continue to demonstrate abnormal dexamethasone (and metyrapone) responses despite a clinical cure and normalization of urinary steroids excretion following treatment. The question that obviously is raised is whether ACTH-secreting adenomas arise de novo or occur as a consequence of stimulation from higher centers. The availability of assays for circulating ACTH level in the plasma has greatly facilitated study of ACTH dynamics in pituitary-dependent CD. The pituitary ACTH secretion in CD responds to a variety of seemingly unrelated agents. Thus, thyrotropin-releasing hormone (TRH),[5] vasopressin,[5,6] and luteinizing-hormone–releasing hormone[7] are all capable of releasing ACTH from pituitary tumors that cause CD, denoting nonautonomy of the tumor and partial preservation of responsiveness.

A major *pharmacological* link of evidence that points to a central nervous system site as the cause of pituitary-dependent CD comes from studies with cyproheptadine. This drug is a serotonin antagonist which in some patients with CD effectively lowers cortisol and ACTH levels, restores normal suppressibility, and can even induce a sustained remission of CD and Nelson's syndrome.[5,8] While the action of this drug is not completely understood, it acts on the central nervous system and is capable of blocking the release of corticotropin-releasing factor (CRF) from the hypothalamus of the rat. More specifically, does cyproheptadine reverse the abnormality in ACTH feedback inherent to patients with CD? Lankford et al.[9] studied the two phases of cortisol feedback suppression of ACTH in nine patients with CD who had been treated either by adrenalectomy or by transsphenoidal microadenectomy. In normal humans cortisol-induced ACTH suppression consists of an early rate-dependent phase and a delayed dose-dependent phase, the two phases being temporally and dynamically distinct. A characteristic abnormality of patients with CD, especially those who undergo adrenalectomy, is an initial paradoxical rise of ACTH feedback abnormality even after removal of the pituitary adenoma. These authors further showed that cyproheptadine reversed the abnormality in all patients, suggesting that higher centers must have an important role in the pathophysiology of CD. Another pharmacological agent, bromocriptine, can also lower ACTH levels in patients with CD.[10] The characterization and isolation of CRF has provided perhaps the strongest support to the notion that hypothalamic (or even more central) influences underlie the pathophysiology of CD. The demonstration of brisk ACTH response to the intravenous administration of ovine CRF clearly implies that the pituitary adenoma cells are not only nonautonomous, but are possibly hypersensitive to the hypothalamic peptide.[11] Finally, ACTH-dependent Cushing's syndrome can result from ectopic secretion of CRH. Carey et al.[12] described a patient with carcinoma of the prostate, metastatic to the median eminence of the hypothalamus. The patient manifested excessive ACTH secretion not suppressible by dexamethasone. The pituitary demonstrated hyperplasia of the corticotropes, while the extracts from the tumor and the metastases contained CRF (but not ACTH) measurable by both radioimmunoassay and bioassay. This was the first documentation of CRF-induced hypercortisolism.

The development of Nelson's syndrome following bilateral adrenalectomy for CD is often cited as evidence to support a hypothalamic etiology for CD. Nelson's syndrome, characterized by markedly increased plasma ACTH levels, despite glucocorticoid replacement, shares several similarities in hormone dynamics with untreated CD. The response to TRH, cyproheptadine, bromocriptine, and CRH administration is strikingly identical in both diseases.

In summary, these lines of evidence form a compelling set of arguments to suggest a hypothalamic etiology in the development of CD (Table 11).

Primary Pituitary Disorder

The two major reasons for considering the pituitary gland as the primary seat of the disease are the high percentage of microadenomas found during surgery and the increasing number of "cures" attained following successful microadenomectomy by transsphenoidal surgery (TPS). In the series of 72 patients reported by Hardy[16] and another series of 100 patients reported by Boggan et al.,[15] the incidence of detecting microadenomas was approximately 80%, and the overall cure rates of tumors confined to the sella turcica were 88% and 87%, respectively. Such figures strongly argue in favor of CD representing a problem arising primarily within the pituitary gland. If this is indeed so, the reason for the partial autonomy of the adenoma cells would require explanation. The

TABLE 11.
Cushing's Disease—A Primary Hypothalamic Disorder

Feature	Reference
Infrequent occurrence of radiologically evident pituitary tumors	Liddle[13]
Presence of abnormal EEG patterns	Krieger and Glick[14]
Persistence of abnormal hypothalamic pituitary axis despite apparent cure	Liddle[3] Orth and Liddle[4]
Failure to find any tumor in a small percentage of patients despite scrutinizing search at surgery	Bogann et al.[15] Hardy[16]
Response to cyproheptadine	Aronin and Krieger[8]
Response to CRF	Chrousos et al.[11]
CD and be mimicked by metastic CRF-secreting neoplasm	Carey et al.[12]
Persistence or recurrence of hypercortisolism despite resection of adenomas	Lamberts et al.[25]

ability to study the behavior of corticostrope adenoma cells in tissue culture has provided insight into this phenomenon. Ludecke et al.[17,18] have shown that isolated corticotropes are only partially suppressible to dexamethasone, strikingly analogous to the partial suppressibility to dexamethasone seen in patients with CD. It is therefore conceivable that the classic response seen in pituitary dependent CD is an inherent feature of the adenoma cells, independent of any "resetting mechanisms" at the level of the hypothalamus.

Several recent in vitro studies have also evaluated the effect of various drugs on the isolated tumor cells grown in tissue culture. Suda et al.[19] have demonstrated that both cyproheptadine and reserpine directly inhibit the release of ACTH and endorphin in vitro from adenomas removed from patients with pituitary-dependent CD. Similar observations have been extended to bromocriptine.[20,21] These powerful data have clearly dealt a blow to the hypothalamic theory of origin of CD, since obviously these drugs can directly act on the pituitary gland, circumventing all mediation by the hypothalamus.

Additional support to the primary pituitary origin of CD is derived from the observation that transient hypoadrenalism develops after successful extirpation of the ACTH-secreting microadenoma. Fitzgerald et al.[22] prospectively evaluated 12 patients who underwent pituitary microsurgery for CD. Eleven of twelve patients developed postoperative hypoadrenalism with deficient adrenal responsiveness to exogenous ACTH administration. Such a response reflects suppression of normal corticotrope function secondary to prior hypercortisolism, reminiscent of the transient postoperative hypocalcemia following removal of a parathyroid adenoma. All patients eventually revealed normal corticotrope function in a time frame that was quite consistent with dormancy of the suppressed population of normal corticotropes. Such a sequence is consistent with a primary pituitary origin for the tumor.

A final argument in favor of the pituitary origin of the disorder is the "corticotroph mass hypothesis" proposed by Jeffcoate et al.[23] These workers reported two patients with pituitary-dependent CD who underwent incomplete transsphenoidal pituitary surgery (TPS) and despite persistence of the disease (and in the presence of residual tumor) demonstrated restoration of full suppression to dexamethasone. It is conceivable that, at least in these two cases, the apparent restoration of suppressibility may merely reflect a reduction in the mass of the tumor cells. Adenoma cells may behave similarly to normal corticotropes, the partial suppressibility merely reflecting their total number. When that number was reduced, as with incomplete TPS, normal suppressibility was restored, despite active disease. Such a hypothesis obviates the need to invoke any hypothalamic mediation or even a resetting of the receptors at either level, the hypothalamus or the pituitary.

From the aforementioned it would appear that a formidable array of data can be lined up to support the notion that CD is a "primary derangement of the pituitary gland" as originally proposed by Harvey Cushing. Table 12 summarizes these data.

While controversy continues regarding the pathophysiology of CD, evidence for the presence of *both* subtypes of CD has been presented by Van Cauter and Refetoff.[24] These workers studied the cortisol pulses, i.e., variations in the circulating cortisol levels in the plasma over a 24-hr period. The "cortisol profile" of normal subjects shows a characteristic jagged pattern with 7–10 pulsatile bursts of secretory activity. The cortisol spikes are due to concomitant ACTH spikes, which in turn are presumed to be due to pulsatile release of CRF by the hypothalamus. Van Cauter and Refetoff delineated two types of patterns in patients with pituitary-dependent CD. One group had a "hypopulsatile" pattern (normal number and absolute height of spikes, but the height relative to the preceding trough was lower than in normal controls); the second group had a "hyperpulsatile pattern" (absolute and relative height of the spikes was greater than in normal controls). The former group may represent patients with CRF-independent ACTH hypersecretion, and the latter group may represent patients with CRF-dependent ACTH hypersecretion.

TABLE 12.
Cushing's Disease—A Primary Pituitary Disorder

Feature	Reference
High incidence of finding microadenoma at surgery	Hardy,[16] Boggan et al.[15]
The extremely high success rate with selective microadenomectomy	Hardy,[16] Boggan et al.,[15] Burch[309]
In vitro response of adenoma cells to dexamethasone	Ludecke et al.[17]
In vitro response of adenoma cells to cyproheptadine, reserpine, and bromocriptine	Suda et al.,[19] Lamberts et al.[10]
Demonstration of dormancy of normal corticotrope population following successful removal of adenoma	Fitzgerald et al.[22]

Intermediate-Lobe Pathology

While indeed most ACTH-secreting microadenomas arise from the anterior lobe, they can also arise from the intermediate lobe, an area of the pituitary that is anatomically not well delineated in the human. Yet, tumors arising from this region assume special importance in CD. Lamberts et al.[25] were the first to point out the differences in ACTH regulation between adenomas originating from the anterior lobe and those arising from the intermediate lobe. The five characteristics of ACTH-secreting adenomas that arise from the intermediate lobe are:

1. The presence of argyrophilic nerve fibers coursing in and around the tumor, suggesting a neural origin.
2. Resistance to dexamethasone suppression, but response to dopamine agonists such as bromocriptine.
3. Frequent association with hyperprolactinemia.
4. Poor visualization by computed tomography of the pituitary.
5. Frequent failure to respond to transsphenoidal microadenectomy, i.e., persistence or recurrence of tumor despite the apparently complete removal of the tumor.

The relatively unsatisfactory response to surgery contrasts sharply with the highly favorable outcome following TPS for adenomas of the anterior pituitary and has tremendous impact when planning the surgical approach for such patients. The reasons for the unique differences that set apart these tumors of the intermediate lobe from conventional microadenomas of the anterior pituitary are less clear.

Lamberts et al.[25] have proposed that depletion of hypothalamic dopaminergic neural input may be the biochemical abnormality that underlies the development of hyperlasia of the corticotropes in the intermediate lobe followed by adenoma formation in some. An identical syndrome occurs in an strikingly parallel animal model in the horse ("equine CD"). Orth et al.[26] described a horse and a pony suffering from myopathy, diabetes, and hirsutism. The animals manifested hypercortisolism with elevated plasma levels of ACTH, disproportionate elevation of α and β melanophore-stimulating hormones, and poor suppressibility to dexamethasone, with responsiveness to intravenous dopamine, bromocriptine, and pergolide.

Tumors of the Adrenal

Adrenal adenomas are benign tumors that originate from the adrenal cortex. These tumors more often tend to remain "nonfunctional" and are detected serendipitously in asymptomatic patients who undergo computerized tomographic evaluation for some other reason. Adrenocortical nodules measuring greater than 1 cm in diameter are found in 1.5–8.7% of unselected postmortem examinations.[27,28] The incidence of detecting nonfunctional asymptomatic benign adrenal adenomas increases with age, especially in patients with hypertension. Although the condition is termed *nonfunctional*, sporadic reports in the literature have suggested that these apparently nonfunctional tumors may secrete small amounts of cortisol, in quantities insufficient to cause Cushing's syndrome.[29–31] It is unclear whether such tumors invariably progress to eventually hypersecrete enough qualities of hormone to result in Cushing's syndrome. The anatomical or functional progression of incidentally discovered "nonfunctional" adrenocortical adenomas has not been convincingly documented in the literature. When adrenal adenomas do become hypersecretory, the hallmark of these tumors is their autonomicity. The reasons that underlie the tumorigenesis of the adrenocortical cells are far from clear. In animal models of adrenal tumors, it has been shown that the adenlyate cyclase of tumor cells can be activated by catecholamines, thyroid-stimulating hormone (TSH), and luteinizing hormone (LH), thereby suggesting the possible presence of multiple receptors for hormones other than ACTH.[32] Matsukura et al.[33] have also shown that the activity of adenylate cyclase in some human adrenocortical adenomas can be stimulated by catecholamines in addition to ACTH. Hirata et al.[34] studied particulate fractions of three cortisol-producing adenomas and convincingly demonstrated the presence of β-adrenergic receptor sites on the cell membrane of these adenomas. Further, production of cortisol from the cultured tumor cells derived from one of the adenomas was significantly stimulated by epinephrine, in addition to ACTH. These data suggest that alterations in cellular membrane characteristics occur in some human adrenal adenomas that render them susceptible to adenylate cyclase stimulation by stimuli other than ACTH.

While adrenal adenomas are autonomous in the strict feedback sense of the term, a substantial number of these tumors respond to exogenous administration of pharmacological doses of ACTH. Further, cultured tumor cells in vitro respond to ACTH, indicating that ACTH responsiveness in these tumors is preserved to some degree. These findings sharply contrast with the inability of *endogenous* ACTH to stimulate cortisol production by adenomas, a characteristic feature of these tumors. Thus, the reasons for the development and the "autonomous" behavior of adrenal adenomas remain largely unresolved.

Nodular Adrenal Hyperplasia

Nodular adrenal hyperplasia is a rare cause of Cushing's syndrome. The disease involves both adrenals, and the terms *micronodular* and *macronodular adrenal hyperplasia* probably represent different ends of the same spectrum. The terminology used in the literature has added to the confusion in describing a disorder of uncertain pathogenesis. Thus, micro- and macronodular hyperplasia,[35-37] bilateral adenomatous adrenal hyperplasia,[38] primary adrenocortical microadenomatosis,[39] and primary adrenal nodular dysplasia[40] have all been used to denote a rare variant characterized by bilateral nodularity of the adrenals in association with Cushing's syndrome. The problems in defining or understanding this entity are further compounded by the impressive variability in steroid dynamics that characterize this entity. While the rarity of this condition defies any attempt to propose a unifying hypothesis for the evolution of nodular hyperplasia, the emerging notion is that the entity of primary adrenocortical nodular dysplasia is probably different from other forms of nodular hyperplasia and therefore should be separately considered. Accordingly, we will first focus on micro- and macronodular hyperplasia and subsequently outline the unique characteristics of primary adrenocortical nodular dysplasia that set this entity apart from the other forms of nodular adrenal disease.

Micro- and Macronodular Adrenal Hyperplasia

The controversial nature of the pathogenesis of bilateral nodular hyperplasia (micro- or macro-) revolves around two theories—the first supposes that the disorder is fundamentally pituitary dependent and that it simply represents a variant of long-standing CD; the second theory claims that it is a primary adrenal malfunction. Added to these is a third, compromise theory that suggests dual control, i.e., semiautonomy of the adrenal nodules, with predominance of the pituitary influence at one time and of the adrenal at another.[41] This theory probably best explains the nature of the disorder.

Nodular hyperplasia as a primary pituitary problem is supported by several lines of evidences.

1. The plasma ACTH level in the syndrome of nodular adrenal hyperplasia is not suppressed to the same extent as in patients with adrenal adenomas. May et al.[42] described a patient with bilateral nodular adrenal hyperplasia with elevated ACTH levels that were suppressible to high-dose dexamethasone. Interestingly, despite ACTH suppression, the high-dose dexamethasone did not cause suppression of plasma cortisol and urinary steroid excretion, suggesting autonomy of the adrenals, in addition to the pituitary problem.

2. Cases of nodular adrenal hyperplasia following long-standing pituitary-dependent Cushing's disease have been described in the literature. Thus, Levin[38] described the development of bilateral "adenomatous adrenal hyperplasia" in a case of Cushing's "syndrome" of 18 years' duration. The most impressive evidence to suggest that nodular hyperplasia may develop in patients with long-standing CD comes from the study of Smals et al.[43] These workers compared the clinical and biochemical findings in 13 patients with macronodular adrenocortical hyperplasia with those of 18 patients with CD and diffuse or micronodular hyperplasia. All patients had undergone bilateral adrenalectomy for their hypercortisolism. The clinical picture, the mean plasma ACTH, cortisol, 17-hydroxycorticosteroid (17-OHCS) excretion, and the incidence of enlargement of sella turcica were nearly identical in both groups. Also, the ACTH response to metyrapone, CRF, TRH, and luteinizing-hormone–releasing hormone (LHRH) showed no significant differences in both groups. The only difference was that patients with macronodular hyperplasia showed less dexamethasone suppressibility, and less responsiveness to exogenous ACTH, indicating a greater degree of adrenal autonomy in comparison to patients with CD. Thus, CD with diffuse adrenal hyperplasia, micronodular hyperplasia, and macronodular hyperplasia may represent entities in the same spectrum of a disease, but in different stages of evolution. As the disease progresses, more autonomy of the adrenal is attained.

3. ACTH dependency of nodular adrenal hyperplasia has also been suggested by studying ACTH levels from pituitary venous effluent. Aron et al.[44] studied two patients with Cushing's syndrome due to nodular adrenal hyperplasia and demonstrated impressive ACTH gradients in the blood obtained from selective catheterization of the inferior petrosal sinus. The combination of elevated as well as suppressible basal plasma ACTH level clearly established pituitary dominance in both patients with nodular adrenal hyperplasia.

4. Finally, the coexistence of pituitary ACTH-dependent Cushing's syn-

drome with a solitary adrenal adenoma has been described.[45] This supports the view that the condition begins as pituitary ACTH-dependent hyperfunction, and that secondary nodules develop consequently, one of which can become autonomous, suppressing the abnormally elevated ACTH. The case reported by Schteingart and Tsao[45] elegantly illustrated the transition from pituitary-dependent CD to adrenocortical adenoma.

While these lines of evidence strongly confer an important role for the pituitary during some time in the evolution of macronodular hyperplasia, some workers have suggested that these nodules arise de novo within the adrenal gland.[37] The two main reasons for this suggestion are the low ACTH levels seen in many patients with adrenal nodular hyperplasia[37] and the marked resistance to dexamethasone suppression seen in the majority of patients with this disorder.[41,46–49] The interpretation of ACTH levels in patients with macronodular disease should take into consideration the fact that significant variation can be encountered if the ACTH in these patients is sampled on multiple occasions. As for resistance to dexamethasone suppression, clearly the nodules are autonomous regardless of whether they arose de novo or secondary to ACTH stimulation in the beginning.

The most plausible explanation for bilateral nodular adrenal hyperplasia involves the existence of dual abnormalities. Abnormalities of the hypothalamic pituitary axis may have initiated the process, with eventual transformation of diffuse hyperplasia into nodular (micro- or macro-) hyperplasia. This is associated with varying degrees of adrenal autonomy of the nodules, which secrete cortisol. Since the pituitary ACTH is suppressible to high concentrations of cortisol, the plasma ACTH becomes suppressible to varying degrees. Since this group of patients represents a heterogenous disorder, and since the steroid dynamic data and ACTH levels depend on the phase of evolution during the testing, the laboratory findings can be extremely confusing and misleading. Thus, in many patients with macronodular hyperplasia the suppression data resemble those of unilateral adenoma, while in others the ACTH level and response to dexamethasone may be identical to pituitary-dependent CD. To confound matters even more, the computerized tomographic appearance can also be confusing, resembling unilateral disease in some patients with bilateral macronodular hyperplasia. Rarely, carcinomatous transformation of macronodular hyperplasia has been reported. Anderson et al.[50] described a patient with long-standing pituitary-dependent CD, who eventually developed secondary nodular hyperplasia followed ultimately by the development of a ("tertiary") virilizing adrenal carcinoma. Such a transformation seems plausible because adrenal carcinoma has been described to occur in glands with congenital adrenal hyperplasia.[51,52] The laboratory profile of such patients can be extremely variegated.

Primary Adrenocortical Nodular Dysplasia

Primary adrenocortical nodular dysplasia (PAND) resembles other forms of macronodular hyperplasia in that Cushing's syndrome develops in the face of adrenal autonomy and the adrenal glands harbor multiple nodules. That the condition was distinct from other forms of macronodular disease of the adrenal was suggested by Ruder et al.[37] These workers pointed out two pathological features that have come to characterize the entity of PAND—the presence of multiple small nodules that were pigmented black or brown, and the presence of internodular adrenal atrophy. Both features are generally absent in the usual varieties of macronodular hyperplasia. It is interesting to note that the combination of pigmented adrenal nodules and Cushing's syndrome had been recognized in 1949 by Chute et al.[53] However, it was not until Ruder et al.[37] delineated this variety that it began to emerge as a distinct subtype. Gradually it became apparent that PAND has recognizable clinical characteristics as well. For instance, its occurrence in children, the mildness of the Cushing's syndrome, and the distinct lack of association with pituitary tumors were recognized, as was the heightened tendency for the development of osteoporosis in this group.[40]

The pathophysiology of PAND became the subject of intense study in the 1980s. McArthur et al.[54] studied two patients with PAND and concluded that it represented a bilateral disease of the adrenal cortex and was a manifestation of functional activity and autonomy of the nodular subpopulation of cells. The impressive atrophy of the internodular adrenocortical tissue is strikingly reminiscent of adrenal tissue deprived of ACTH. The common occurrence of this disease in children and the presence of discoid lesions in infants with the syndrome suggested to the authors that the overactive subpopulation of cells may evolve due to a defect in the involution of the fetal adrenal cortex. While the term *dysplasia* is used to denote this entity, it is by no means a precancerous condition.

Larsen et al.,[55] in an impressive review of the 30 reported cases in the literature that met the criteria of PAND, including one case of their own, succinctly summarized the features of PAND that delineate this form of Cushing's syndrome:

1. The disease occurs predominantly in infants, children, and young adults, the youngest being a 7-day-old infant.
2. A familial occurrence has been noted in some cases.[39,56–58]
3. The clinical severity of the hypercortisolism is usually mild.
4. Osteoporosis tends to occur frequently.
5. The cortisol elevation is mild to moderate and fails to suppress to high-dose dexamethasone administration.
6. The ACTH levels are suppressed, and the sella turcica is nonenlarged.
7. The histopathological hallmark is the presence of multiple, small, black or brown pigmented nodules in both adrenals, with remarkable atrophy of adrenal tissue between these nodules. The consistency with which pigmentation occurs has earned it the name "primary pigmented nodular adrenocortical disease."[59]
8. The condition can be cured only by bilateral adrenalectomy, with almost no incidence of development of Nelson's syndrome.
9. Other coexisting anomalies have been described in association with PAND. Somatic features such as unusual facies with hyperteleorism, small heads, large fontanelles, and dark skin pigmentation were present in a brother and sister with PAND.[58] Other congenital abnormalities

such as the Peutz–Jeuger type of pigmentation and cardiac myxomas have been reported in patients with PAND.[39] The triad of myxomas, spotty pigmentation, and adrenal hyperfunction has been reported by Carney et al.[60,61] in kindreds, suggesting an autosomal dominant pattern of inheritance. In fact, death from embolic disease has been described in a 16-year-old who demonstrated PAND at autopsy.[62] In addition, large-cell, calcifying Sertoli cell tumors of the testis have also been described in patients with PAND.[62]

The aforementioned unique characteristics of PAND have set this syndrome apart from other forms of nodular hyperplasia of the adrenal glands. While the etiology of this disorder, at best, remains speculative, Van Berkhout et al.[63] have proposed that it may represent a putative receptor–antibody disease. The serum from two sisters with PAND was shown to contain immunoglobulins that stimulated adrenocortical cell growth in a cytochemical bioassay system. Thus, circulating growth factors may have been involved in the pathogenesis of the adrenal hyperfunction seen in the siblings. If confirmed in larger series of patients, these findings would place PAND in the realm of immunoendocrinopathies (such as Graves' disease and sporadic euthyroid goiter) where receptor antibodies stimulate growth and function of target glands by competing and displacing the physiological stimulator, in this case ACTH.

Ectopic ACTH Secretion

ACTH was once believed to be exclusively produced by the anterior pituitary gland. However, in the 1960s several workers[64–67] found that certain non-pituitary carcinomas produced ACTH, causing the "ectopic ACTH syndrome." Such a phenomenon had been postulated as early as 1928 by Brown.[68] The literature of the past two decades is replete with a proliferation of reports of the ectopic ACTH syndrome, rendering it one of the most prototypical examples of ectopic hormonogenesis. Diverse tumors—mostly tumors that originate from neural crest derivatives (Table 13)—have been reported to cause ectopic ACTH hypersecretion. Thus, ectopic ACTH secretion has been reported in association

TABLE 13.
Ectopic ACTH-Secreting Tumors

Oat cell carcinoma of lung
Thymoma
Pancreatic carcinoma
Carcinoid tumors
   Bronchial carcinoids
   Pancreatic carcinoids
   Prostatic carcinoids
   Thymic carcinoids
Medullary cancer of the thyroid
Pheochromocytoma
Paraganglioma
Hepatocellular carcinoma

with oat cell carcinoma of the lung, carcinoma of the pancreas and thymus, carcinoid tumors (from the bronchus, pancreas, thymus, or prostate), medullary thyroid cancer, gastrinomas, pheochromocytomas, and paragangliomas.[69-92] The ability of these cells to uptake and decarboxylate biogenic amine precursors is largely responsible for their proclivity to secrete ACTH, among other peptide hormones.

Small cell carcinoma of the lung remains as the most common neoplasm associated with ectopic ACTH secretion. In fact, secretion of "ACTH-like peptides" may be a ubiquitous phenomenon in this cancer. Using radioimmunoassay techniques, Gerwirtz and Yalow[93] have demonstrated "ACTH immunoreactivity" in nearly all patients with small cell lung cancer, yet clinical or biochemical expression of hyperadrenocorticism evolves in less than 5% of such patients. This is a reflection of the inability of these tumors to cleave the precursor peptides of ACTH into biologically active ACTH. Patients with small cell lung cancer who do develop the syndrome of hyperadrenocorticism tend to have a poorer prognosis.[94]

Studies that have attempted to characterize the ACTH secreted by neoplastic tissue have suggested that ectopic ACTH, when compared with pituitary ACTH, has a larger molecular weight. Thus, patients with ectopic ACTH secretion tend to demonstrate a 22,000-dalton ACTH (22-K ACTH) in addition to the native 1-39 form in their circulation.[95] This larger-molecular-weight form of ACTH contains the N-terminal region of Pro-opio-cortin (N-POC), a fragment that has been shown to potentiate ACTH-induced steroidogenesis by adrenocortical cells.[96,97] It has been suggested, but not conclusively proven, that this fragment is responsible for causing hypokalemia, a biochemical hallmark of the ectopic ACTH hypersecretion, by increasing synthesis of desoxycorticosterone by the zona glomerulosa. Ratter et al.,[95] comparing the immunoreactivity with the bioactivity of "ectopic ACTH peptide," found that it possessed much lower bioactivity relative to its immunoreactivity.

In addition to causing Cushing's syndrome by ectopic secretion of ACTH, certain neoplasms can cause the same phenomenon by secreting CRF. Following the original report by Upton and Amatruda[98] of CRH-like material extractable from two tumors that caused a syndrome resembling ectopic ACTH syndrome, several reports have appeared in the literature.[99-102] Carey et al.[12] reported a patient with metastatic carcinoma of the prostate presenting with ACTH-dependent Cushing's syndrome. At autopsy, large areas of the median eminence and the pituitary stalk were replaced by tumor that contained cells which stained positive for CRH immunoreactivity. Ectopic hypersecretion of CRH has been documented in patients with diverse neoplasms such as lung cancer,[103] nephroblastoma,[101] pheochromocytoma,[104] metastatic medullary thyroid carcinoma,[105] bronchial carcinoids,[106] and intrasellar gangliocytomas.[107] Occasionally tumors have been known to ectopically secrete both ACTH and CRH.[106,108] This phenomenon is best illustrated in the case reported by Schteingart et al.,[108] who identified two distinct cell populations in a bronchial carcinoid, one secreting ACTH/β-endorphin, and the other secreting CRH. The importance of recognizing ectopic CRH secretion by neoplastic tissue lies in the fact that such a phenomenon can result in the development of a syndrome identical to pituitary-dependent CD, since ectopic CRF stimulates the corticotropes to secrete endogenous ACTH.

Finally, Howlett et al.[109] have described yet another fascinating mechanism for the development of the "ectopic" ACTH syndrome. These workers reported a patient with metastatic medullary thyroid carcinoma due to ectopic production of a bombesinlike peptide that in turn stimulated excessive secretion of pituitary ACTH. The patient, a 41-year-old man, presented with Cushing's syndrome and the biochemical features of ectopic ACTH production, as well as mediastinal metastases from medullary carcinoma of the thyroid. The peripheral plasma contained markedly elevated levels of immunoreactive bombesin as well as calcitonin; selective venous catheterization revealed a gradient of immunoreactive bombesin, but not of ACTH, in the mediastinal vein. The tumor extracts also contained bombesin, but not vasopressin or CRH. Thus, stimulation of endogenous ACTH by the ectopically secreted bombesin was postulated to be the cause of hypercortisolism.

Histopathology

The histopathology of the pituitary gland and the adrenals in patients with Cushing's syndrome varies according to the etiology.

Pituitary-Dependent Cushing's Disease

The most common anatomical lesion responsible for pituitary-dependent CD is a tumor in the anterior pituitary, usually under 10 mm in size (microadenoma). Less commonly, CD may result from macroadenomas (tumors larger than 10 mm), invasive macroadenomas or diffuse hyperplasia. It should be realized that in approximately 15–20% of patients with hormonally documented pituitary-dependent CD, no abnormalities may be found despite the most careful search during surgery. The location of the microadenomas is usually lateral. In the series reported by Boggan et al.,[15] the tumor arose laterally in 60 of 62 patients operated on. Hardy[16] reported the gross anatomical findings in 72 patients subjected to transsphenoidal exploration for pituitary-dependent CD and found microadenomas in 52. The likelihood of finding adenomas that were located deeply in the central wedge of the pituitary was more probable in the case of the smallest adenomas.

Histopathologically, the typical pituitary corticotrope adenoma consists of cells that are well granulated and basophilic, staining positive with periodic acid–Schiff stains. The normal acinar architecture of the pituitary is replaced by these cells that form nodular masses. A characteristic feature of these cells is the presence of microfilaments measuring 70 Å, considered to represent cytokeratin.[110] When the accumulation of these filaments is extensive, the appearance resembles the characteristic "Crooke's hyalinization."[111] This histological hallmark, best brought out by the staining with the Alcian blue–PAS–orange G stain, is specific enough for corticotrope adenoma to obviate need for ultrastructural or immunocytochemical techniques. When in doubt, definitive diagnosis of

corticotropin-secreting adenoma can be made by employing specific immunoperoxidase stains. When intermediate lobe tumors are suspected, silver stains
need to be employed to demonstrate the argyrophilic strands that traverse the
tumor.

The adrenal glands in pituitary-dependent CD are enlarged and demonstrate diffuse hyperplasia of the cells of the zona fasciculata and the zona reticularis. In some cases micronodularity may be evident in microscopic sections.

Nodular Adrenal Disease

The term *macronodular hyperplasia* is used to denote the presence of one or
more yellow nodules visible to the naked eye and often 2–3 cm in diameter.[112]
The size of these nodules generally correlates with the adrenal weight. The
macronodules are usually composed of clear cells arranged in acini and cords,
often with simple or micronodular hyperplasia being present in the adjacent
cortical tissue. This finding sharply contrasts with the internodular atrophy that
characterizes primary adrenocortical nodular dysplasia. The macronodules are
generally encapsulated. Although the nuclei of the cells of macroadenoma may
occasionally demonstrate atypia and polymorphism, this is a low-frequency
occurrence.

The histopathological appearance of PAND is typical. Both the adrenal
glands are characteristically studded with multiple, small, black or brownish
nodules. These nodules are smaller in size, when compared to the size of the
nodules in macronodular hyperplasia, and measure between 3 and 5 mm. The
unique pigmentation is due to the presence of lipofuschin pigment within the
nodules. A universal finding in these adrenals is the presence of atrophy of
adrenals between the nodules. Although invasion of pericapsular fat and even
vascular structures by these small nodules has been described,[39,113] these lesions
are neither malignant nor premalignant.

Adenoma

The cortisol-secreting adrenal adenoma is usually solitary, and occasionally
multiple. On gross appearance, adrenal adenomas appear as round, capsulated
tumors originating from the adrenal cortex. The color of these tumors can be
purely yellow, purely black, or very between yellow and dark brown. This colora-
tion contrasts with aldosterone-secreting adenomas, which are usually purely
yellow. Microscopically the yellow adenomas are composed of predominantly
clear cells, while the black adenomas are comprised of compact cells, and the
brown adenomas consist of both compact and clear cells, but with a predominance of the former. The black and brown adenomas of the adrenal have been
considered rather rare.[114–116] However, in one recent report, Komiya et al.[117]
found black or brown adenomas in 71% of 17 patients with Cushing's syndrome
caused by an adrenal tumor. In the same study the authors made the interesting
observation that black or brown adenomas appeared to have higher radiological
densities on computed tomography and functionally secreted lower amounts of
androgens and aldosterone in comparison to the group with purely yellow adenomas. The significance of these interesting observations is not known.

Carcinoma

Adrenocortical carcinoma is a highly malignant tumor. While the size of these tumors is variable, in general they tend to be large, sometimes weighing as much as 4 kg. On gross appearance adrenal carcinoma has a mottled appearance, with signs of invasion of adjacent tissue or vascular structures. This is a highly vascular tumor. The cut section is usually soft, in contrast to the firmness of adenoma, and the tumor is composed of uniform cells arranged in an alveolar pattern as a compact mass, or it may appear syncytial. The cells demonstrate vesicular nuclei, pleomorphism, and mitoses. Pyknotic nuclei set in an intensely pink cytoplasm with little or no lipid is a characteristic histological appearance when present.

Clinical Features

The clinical features of CD are a result of prolonged overproduction of cortisol. The features of hypercortisolism can be striking enough to be recognized at a glance, or subtle enough to be missed by the untrained eyes and mind. The classic textbook manifestations of fat deposition in the face, neck, supraclavicular area, and neck resulting in the familiar moon face, buffalo hump, and truncal obesity with relative sparing of extremities are less common these days. This is probably because sensitive tests have made it possible to diagnose the disease at an early stage. For descriptive purposes the clinical features can be viewed from the following vantage points: the "classic" presentation, the early manifestations, the atypical manifestations, and the clinical differences between the various types of hypercortisolism. Finally, the section will be concluded by correlating the physical findings to the well-known (and the not so well-known) effects of glucocorticoid hormones.

The "Classic" Presentation

The detailed descriptions of the first 12 reported patients with CD by Harvey Cushing in 1932[1] rank among the best descriptions of a new syndrome in the archives of medical literature. The reiteration of the same description half a century (and several hundred patients) later by numerous reviewers is attestation to that fact.

The evolution of pituitary-dependent CD is a slow process, gradually evolving into the plethoric features that have become strongly associated with that disease. Table 14 outlines the most frequently encountered features in patients with hypercortisolism.

Several series have analyzed the incidence of the numerous physical findings encountered in patients with hypercortisolism in general.[31,118,119] The recurring themes that emerge upon reviewing the literature are that (1) a high index of suspicion is required to make an early diagnosis, and (2) although several features of hypercortisolism can be caused by diverse disorders, when they occur collectively in the same patient CD is more likely to be present. The old adage that "the presence of thin skin, thin muscle, and thin bones in a fat

TABLE 14.
Frequency of Clinical Features
in Cushing's Syndrome

Feature	Incidence (%)
Obesity	88–95
Plethora	60–90
Thin skin	80
Diastolic hypertension	76–87
Striae	50
Muscle weakness	60–90
Easy bruisability	42–65
Menstrual disorders	65–85
Hirsutism in women	64–80
Psychiatric problems	42–60
Backache (osteoporosis), edema	40–48

person should raise the possibility of Cushing's" is an excellent one, since atrophy of the skin (84%), myopathy (90%), osteopenia (50%), and obesity, defined here as >115% IBW (80%), are frequent features of chronic hypercortisolism. One recent review by Ross and Linch[120] analyzed the presenting features in 70 patients with documented CD seen in the past 30 years, with a view to identifying the discriminatory features that would aid in making an early diagnosis. The authors noted that the three most discriminatory findings were the presence of ecchymoses, myopathy, and hypertension. Three-fourths of the patients had diastolic blood pressure in excess of 90 mm Hg. Myopathy, evident objectively, was present in 56% of patients, and psychiatric disturbances figured prominently in two-thirds of the patients. While only 3% of patients were not obese, it was noted that the distribution of fat was not a particular discriminatory factor.

Hypertension, one of the most common features in patients with pituitary-dependent CD, is modest, often responding excellently to therapy. The mechanism(s) that underlie the hypertension of CD are not well delineated. Biglieri et al.[121] studied adrenal secretion in vivo and in vitro in patients with Cushing's syndrome and concluded that mineralocorticoid mediation is not a factor in the causation of hypertension in this disease. Although an increase in plasma renin activity has been occasionally reported in patients with Cushing's syndrome[122] and an increase in vascular response to pressor agents has been documented,[123] in general, the renin–angiotensin–aldosterone axis in patients with Cushing's syndrome shows no significant alterations.[124]

In a more recent student, Saruta et al.[125] evaluated the kallikrein–kinin system in 12 patients with Cushing's syndrome and were able to demonstrate a significant decrease in urinary kallikrein and prostaglandin E in patients with the syndrome. In addition, the oral administration of captopril, an angiotensin I–converting enzyme inhibitor, reduced the blood pressure effectively. The possibility that the blood pressure elevation in patients with Cushing's syndrome might be secondary to suppression of depressor systems, coupled with enhanced pressor response to vasoactive substances, awaits further elucidation.

The *psychopathy* of Cushing's syndrome deserves special emphasis since it often highlights the clinical presentation. Mental changes are reported to occur in 50–60% of patients with Cushing's syndrome.[126,127] These include marked emotional lability, paranoid ideation, confusion, and disorientation. In severe cases delusions, hallucinations, depression, and delirium may occur. Glaser[128] reported a 10% incidence of suicide attempts in patients with Cushing's syndrome. It is interesting to note that euphoria is an uncommon facet of Cushing's psychopathy, whereas patients placed on steroids are often euphoric. It is known that steroids may have a direct effect on neuronal tissue. It has been shown that steroids directly induce electrophysiological changes in the neurons, such as diminished synaptic delay and slowing of axonal conduction. Historically, Harvey Cushing recognized that emotional disturbances were prominent features of the disease; in fact, one of his patients was found in a mental hospital with depression, irritability, and memory loss.[1] The spectrum of psychiatric abnormalities encountered in patients with Cushing's syndrome and the relief of these abnormalities with lowering of cortisol have been amply documented by Jeffcoate et al.[129] Also, the observation that acute psychosis can be the sole manifestation of "occult" CD has been emphasized in the literature.[130] Recently, Starkman and Schteingart[131] demonstrated a statistically significant correlation between the neuropsychiatric disability in Cushing's syndrome and the levels of cortisol and ACTH. Their study suggests that the incidence and severity of psychiatric manifestations are greater in patients with hypercortisolism due to elevated pituitary ACTH levels, i.e., central or pituitary-dependent CD. They observed that patients with adrenal adenomas (low ACTH levels) did not have as severe a psychiatric abnormality as patients with central CD (high ACTH levels) in spite of having comparable cortisol levels. The obvious assumption is that ACTH contributes more to the psychopathy of Cushing's syndrome than the cortisol levels. The implication that beta endorphins may be related to the Cushing's psychopathy has generated considerable interest in recent years. Beta endorphin is an opiatelike peptide extracted from the normal pituitary gland. Beta endorphin and ACTH are often secreted concomitantly to various stimuli that normally stimulate ACTH. Therefore, it seems plausible that when excessive ACTH secretion is present (central CD), a concomitant hypersecretion of beta endorphins may also occur. A single report demonstrating high endorphin levels in the cerebrospinal fluid (CSF) of psychotic patients has attempted to link mental disease with endorphins.[132] The significance of ACTH, endorphins, and the mental illness seen in Cushing's disease awaits elucidation.

The *myopathy* associated with hypercortisolism is primarily due to protein catabolism caused by excess glucocorticoids. Like most endocrine myopathies, the proximal group of muscles are involved more commonly than the distal, subjective features are more striking than objective evidence, and the biochemical abnormalities are even less striking. Rarely, overt severe wasting of the muscles may be encountered. Khaleeli et al.[133,134] have recently characterized the clinical, biochemical, functional, and structural abnormalities in six patients with Cushing's syndrome, three of whom suffered from pituitary-dependent CD. Force measurements, using both the myometer and strain gauge techniques, demonstrated quadriceps weakness in every patient. The histological appearance by light microscopy was consistent with type II fiber atrophy in some, but not all, patients with endogenous hypercortisolism. Electromyographic evidence of myopathy was common to both endogenous and exogenous hypercortisolism. The skeletal muscle content of sodium was high, but that of creatine was low. Notably, the skeletal muscle content of potassium was normal. Several of these abnormalities were reversible upon restoration of eucortisolemia.

The *thinning* of the skin, in many an experienced endocrinologist's view, is a remarkably good clue to hypercortisolism, which is the only endocrine cause for this phenomenon.[135,136] The easy *bruisability* is not due to any abnormalities in coagulation, but merely represents increased fragility of the capillaries in the thinned-out skin. The *striae*, which can be quite impressive when classic, assume a violaceous hue, and are due to loss of collagen caused by glucocorticoid excess (Fig. 3).

Hyperpigmentation, a finding that was noted in five of the original 12 patients described by Cushing,[1] is due to the pigmentary effects of lipotropin. This finding is more frequent in patients with ectopic ACTH secretory syndrome.[137]

Hirsutism, glucose intolerance, and hypertension is an impressive triad seen in a high percentage of patients with non-Cushingoid obesity as well. In many obese patients, these features may even be associated with subjective fatigue, mild depression, and menstrual irregularities, all secondary to obesity per se. Therefore, it is easy to miss CD in a patient with obesity. The occurrence of easy bruisability, puffy facies, significant psychopathy, or edema should increase the index of suspicion to detect hypercortisolism. As pointed out by Gold,[138] diversity merits more emphasis, because the incidence of the occurrence of any single feature alone varies so widely among reported series that no single finding is a requisite for the diagnosis. On the other hand, the commonplace nature of

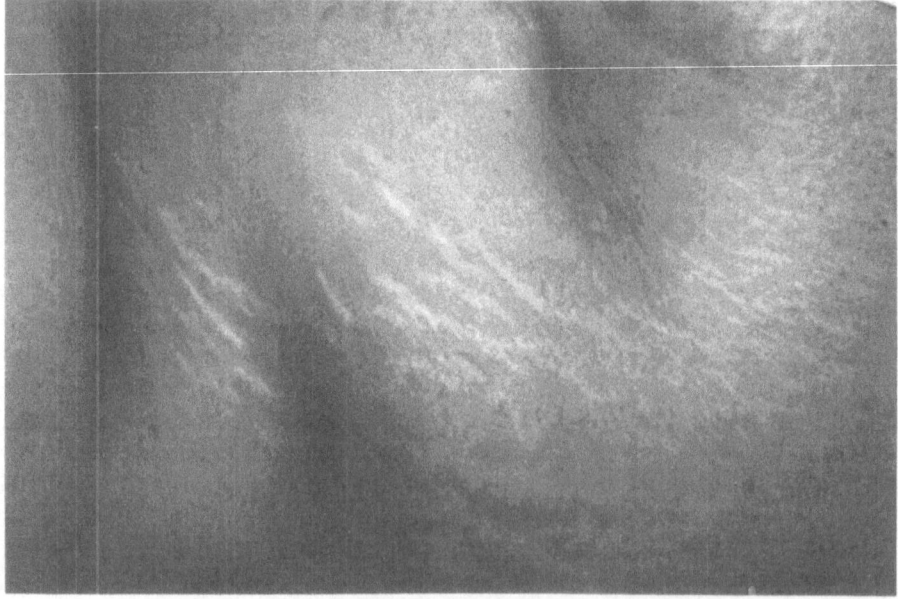

FIGURE 3. Cutaneous striae with a violaceous hue in a 30-year-old patient with hypercortisolism.

several features of this syndrome are such that, if surveyed individually, half the population of the United States would be encompassed.[138]

Early Manifestations

As alluded to earlier, a high index of suspicion is required to recognize the disease in its early stages. Since pituitary-dependent CD evolves slowly and progresses gradually, the changes in appearance may be so insidious as to not concern the patient. In this regard pituitary-dependent CD resembles acromegaly, a closely related disorder of the pituitary gland. The following clues may be helpful in suspecting the disease in its early phase.

1. The development of "puffiness" (plethora) of the face in a patient with hypertension of recent onset (Fig. 4).
2. The combination of edema of the feet and recent-onset hypertension in an obese patient.
3. Any patient with a history of recent weight gain, easy bruisability, and hypertension.

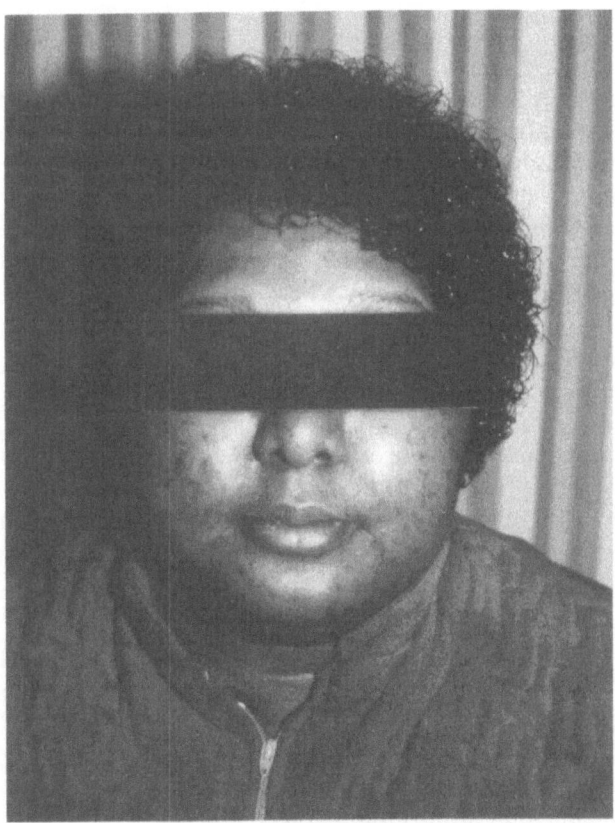

FIGURE 4.   Puffiness of face with acne and facial hirsutism in a patient with Cushing's syndrome.

4. The presence of spontaneous hypokalemia in patients with recent-onset hypertension.
5. The combination of hirsutism and menstrual irregularities in a patient with recent weight gain.
6. Obesity with significant muscle weakness.
7. Unexplained psychiatric distrubances, especially depression, in association with any of the other factors.
8. Growth retardation in an obese child.

Only a small percentage of obese patients turn out to harbor CD, and indeed, Cushing's syndrome seldom if ever causes morbid obesity. Nevertheless, CD, a disorder that can kill if unrecognized, but can be cured if detected, is an important etiology to be kept in mind when evaluating any patient with recent onset of obesity. If the question "Can this be due to hypercortisolism?" is asked in every instance, the diagnosis is less likely to be missed. The five clues that may have discriminatory value, in no particular order of importance, are easy bruisability (ecchymoses), puffiness of face, muscle weakness, hirsutism, and associated psychiatric problems.

Atypical Features

The term *atypical* may not be quite appropriate in discussing certain unusual findings in a disease that is as rare, as unpredictable, and as fascinating as Cushing's syndrome. Unusual deposition of fat in uncommon areas, "lipomatosis," may be seen in this syndrome. The most well-known example of this phenomenon is the mediastinal widening that occurs as a consequence of mediastinal lipomatosis.[139,140] Rarely fat deposition may occur in the paracardiac space.[141] A more dangerous location for fat accumulation is the epidural space. Such epidural lipomatosis has been reported to be severe enough to cause compression of the spinal cord.[142-144] Very rarely the rectum can be displaced anteriorly by the presacral deposition of fat[145] or a fatty mass can be seen in the liver.[146] Although these phenomena occur more often with exogenous steroid administration, they are recognized features of endogenous hypercortisolism of any etiology.

Nephrolithiasis, once considered to be unusual in hypercortisolism, is being noted with increasing frequency. In the series of Ross and Linch[120] renal calculi were present in 15% of patients with endogenous hypercortisolism. In the same series *loss* of *scalp hair* was present in a small, but significant percent (13%) of patients.

Since the vast majority of patients with pituitary-dependent CD harbor only small (micro-) adenomas, headaches, visual field cuts, and other parasellar phenomena are highly unusual. These features, when present, strongly point to the presence of invasive macroadenomas. However, headaches are extremely common in patients with hypercortisolism regardless of the etiology, even in the absence of any increase in the sellar pressure.

Other unusual but important features include the occurrence of cyclical edema,[147] increased tendency for developing superficial cutaneous fungal infections,[148-150] and thromboembolic phenomena.[151,152]

Etiological Differentiation

This may well be impossible on clinical grounds alone. Although the syndrome evolves slowly when pituitary disease underlies the problem and may evolve rapidly when caused by adrenal neoplasms, this may not be useful in a given case. The presence of *galactorrhea* and generalized *hyperpigmentation* may point to the pituitary as the site of CD. However, these findings are not consistently present. The occurrence of *virilization* in association with features of hypercortisolism is a strong sign of adrenocortical carcinoma.[153,154]

While virilization in conjunction with Cushing's syndrome raises the ominous specter of adrenal carcinoma, this need not always be the case. Rarely, a benign adrenal adenoma can result in virilizing syndromes. ("pure virilizing adrenal adenoma"). While such a phenomenon is extremely rare, Gabrilove et al.[155] were able to collect 34 such instances of pure virilizing adrenal adenoma of adult females in the literature, including three cases of their own. The evolving clinical syndrome is identical to that caused by pure virilizing adrenal carcinoma and virilizing ovarian tumors. A notable observation in these patients is the frequent elevation of plasma testosterone levels, a finding that was often conducive to the mistaken initial diagnosis of ovarian tumor in many instances. Virilizing adrenal adenomas have also been described in girls,[156–159] in boys,[160–162] in postmenapausal females,[163] and during pregnancy, resulting in virilization of mother and infant.[164]

In addition to secreting androgens, adrenal adenomas can also secrete estrogens, resulting in isosexual precocity in girls[165,166] and feminization in boys.[167,168]

When Cushing's syndrome is caused by ectopic ACTH secretion, three distinct syndromes evolve—cacehetic, pigmentary, and plethoric. The most frequent are cachectic and pigmentary. The *cachectic* form of ectopic ACTH-secreting syndrome, the prototype of which is lung cancer, is characterized by severe weight loss, darkening of skin, and metabolic changes of profound cortisol excess (hypokalemic alkalosis and hyperglycemia). In the *pigmentary* form, the clinical picture is dominated by increased skin pigmentation, with little or no clinical evidence of Cushing's syndrome. The reason for the paucity of plethoric findings in these circumstances is that the malignancy does not allow the patient enough time to develop Cushingoid features. Occasionally, however, patients with ectopic ACTH secretion may indeed manifest the *plethoric* features of CD. The four classic examples of such a phenomenon are represented by ectopic ACTH secretion by carcinoids of the bronchus,[169,170] carcinoid tumors of the pancreatic islets,[171] medullary carcinoma of the thyroid,[88,172] and pheochromocytoma.[79,173] The ectopic ACTH secreted by these tumors which are benign, or of a low-grade malignancy, results in chronic stimulation of both adrenal glands. In all cases of ectopic ACTH-secreting syndrome, the disease results in stimulation of *both* adrenal glands. To this extent, the disorder resembles pituitary-dependent CD. The majority of tumors that secrete ACTH ectopically are indeed autonomous. A rare exception of this rule is bronchial carcinoid, which may demonstrate some degree of responsiveness to suppression tests. Whether these cases represent ectopic secretion of a CRF-like peptide is a controversial issue.

When Cushing's syndrome is secondary to iatrogenic causes—the most common form of CD—the clinical features are again identical to other forms of plethoric CD. There is a higher incidence of certain physical findings when Cushing's syndrome is caused by exogenous steroid administration. These include a higher incidence of ocular findings such as cataracts and papilledema, myopathy, ischemic necrosis of the femoral head, and pancreatitis. The reasons for these phenomena are not clear. It is also recognized that the incidence of fractures and opportunistic infections is greater in iatrogenic Cushing's syndrome. Table 15 outlines the clinical presentations of the four forms of hypercortisolism.

Clinicopathological Correlations

The overproduction of cortisol in patients with Cushing's syndrome is clearly correlated with the five experessions of such a phenomenon—the catabolic effects on the skin, muscle, and bone; the antiinsulin effects; the central nervous system effects; the antiinflammatory effects; and the consequences on growth.

The catabolic effect on the skin and the muscle have already been alluded to. It is noteworthy that myopathy, a catabolic, protein-losing effect of glucocorticoids, can be offset by the increase in anabolic androgens seen in association with adrenocortical carcinoma. The catabolic effects of glucocorticoids on bone metabolism deserve special mention, since this contributes to significant morbidity. Most patients with iatrogenic or spontaneous Cushing's syndrome suffer from a severe and significant form of bone loss. The changes in the skeletal system that occur in association with hypercortisolism have been extensively reviewed in the radiology literature.[174] In addition to osteopenia, hypercalciuria (urinary calcium excretion greater than 250 mg/day) has been reported in approximately 45% of patients with hypercortisolism.[118] A great deal of controversy revolves around the mechanisms that cause bone loss. Although it is generally accepted that glucocorticoid excess suppresses intestinal calcium absorption, the mechanism for such a phenomenon is far from clear. Most investigators have described a secondary increase in parathyroid hormone (PTH) levels to counteract the decrease in intestinal absorption.[175] The effects of short-term administration of glucocorticoids on intestinal calcium absorption and its relationship to circulating PTH and vitamin D metabolite concentrations have been studied by Hahn and his co-workers.[176] They concluded that the reduced intestinal calcium absorption following glucocorticoid administration could not be attibuted to changes in immunoreactive PTH or circulating concentrations of the major known metabolites of vitamin D. To help resolve the difference cited in the literature, Findling et al.[177] studied the relationship of vitamin D metabolites and PTH to calcium and phosphorus homeostasis in seven patients with spontaneous Cushing's syndrome. Their data suggest that endogenous hypercortisolism decreases tubular reabsorption of phosphorus and increases PTH, which in turn results in an increase in 1,25-(OH) D. These effects may contribute to the bone loss caused by the direct action of cortisol on bone. Histological and biochemical studies have demonstrated that glucocorticoids are not only capable of inhibiting bone formation, but can also *directly* stimulate bone-resorbing cells.[178] The increased PTH secretion assumes importance as a contributory

TABLE 15.
Hypercortisolism: A Clinical Perspective[a]

Pituitary-dependent CD	Adrenal adenoma	Adrenal carcinoma	Ectopic ACTH syndrome	Iatrogenic
Slow evolution	Slow evolution	Rapid evolution	Cachexia	History of steroid therapy
Plethoric features	Plethoric features	Virilization	Pigmentation	Cataracts
Headaches	Unilateral disease	Absence of myopathy	Metabolic ($\downarrow$ K, $\uparrow$ glucose)	Papilledema
Field cuts		Dissemination to lung, bone	Rarely cushingoid	Femoral head necrosis
Galactorrhea			(carcinoids of bronchus,	
Pigmentation			islet cell tumors, MCT,	
Both adrenals enlarged			tumors, MCT, pheo)	

[a] From Kannan.[351]

factor, especially in view of data that support the notion that parathyroidectomy abolishes the osteoclastic effect of steroids on bone in animals.[179] These facts may play an important role in the prevention and treatment of steroid-related osteopenia.

The anti-insulin effect of hypercortisolemia is considered to be exerted at the postreceptor level. Nosadini et al.[180] studied five women with Cushing's syndrome and impaired oral glucose tolerance tests. By plotting the insulin-induced, disposal dose–response curves obtained with the use of the "euglycemic clamp procedure," these workers demonstrated a marked lowering of maximal glucose disposal (MGD). There were no significant differences in the insulin-binding capacity of erythrocytes and monocytes between normals and patients with Cushing's syndrome. It is believed that glucocorticoid excess impairs tissue glucose disposal through a postreceptor mechanism.

The consequences of hypercortisolism on growth assume importance when the disease strikes youngsters.[181] Although the dynamics of growth hormone in adults with hypercortisolism may demonstrate a blunting in provoked secretion,[182] this is not the case in children; growth hormone dynamics in children with iatrogenic, or spontaneous steroid excess is generally considered to be normal.[183,184] These observations, coupled with the fact that growth hormone therapy fails to induce linear growth in hypercortisolemic children,[185,186] suggest that glucocorticoids may interefere with somatomedin generation or action. Indeed, several studies have demonstrated low levels of net circulating somatomedin activity following glucocorticoid administration.[187–189] There is some controversy as to whether this reflects decreased production or indicates antagonism to the activity of these peptides. The results from studies using bioassay systems that test the direct effect of glucocorticoids on cartilage growth have yielded conflicting results.[190,191] A recent concept gaining favor is that hypercortisolemia can lower somatomedin activity by inducing changes in somatomedin inhibitors. These inhibitors of somatomedin activity are present in normal subjects.[192] Unterman and Phillips[193] have demonstrated that glucocorticoid administration is followed by an increase in circulating somatomedin inhibitors, which can account for the fall in the net somatomedin activity in the sera of patients with hypercortisolemia. Regardless of the mechanism, hypercortisolemia leads to a significant impairment of linear growth, resulting in retarded skeletal maturation. The vertebral "growth arrest lines" (zones of increased density corresponding with the vertebral plates) may linger as residual radiological changes, years after cure, serving as a grim reminder of the childhood CD.[194]

The Laboratory Diagnosis of Cushing's Disease

The diagnostic studies involved in the evaluation of patients with Cushing's syndrome fall into three categories. The first phase of workup involves establishing the presence of true hypercortisolism. The second phase consists of tests that are aimed at establishing a primary pituitary etiology for the hypercortisolism in contrast to a primary adrenal or ectopic source. The third and final phase consists of performing the appropriate localizational procedures as indicated.

During all phases of the workup, the yield from the various diagnostic studies ranges from being conclusive and straightforward to frustratingly equivocal. Although the availability of newer tools such as assays for measurement of urinary free cortisol and plasma ACTH, high-resolution computerized tomography, and the corticotropin releasing hormone (CRH) tests have all enhanced our diagnostic armamentarium, each phase of workup for this disease can be fraught with uncertainty. The following discussion focuses on the step-by-step evaluation of the patient suspected to have endogenous hypercortisolism.

Establishing "True" Hypercortisolism

A basic understanding of the synthesis, secretion, and circulation of glucocorticoids is essential to elucidate the best screening tests for hypercortisolism. Figure 5 outlines the pathways and enzymes involved in adrenal steroidogenesis.

The major stimulus for glucocorticoid *secretion* is ACTH. Under the influence of this hormone, adrenal steroidogenesis takes place through all the necessary enzymatic steps. After secretion and release, the cortisol becomes bound to a plasma globulin called *transcortin*. Transcortin has an extremely high affinity for cortisol (75% of all cortisol). Transcortin also binds to a variable extent with other steroids such as progesterone, corticosterone, deoxycorticosterone, and deoxycortisol. The binding of cortisol to transcortin is reversible. Cortisol is also bound to albumin (15%). Approximately 10% of cortisol is "free." It is this free fraction that moves into the cell to exert its action. Measurement of plasma cortisol by radioimmunoassay reflects only the cortisol bound to transcortin. Conditions that alter the transcortin would alter cortisol measurements. Thus, pregnancy, estrogen therapy, and obesity are associated with elevated cortisol levels due to elevated transcortin levels.

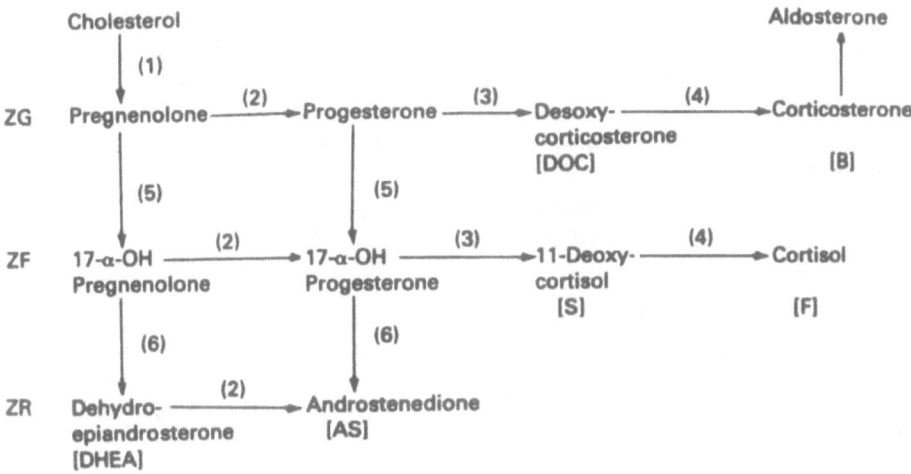

FIGURE 5. Adrenal steroidogenesis. Enzymes: (1) desmolase, (2) 3-β-OH-dehydrogenase, (3) 21-hydroxylase, (4) 11-hydroxylase, (5) 17-α-hydroxylase, (6) 17, 20-desmolase. ZG, zona glomerulosa; ZF, zona fasciculata; ZR, zona reticularis.

The glucocorticoids are mainly *inactivated* in the liver. By a series of reduction reactions cortisol is converted to dihydro- and tetrahydrocortisol, which is then processed through the glucuronyl transferase system to become tetrahydrocortisol glucuronide. Similarly, deoxycortisol is excreted as tetrahydrodeoxycortisol. These two metabolites are measured *17-hydroxycorticosteroids*, in the urine (Porter–Silber chromogens). Approximately 30% of cortisol secreted is metabolized to tetrahydrocortisol glucuronide. The collective measurement of 17-hydroxycorticosteroids in the urine in general reflects glucocorticoid activity, even though tetrahydrodeoxycortisol is not a biologically active compound (compound S). If one wishes to extract or fractionate the amount of tetrahydrocortisol from the tetrahydrodeoxycortisol, this can be done by extractions utilizing carbontetrachloride. In practice, however, this is not generally necessary, since the contribution of 11-deoxycortisol to the urinary 17-hydroxycorticoids is minimal. The 17-hydroxycorticoids can be affected by drugs such as spironolactone, dilantin, estrogens, and exogenous steroid administration.

In normal individuals, cortisol is secreted by the adrenal cortex in a series of "bursts" rather than in a steady continuous fashion.[195] If 24-hr profiles of serum cortisols are obtained by measuring the cortisol level in the plasma every 20–30 min, 7–10 "secretory spikes" can be identified. This phenomenon of "pulsatile" or "episodic" secretion should be taken into account when interpreting a sample of plasma cortisol drawn in an isolated period of time. In general, however, the amplitude of the spikes is greater in the early-morning hours and lowest during the late night, with a noticeably protracted quiescent period around midnight. The cortisol spikes are a reflection of ACTH spikes that precede them.

When screening for a disorder that is potentially lethal, but that if diagnosed is mostly curable, it is desirable to have a screening test with the highest sensitivity; i.e., the test should yield very low or negligible false negatives; if a false positive diagnosis were made, the test could be repeated or other tests performed for confirmation. As screening tests, the combination of the overnight dexamethasone test and the collection of urine for 24-hr free cortisol is a reasonable standard of screening. Unfortunately, no single screening procedure for hypercortisolism is 100% free of false negatives. The following tests are discussed in this section.

1. Overnight dexamethasone suppression test.
2. 24-Hr urinary free cortisol (UFC).
3. Timed integrated concentration of cortisol (IC).
4. 24-Hr urinary 17-hydroxycorticosteroids (17-OHCS).
5. Random plasma cortisol.
6. Evaluation of the circadian rhythm.
7. Urinary steroid profiles.

The Overnight Dexamethasone Suppression Test

The test is based on the principle that when 1 mg of dexamethasone is administered by mouth to normal individuals at night, before sleep, it abolishes the nyctohemeral ACTH rise. As a consequence, the 8 A.M. cortisol level obtained the following morning will be suppressed below a defined limit of 5 μg%.

The dexamethasone administered is potent enough to induce suppression of the hypothalamic pituitary axis, but does not interfere in the assay of plasma cortisol. Often 15–30 mg of flurazepam by mouth is given to ensure a rested night. The unique features of the overnight dexamethasone suppression test are the ease with which it can be performed, the inexpensiveness, and, above all, the extremely low incidence of false negative results. Crapo[196] examined the data from several series and pointed out the following important observations. First, the number of patients with Cushing's syndrome who suppressed normally was negligible; only 3 of 154 patients (1.9%) with Cushing's syndrome demonstrated normal suppression, one of whom on repeat testing demonstrated nonsuppression; the other two were not retested. Second, the incidence of false positives (i.e., failure to suppress in the absence of hypercortisolism, was very low in normal, nonobese, nonmedicated ambulatory patients (5 of 466, or 1.1%). Third, the incidence of false positives was high in obese controls (13% of 173), in hospitalized patients, and in those with chronic illness. Table 16 outlines the conditions characterized by nonsuppression to the overnight dexamethasone test.

As a screening test, when cost effectiveness is considered, the overnight dexamethasone test is highly recommended as the first line of screening.[197–200] Meikle[201] has suggested that the discriminatory value of the test can be enhanced by the simultaneous measurement of plasma level of dexamethasone along with plasma cortisol level at 8 A.M. The demonstration of plasma dexamethasone levels high enough to exert suppression clearly aids in interpretation of the overnight dexamethasone suppression tests. The assay of dexamethasone in plasma will permit identification of these patients with inappropriately low plasma levels of the drug, due to either delayed or altered drug metabolism. This is particularly the case in patients receiving anticonvulsant therapy. The clinical importance of

TABLE 16.
Nonsuppression to Overnight Dexamethasone Test

Condition	Mechanism
Pituitary-dependent Cushing's syndrome	Partial pituitary autonomy; reset hypothalamus
Adrenal adenoma or carcinoma	ACTH is already maximally suppressed by autonomous secretion of steroids by the adrenal tumor
Ectopic ACTH-producing neoplasma	Tumor elsewhere secretes ACTH autonomously
False positive results	
Obesity	Elevated transcortin (which binds the dexamethasone and decreases bioavailability)
Estrogen therapy (conjugated equine estrogen >1.25 mg)	Elevated transcortin
Diphenylhydantoin	Metabolism of dexamethasone (increased hepatic mitochondrial inactivation to less polar metabolites)
Alcohol	? Increased clearance
Malabsorption	Poor absorption of the administered drug
Depression and other psychiatric diseases	Abnormal hypothalamic–pituitary axis
Stress (e.g., chronic illness, hospitalization, surgery)	Activation of hypothalamic–pituitary axis

concomitantly assaying the plasma dexamethasone level has been highlighted by two reports, one by Meikle et al.[202] and another by Caro et al.,[203] where the diagnosis of Cushing's syndrome would have been difficult to establish without the information relating to the plasma level of dexamethasone or its metabolism. The assay, however, is not routinely available. Therefore, in clinical practice, when the overnight dexamethasone test is equivocal or abnormal, the next step in the approach to the problem is measurement of free cortisol in 24-hr urine.

24-Hr Urinary Free Cortisol

Regarded by many as the initial choice for screening, the measurement of "free" cortisol in the urine has several advantages.[204,205] First, and most important, it measures the amount of cortisol that is "free," i.e., unbound to transcortin, and hence metabolically active. The binding of cortisol to its globulin (transcortin or cortisol binding globulin) typically begins to reach saturation at levels of serum cortisol of approximately 25 µg.[206] When the concentrations of plasma cortisol exceed this level, there is a disproportionate increase in the level of the unbound or free cortisol. Since the bound cortisol is not filtered by the kidneys, measurement of the free cortisol reflects the amount of cortisol above and beyond the saturation of transcortin; therefore, the UFC measures the integrated glucocorticoid secretory activity in a 24-hr period; second, the UFC is not affected by increases in transcortin, unlike plasma cortisol which can be elevated whenever transcortin is increased (pregnancy, obesity, etc.). Third, and more important, under circumstances of hypersecretion, because of the saturation of transcortin, UFC increases exponentially, while metabolite secretion (17-OHCS) increases only linearly. This is the reason why measurement of 24-hr UFC is clearly superior to measurement of 17-OHCS in the 24-hr urine, in screening for hypercortisolism. The value of UFC in the diagnosis of Cushing's syndrome has been reviewed by Trecan et al.[207] They studied the sensitivity, specificity, and predictive value for Cushing's syndrome based on measurement of basal and postdexamethosone levels of free cortisol in 24-hr urines. The subjects consisted of 26 patients proved to have Cushing's syndrome (20 with CD) and 93 normal subjects who were "suspected" of harboring hypercortisolism, but turned out not to have the syndrome. Based on their results both the sensitivity and the specificity of UFC measurement approached 96%. The value of the basal collection in predicting the presence of Cushing's syndrome using 95 µg as cutoff was 80%. This increased to 100% when the postsuppression cutoff value of 10 µg was employed. The predictive value for the absence of Cushing's syndrome, based on a basal level of 95 µg or less, was 99%. The authors also noted that measurement of 24-hr urinary 17-OHCS provided only partial separation of the two groups, with considerable overlap.

Excellent as it may be, the 24-hr UFC suffers from the banal problems of completeness of urine collection. It is generally believed that the UFC is not significantly affected by modest decreases in glomerular filtration rate (GFR). However, occasionally the UFC can increase when the GFR is increased. This is particularly likely to happen when the UFC is measured immediately following withdrawal of diuretic therapy. Haigh and Tevaarwerk[208] described a hypertensive patient with Cushingoid features in whom the UFC was markedly increased when measured after withdrawal of her diuretic therapy; the UFC continued to rise despite dexamethasone, simulating a "paradoxically increased" response pattern to dexamethasone administration. The creatinine clearance had also markedly increased, presumably owing to diuretic withdrawal. When the UFC was expressed per 100 ml of filtered plasma, the authors found no differences in the daily excretion pattern. Also, UFC can be elevated in patients with psychiatric illness, particularly depression, and in stress. Most workers agree that the UFC serves as an excellent diagnostic tool in the obese patient suspected of having Cushing's syndrome, especially in the face of an equivocal overnight dexamethasone suppression test.

Timed Integrated Concentration of Cortisol (IC)

Measurement of the IC of cortisol in the plasma is distinctly advantageous in comparison to measurement of the hormone in a single sample drawn at an isolated point in time. This is because of the "episodic" or "pulsatile" nature of cortisol secretion by the adrenal cortex. The term *integrated* denotes the concentration of hormone in pooled blood collected at a constant rate over a period of time, the advantage being that it has a smaller variability and reflects the overall prevailing blood level better than a single sample drawn at a given point in time. Several studies[195,209] have characterized the normal episodic pattern of cortisol secretory activity. Patients with CD are no exception to the fluctuating pattern,[210] a phenomenon that may greatly minimize the value of a single random cortisol level. Studies that determine the 24-hr integrated mean cortisol concentrations are highly representative of adrenal activity,[211] but are impractical and cumbersome to employ as screening tests. Therefore, attempts have been made to evaluate the value of 3-hr or 6-hr IC of cortisol for diagnosing Cushing's syndrome. Zadik et al.[212] studied the clinical usefulness of a shortened, practical version of the 24-hr integrated measurement, in its ability to detect Cushing's syndrome; 68 normal subjects and 13 patients with Cushing's syndrome constituted the study. When the IC of cortisol in different time slots was compared between the two groups, they noted that there was no overlap in the 24-hr integrated plasma cortisols between the two groups. More important, the IC of cortisol drawn between 8 P.M. and 2 A.M. clearly separated controls from patients with Cushing's syndrome. The conclusion drawn was that the test, when performed between these hours (a period during which IC levels in normal subjects are at a nadir), demonstrated a discriminatory value that was as good as more-protracted 24-hr sampling. Halbreich et al.[213] studied an even shorter version of the test and evaluated the mean or integrated cortisol concentration between the afternoon hours of 1 and 4 P.M. and concluded that it was a reliable and powerful indicator of cortisol hypersecretion. A present measurement of IC has not replaced the more conventional screening tests. At best, it may be used when the clinical suspicion is strong and the other *screening* tests are equivocal.

24-Hr Urinary 17-Hydroxycorticosteroids

Although the 24-hr urinary 17-OHCS measures the metabolites of glucocorticoids, the amount of 17-OHCS excreted in a 24-hr period does not clearly separate normals from hypercortisolemic patients. This is not surprising since the Porter–Silber reaction (the color reaction between corticostroids and phenylhydrazine) measures only 50–60% of extractable glucocorticoid metabolites in the urine. It is well known that a significant number (as high as 10%) of patients with hypercortisolism may show normal 24 hr 17-OHCS levels in the urine, while as many as 25% of non-Cushingoid obese subjects may demonstrate an elevated 24-hr urinary 17-OHCS.[214–216] Although it is true that the discriminatory value of urinary 17-OHCS can be improved by expressing the 17-OHCS in terms of the urinary creatinine, the false negatives are still too high for it to be a useful screening test. In general, most but not all patients with hypercortisolism demonstrate urinary 17-OHCS levels greater than 9 mg/day per gram of urine creatinine.[217] Although the test is not suitable for screening, the measurement of 17-OHCS plays a dominant role in the standard dexamethasone suppression tests. This is because standardization of this test is based on comparisons of 17-OHCS values before and after oral dexamethasone administration (the "Liddle test").

Random Cortisol Level

Measurement of a random cortisol level is an extremely unreliable screening method for hypercortisolism. The plasma cortisol, measured by radioimmunoassay, is affected by multiple variables, such as increase in the cortisol binding globulin, drugs, and stress. The physiological range of cortisol in normal plasma is wide and variable and markedly fluctuates during the 24-hr cycle as a result of episodic secretion. The variable between the highest and lowest values in a single 24-hr period can be as much as a 10-fold difference. This magnitude of fluctuation undermines the meaningful interpretation of a single cortisol determination drawn at random. For this reason the "spot" cortisol determinations in the plasma are poor discriminators between normals and patients with hypercortisolism. Moreover, the rise in plasma cortisol associated with increased cortisol binding globulin overlaps greatly with the levels seen in hypercortisolemic subjects.

Circadian Rhythm of Plasma Cortisol

A loss of circadian variation of the plasma cortisol, reported for the first time in 1960,[218] has been regarded as a time-honored test for detecting hypercortisolism. Although it is true that in normal subjects the ACTH–cortisol secretory pattern demonstrates a circadian rhythm, the precise demonstration of loss of rhythm would depend on performing multiple determinations for cortisol in plasma during crucial periods of the day. The fact that cortisol is secreted by the adrenal cortex in bursts with variable amplitudes may mask the phenomenon of circadian rhythm, merely because of the timing of blood sampling. Thus, even in normal subjects, depending on the precise moment at which blood is sampled, the circadian pattern of cortisol may not be readily evident. Furthermore, in contrast to the commonly held belief, some patients with pituitary-dependent CD may indeed demonstrate preservation of the circadian rhythm. Glass et al.[219] reported two such patients who underwent hourly blood sampling for 24 hr. Analysis of the cortisol patterns revealed statistically significant circadian rhythms in both. The amplitude of the rhythms fell within the range for the amplitude of the diurnal rhythm of cortisol seen in normal subjects. Similarly, Van Cauter et al.[220] reported the preservation of diurnal pattern of cortisol rhythm in two patients with CD. Despite these interesting observations, it is still widely believed that the majority of patients with CD lack diurnal variation. The conflicting data may be due to the fact the pituitary-dependent CD is a heterogeneous disorder characterized by CRH-independent as well as CRH-dependent forms.[24] If so, the lack of circadian rhythm would be expected to be a striking feature of the CRH-independent form while preservation of the diurnal rhythm would be anticipated in the CRH-dependent variety of CD. Improved methods for interpreting hormone rhythms, such as statistical analyses by periodograms, may permit better understanding of the phenomenon. Rarely, adrenal adenoma, with consistent, but temporally reversed circadian rhythym of cortisol, has been described.[221]

Urinary Steroid Profile

The metabolites of cortisol and androstenedione undergo a unique alteration in patients with hypercortisolism. Physiologically, both these steroids are metabolized in the liver by enzymes containing 5-α-reductase and 11-β-hydroxysteroid dehydrogenase activity. These enzymes are highly sensitive to circulating T_3 levels. For instance, patients with hypothyroidism have tremendous impairment in the activity of these enzymes. Since hypercortisolism of any cause inhibits the peripheral deiodination (outer-ring deiodination) of T_4 to T_3, an analogous situation develops. As a result of this, the hepatic metabolism of glucocorticoids and androgen shifts in a different direction, yielding a singular pattern, characterized by 5-β- and 11-β-hydroxysteroid metabolites in the urine (see Table 17). The only other condition characterized by such a shift in metabolism is hypothyroidism. Since these metabolites can be readily measured by gas chromatography, several workers have evaluated the discriminative value of this "neutral steroid profile" in correctly identifying the presence of Cushing's syndrome.[222,223] Recently, Phillipou[224] studied seven patients with Cushing's syndrome and noted that in every case there was a striking alteration in the hepatic metabolism of cortisol and androstenedione. Thus, the ratios of etiocholanalone/androstenedione and THF/Allo-THF were increased, while the ratio of THE/THF was decreased. There were no false negatives in any patient with Cushing's syndrome in all three series. The only condition that mimics the profile seen in hypercortisolemic patients is hypothyroidism; rarely, patients with 5-α-reductase deficiency or porphyria may have similar profiles. Until a large number of patients are evaluated, the urinary steroid profile is not likely to replace the UFC or the overnight dexamethasone test as the screening test of choice for Cushing's syndrome.

Establishing the Etiology of Hypercortisolism

The second step in the approach to the patient with hypercortisolism is determining the source of the disorder, i.e., whether it is originating from the pituitary, the adrenal, or an ectopic source. The three diagnostic procedures that

TABLE 17.
Neutral Steroid Profiles in Normal and Hypercortisolemic Patients

5-α Compounds	5-β Compounds	11-OH compounds	11-Keto compounds	Normal pattern	CD
Androstenedione (AS)	Etiocholanolone (EC)	Tetrahydrocortisol (THF)	Tetrahydrocortisone (THE)	AS > EC, THE > THF	EC > AS, THF > THE
Allotetrahydrocortisol (allo-THF)	Tetrahydrocortisol (THF)			allo-THF > THF	THF > allo-THF

help in this exercise are the standard dexamethasone suppression test, the plasma ACTH level, and the metyrapone test. When results of all three tests point in the same direction, the etiological diagnosis can be established with a high confidence limit. If all three do not point to the same site, one or more need to be repeated, and other diagnostic avenues must be pursued. The introduction of a fourth test—the CRH test—has added a new dimension to the diagnostic approach for hypercortisolism. It is too soon to speculate if the new is going to replace the old, conventional methods of testing. The principles, the classic responses, the atypical responses, and the diagnostic accuracy of the following tests will be discussed below:

1. The standard dexamethasone suppression test
2. The plasma ACTH assay
3. The metyrapone test
4. The ovine CRF test

The Standard Dexamethasone Suppression Test

The pioneering work by Grant Liddle[225] in 1960 laid the framework for what is still considered by many as the "gold standard" test in the etiological diagnosis of hypercortisolism. The passage of a quarter-century has seen the emergence (as well as the exit) of several "newer" tests claiming superior diagnostic accuracy. The standard dexamethasone test, or the "Liddle test," as it is appropriately called, has withstood the test of time, mostly because of the solid principles on which it is founded. Liddle's thorough analysis[225] of the test results that led to the criteria he established for normals ranks among the best investigative work ever reported in the literature. A brief outline of these criteria is necessary for further understanding of the abnormal states. The standard dexamethasone test consists of 2-mg (low dose) and 8-mg (high-dose) dexamethasone suppression tests. The test involves measuring the 24-hr urine for 17-OHCS before, during, and after the low and high doses of dexamethasone administered orally. The normal response to the low dose (2 mg/day for 2 days) is a decline in 24-hr urinary 17-OHCS to below an absolute value of 3.5 mg/g of creatinine per day. All obese, non-Cushingoid patients will normally suppress to the low dose, while regardless of the etiology, patients with Cushing's syndrome fail to do so. The high-dose dexamethasone, which extends the test for 2 more days, involves administering 8 mg/day for 2 days and comparing the urinary 17-OHCS of the second day of high dose with the basal level. A decline in the 17-OHCS or the UFC by 50% or more of the baseline is defined as suppression. Most patients with pituitary-dependent CD demonstrate suppression to the high dose, while patients with adrenal tumors and ectopic ACTH-secreting tumors do not. The discriminatory value of the high-dose dexamethasone suppression test is based on the autonomy of these tumors as well as the preservation of partial responsiveness in pituitary-dependent CD. To reiterate, the classic "patterns" for the various types of hypercortisolemic states are:

1. *Central CD* (pituitary-dependent) demonstrates no suppression to the low dose, but adequate suppression to the high dose. In this entity, the servomechanism for feedback at the pituitary (or hypothalamic) level is set at a higher threshold.

2. *Adrenal tumors* (adenoma or carcinoma) fails to demonstrate suppression to a low or high dose. This type is classically autonomous, since the ACTH is already maximally suppressed by the increase in circulating glucocorticoids.

3. *Ectopic ACTH-secreting tumors* again fails to demonstrate suppression to any dose of dexamethasone, since they are independent of pituitary ACTH.

There are, however, notable exceptions to the classic pattern in each type of CD.

Cushing's Disease.   When there is a strong clinical or radiological suspicion of pituitary-dependent CD, but the standard dexamethasone test reveals no suppression to a high dose, the following dynamic phenomena must be kept in mind and appropriate testing undertaken.

1. Pituitary ACTH-dependent CD nonsuppressible to high-dose dexamethasone. As many as 15–20% of patients with pituitary-dependent CD may not show the classic suppression to high-dose dexamethasone.[215,226] These patients may require a "high-high-dose dexamethasone suppression test" (4–8 mg q.i.d.) since their hypothalamic–pituitary threshold for suppression is greatly elevated.[227,228]

2. Pituitary ACTH-dependent CD nonsuppressible to dexamethasone, but suppressible to intravenous cortisone. This unusual but important phenomenon, in which the receptors at the hypothalamic–pituitary level do not recognize dexamethasone but do "see" cortisol, was first reported by Carey.[229]

3. Patients with pituitary-dependent CD caused by intermediate-lobe tumors are often resistant to suppression by high-dose dexamethasone, although they may suppress to bromocriptine.[25]

4. Pituitary ACTH-dependent CD with paradoxical increase in ACTH during dexamethasone suppression test.[230]

5. Central CD (pituitary ACTH-dependent) with periodic hormonogenesis.[231]

The latter two phenomena are recognized to be part of the same entity, where pituitary tumors secrete ACTH on a cyclical basis. Variable cycles have been reported, i.e., once in 12 days, or as long as once in 80 days. In between the "bursts," all tests may be normal, or slightly elevated, but during the secretory "bursts" the ACTH and cortisol secretions are totoally autonomous, occuring randomly and independently of the fortuitous administration of dexamethasone. In these cases, it is often impossible to make the correct diagnosis by the dexamethasone test. The patient should be retested at a later date.

Occasionally, patients with pituitary-dependent CD may show "normal" suppression to dexamethasone. This is particularly likely to happen in the presence of an abnormally decreased clearance of dexamethasone. Thus, the "low"-dose dexamethasone test assumes the equivalence of a "high" dose owing to the decreased clearance of the administered dose. Kapcala et al.[232] described a case of proven CD where the patient "suppressed" to the low-dose dexamethasone test. Simultaneous determination of plasma dexamethasone levels in this case demonstrated decreased clearance of the drug and offered an explanation for the "normal" suppression to high dose. Table 18 highlights the various atypical and unusual patterns that may be encountered in patients with pituitary-dependent CD in response to the standard dexamethasone suppression test.

Adrenal Tumors. The phenomenon of adrenal tumors that retain partial suppressibility of ACTH to high-dose dexamethasone may be seen in an entity called macronodular hyperplasia of the adrenal glands.[44] These are basically cases of long-standing hyperplasia of the adrenals, resulting in a single or multiple "autonomous" adenoma(s) with varying degrees of autonomy. More recently, Smals et al.[43] systematically compared the biochemical and pathological features of 13 patients with macronodular hyperplasia; their study reconfirms the notion that the entity may be a result of long-standing CD with varying degrees of pituitary dependence and adrenal autonomy.

Ectopic ACTH-Secreting Tumors. While the characteristic response of these tumors is nonsuppression to any dose of dexamethasone, some such tumors demonstrate suppression to high-dose dexamethasone. A notable example of this phenomenon is the ACTH-secreting carcinoid tumors, usually of the bronchus. These may mimic pituitary-dependent CD to a remarkable degree[74,233,234] with the following similarities:

1. Nearly a third of these tumors demonstrate some preservation of suppression to high-dose dexamethasone, yielding dynamic data identical to pituitary-dependent CD.
2. The plasma ACTH level, in contrast to most cases of ectopic ACTH syndrome, is often not strikingly high, overlapping ranges that can be seen in pituitary-dependent CD.
3. The tumor can be clinically, radiologically, and biochemically silent, explaining the asymptomatic nature, negative chest x-rays, and normal serotonin metabolites encountered.

TABLE 18.
Pituitary-Dependent Cushing's Disease:
Response Patterns to the Standard Dexamethasone Test

I. Classic response
    Nonsuppression to low-dose dexamethasone
    Suppression to high-dose dexamethasone
II. Atypical response
    A. Suppression to low dose, due to decreased clearance of dexamethasone
    B. Nonsuppression to high dose
        1. Due to higher threshold for feedback suppression
        2. Due to receptor insensitivity to dexamethasone
        3. Due to intermediate-lobe tumor secreting ACTH
        4. Due to periodic hormonogenesis

4. Owing to its benign nature, the patient may evolve into plethoric CD, heightening the similarity to pituitary-dependent Cushing's syndrome.

It is, therefore, not surprising that these features render the situation conducive to missing the real cause of hypercortisolism in "silent" ACTH-secreting carcinoids. In some cases, the only definitive method for establishing or excluding a pituitary source of ACTH is selective venous sampling of the inferior petrosal sinus.[227,235,236]

Finally, it must be realized that abnormal test results may occur in the absence of true hypercortisolism.[237] Patients with severe endogenous depression may show persistent abnormalities in ACTH dynamics, demonstrating failure to suppress with a low dose, but good suppression with a high dose. In such patients even the diagnosis of CD becomes dubious, much less the etiological separation.[238] These exceptions in no way mitigate against the overall value of the standard dexamethasone test.

Attempts to simplify the standard dexamethasone test by obviating the need for collection of 24-hr urine have resulted in the emergence of simpler protocols. For instance, Kennedy et al.[239] have reported that measurement of serum cortisol concentrations after low-dose dexamethasone administration is as accurate and reliable as measurement of urinary steroids. In a series consisting of 23 obese controls, 11 patients with possible Cushing's syndrome, and 10 with definite Cushing's, the serum cortisol determinations at 8 hr after the low-dose dexamethasone was in concordance with the UFC measurements. Thus, in 97% of patients without Cushing's syndrome the serum cortisol declined below 2.2 μg% after the low dose. False positive responses were seen in only one patient, while all patients with Cushing's syndrome failed to show suppression below the cutoff serum cortisol level of 2.2 μg%.

In another impressive study aimed at simplification of the high-dose dexamethasone test for the etiological diagnosis of Cushing's syndrome, Tyrell et al.[240] evaluated the plasma cortisol concentration following a single 8-mg dose of dexamethasone administered orally at 11 P.M. Suppression to the single dose of 8 mg dexamethasone was defined as a 50% drop in the serum cortisol. This simplified test, in terms of predicting the presence of CD, had a sensitivity of 92%, a specificity of 100%, and an accuracy of 93%. These values were equal to or exceeded those of the standard test using measurement of urinary 17-OHCS as a parameter to define suppression. Of course, a nonsuppressed postdexamethasone value would not delineate tumors of the adrenal causing the Cushing's syndrome from the ectopic ACTH syndrome. Since this study involved a large number of patients (60 with Cushing's disease, nine with adrenal tumors, and seven with the ectopic ACTH syndrome), and since in some cases this version of the test proved to be better than the conventional high-dose test, the authors concluded that this overnight high-dose test was a practical, simple, and reliable maneuver that helps in the etiological separation of CD from other causes of hypercortisolism.

Regardless of the format used, the standard dexaemthasone suppression test has played, and will continue to play, a major role in the differential diagnosis of Cushing's syndrome. The results of the standard dexamethasone test assume greater significance when viewed in conjunction with the plasma ACTH level, the second test to be performed in the etiological diagnosis of hypercortisolemic patients.

The Plasma ACTH Assay

Since the concentration of circulating ACTH levels in the plasma of normal humans is highly responsive to the ambient concentrations of cortisol in the plasma, measurement of ACTH should permit distinction between the various causes of hypercortisolism, and indeed it does so in a significant number of patients. Theoretically, patients with hypercortisolemia due to adrenal tumor should have very low, i.e., suppressed, levels of ACTH, while patients with CD should demonstrate ACTH levels inappropriate to the level of circulating cortisol level, i.e., a mildly elevated or even "normal level. Patients with hypercortisolemia secondary to ecotpic ACTH-secreting syndrome generally demonstrate markedly elevated levels of ACTH. However, several important issues may cloud the interpretation of the ACTH assay:

1. The assay for ACTH in plasma is very "delicate." Blood has to be collected with meticulous care using chilled EDTA-containing tubes on ice; the blood must be centrifuged and the plasma separated within 2 hr of collection. Failure to properly collect the sample could spuriously elevate or decrease the plasma ACTH level.
2. The ACTH level in plasma is subject to a great deal of fluctuation due to the inherent "pulsatile" nature of its secretion from the pituitary. Therefore, the level of hormone in a randomly drawn sample of blood is not reflective of the overall prevailing concentration of the hormone. The same diagnostic limitations that applied to the random serum cortisol are also applicable to a single ACTH level drawn at random.
3. The range of basal ACTH in the plasma of patients with pituitary-dependent CD is frustratingly wide (from a low as 30 pg to as high as 180 pg). Most, but not all patients with hypercortisolism due to adrenal tumor demonstrate basal ACTH levels below 10 or 20 pg. Horrocks and London[241] evaluated the diagnostic value of the 9 A.M. plasma ACTH in 58 normal subjects and seven patients with pituitary-dependent CD and found no overlap between the two groups. However, it is crucial for each laboratory to accurately define the 9 A.M. normal range in order to avoid spurious results.
4. Finally, the differences in assay systems employed by various reference laboratories make it impossible to interpret the ACTH assay by merely looking at a number.

The plasma ACTH assay viewed in conjunction with the standard dexamethasone test provides additional information (Table 19).

The Metyrapone Test

One of the main disadvantages of the standard dexamethasone suppression test is the inherent difficulty of ensuring a proper and complete 24-hr urine collection. Regardless of whether the parameter compared is free cortisol or the

TABLE 19.
Interpretative Value of DXM Test Coupled with ACTH Assay

Suppression to high-dose dexamethasone	Plasma basal ACTH	Diagnosis
Suppressible	Normal or mildly elevated (50–100)	CD
Nonsuppressible	Low (<10)	Cushing's syndrome secondary to adrenal tumor
Nonsuppressible	High (>200)	Cushing's syndrome secondary to ectopic ACTH
Suppressible	Low (<20)	Macronodular hyperplasia
Suppressible	Modest elevation (150–200)	Ectopic ACTH secondary to carcinoid (or) CD

more conventional 17-OHCS, collection for 7 days (two basal, four for the low- and high-dose days, and one postcontrol) is difficult even under the best of circumstances. For this reason, it would be desirable to have a dynamic test that does not involve urine collection and yet has a discriminatory value that equals (or excels) the standard dexamethasone test; the metyrapone test of the 1980s seems to satisfy both these objectives.

Metyrapone is a chemical that blocks the 11-hydroxylation of 11-deoxycortisol to cortisol. When administered at a dose of 750 mg by mouth every 4 hr for six doses, metyrapone almost completely blocks the 11-hydroxylation of 11-deoxycortisol. As a result, the concentration of cortisol rapidly declines, causing the hypothalamic pituitary unit to respond to releasing more ACTH. This in turn would lead to an increase in the level of 11-deoxycortisol, the immediate precursor steroid proximal to the block. The availability of a sensitive radioimmunoassay for the measurement of 11-deoxycortisol in the plasma has greatly simplied the metyrapone test, obviating the nuisance of urine collection. A number of smaller series have evaluated the diagnostic utility of metyrapone.[242,243] The largest series that has propsectively compared the usefulness of the metyrapone test versus the standard high-dose dexamethasone test in 25 unselected patients with hypercortisolism is that of Sindler et al.[244] These workers compared the ability of the two tests to accurately determine the etiology of Cushing' syndrome. The results of their study indicated that the metyrapone test was more accurate than the standard high-dose dexamethasone test in differentiating pituitary-dependent CD from adrenal tumors causing hypercortisolism. All patients with CD demonstrated a postmetyrapone 11-deoxycortisol level greater than 10 μg/dl, while all patients with adrenal adenomas had a suppressed 11-deoxycortisol level (below 10 μg/dl) postmetyrapone. The reasons for these behavior patterns are straightforward. The normal response to metyrapone administration is an increase in the plasma level of 11-deoxycortisol. The response is preserved in patients with pituitary-dependent CD and is abolished in adrenal tumors causing CD since the ACTH is maximally suppressed. Metyrapone has more far-reaching effects than just blocking 11-beta-hydroxylase. It is felt that metyrapone has a dual action on cortisol biosynthesis—a distal block on 11-beta-hydroxylase, and a proximal block early in the steroidogenesis that involves cholesterol cleavage.[245] The latter block is not significant in normal subjects since it can be overcome by ACTH, but in patients who have no ACTH in plasma (i.e., patients with adrenal tumors), metyrapone leads to significant lowering of 11-deoxycortisol. The response of ectopic ACTH-secreting syndrome to metyrapone is mostly similar to that of adrenal tumors, resulting in very little or no increase in 11-deoxycortisol following metyrapone. The convenience and accuracy of the metyrapone test have made it an important dynamic test in determining the etiology of CD. However, the predictive value of the metyrapone test is not exempt from the bizarre exceptions that pervade all diagnostic tests for Cushing's syndrome. There have been reported instances of unilateral adrenal neoplasms responding to metyrapone administration.[246]

The Ovine CRH Stimulation Test

In a sense all the aforementioned three diagnostic studies—the dexamethasone suppression tests, the plasma ACTH level, and the metyrapone study—are focussed on the responsiveness of the corticotrope. Therefore, when in the early 1980s the structure of CRF was characterized, a new method to study aberrant corticotrope behavior was developed. Vale et al.[247] characterized the 41-residue ovine hypothalamic peptide CRH and showed that it effectively stimulates the secretion of corticotropin and β-endorphin. The in vivo effectiveness of synthetic ovine CRH was established by Rivier et al.,[248] who demonstrated the selective effects of this peptide on human corticotropes. As expected, numerous studies from international centers began reporting on the diagnostic use of ovine CRH in altered states of corticotrope function. Orth et al.[249] were among the earliest to study the response pattern of ACTH to CRH stimulation. They reported two patients with mild CD who underwent test with CRH shortly before and 1 week after successful transsphenoidal microadenectomy. When CRH was given in a dose of 1 µg/kg body weight, a brisk, exaggerated ACTH response was seen in the first patient. This "hyperresponsive" ACTH pattern disappeared following removal of the microadenoma. Similar, but less impressive results were seen in the second patient. The authors suggested that the response pattern in these two patients with ACTH-secreting microadenomas was consistent with the fact that CRH may be involved in the pathogenesis of the CD seen in these two patients. Muller et al.[250] from Germany reported on the effects of intravenous CRH on ACTH concentrations in seven patients with pituitary-dependent CD, two patients with adrenal tumors, and one with the ectopic ACTH-secreting syndrome. All seven patients with pituitary-dependent CD demonstrated a brisk, hyperresponsive ACTH pattern, despite the fact that four had normal baseline ACTH levels. In contrast, the ACTH response to CRH in the other two groups of patients remained unaltered.

Chrousos et al.[11] presented the hitherto largest series to date in 1984; they studied the response of ACTH to CRH in 22 patients with various forms of hypercortisolism. Again, all 13 patients with pituitary-dependent CD showed a robust increase in the already elevated basal ACTH level. They confirmed prior data that following microadenomectomy there was restoration of the CRH-induced ACTH (and cortisol) responses to near normalcy or normalcy. Once again, there was confirmation of the fact that patients with ectopic ACTH syndrome (with very high basal ACTH levels) and those with adrenal neoplasms (with very low basal ACTH levels) behaved identically with absolutely no increase in ACTH level when CRH was administered. The results of these studies clearly seemed to emphasize one fact—the failure of ACTH to respond to a bolus of CRH in the setting of CD clearly excludes the pituitary as the source of the problem.

While the enthusiasm for the CRH test was gaining momentum, a few studies began to raise some doubt as to whether all patients with pituitary-dependent CD behave in a homogenous fashion. Pieters et al.[251] studied the ACTH response to intravenous CRH administration in five patients with pituitary-dependent CD. While they noted that the absolute increments of ACTH (and cortisol) levels of these patients following CRH administration were significantly higher than in normals, the CRH responses in individual patients were variable. For instance, in three patients, the ACTH increase following CRH was within the mean ±1 SD of the response seen in normal subjects. More interestingly, one patient with CD with bilateral micronodular hyperplasia demonstrated a negligible ACTH response to CRH. The authors concluded that the CRH–ACTH axis in patients with pituitary-dependent CD may reveal hyperresponsive, normoresponsive, and even hyporesponsive patterns when tested with exogenous CRH. When the behavior of pituitary adenomas removed surgically and maintained in culture were studied, somewhat unexpected data emerged. Suda et al.[252] evaluated the ACTH responsiveness in vitro by using superfusion of pituitary adenoma tissue removed from 16 patients with CD. Their data indicate that adenoma tissue, in some patients, demonstrated a strikingly low sensitivity to CRH. Similar data had been previously reported to Shibasaki et al.,[253] who demonstrated that the pituitary adenomatous tissue removed from two patients with CD responded to a variety of agents with great variability. In particular, CRH failed to stimulate the secretion of ACTH or β endorphin from adenoma tissue of both patients, while vasopressin, 3-isobutyl-methylxanthine, and high concentrations of potassium caused a prompt increase in the secretion of these POMC-derived peptides. These findings suggested that these corticotrope adenomas had lost their receptors to CRH or the postreceptor mechanisms were nonfunctional. The reasons(s) for the dichotomy between the in vivo and the in vitro ACTH responses to CRH in some patients with pituitary-dependent CD is not clear. Perhaps, the in vivo ACTH response to CRH is modulated by other factors such as vasopressin or angiotensinogen.[252]

Regardless of these discrepanies in in vitro studies, the ovine CRH stimulation test is rapidly becoming an important tool in the differential diagnosis of Cushing's syndrome.[11,254–256] Studies that compare the CRH test with the standard dexamethasone suppression test have quite favorably appraised the CRH test. In one recent report, Nieman et al.[257] performed both tests in 33 patients with pituitary ACTH-mediated CD and eight with the ectopic ACTH syndrome; 29 of 33 patients with CD responded to CRH with an increase in cortisol, while none of eight with the ectopic ACTH syndrome did. A response to the CRH test was defined as an increase in the mean response variable (ACTH or cortisol) of more than four times the intraassay coefficient of variation at the mean baseline concentration.[257] Comparison of the results of the dexamethasone suppression test revealed that 29 of 33 patients with CD suppressed to the high-dose dexamethasone, as did one patient with the ectopic ACTH syndrome. The authors concluded that the ovine CRH test works as well as the standard dexamethasone test in differentiating CD from the ectopic ACTH syndrome. The diagnostic power of each test was enhanced when the two were combined. The concordance between these two tests is not surprising, since the two phenomena, i.e., suppressibility to high-dose dexamethasone and responsivity of ACTH to CRH, are believed to be related.[258] Since the distinction between pituitary ACTH-dependent CD and ectopic ACTH syndrome can pose considerable difficulty, it is hoped that the CRH stimulation test would aid in making this distinction. While the general consensus is that ectopic ACTH syndrome is characterized by a leak of ACTH (or cortisol) response to the administration of ovine CRH, reports of ectopic ACTH syndrome that "responded" to CRH continue to appear in the literature.[103,259] Given the heterogeneity of ectopic ACTH/CRH-secreting tumors, coupled with the lack of a precise definition of what constitutes a response to CRH, the results of this test in an individual patient should be interpreted with caution.

These conflicting data regarding the responsiveness of ACTH to CRH in pituitary-dependent CD and in ectopic ACTH syndrome clearly indicate the need for more studies. It is unlikely that the CRH test will wholly replace the need for the conventional dynamic tests that have been in vogue for decades. In a resounding editorial article resonating with clarity, David Orth[260] has placed in perspective the new and the old in CD. Although the CRH test may separate most patients with pituitary-dependent CD from the ectopic ACTH syndrome, patients in the latter group with recent-onset hypercortisolemia and incomplete pituitary suppression may respond to CRH, mimicking the dynamics of CD.[260] The author's conclusion was that the demonstration of relative versus absolute resistance to glucocorticoid inhibition, as shown by the dexamethasone suppression test, will continue to remain the best single method to differentiate CD from Cushing's syndrome due to adrenal tumor or ectopic ACTH secretion.

Table 20 outlines the important diagnostic tests in Cushing's syndrome.

In summary, the following diagnostic approach is followed at Cook County Hospital in the evaluation of hypercortisolism. As a first step the *overnight dexamethasone* test is performed on an outpatient basis. If the postdexamethasone 8 A.M. cortisol is below 5 μg, no further workup is indicated unless the clinical findings are extremely strong. If the patient fails to suppress to the overnight dose, a 24-hr urine is collected for measurement of *free cortisol*. If the values exceed 90 μg, a *high-dose dexamethasone* suppression test is performed. If the patient shows suppression (50% decline in basal 17-OHCS or urinary free cortisol), pituitary-dependent disease is suspected; if the patient fails to show suppression, an adrenal tumor or ectopic source of ACTH is suspected. Further hormonal confirmation of either entity is provided by performing a *basal ACTH* level in plasma and the *metyrapone* test using 11-deoxycortisol level as the parameter. If *all* tests point in the direction of the pituitary, the sella turcica is evaluated by *computerized tomography* (CT); if all tests, on the other hand, point toward an adrenal source, both adrenals are evaluated by CT. When the CT of sella is normal, but the hormonal data point to the pituitary, the only definitive method

TABLE 20.
Diagnostic Studies in Evaluation of Cushing's Syndrome

Test	Comment
Screening	
The overnight dexamethasone suppression test	Limited by the high percentage of false positive results
The 24-hr urinary free cortisol	Limited by the problem of complete urine collection
Etiological diagnosis	
The standard dexamethansone suppression tests	High-dose test distinguishes pituitary from non-pituitary dependency in 60–80% of cases
Plama ACTH level	Serves as a valuable adjunct to other dynamic tests
Metyrapone test	Separates pituitary dependent from nondependent Cushing's; discriminatory value equals, perhaps excels, the standard dexamethasone test
CRF test	New technique to view an old principle (i.e., ACTH responsiveness in hypercortisolism)
Localizing procedures	
CT scan of adrenals	Outlines abnormalities in nearly all cases of adrenal tumors
CT scan of sella	Abnormal in nearly 20–70% of pituitary-dependent CD
Selective venous catheterization of inferior petrosal sinus for ACTH gradient	When performed properly, establishes or excludes the pituitary origin of ACTH excess

to document the source of the ACTH is by selective venous sampling of the inferior petrosal sinus; when ectopic ACTH syndrome is suspected, appropriate radiological studies are undertaken to detect the primary tumor. The algorithm is outlined in Figure 6.

Localization of the Etiology

The anatomical localization of the source of hypercorticolism depends on the information derived from the steroid dynamics. When all data point to a pituitary source, the localizational procedures are aimed at imaging the pituitary by CT. When the hormonal data are indicative of an adrenal neoplasm, the focus is shifted on CT of the adrenal. When the data are inconclusive, CT of both the adrenals and the pituitary is indicated. An additional procedure, selective venous sampling of the inferior petrosal sinus, has added a new dimension for the diagnosis of "difficult" cases. The advent of CT of the adrenal glands has diminished the need for adrenal venography, selective catheterization of adrenal veins, and imaging procedures with iodocholesterol. The focus in this section will be the following:

1. Computed tomography of the adrenals
2. Computed tomography of the pituitary
3. Selective venous sampling of the inferior petrosal sinus
4. Other localizational procedures

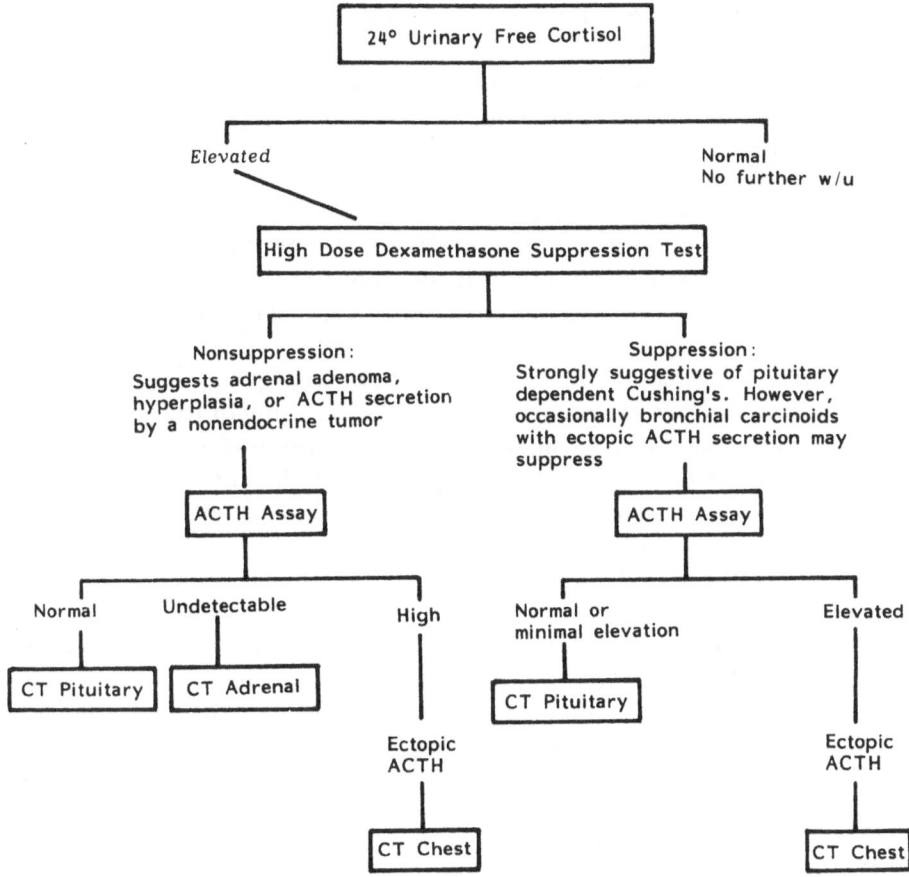

FIGURE 6. Algorithmic approach for evaluating hypercortisolism.

Computed Tomography of the Adrenal Gland

The adrenal glands, located retroperitoneally, amid the abundant fat that is always present in patients with Cushing's syndrome, lend themselves to excellent anatomical depiction by CT. The reader is referred to standard treatises in radiology[261] for the characteristics of the normal and abnormal adrenal by CT. However, some basic observations merit emphasis. First, the likelihood of delineating the adrenal gland by CT depends on the technique used. With 2-cm sections both glands can be defined in only 54% of normal individuals, while 1-cm sections reveal the adrenal glands in approximately 90%.[262] This figure can be further improved to 97%, by the use of fast scans done at narrower intervals.[263] Second, the shape of adrenal glands is crucial in disclosing underlying disease. Although some degree of anatomical variability is to be expected, the margins are generally straight or concave. This is particularly so for the outer margin. The "limbs" of the adrenals measure 3–4 mm in thickness. The length is generally 2–4 cm. Alteration in thickness is the most consistent observation when underlying adrenal disease is present. Third, when the size and shape of the adrenals are normal, adrenal tumors are unlikely. However, tumors can be present even when the size is normal; in such cases the margins assume a convex or irregular configuration.

Computed tomography of the adrenals is a high-yield procedure for detection of cortisol-secreting adrenal tumors. The chance of detecting such tumors by CT is clearly higher than the yield of CT in detecting aldosteronomas or pheochromocytomas. Cortisol secreting adrenal tumors are almost 2 cm or larger in diameter and easily visualized by tomography. Since the cholesterol content of these adrenal tumors is abundant, they are 10–20 Hounsfield units lower in density when compared to the adjacent soft tissue. White et al.[264] noted that CT correctly identified all 15 patients in their series with adrenal tumors, distinguishing five carcinomas from the 10 adenomas. The appearance of both adrenals in pituitary-dependent CD ranges from "normal" to slight enlargement bilaterally. The classic appearance, when present, is generalized thickening of the adrenal limbs bilaterally. In general, the bilateral increase in size of the adrenal glands tends to be greater when ectopic ACTH syndrome underlies the etiology of hypercortisolism. In the series of White et al.[264] CT was also valuable in identifying small lesions in the lung, mediastinum, and pancreas—areas that are strongholds for neoplasms secreting ACTH. In a study reported by Dunnick et al.[265] CT had a 100% accuracy in localizing cortisol-secreting adenomas. More important, no false positive diagnoses were encountered in patients with adrenal adenomas causing Cushing's syndrome. Based on the CT appearance, the authors were able to separate CD from adrenal tumors with a high degree of accuracy. In each case the findings on CT correlated with the findings on adrenal venous sampling.

The singular advantage of CT of the adrenals in Cushing's syndrome is that it is *always* abnormal when intrinsic adrenal pathology (adenoma or carcinoma) is causing the Cushing's syndrome. The view that CT of the adrenals should be the first imaging procedure regardless of the hormone data is shared by several authorities for the following reasons. If the study reveals unilateral, intrinsic adrenal disease (which it does with 100% accuracy), then no further localizational procedures would be needed. If on the other hand the adrenal CT reveals bilateral disease (enlargement) or normal adrenals, then the appropriate studies to differentiate pituitary from ectopic source of ACTH secretion may be undertaken. Patients with ACTH-dependent CD demonstrate normal or prominent adrenals with a normal configuration, while a significant portion of patients with ectopic ACTH syndrome demonstrate bilateral adrenal enlargement.

Despite the ease with which adrenal CT discloses the presence of unilateral tumor, occasionally the cross-sectional appearance of an adrenal tumor can be similar to the cross-sectional appearance of the upper pole of the unenhanced kidney and the inferior vena cava. Goldman et al.[266] have pointed out that enlargement of the adrenals usually assumes an ovoid configuration. Adrenal masses follow the principle that soft tissue structures show a tendency to conform to local limitations of space. This is especially so for the right adrenal, which has a very limited narrow space available between the liver and crus of the diaphragm. Therefore, tumors that arise from the right adrenal assume an oval shape when they enlarge upward. The recognition of this oval configuration may minimize observer error in interpretation. Of course, depending on their size and directional growth pattern, adrenal tumors can also be round.

Computed Tomography of the Pituitary

The sensitivity and specificity of CT in depicting pituitary microadenoma is much lower than that of detecting adrenal tumors. The incidence of detecting pituitary microadenomas can range from as low as 20% to as high as 70%.[162,163] Microadenomas smaller than 1–2 mm are not within the resolution of CT of the pituitary. When the CT of the pituitary is completely normal, and the CT of adrenals shows "normal-sized" or slightly enlarged adrenal glands, an ACTH source for the hypercortisolism can be presumed. Unequivocal documentation of a pituitary origin would have to rest on selective sampling of the inferior petrosal sinuses bilaterally, to demonstrate an ACTH gradient.

Selective Venous Sampling of the Inferior Petrosal Sinus

The venous anatomy of the pituitary is such that each half of the gland is drained by a venous plexus into either the ipsilateral inferior petrosal sinus or the intercavernous sinus crossing the floor of the pituitary fossa.[267] Figure 7 diagrammatically outlines the venous drainage of the pituitary gland. The pituitary is drained via the cavernous sinus posteriorly into the superior and inferior petrosal sinuses, and from there into the jugular bulb and vein. The blood from the transverse, sigmoid, and occipital sinuses also drains into the jugular bulb. Since there is increased admixture of blood in the jugular bulb and distal to it, sampling of the petrosal sinus is critical to determine the presence or absence of a hormone gradient. Although experimental data in humans are lacking, experiments in the rhesus monkey suggest that pituitary venous drainage is laternized. Oldfield et al.[268] determined the extent of intermixing of blood between the

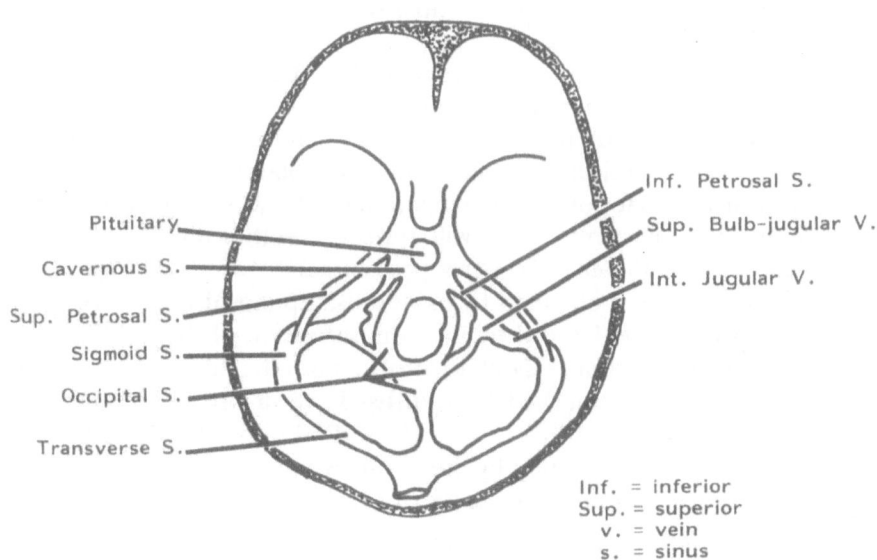

FIGURE 7.   Venous drainage of the pituitary gland.

cavernous sinuses and between both inferior petrosal sinuses by injecting Tc colloidal sulfur into the superior orbital vein of rhesus monkeys. By comparisons of radioactivity in the inferior petrosal sinuses, they showed that the mean relative radioactivity was negligible in the contralateral inferior petrosal sinuses. Their findings suggest that mixture of blood between the cavernous sinuses and the inferior petrosal sinuses is insignificant. Thus, the bulk of evidence in humans favors the notion that pituitary venous drainage is lateralized. Therefore, a laterally placed microadenoma will result in increased ACTH levels in the ipsilateral petrosal sinus. This premise has been confirmed by several studies.[227,269–272] Accordingly, measurement of the ACTH gradient (the ratio of central to peripheral ACTH concentration) by catheterization of the inferior petrosal sinus is accurately representative of the pituitary origin of the tumor. Several important aspects of this procedure deserve emphasis.

1. The central-to-peripheral ACTH concentration must be interpreted cautiously, owing to the short half-life of ACTH and the intrinsically fluctuant nature of ACTH secretion.
2. Failure to properly collect and handle the sample for a difficult assay may result in misleading numbers.
3. The need for *bilateral* petrosal sinus sampling has not been sufficiently emphasized in the literature. Such a procedure will clearly demonstrate an ipsilateral gradient on the side of the microadenoma, if it is laterally situated. A midline microadenoma will demonstrate equal gradients on both sides, but that are higher than normal, information that is likely to help the neurosurgeon. If only one side is sampled, and if it were not the side at which the tumor was lateralized, the information yielded would not be clearly conclusive in localization of the source, or the side.
4. Lateralizing gradients are not consistently obtained with jugular venous sampling. It is, therefore, essential that the catheter be engaged in the inferior petrosal sinus.
5. The catheters should not be advanced to the cavernous sinus, owing to the risk of cavernous sinus thrombosis.
6. The correct positioning of the catheter can be verified by gentle retrograde contrast injection immediately after placement of cathether in the inferior petrosal sinus. Following unilateral injection the contrast will fill both cavernous sinuses and will appear as reflux down the contralatral side, if the catheter is properly positioned.
7. The information obtained by bilateral sampling of the inferior petrosal sinuses can be enhanced by the administration of ovine CRF, 1 μg/kg. After the basal sample is obtained from both the inferior petrosal sinuses and peripheral vein, the CRF is infused over a period of 1 min. Petrosal samples are obtained at 1, 3, 5, 10, and 30 min after the infusion.
8. Because ACTH secretion by the pituitary microadenoma is pulsatile, it is absolutely essential that both petrosal sinuses are sampled *simultaneously.*

The results of the study should be interpreted with all these factors in mind. Obviously, when the CT scan of the pituitary has disclosed microadenoma, there is no need to resort to selective venous sampling. But in the patient with hypercortisolism and a normal sella, with dynamic data that suggest a pituitary source, a properly performed selective venous sampling of the inferior petrosal sinus greatly helps in confirming or excuding a pituitary origin. Typical findings that characterize a laterally placed tumor are the following:

1. A large ACTH gradient on the ipsilateral petrosal sinus.
2. A less impressive, but mildly increased ACTH gradient on the contralateral side.
3. A prompt and dramatic rise in ACTH gradient on the ipsilateral side after administration of ovine CRF.
4. A less marked gradient on the contralateral side following CRF administration.

It is believed that the study, when performed correctly, can accurately predict the site of the microadenoma in nearly every case. The information is especially helpful for the neurosurgeon if microsurgery fails to show a tumor. If the side of the ACTH secretion is known, a hemipituitectomy done on the side of gradient is curvative.

Other Localizational Procedures

The proven track record of CT in documenting adrenal tumors has obviated the use of other radiological procedures for localizing adrenal tumors. Three such procedures that were once very much in vogue, but currently are used less frequently, are iodocholesterol scans, adrenal venography, and adrenal venous sampling.

Iodocholesterol Scan. Following the first reported visualization of the adenal glands in a patient with Cushing's syndrome,[273] adrenal imaging with [131I]19-iodocholesterol became a frequently employed test in selected centers. The bilateral uptake by hyperplastic tissue contrasts with the scintigraphic appearance of unilateral adenoma, which is characterized by unilateral uptake on the side of the tumor with no uptake on the contralateral side. Although scintigraphic imaging with [131I]19-iodocholesterol has enjoyed a good reputation,[274–276] the limited availability of the test and the superior results obtained with CT have resulted in a declining use of iodocholesterol scanning.

Adrenal Venography.[277–280] Adrenal venography, obviously an invasive localizational procedure, is seldom performed today for identifying adrenal tumors. In addition to the technical problems involved with the cannulation of adrenal veins, the procedure can result in complications. These usually include thrombosis of the adrenal veins[281] and less commonly adrenal infarction. Remission of Cushing's syndrome due to infarction of the adrenal tumor following venography has been described.[282] The accuracy of CT in correctly localizing cortisol-secreting adrenal tumors has obviated the use of venography.

Adrenal Venous Sampling. Although adrenal vein sampling continues to remain an important procedure for localizing aldosterone-secreting adenomas, this procedure, like venography, is being employed less frequently in the localizational diagnosis of Cushing's syndrome caused by adrenal adenomas.

Again, the diagnostic superiority of CT in detecting cortisol-secreting adrenal tumors is unsurpassed, rendering the need for invasive procedures unjustifiable.

Differential Diagnosis

The differential diagnosis of Cushing's syndrome can be viewed from two aspects: First is the differentiation of true hypercortisolism from other conditions that mimic the clinical, hormonal, and dynamic features of Cushing's syndrome. Included in this category are obesity, drug interference, endogenous depression, alcohol abuse, and chronic renal failure. The second aspect of the differential diagnosis of Cushing's syndrome involves etiological separation of the various causes of hypercortisolism. Accurate diagnosis of the etiology is obviously crucial for therapy.

Conditions That Mimic Cushing's Syndrome

Several "non-Cushingoid" conditions can result in alterations in the circulating levels and dynamics of cortisol secretion. These alterations can involve binding proteins, can affect the renal excretion of cortisol and its metabolites, and can even affect the secretory activity of the hypothalamic pituitary unit. In fact, these conditions often form the bulk of consultations requested to exclude Cushing's syndrome.

Obesity

Obesity can be differentiated from true hypercortisolism by the measurement of UFC levels in a 24-hr sample. Unless complicated by psychiatric illness, obesity is characterized by normal 24-hr UFC. When UFC measurements are not available, the low-dose dexamethasone test can be performed, the results of which are nearly always normal in uncomplicated obesity.

Drug Interference

Various states of drug interference can cause alterations in the dynamic tests employed in the evaluation of hypercortisolism; of particular importance is diphenyl hydantoin. Patients on this anticonvulsant drug often show failure to suppress to the overnight and low-dose dexamethasone tests. This is due to accelerated clearance of dexamethasone occasioned by the effect of diphenylhydantoin on hepatic microsomal induction. An alternate method of studying steroid dynamics in patients receiving the drug is by administering hydrocortisone and measuring the plasma level of corticosterone.[283] Non-Cushingoid patients on diphenylhydantoin decrease their 8 A.M. plasma corticosterone level to below 270 ng/dl and below 50% of the baseline following 50 mg of hydrocortisone at midnight.

Psychiatric Illness

Psychiatric illness, especially depression, can mimic several facets of the abnormal steroid dynamics of Cushing's syndrome. Thus, increased cortisol secretion, more secretory spikes,[284] and even increased UFC levels[285] are often encountered in patients with depression. Absence of diurnal variation and abnormalities in the overnight and even the low-dose dexamethasone suppression tests can occur in patients with depression. The combination of an elevated UFC, abnormal suppression to low-dose dexamethasone, and preservation of suppression to high-dose dexamethasone test closely resemble the dynamics of pituitary-dependent CD. The frequency with which abnormal steroid dynamics are encountered in endogenous depression, reverting to normal with treatment, has rendered the dexamethasone suppression test a controversial parameter in the diagnosis and followup of patients with unipolar and bipolar depressive illness. Since depression is often a prominent aspect of Cushing's syndrome, and shares several facets of hormone dynamics with that syndrome, it may be difficult to establish the presence of "true hypercortisolism" in patients who are depressed. The following may help in identifying Cushing's syndrome in a depressed patient.

1. The cortisol response to insulin-induced hypoglycemia is preserved in most cases of endogenous depression.[286]
2. The steroid dynamics revert to normal upon treatment of the depressive disorder, in contrast to Cushing's syndrome, where the abnormal suppression data persist despite adequate treatment with antidepressants.
3. Recently, the CRF test has been used as a tool in the diagnostic evaluation of depression.[287,288] The ACTH response to CRF in patients with depression is appropriate for a pituitary gland exposed to the negative feedback of chronic hypercortisolemia, resulting in the restraining effect to CRF stimulation. Such a response separates the depressed patient from one with pituitary-dependent CD.

Alcoholic Pseudo-Cushing's Syndrome

Chronic alcoholism shares several similarities to the clinical and hormonal features of hypercortisolism:

1. The clinical syndrome of "pseudo-Cushing's" caused by chronic alcoholism includes weakness, plethora, rounding of face, fatigue, peripheral muscle wasting, thin skin, and even purple striae. The picture may be further complicated by the occurrence of hypertension, glucose intolerance, and osteopenia in chronic alcoholism. In an individual patient, however, only a few of the above features may be present.
2. The serum cortisol levels are often elevated, frequently demonstrating failure to suppress with low-dose dexamethasone.[289]
3. Abnormal diurnal rhythms may be associated with the high cortisol levels.[289,290]
4. Rarely, hyperresponsiveness of the hypothalamic pituitary axis to

metyrapone and an elevated secretory rate of cortisol may be encountered.[291] The similarity to pituitary-dependent Cushing's syndrome may be further accentuated by the abnormally blunted plasma cortisol response to hypoglycemia.[292]

The abnormal steroid dynamics seen in chronic alcoholism could be a result of stress, alcohol withdrawal, intercurrent illness, or concomitant psychiatric disease. Elias et al.[293] evaluated the effect of acute alcohol administration on the HPA axis in five chronic alcoholic patients and compared the effects with those in normal controls. The mean plasma ACTH concentration, as well as effect of exogenous ACTH on the release of cortisol and aldosterone, were similar in both groups. The authors speculated that the pseudo-Cushing's syndrome of alcoholics might represent a state of stress-induced hypercortisolemia caused by multiple episodes of subacute alcohol withdrawal.

There is a paucity of studies that have evaluated the steroid dynamics in the nonimbibing phase of chronic alcoholics. It is generally believed that these abnormalities resolve during abstinence. Measurement of UFC in the nonimbibing phase is probably the best means of screening for hypercortisolism.

Chronic Renal Failure

The diagnosis of Cushing's syndrome can be obfuscated by the presence of chronic renal failure.[294] The reasons for this are multifactorial. Inadequate urine collection, erratic absorption of dexamethasone,[295] spurious elevation of plasma cortisol due to accumulation of interfering substances, prolongation of the half-life of circulating cortisol, and the stress of a chronic illness can all contribute to difficulties in the diagnosis of true hypercortisolism when renal failure is present. Nolan et al.[296] have pointed out that the plasma of patients with chronic renal failure contains water-soluble glucuronides and unconjugated glucocorticoids that interfere with measurement of plasma cortisol. This cross-reactivity can be eliminated by extraction with dichloromethane, a procedure employed to assess the true cortisol activity in patients with chronic renal failure. This procedure, coupled with performance of dexamethasone suppression tests by the intravenous, rather than the oral, route, may help establish the diagnosis of true hypercortisolism in patients with chronic renal failure.

Factitious Cushing's Syndrome

Factitious Cushing's syndrome is extremely rare. If dexamethasone is being abused, the diagnostic problems are relatively simple, because the hormonal profile would reveal low UFC with an appropriately suppressed plasma ACTH level. The problem can become confounding with surreptitious abuse of cortisone or hydrocortisone, which obviously cross-reacts with the serum cortisol assay. The profile in such cases would closely resemble that of Cushing's syndrome caused by adrenal tumors. The exogenous source for the elevated cortisol level can be demonstrated by the measurement of corticosterone level in the plasma. Cook and Meikle[297] employed measurement of plasma corticosterone levels to identify a patient with factitious Cushing's syndrome. Endogenous hypercortisolemia should be accompanied by simultanteous corticosterone secretion, while exogenous hypercortisolism—which physiologically suppresses the adrenals—is associated with low corticosterone levels even after ACTH stimulation. This novel role of corticosterone concentrations can be a vital key to provide data for confronting the patient with surreptitious cortisone (or hydrocortisone) abuse.

Etiological Diagnosis of Cushing's Syndrome

As indicated earlier, the distinction between the various etiologies for Cushing's syndrome is crucial for proper therapy. The individual features of each variety of Cushing's syndrome is outlined next, underscoring the fact that these are broad generalizations.

Pituitary-Dependent Cushing's Disease

The typical patient with pituitary-dependent CD demonstrates preservation of suppression to high-dose dexamethasone and responds to metyrapone with an increase in 11-deoxycortisol; the basal plasma ACTH concentrations are normal to minimally elevated, with further increments with the administration of ovine CRH. The plasma dehydroepiandrosterone (DHEA) sulfate levels are normal to mildly elevated. The CT of the adrenals reveal normal-sized to slightly enlarged glands bilaterally, while the pituitary gland may reveal an abnormality in 35–60% of patients. Some patients with CD may show a paradoxical ACTH response to TRH or LHRH.

Adrenal Adenoma

When Cushing's syndrome is caused by adenomas of the adrenal cortex, the patient fails to suppress with high-dose dexamethasone and fails to respond to metyrapone administration. The plasma ACTH levels are low and show no response to exogenous administration of CRH. The adrenal steroids respond, in a significant number of patients, to exogenous ACTH administration. The level of the adrenal androgen DHEA sulfate is depressed in patients with adrenal adenomas that cause Cushing's syndrome, often remaining low for as long as 2 years following surgical extirpation of the tumor.[298] This feature, a reflection of chronic suppression of endogenous ACTH levels, is characteristic of adrenal adenoma. Computed tomographic examination almost always demonstrates a unilateral mass with a "shrunken" adrenal on the contralateral side. Table 21 outlines the differences between Cushing's syndrome secondary to CD and adrenal tumors.

Macronodular Hyperplasia

Macronodular adrenocortical hyperplasia may be associated with conflicting dynamic data since this entity is characterized by varying degrees of adrenal auton-

TABLE 21.
Differences between Pituitary-Dependent Cushing's Disease and Adrenal Neoplasms

Test	Pituitary-dependent Cushing's disease	Adrenal neoplasm
Response to 24-hr 17-OHCS or UFC to high-dose dexamethasone	Preserved	Lost
Response of 11-deoxycortisol to metyrapone	Present	Absent
Plasma ACTH level	Normal to minimally elevated	Suppressed
Plasma DHEA-S level	Normal to minimally elevated	Suppressed
ACTH response to CRF	Preserved; may be exaggerated	Absent
Pardoxic ACTH response to TRH, LHRH	Often present	Absent
CT scan of pituitary	May be abnormal	Normal
CT scan of adrenals	Both adrenal enlarged	Unilateral neoplasm

omy coupled with varying degrees of pituitary dependency; suppression tests, basal ACTH levels, and ACTH response to CRF may not always provide clear separation of this entity from CD or from unilateral tumors of the adrenal. Computerized tomography may or may not reveal the macronodular disease, but usually bilateral disease.

Adrenal Carcinoma

Adrenal *carcinoma* should be suspected when virilization occurs in association with hypercortisolism. This rare disease is a devastating disorder affecting younger patients, although it can occur at any age. The most common presentations of adrenal carcinomas are the development of Cushing's syndrome and virilization. In most instances these two syndromes occur together. Adrenal carcinomas causing virilization alone, without a concomitant increase in glucocorticoids, have been described, but the reverse has not; i.e., whenever an adrenal carcinoma causes Cushing's syndrome, there is invariably an elevation of 17-ketosteroids. The hormonal hypersecretion by an adrenal carcinoma is a rather paradoxical phenomenon because when adrenal cells become malignant they lose many enzyme systems needed to produce hormones. Therefore, these tumors are regarded as rather inefficient hormone producers. The implications of this observation are twofold. First, when there is evidence of hormone hypersecretion at the time of diagnosis, it means that the tumors have been present for a long time and have grown large enough to produce hormone hypersecretion. Second, because of the enzyme deletions several precursors are released in the circulation. Thus, DHEA, estrone, estradiol, and 11-deoxycortisol may be detected in the urine before virilization has occurred. The rapid development of Cushing's syndrome, the presence of virilizing signs, and a palpable mass would clearly indicate the clinical diagnosis of adrenocortical carcinoma. In addition to virilization, adrenal carcinomas can result in precocious puberty in both sexes and feminization of males.

Ectopic ACTH Syndrome

The recognition of *ectopic ACTH syndrome* in its characteristic presentation of weight loss and pigmentation with metabolic phenomena (hypokalemia, alkalois, hyperglycemia, and so forth) is relatively simple. In most patients the underlying malignancy is apparent and the plethoric features of Cushing's syndrome are absent. The cortisol levels are usually quite high (>50 μg) as is the plasma ACTH level (>250 pg). These patients are usually nonresponsive to the administration of dexamethasone, metyrapone, and CRF. The exceptional features of the ectopic secretion of ACTH by carcinoids have already been alluded to. The occurrence of plethoric Cushing's syndrome in the setting of suppressibility to high-dose dexamethasone, with only modest increments in plasma ACTH levels closely resembles pituitary-dependent CD. The metyrapone test or the CRF test has not been evaluated in a large series of patients with dexamethasone-suppressible ectopic ACTH syndrome. The only definite method of excluding a pituitary source of ACTH in such patients with an occult primary is by measuring the ACTH gradient by simultaneous bilateral inferior petrosal catheterization, preferably coupled with administration of CRF just prior to sampling.[235,236]

Periodic Hormonogenesis

It is difficult to establish the diagnosis (much less the etiology) of hypercortisolism when it occurs in the setting of *periodic hormonogenesis.* Apparently these represent examples of a "tumor on a timer." The responses to dexamethasone administration are highly variable, ranging from no response to a "paradoxical" response. Several reports have highlighted the entity. The first documentation of such a phenomenon was by Bailey[299] in a patient with a malignant bronchial carcinoid, which secreted ACTH cyclically with dramatic "steroid cycles" every 18 days. Brown et al.[231] described cycles of 11 days in a patient with a chromophobe adenoma causing CD. The cyclical change in steroid levels in these cases was unaffected by dexamethasone, and the apparent paradoxical response to dexamethasone was fortuitous. The obvious difficulty is that one cannot interpret the steroid levels as an effect of drug administration, and the increase or decrease in steroid levels is totally a reflection of periodic hormogenesis. The most remarkably example of such a phenomenon is the case reported by Lieberman et al.[230] These workers studied a patient with CD secondary to a chromophobe adenoma for 243 days and presented evidence for periodicity in cortisol production, with cycles occurring every 85 days associated with "laboratory remissions" in between, and paradoxical responses to dexamethasone administration. Thus, paradoxical responses to dexamethasone are best explained by the presence of an autonomous rhythm and periodic changes in hormone levels with nothing more than a coincidental temporal relation to the administration of dexamethasone.

Since it has been implied that cortisol secretion in such cases is independent of stimulation or feedback, it is interesting to note that Jordan et al.[300] studied a patient with cyclic Cushing's syndrome of pituitary origin, apparently set at a 35-day cycle. The fact that the pituitary did respond to feedback was evidenced by a decline in UFC level with cyproheptadine and a dramatic rise in ACTH in response to the stress of anesthesia. Also, cyclical CD with two distant and independent rhythms has been described.[301]

The phenomenon of periodic hormonogenesis (cyclical Cushing's syndrome) is not restricted to pituitary-dependent CD. Even though pituitary and ACTH-secreting tumors predominate in the display of this astonishing colorful spectable of periodic hormonogenesis,[302–304] adrenal tumors are not exempt.[305,306] The importance of recognizing the phenomenon of periodic hormonogenesis is fourfold. First, the results of a high-dose dexamethasone test could be misinterpreted as nonsuppressible when, in fact, they are due to the periodic hormonogenesis; second, between cycles the hormone data may be "normal," and hence even establishing the hormonal diagnosis could be difficult; third, intermittent hypokalemia, periodic mental changes, intermittent loss of diabetic control, and osteopenia can elude diagnosis if this entity is not kept in mind. Finally, intermittent or cyclical hypercortisolism may result in "spontaneous remissions" of the disease.[307,308]

Treatment

The treatment of Cushing's syndrome is based on the etiology of hypercortisolism. Although the adrenal glands are the ultimate target regardless of the etiology, therapy is directed against the primary pathogenic mechanism that stimulated the adrenals in the first place. Thus, pituitary-dependent CD is treated by pituitary surgery, pituitary radiation, and adjunctive drug therapy with neuropharmacotherapeutic agents. Adrenal adenoma is simply treated by surgical removal of the tumor, a procedure that involves unilateral adrenalectomy. Treatment for adrenal carcinoma involves surgery, when indicated, along with chemotherapy with o p′ DDD. Cushing's syndrome caused by ectopic ACTH secretion is managed by therapy directed against the primary neoplasm, coupled with drug therapy that lowers cortisol (metyrapone, ketoconazole) or newer agents that antagonize the peripheral effects of glucocorticoids (RU 486).

Cushing's Disease

The ideal treatment for pituitary-dependent CD should be one that will completely remove the source of ACTH hypersecretion, restore cortisol secretion to normal, avoid hypopituitarism, and would be associated with negligible morbidity and mortality. These ideals, at least theoretically, are best fulfilled by selective microadenectomy by transsphenoidal pituitary surgery (TPS). Other forms of therapy—bilateral adrenalectomy, pituitary radiation, and therapy with drugs that lower ACTH–cortisol hypersecretion—fall short of these ideal requisites of therapy. The most satisfactory results for a "cure" are obtained when primary treatment is directed at the pituitary gland. Attempts at treating the hypothalamus with chemical therapy are fraught with unpredictable and transitory responses, while therapy aimed at the adrenals is drastic, with the potential for trading off two diseases (Addison's and Nelson's) for one disorder. With the increasing expertise that is emerging for pituitary microsurgery, this is clearly becoming the first and in many cases the only choice. The options for treating patients with pituitary-dependent CD are several and include the following:

1. Transsphenoidal pituitary microsurgery.
2. Pituitary radiation.
3. Bilateral total adrenalectomy.
4. Neuropharmacotherapeutic drugs.

Transsphenoidal Pituitary Microsurgery

The procedure of transsphenoidal pituitary surgery has gained immense popularity in the past decade. Several workers have published their experience with this procedure in the treatment of pituitary-dependent CD, the two largest studies being that of Boggan et al.[15] from San Francisco, who treated 100 patients with CD between 1974 and 1981, and the series of 75 patients of Hardy[16] from Montreal. A smaller, but equally good series by Burch[309] has also outlined the results of this procedure in 19 patients who underwent pituitary microsurgery for CD in a 6-year period. The important aspects to focus on are the success rate of the procedure, the definition of "cure," the adverse effects, and the reasons for failure.

It is now generally accepted that the cure rate of TPS far exceeds that of every other available therapeutic modality for CD. The overall cure rate was noted to be 88% in the series by Hardy and 87% in Boggan's series; of course, these figures are applicable only to tumors confined to the sella turcica. The success rate drops significantly when the tumor is no longer confined to the sella (25–48%). It can therefore be emphatically stated that when the tumor is confined to the sella, TPS offers the patient the best chance of a cure. It is interesting to note that in all three series, the incidence of finding no obvious tumor during surgery was approximately 20%. Before discussing the neurosurgeon's options in such a situation, it is necessary to understand the exploratory strategy practiced by most neurosurgeons who perform microsurgery.[310] The initial incision is in the midline carried into the neurohypophysis. If an adenoma is not visualized, both the lateral wings should be inspected to detect a resectable adenoma. If no adenoma is seen in either location, a final surgical sweep of the pars intermedia is made to detect a tumor in that region. If the exploration is completely negative, and the surgeon has data from selective inferior petrosal sinus cathetherization that denote a unilateral source, then a hemipituitectomy on the side of the increased gradient is performed. If no catheterization data are available, the choice between a medial wedge resection and a total hypophysectomy is difficult, often resting with the individual neurosurgeon's prior experience. Since the incidence of "negative exploration" is approximately only 20%, in the majority of patients an adenoma will be found (even with no abnormalities on CT), which can be successfully resected. It is in these situations that a cure rate greater than 80% can be offered to patients by the experienced and expert neurosurgeon.

The efficacy of TPS in children and adolescents, a group normally treated with conventional radiation, has been established by the report of Styne et al.[311]

They treated 15 children and adolescents with CD with TPS and were able to achieve a cure in 14, with a negligible incidence of hypopituitarism.

The success of TPS can be evaluated within a few days after surgery. If the source of the pituitary ACTH was completely removed, the adrenals rapidly become hypofunctional. This is because the normal corticotropes have been rendered dormant owing to the suppressive effect of chronic hypercortisolism. It follows, then, that secondary hypoadrenalism (transient as it may be) is an anticipated reflection of successful resection of an ACTH-producing pituitary microadenoma.[22] In fact, if hypocortisolism does not develop imediately after TPS, it is unlikely that the patient is cured, despite normalization of UFC and restoration of normal suppressibility to dexamethasone.[309] Such patients tend to show a high rate of recurrence of hypercortisolism.[312] In anticipation of transient secondary hypoadrenalism patients undergoing TPS are covered with hydrocortisone starting on the day of surgery. The transient secondary hypoadrenal state caused by prolonged suppression of normal corticotropes lasts for a variable period of time, ranging from weeks to even months. Hypocortisolism (with a low ACTH level) in the postoperative period is the best prognostic parameter for predicting cure.[309]

The adverse effects of TPS are few. The morbidity associated with the procedure is very low. Transient diabetes insipidus is the most frequent complication. Less commonly cerebrospinal fluid leak, meningitis, and optic nerve damage may occur. The mortality, when it occurs, is more likely to be seen in patients with invasive tumors. The single most important factor that determines the outcome of surgery is the experience of the neurosurgeon. Such experience is invaluable in localizing the adenoma and in properly exploring the entire pituitary. When the exploration is negative, the correctness of the further decision (medial wedge resection, versus selection partial central hypophysectomy, versus total hypophysectomy) is entirely a reflection of the neurosurgeon's prior experience. It is, therefore, essential that the patient be referred to a center with an established track record.

There are several reasons for the persistence or recurrence of hypercortisolism following TPS. The three major reasons for such failure are when corticotrope hyperplasia underlies the disorder; when the tumor is invasive; and when tumors originate from the intermediate lobe to cause CD. There is a fourth reason—erroneous diagnosis, the most notorious example being ectopic ACTH secretion caused by an occult, relatively benign neoplasm mimicking the dynamics of pituitary-dependent CD.[235] *Corticotrope hyperplasia* can be the cause of pituitary-dependent CD in a minority of patients. There are no definite preoperative methods for making such a diagnosis. The existence of such an entity has been clearly documented in the literature.[2,313–315] The fact that corticotrope hyperplasia can coexist in addition to a discrete adenoma of the adenohypophysis can further confound the issue.[135] Although corticotrope hyperplasia can occur in a diffusely scattered form, clusters of corticotrope nests are especially found in the central posterior area of the adenohypophysis. The therapeutic choice for corticotrope hyperplasia has not been well established, the recommended options ranging from total hypophysectomy[315] to selective central partial hypophysectomy.[16]

The poor response rate of *intermediate-lobe tumors* to pituitary microsurgery has been recognized since Lamberts et al.[25] proposed the existence of two types of ACTH-secreting microadenomas; one from the adenohypophysis, and one originating from the intermediate lobe. Tumors that originate from the intermediate lobe are probably of neural origin and are histologically characterized by argyrophilic nerve fibers traversing the tumors, as well as by the hormonal triad of poor suppression to dexamethasone, responsiveness to bromocriptine, and hyperprolactinemia. If such patients could be identified preoperatively, the option for treatment with bromocriptine is a valid one to offer the patient, since surgical cure for this subset of patients is not gratifying. If the condition is identified during surgery, by the characteristic neurofibrillary fibers within the adenoma, the choice between extensive surgery and partial surgery with adjunctive bromocriptine therapy postoperatively is difficult to make with the current state of our knowledge.

The third reason for therapeutic failure with TPS is when the tumor is invasive. Lateral invasion of tumor is a well-recognized situation for noncure.[15] Under such circumstances the choices are between complete hypophysectomy and bilateral adrenalectomy. Each case should be individualized since the price to be paid here for the cure of CD is the exchange for one or more deficiency states.

Despite these factors, selective adenomectomy provides the best chance for a permanent cure. As our understanding of the disease progresses, and newer methods of identifying the "poor responders" become available, the success rate is likely to improve above the excellent chances of a cure for this disease.

Bilateral Adrenalectomy

This surgical method reverses hypercortisolism immediately, and for the most part permanently. The only instances of recurrence are when some adrenal tissue is inadvertently left behind; when the capsule is ruptured and clusters of cells are released into the retroperitoneal bed, where they "germinate" and gradually begin to hyperfunction ("retroperitoneal adrenal remnants"); or when adrenal rest cells in the gonads (usually testes) become functional under the chronic stimulatory influence of ACTH. The objections to bilateral total adrenalectomy as the *first* line of therapy for Cushing's disease are the following:

1. Removing the target organ(s) without removing the obvious seat of primary pathology is scientifically unsound strategy.
2. Permanent adrenal failure, with the need for lifelong replacement therapy, is a high price to pay at an age when selective microadenectomy of the pituitary is an established procedure.
3. The morbidity and mortality of bilateral adrenalectomy are significant. Most centers acknowledge a 5% mortality with this procedure. The postoperative complications include infection, bleeding, thromboembolic phenomena, and pancreatitis.
4. The most important objection to bilateral total adrenalectomy is the possibility of developing Nelson's syndrome.[316] This entity is characterized by the development of generalized hyperpigmentation, progressive enlargement of the sella turcica, and the development of visual-field cuts.

Hormonally, marked elevations of ACTH and β lipotropin–endorphin peptides are seen. Nelson's syndrome represents growth of a preexisting tumor, which is enhanced by adrenalectomy. Cook et al.[317] have demonstrated the restraining effect of glucocorticoids on ACTH during all stages of pituitary-dependent CD. When the suppressive effect of glucocorticoids is eliminated by removal of the adrenal, the ACTH secretion and the tumor growth are enhanced. In some cases the tumor growth is considerably accelerated, resulting in rapidly progressive, invasive tumors. The dynamic data in Nelson's syndrome are almost identical to those in pituitary-dependent CD. The only notable difference is that the ACTH of patients with Nelson's syndrome is often responsive to hypoglycemia.[5] The incidence of Nelson's syndrome following bilateral adrenalectomy for CD ranges from 8 to 35% and higher in children.[318] The incidence and the temporal aspects of the development of Nelson's syndrome have recently been reviewed.[319] Nelson's syndrome developed in 28% of 50 patients who had undergone bilateral adrenalectomy for CD. While the syndrome usually developed within 5 years, occasionally it was noted to develop as long as 12 years following surgery. More important, anaplastic changes in the tumor cells were noted in two patients.

While the triad of hyperpigmentation, marked ACTH elevations, and abnormal sellar dimensions is the usual presentation, occasionally Nelson's syndrome can manifest with recurrent hypercortisolism,[320–322] testicular tumors in males,[323,324] or paraovarian tumors in females[325,326] with virilization. These phenomena are a result of chronic stimulation of adrenal rests by the excessively produced ACTH. It is not clear whether Nelson's syndrome can be prevented by prior radiation to the pituitary. Once the syndrome has developed, pituitary surgery becomes mandatory.

The indications for bilateral adrenalectomy are

1. When the hypercortisolism is life threatening, and a rapid 100% effective modality is imminently required to correct it.
2. When pituitary surgery has failed to cure the patient's disease.
3. When hypercortisolism recurs after a period of remission following pituitary surgery.
4. When macronodular hyperplasia is suspected on the basis of hormonal studies or CT of the adrenals.

Pituitary Radiation

Conventional radiation to the pituitary gland is a time-honored therapeutic modality for CD. About 5000 rads are delivered to the pituitary gland over a 4–5-week period. The main advantage of the procedure is that it is relatively free from major side effects. Conventional radiation has enjoyed the reputation of being an innocuous form of therapy since it is devoid of complications such as hypopituitarism or optic nerve damage. The major drawback of conventional pituitary radiation, particularly in adults, is the relatively low cure rates attained. Orth and Liddle[4] evaluated the results of conventional radiation given to a large group of patients with pituitary-dependent CD. Of the 50 patients treated, complete cure was attained in only 10. Improvement was seen in another 13 patients. Results of therapy were seen from 6 to 18 months following radiation. The low cure rate, and the long delay in attaining a remission have been compelling reasons to minimize the enthusiasm for this therapy. Besides, in a long-term followup of pituitary function following megavoltage therapy for CD, Sharpe et al.[327] demonstrated the development of overt or subtle hypopituitarism in a small, but significant number. The results with conventional radiation to the pituitary are more encouraging in children and adolescents with the disease. In the report by Jennings et al.,[328] a cure was seen in 13 of 15 children with CD treated by conventional radiation. The lack of adverse effects, particularly hypopituitarism, is especially important in this group of patients where preservation of growth and sexual maturation are vital concerns while planning therapy. Considering this factor, and coupled with its high cure rate, conventional radiation is favored by many as the first option in children and adolescents with CD.

Proton beam (or heavy-particle) radiation is a highly effective method to treat pituitary-dependent CD. The major limitation of this modality is the nonavailability of the procedure in most centers and the higher incidence of complications such as hypopituitarism and oculomotor nerve palsies.

The value of prophylactic radiation to the pituitary gland (prior to bilateral adrenalectomy) in the prevention of Nelson's syndrome is controversial and unsettled.[4,318,329]

Neuropharmacotherapy

The role of neuropharmacological agents in the treatment of CD is at best adjunctive. The basis for the use of drugs that modulate ACTH secretion rests on the ability of certain drugs to impair or deplete monoamine concentrations in the hypothalamus. Hypothalamic release of CRH can be affected by neurotransmitter concentration in the hypothalamus. Thus inhibitors of serotonin (and in some instances dopamine agonists) are capable of lowering CRH and indirectly ACTH concentrations. The notion that these drugs exert their action solely on the hypothalamus must be reexamined in view of in vitro data that show a direct inhibitory effect of these drugs on the adenoma cells maintained in culture.[252] In addition to serotonin antagonists and dopaminergic agonists, noreprinephrine depletors as well as agonists to gamma-aminobutyric acid (GABA) have extended the neuropharmacological armamentarium. Several general observations are pertinent in this regard:

1. Although the basis for drug therapy is solidly based on the abnormal hormone dynamics that characterize pituitary-dependent CD, these data cannot be transposed to all patients with CD. The heterogeneity of this disease precludes predictability of response in an individual case.
2. Even when drug therapy works, as exemplified by cyproheptadine, discontinuation of the drug is usually attended by a return of hypercortisolism. Recommendation of protracted, possibly lifelong, therapy is not justified when more definitive cures are available.
3. Side effects of the drugs employed should be considered when long-term therapy is contemplated.

4. Definitive therapy should not be deferred while waiting for a drug-induced remission that may never come.

Cyproheptadine is the most notable drug used to treat pituitary-dependent CD. Much of our knowledge regarding this drug is due to the pioneering studies of Krieger et al.[330,331] Administration of cyproheptadine is attended by lowering of cortisol and ACTH levels, restoration of suppressibility to dexamethasone, and even a clinical remission in some patients with pituitary-dependent CD. The drug has to be administered in large doses (24 mg/day), with the expectation of no more than a 50% chance of remission, which lasts only as long as the drug is continued. There are no definite methods of prospectively identifying the patients who are unlikely to respond. The most serious and undesirable effect of cyproheptadine therapy is weight gain.

The use of bromocriptine is likely to benefit patients with CD originating from the intermediate lobe. Although the suppressive effect of dopamine agonists is well established in such patients,[20,25,332] the success rate of chronic dopamine agonist therapy has not been clearly established. Similarly, the use of reserpine[333] has not been well established.

Several newer drugs have entered the scene in the recent past. Valproate sodium, a GABA agonist, has been shown to be effective in lowering the ACTH level in some patients with CD and Nelson's syndrome.[334,335] The effectiveness of treatment of CD with sodium valproate has been noted by Jones et al.[336] The drug has also been successfully employed to lower ACTH levels in patients with Nelson's syndrome.[337] Adrenolytic agents (e.g., o p' DDD, aminoglutethimide, trilostane) are used in the preoperative preparation of the patient when the clinical expression is severe. Most authorities who perform TPS do not employ these agents routinely in the preoperative management.[309] Prolonged use of ketoconazole in the management of CD has been described by Sonino et al.[338] to correct the hypercortisolism quickly, following unsuccessful TPS.

Adrenal Adenoma

The treatment for patients with Cushing's syndrome caused by a unilateral solitary adrenal adenoma is unilateral adrenalectomy. Adrenal surgery is usually followed by a complete recovery of the hormonal and clinical features of Cushing's syndrome, with gradual functional restoration of the contralateral atrophied adrenal cortex. Välimäki et al.[339] have demonstrated that recovery of the contralateral suppressed adrenal cortex can be a slow process in some patients who undergo adrenal surgery for adenoma, taking as long as 28 months. During this period replacement with hydrocortisone is essential. Testing of adrenal–pituitary function often reveals that recovery of ACTH antedates recovery from adrenal dormancy. It is also noteworthy that while most of the expressions of hypercortisolism disappear, obesity, hypertension, and the bone disease of hypercortisolism may remain as residual stigmata even after successful removal of the adenoma.

When the hypercortisolism is severe, preoperative reduction in the cortisol levels may permit surgical ease during the adrenal surgery. This can be attained by one of several drugs that inhibit steroidogenesis. While o p' DDD, aminoglutethimide, or metyrapone has traditionally been employed for the purpose, newer agents such as ketoconazole, trilostane, and the glucocorticoid-receptor antagonist RU 486 merit investigation for this purpose.

Bilateral Macronodular Hyperplasia

The treatment for this variant of Cushing's syndrome is controversial, because its pathogenesis is not well understood. Since varying degrees of adrenal autonomy are usually present, bilateral adrenalectomy has been advocated. This is especially so in patients with primary adrenocortical nodular dysplasia, a condition that responds only to bilateral adrenalectomy. Patients with the more common variety of macronodular hyperplasia deserve close follow-up to detect the evolution of Nelson's syndrome following bilateral adrenalectomy. It is not certain whether patients with macronodular hyperplasia need pituitary radiation or surgery following bilateral adrenalectomy. The relative rarity of this entity, coupled with its ambiguous pathogenesis, has made bilateral adrenalectomy the initial mode of therapy.

The options for the patient with Cushing's syndrome due to unilateral adenomas on one side with hyperplasia on the other are even more uncertain. The choices range between bilateral adrenalectomy alone, unilateral adrenalectomy coupled with additional treatment directed at the pituitary, and bilateral adrenalectomy with therapy for pituitary ACTH hypersecretion. The choice depends on the certainty with which hormonal and radiological evidence for excess ACTH secretion can be demonstrated. When excess or inappropriate ACTH secretion is evident, the possibility that the adrenal adenoma represents a "tertiary" phenomenon is more plausible. If unilateral adrenalectomy with treatment directed to the pituitary is chosen as the option, careful monitoring of glucocorticoid function is mandatory. Should the hypercortisolemia persist or recur, a second adrenal surgery would become necessary to remove the source of hypercortisolism from the hyperplastic gland.

Adrenal Carcinoma

The treatment for this devastating disease is discussed separately in Chapter 10.

Adrenolytic Therapy

The successful treatment of Cushing's syndrome depends on specific therapy directed against the etiology of hypercortisolism. However, even when the cause can be clearly identified, it is desirable, in many instances, to lower cortisol levels by drug therapy in preparation for surgery, or while awaiting the response to pituitary radiation. Most of the drugs used for this purpose exert their antisteroidal effects by blocking one or more enzymes involved in the biosynthetic steps in adrenal steroidogenesis. The three prototypical drugs that have traditionally been employed for such a purpose are o p' DDD, aminoglutethimide, and metyrapone. Although "medical adrenalectomy" by means of these drugs effectively interrupts adrenal steroidogenesis, this does not supplant the use of definitive therapy aimed at correcting the etiology of Cushing's syndrome. In addition to these three drugs, the arrival of newer drugs on the scene has facilitated pharmacological control of hypercortisolism. These include ketoconazole, which has been impressively effective; trilostane, which has received mixed reviews; and the steroid-receptor antagonist RU 486, which is still experimental at the time of writing.

o p' DDD (o p'-Dichlorodiphenyl Dichloroethane).   o p' DDD (Mitotane) in a daily dose of 0.5–3 g blocks adrenal 11-β-hydroxylation and thereby reduces production of cortisol. In addition, the drug has a cytolytic action and causes alterations in the mitochondrial morphology of the zona fasiculata. In doses exceeding 3 g daily, o p' DDD causes mineralocorticoid deficiency in addition to glucocorticoid deficiency; chronic therapy with o p' DDD results in adrenal atrophy. In addition to its effects on adrenal steroidogenesis, the drug affects the peripheral metabolism of cortisol and androgens, resulting in increased excretion of unconjugated 6-β-OH derivatives. This is the reason for the impressive reduction in the urine 17-OHCS during the early weeks of o p' DDD therapy despite a lack of changes in the cortisol secretory rate or the plasma cortisol level.

Although o p' DDD deposits in several tissues (such as subcutaneous fat, liver, brain, and ovaries) following absorption, it usually does not affect the function of these organs. This broad specificity of binding may be responsible for the long-lasting action of the drug even after withdrawal. The usual side effects of the drug include nausea, vomiting, somnolence, weakness, and dermatitis. The gastrointestinal adverse effects are the main reason for poor patient acceptance. Owing to these side effects, the drug is started in lower doses and gradually increased. The maximal adrenolytic effect of o p' DDD is usually attained in 8 weeks following high-dose therapy (6–12 g/day). At this point most patients on therapy require substitutive gluco- and mineralocorticoid therapy.

While o p' DDD has mostly been used in the palliative therapy of adrenal carcinoma, especially metastatic, it has also been used to treat pituitary-dependent CD. Luton et al.[340] used high-dose o p' DDD therapy in 62 patients with CD for an average of 8 months and attained an impressive remission in 38 of the 46 patients who received the drug alone and in all 16 who received the drug in combination with cobalt irradiation to the pituitary. Of course, 60% of these patients subsequently relapsed, but responded to an additional course of o p' DDD. Couzinet et al.[341] have reported durable remission of CD after treatment with o p' DDD. When tolerated, o p' DDD can serve as excellent adjunctive therapy to lower adrenal glucocorticoid production regardless of the etiology of Cushing's syndrome.

Aminoglutethimide.   Aminoglutethimide (α-ethyl-α-p-aminophenyl glutarimide) decreases cortisol production by inhibiting conversion of cholesterol to pregnenolone. However, this effect can be overridden and overcome by ACTH. Thus, when subjects with an intact hypothalamic–pituitary–adrenal axis are given aminoglutethimide, the compensatory increase in ACTH prevents the development of hypocortisolemia. In contrast, patients with Cushing's syndrome caused by tumors demonstrate significant blocking of adrenal steroidogenesis by aminoglutethimide, since their pituitary ACTH level is suppressed. Interestingly, aldosterone secretion declines in all patients placed on aminoglutethimide, suggesting that the drug has multiple loci of action. It is believed that aminoglutethimide also inhibits 11-β-hydroxylase and perhaps inhibits 3-β-hydroxysteroid dehydrogenase and 18-hydroxylase as well. In addition to its inhibitory effects on adrenocortical biosynthesis, aminoglutethimide also affects testicular (and possibly ovarian) steroidogenesis.

Aminoglutethimide can be used as an interim measure for reducing cortisol levels in patients with Cushing's syndrome. A 50–60% reduction in baseline cortisol levels can be attained with the use of aminoglutethimide in patients with all forms of Cushing's syndrome. However, an attenuation in the effectiveness of the drug is often seen in patients with pituitary ACTH-dependent CD ("escape phenomenon"). The adverse side effects of the drug include drowsiness, anorexia, skin rash, and dizziness. These effects often spontaneously resolve with continuation of therapy. Patient acceptance of aminoglutethimide is more than with o p′ DDD. Supplementation of glucocorticoid and mineralocorticoid therapy is essential as adrenal function declines with aminoglutethimide treatment. Unlike o p′ DDD, which has the potential for causing permanent adrenal failure, aminoglutethimide treatment results in only transient hypoadrenalism.

Metyrapone.   Since its introduction in 1958, metyrapone has been widely used as a diagnostic agent to test the ability of the pituitary to respond to declining concentrations of cortisol. The drug has also found application in lowering the metabolically active cortisol concentrations in patients with Cushing's syndrome caused by adrenal tumors and ectopic ACTH-secreting tumors. Amelioration of the hypokalemia and hyperglycemia of patients with the ectopic ACTH syndrome treated with metyrapone is well known.[342] The dose employed is usually 500–750 mg three to four times a day. Metyrapone therapy is especially useful for palliative purposes in this setting, since many patients with the metabolic abnormalities of the ectopic ACTH syndrome are too ill for other forms of treatment. Metyrapone has also been used in the treatment of pituitary-dependent CD, a paradoxical observation in light of the fact that ACTH levels would be expected to increase and overcome the block in 11-β-hydroxylation induced by metyrapone. A possible explanation may be that in addition to causing a block in 11-β-hydroxylation the drug also blocks the side-chain cleavage of cholesterol, inhibiting the formation of pregnenolone. Regardless of the mechanism(s), metyrapone has been used in pituitary-dependent CD with varying degrees of success.[343,344]

Newer Drugs Employed in Treatment of Cushing's Syndrome.   Ketoconazole, a broad-spectrum antifungal agent, can cause inhibition of adrenal steroidogenesis. The mechanisms involved in ketochonazole-induced hypoadrenalism are outlined in chapter 2. This adverse effect of ketoconazole has been exploited to treat hypercortisolism. Loli et al.[345] treated seven patients with CD and one with adrenal adenoma with 600–800 mg/day of ketoconazole for 3–13 months. Rapid and persistent improvement in the clinical and hormonal features of hypercortisolism was noted. An interesting observation was that there were no significant differences in ACTH levels or in the response of ACTH to CRH before or after treatment with ketoconazole. The beneficial effects of prolonged treatment of recurrent, severe CD with ketoconazole have been described by other workers.[338] The palliative use of ketoconazole in patients with ectopic ACTH syndrome has broadened the indications for the use of ketoconazole in patients with hypercortisolism.[71]

Trilostane is an agent that, in vitro, effectively blocks the enzyme 3-β-hydroxysteroid dehydrogenase. Its efficacy in patients with Cushing's syndrome has not been satisfactorily evaluated in large numbers of patients. The variability in the reports in the literature[346,347] regarding the efficacy of trilostane may be related to individual variabilities in patient response caused by in vivo metabolism of trilostane. The paucity of clinical trials has limited the widespread use of trilostane in Cushing's syndrome.

Finally, a clinically applicable antagonist of glucocorticoids can be a useful modality to block the peripheral efects of cortisol. A new experimental drug, RU 486, has been shown to antagonize the peripheral actions of glucocorticoids in humans.[348,349] Neiman et al.[350] treated a patient with ectopic ACTH syndrome by using oral RU 486 for 9 weeks. In addition to causing marked subjective and objective improvement, the drug led to complete normalization of all glucocorticoid-sensitive parameters. The drug was well tolerated, with very few adverse effects. If proven in larger series, competitive inhibition of glucocorticoids may become an important therapeutic strategy in the management of patients with Cushing's syndrome.

References

1. Cushing H: The basophillic adenomas of the pituitary body and their clinical manifestations. *Bull Johns Hopkins Hosp* **50:**137, 1932.
2. Taylor HC, Velasco ME, Brodkey JS: Remission of pituitary dependent Cushing's disease after removal of nonneoplastic pituitary gland. *Arch Intern Med* **140:**1366, 1980.
3. Liddle GW: Tests of pituitary–adrenal suppressibility in the diagnosis of Cushing's syndrome. *J Clin Endocrinol* **20:**1539, 1960.
4. Orth DN, Liddle GW: Results of treatment in 108 patients with Cushing's syndrome. *N Engl J Med* **285:**243, 1971.
5. Krieger DT, Luria M: Plasma ACTH and cortisol responses to TRF, vasopressin or hypoglycemia in Cushing's disease and Nelson's syndrome. *J Clin Endocrinol Metab* **44:**361, 1977.
6. Krieger DT: Medical treatment of Cushing's disease. In Tolis G (ed): *Clinical Neuroendocrinology: A Pathophysiological Approach.* Raven Press, New York, 1979, pp. 423–427.
7. Pieters GFFM, Smals AGH, Benraad TJ, et al: Plasma cortisol response to thyrotropin-releasing hormone and luteinizing hormone-releasing hormone in Cushing's disease. *J Clin Endocrinol Metab* **48:**874, 1979.
8. Aronin N, Krieger DT: Sustained remission of Nelson's syndrome after stopping cyproheptadine treatment. *N Engl J Med* **302:**453, 1980.
9. Lankford HV, Tucker HSG, Blackard WG: A cyproheptadine-reversible defect in ACTH control persisting after removal of the pituitary tumor in Cushing's disease. *N Engl J Med* **305:**1244, 1981.
10. Lamberts SWJ, Klijn JGM, de Quijada M, et al: The mechanism of the suppressive action of bromocriptine on adrenocorticotropin secretion in patients with Cushing's disease and Nelson's syndrome. *J Clin Endocrinol Metab* **51:**307, 1980.
11. Chrousos GP, Schulte HM, Oldfield EH, et al: The corticotropin-releasing factor stimulation test—An aid in the evaluation of patients with Cushing's syndrome. *N Engl J Med* **310:**622, 1984.
12. Carey RM, Varma SK, Drake CR Jr, et al: Ectopic secretion of corticotropin releasing factor as a cause of Cushing's syndrome. *N Engl J Med* **311:**13, 1984.
13. Liddle GW: Pathogenesis of glucocorticoid disorders. *Am J Med* **53:**638, 1972.

14. Krieger DT, Glick SM: Sleep EEG stages and plasma growth hormone concentration in states of endogenous and exogenous hypercortisolemia or ACTH elevation. *J Clin Endocrinol Metab* **39:**986, 1974.
15. Boggan JE, Tyrrell JB, Wilson CB: Transsphenoidal microsurgical management of Cushing's disease. *J Neurosurg* **59:**195, 1983.
16. Hardy J: Cushing's disease: 50 year later. *Can J Neurol Sci* **9:**375, 1982.
17. Ludecke DK, Schabet M, Saeger W: In vitro secretion of adenoma and anterior lobe cells in two typical cases of Cushing's disease. *Neurosurgery* **12:**549, 1983.
18. Ludecke DK, Westphal M, Schabet M, et al: In vitro secretion of ACTH, β-endorphin and β-lipotropin in Cushing's disease and Nelson's syndrome. *Hormone Res* **13:**259, 1980.
19. Suda T, Tozawa F, Mouri T, et al: Effects of cyproheptadine, reserpine, and synthetic corticotropin-releasing factor on pituitary glands from patients with Cushing's disease. *J Clin Endocrinol Metab* **56:**1094, 1983.
20. Lamberts SWJ, Klijn JGM, Quijada M, et al: The mechanism of the suppressive action of bromocriptine on adrenocorticotropin secretion in patients with Cushing's disease and Nelson's syndrome. *J Clin Endocrinol Metab* **51:**307, 1980.
21. Ishibashi M, Yamaji T: TRH stimulation and dopaminergic and antiserotonergic inhibition of ACTH release from cultured pituitary adenoma tissues of Nelson's syndrome. Program of the 6th Internal Congress of Endocrinology, Melbourne, Australia, 1980, p. 407.
22. Fitzgerald PA, Aron DC, Findling JW: Cushing's disease: Transient secondary adrenal insufficiency after selective removal of pituitary microadenomas; evidence for a pituitary origin. *J. Clin Endocrinol Metab* **54:**413, 1982.
23. Jeffcoate WJ, Dauncey S, Selby C: Restoration of dexamethasone suppression by incomplete adenomectomy in Cushing's disease. *Clin Endocrinol* **23:**193, 1985.
24. Van Cauter E, Refetoff S: Evidence for two subtypes of Cushing's disease based on the analysis of episodic cortisol secretion. *N Engl J Med* **313:**1343, 1985.
25. Lamberts SWJ, De Lange SA, Stefanko SZ: Adrenocorticotropin-secreting pituitary adenomas originate from the anterior or the intermediate lobe in Cushing's disease: Differences in the regulation of hormone secretion. *J Clin Endocrinol Metab* **54:**286, 1982.
26. Orth DN, Holscher MA, Wilson MG, et al: Equine Cushing's disease: Plasma immunoreactive proopiolipomelanocortin peptide and cortisol levels basally and in response to diagnostic tests. *Endocrinology* **110:**1430, 1982.
27. Symington T: *Functional Pathology of the Human Adrenal Gland.* Williams & Wilkins, Baltimore, 1969, p. 151.
28. Hedeland H, Östberg G, Hökfelt B: On the prevalence of adrenocortical adenomas in an autopsy material in relation to hypertension and diabetes. *Acta Med Scand* **184:**211, 1968.
29. Carbonnel B, Chatal JF, Ozanne P: Does the corticoadrenal adenoma with "pre-Cushing's syndrome" exist? *J Nucl Med* **22:**1059, 1981.
30. Beyer HS, Doe RP: Cortisol secretion by an incidentally discovered nonfunctional adrenal adenoma. *J Clin Endocrinol Metab* **62:**1317, 1986.
31. Bertagna C, Orth DN: Clinical and laboratory findings and results of therapy in 58 patients with adrenocortical tumors admitted to a single medical center (1951 to 1978). *Am J Med* **71:**855, 1981.
32. Schorr I, Ney RL: Abnormal hormone responses of an adrenocortical cancer adenyl cyclase. *J Clin Invest* **50:**1295, 1971.
33. Matsukura S, Kakita T, Sueoka S, et al: Multiple hormone receptors in the adenylate cyclase of human adrenocortical tumors. *Cancer Res* **40:**3768, 1980.
34. Hirata Y, Uchihashi M, Sueoka S, et al: Presence of ectopic β-adrenergic receptors on human adrenocortical cortisol-producing adenomas. *J Clin Endocrinol Metab* **53:**953, 1981.
35. Joffe SN, Brown C: Nodular adrenal hyperplasia and Cushing's syndrome. *Surgery* **94:**919, 1983.
36. Donaldson MDC, Grant DB, O'Hare MJ, et al: Familial congenital Cushing's syndrome due to bilateral nodular adrenal hyperplasia. *Clin Endocrinol* **14:**519, 1981.
37. Ruder HJ, Loriaux DL, Lipsett MB: Severe osteopenia in young adults associated with Cushing's syndrome due to micronodular adrenal disease. *J Clin Endocrinol Metab* **39:**1138, 1974.
38. Levin ME: The development of bilateral adenomatous adrenal hyperplasia in a case of Cushing's syndrome of eighteen years' duration. *Am J Med* **40:**318, 1966.

39. Schweizer-Cagianut M, Froesch ER, Hedinger C: Familial Cushing's syndrome with primary adrenocortical microadenomatosis (primary adrenocortical nodular dysplasia). *Acta Endocrinol* **94:**529, 1980.

40. Meador CK, Bowdoin B, Owen WC, et al: Primary adrenocortical nodular dysplasia: A rare cause of Cushing's syndrome. *J Clin Endocrinol Metab* **27:**1255, 1967.

41. Choi Y, Werk EE Jr, Sholiton LJ: Cushing's syndrome with dual pituitary-adrenal control. *Arch Intern Med* **125:**1045, 1970.

42. May PB, Sobel H, Schneider G, et al: Nodular adrenal hyperplasia with elevated adrenocorticotropic hormone levels. *Arch Intern Med* **143:**136, 1983.

43. Smals AGH, Pieters GFFM, van Haelst UJG, et al: Macronodular adrenocortical hyperplasia in long-standing Cushing's disease. *J Clin Endocrinol Metab* **58:**25, 1984.

44. Aron DC, Findling JW, Fitzgerald PA, et al: Pituitary ACTH dependency of nodular adrenal hyperplasia in Cushing's syndrome: Report of two cases and review of the literature. *Am J Med* **71:**302, 1981.

45. Schteingart DE, Tsao HS: Coexistence of pituitary adrenocorticotropin-dependent Cushing's syndrome with a solitary adrenal adenoma. *J Clin Endocrinol Metab* **50:**961, 1980.

46. Burke CW, Beardwell CG: Cushing's syndrome. *Q J Med* **42:**175, 1973.

47. Josse RG, Bear R, Kovacs K, et al: Cushing's syndrome due to unilateral nodular hyperplasia: A new pathophysiological entity. *Acta Endocrinol (Copenh)* **93:**495, 1980.

48. Kirschner MA, Powell RD, Lipsett MB: Cushing's syndrome: Nodular cortical hyperplasia of adrenal glands with clinical and pathological features suggesting adrenocortical tumor. *J Clin Endocrinol Metab* **24:**947, 1964.

49. Katz J: Failure of dexamethasone suppression in adrenal hyperplasia. *Arch Intern Med* **118:**265, 1966.

50. Anderson DC, Child DF, Sutcliffe CH, et al: Cushing's syndrome, nodular adrenal hyperplasia and virilizing carcinoma. *Clin Endocrinol* **9:**1, 1978.

51. Dluhy RG, Barlow JJ, Mahoney EM, et al: Profile and possible origin of an adrenocortical carcinoma. *J Clin Endocrinol Metab* **33:**312, 1971.

52. Hamwi GJ, Serbin RA, Kruger FA: Does adrenal cortical hyperplasia result in adrenocortical carcinoma? *N Engl J Med* **257:**1153, 1957.

53. Chute AL, Robinson GC, Donahue WL: Cushing's syndrome in children. *J Pediatr* **34:**20, 1949.

54. McArthur RG, Bahn RC, Hayles AB: Primary adrenocortical nodular dysplasia as a cause of Cushing's syndrome in infants and children. *Mayo Clin Proc* **57:**58, 1982.

55. Larsen JL, Cathey WJ, Odell WD: Primary adrenocortical nodular dysplasia, a distinct subtype of Cushing's syndrome. Case report and review of the literature. *Am J Med* **80:**976, 1986.

56. Arce B, Licea M, Hung S, et al: Familial Cushing's syndrome. *Acta Endocrinol* **87:**139, 1978.

57. Bohm N, Lippman-Grob B, Pitrykowski WV: Familial Cushing's syndrome due to pigmented multinodular adrenocortical dysplasia. *Acta Endocrinol* **102:**428, 1983.

58. Donaldson MDC, Grant DB, O'Hare MJ, et al: Familial congenital Cushing's syndrome due to bilateral nodular adrenal hyperplasia. *Clin Endocrinol* **14:**519, 1981.

59. Shenoy BV, Carpenter PC, Carney JA: Bilateral primary pigmented nodular adrenocortical disease: Rare cause of the Cushing's syndrome. *Am J Surg Pathol* **8:**335, 1984.

60. Carney JA, Gordon H, Carpenter P, et al: The complex of myxomas, spotty pigmentation, and endocrine overactivity. *Medicine (Baltimore)* **64:**270, 1985.

61. Carney JA, Hruska LS, Beauchamp GD, et al: Dominant inheritance of the complex of myxomas, spotty pigmentation, and endocrine overactivity. *Mayo Clin Proc* **61:**165, 1986.

62. Proppe KH, Scully RE: Large cell calcifying Sertoli cell tumor of the testis. *Am J Clin Pathol* **74:**607, 1980.

63. Van Berkhout FT, Croughs RJM, Kater L, et al: Familial Cushing's syndrome due to nodular adrenocortical dysplasia. A putative receptor-antibody disease? *Clin Endocrinol* **24:**299, 1986.

64. Christy NP: Adrenocorticotrophic activity in the plasma of patients with Cushing's syndrome associated with pulmonary neoplasms. *Lancet* **1:**85, 1961.

65. Liddle GW, Givens JR, Nicholson WE, et al: The ectopic ACTH syndrome. *Cancer Res* **25:**1057, 1965.

66. Meador CK, Liddle GW, Island DP, et al: Cause of Cushing's syndrome in patients with tumors arising from "nonendocrine" tissue. *J Clin Endocrinol Metab* **22:**693, 1962.

67. Holub DA, Katz FH: A possible etiologic link between Cushing's syndrome and visceral malignancy. *Clin Res* **9:**194, 1961.

68. Brown WH: A case of pluriglandular syndrome: "Diabetes of bearded women." *Lancet* **2:**1022, 1928.

69. Bornstein P, Nolan JP, Bernanake D: Adrenocortical hyperfunction in association with anaplastic carcinoma of the respiratory tract. *N Engl J Med* **264:**363, 1961.

70. Egan RT, Maurer LH, Forcier RJ, et al: Small cell carcinoma of the lung. *Cancer* **37:**527, 1974.

71. Shepherd FA, Hoffert B, Evans WK, et al: Ketoconazole: Use in the treatment of ectopic adrenocorticotropic hormone production and Cushing's syndrome in small-cell lung cancer. *Arch Intern Med* **145:**863, 1985.

72. Allott EN, Skelton MO: Increased adrenocortical activity associated with malignant disease. *Lancet* **2:**278, 1960.

73. DeStephano DB, Lloyd RV, Schteingart DE: Cushing's syndrome produced by a bronchial carcinoid tumor. *Hum Pathol* **15:**890, 1984.

74. Strott CA, Nugent CA, Tyler FH: Cushing's syndrome caused by bronchial adenomas. *Am J Med* **44:**97, 1968.

75. Morse WI, Kerenyi N, Nelson DH: Prolonged hyperadrenocorticortrophism and pigmentation associated with bronchial carcinoid tumour. *Can Med Assoc J* **96:**104, 1967.

76. Spirn PW, Godleski JJ, Dluhy RG, et al: Young woman with Cushing's syndrome. *Invest Radiol* **17:**336, 1982.

77. O'Riordan JLH, Blanshard GP, Moxham A, et al: Corticotrophin-secreting carcinomas. *Q J Med* **35:**137, 1966.

78. Brown LR, Aughenbaugh GL, Wick MR, et al: Roentgenologic diagnosis of primary corticotropin-producing carcinoid tumors of the mediastinum. *Radiology* **142:**143, 1982.

79. Melone CR, Tucci J, Canary JJ, et al: Cushing's syndrome due to bilateral adrenocortical hyperplasia caused by a benign adrenal medullary tumor. *J Clin Endocrinol Metab* **26:**1192, 1966.

80. Schteingart DE, Conn JW, Orth DN, et al: Secretion of ACTH and B-MSH by an adrenal medullary paraganglioma. *J Clin Endocrinol Metab* **34:**676, 1972.

81. Apple D, Kreines K: Case report. Cushing's syndrome due to ectopic ACTH production by a nasal paraganglioma. *Am J Med Sci* **283**(1):32, 1982.

82. Huntrakoon M, Lin F, Heitz PU, et al: Thymic carcinoid tumor with Cushing's syndrome: Report of a case with electron microscopic and immunoperoxidase studies for neuron-specific enolase and corticotropin. *Arch Pathol Lab Med* **108:**551, 1984.

83. Thorner MO, Martin WH, Ragan GE, et al: A case of ectopic ACTH syndrome: Diagnostic difficulties caused by intermittent hormone secretion. *Acta Endocrinol* **99:**364, 1982.

84. Ghali VS, Garcia RL: Prostatic adenocarcinoma with carcinoidal features producing adrenocorticotropic syndrome. Immunohistochemical study and review of the literature. *Cancer* **54:**1043, 1984.

85. Lyons DF, Eisen BR, Clark MR, et al: Concurrent Cushing's and Zollinger–Ellison syndromes in a patient with islet cell carcinoma. Case report and review of the literature. *Am J Med* **76:**729, 1984.

86. Maton PN, Gardner JD, Jensen RT: Cushing's syndrome in patients with the Zollinger–Ellison syndrome. *N Engl J Med* **315:**1, 1986.

87. Melvin KEW: Tashjian AH Jr, Cassidy CE, et al: Cushing's syndrome caused by ACTH- and calcitonin-secreting medullary carcinoma of the thyroid. *Metabolism* **19:**831, 1970.

88. Rosenberg EM, Hahn TJ, Orth DN, et al: ACTH-producing medullary carcinoma of the thyroid presenting as severe idiopathic osteoporosis and senile purpura: Report of a case and review of the literature. *J Clin Endocrinol Metab* **47:**255, 1978.

89. Lamberts SWJ, Hackeng WHL, Visser TJ: Dissociation and association between calcitonin and adrenocorticotropin secretion. *J Clin Endocrinol Metab* **50:**565, 1980.

90. Takai S, Ogihara T, Miyachi A, et al: A case of medullary thyroid carcinoma with ectopic ACTH syndrome. *Folia Endocrinol Jpn* **53:**1279, 1977.

91. Shimatsu A, Kato Y, Tanaka I, et al: Plasma calcitonin and ACTH responses to lysine vasopressin, calcium and pentagastrin in a patient with medullary thyroid carcinoma associated with Cushing's syndrome. *Clin Endocrinol* **18:**119, 1983.

92. Himsworth RL, Bloomfield GA, Coombes RC, et al: "Big ACTH" and calcitonin in an ectopic hormone secreting tumour of the liver. *Clin Endocrinol* **7**:45, 1977.

93. Gerwirtz G, Yalow RS: Ectopic ACTH production in carcinoma of the lung. *J Clin Invest* **53**:1022, 1974.

94. Bishop MC, Ross EJ: Adrenocortical activity in disseminated malignant disease in relation to prognosis. *Br J Cancer* **24**:719, 1970.

95. Ratter SJ, Gillies G, Hope J, et al: Pro-opiocortin related peptides in human pituitary and ectopic ACTH secreting tumours. *Clin Endocrinol* **18**:211, 1983.

96. Pedersen RC, Brownie AC: Adrenocortical response to corticotropin is potentiated by part of the amino terminal region of pro-corticotropin-endorphin. *Proc Natl Acad Sci USA* **77**:2239, 1980.

97. Pedersen RC, Brownie AC: Proadenocorticotropin/endorphin derived peptides: Coordinate action on adrenal steroidogenesis. *Science* **208**:1044, 1980.

98. Upton GV, Amatruda TT Jr: Evidence for the presence of tumor peptides with corticotropin-releasing-factor-like activity in the ectopic ACTH syndrome. *N Engl J Med* **285**:419, 1971.

99. Suda T, Demura H, Demura R: Corticotropin-releasing factor-like activity in ACTH-producing tumors. *J Clin Endocrinol Metab* **44**:440, 1977.

100. Yamamoto H, Hirata Y, Matsukura S, et al: Studies on ectopic ACTH-producing tumors. *ACTA Endocrinol (Copenh)* **82**:183, 1976.

101. Hashimoto K, Takahara J, Ogawa N, et al: Adrenocorticotropin, β-lipotropin, β-endorphin and corticotropin-releasing factor-like activity in an adrenocorticotropin-producing nephroblastoma. *J Clin Endocrinol Metab* **50**:461, 1980.

102. Wakabayashi I, Ihara T, Hattori M, et al: Presence of corticotropin-releasing factor-like immunoreactivity in human tumours. *Cancer* **55**:995, 1985.

103. Suda T, Kondo M, Totani R, Hashimoto N, et al: Ectopic adrenocorticotropin syndrome caused by lung cancer that responded to corticotropin-releasing hormone. *J Clin Endocrinol Metab* **63**:1047, 1986.

104. Suda T, Tomori N, Tozawa F, et al: Immunoreactive corticotropin-releasing factor in human hypothalamus, adrenal, lung cancer and pheochromocytoma. *J Clin Endocrinol Metab* **58**:919, 1984.

105. Belsky JL, Cuello B, Swanson LW, et al: Cushing's syndrome due to ectopic production of corticotropin-releasing factor. *Clin Endocrinol Metab* **60**:496, 1985.

106. Zarate A, Kovacs K, Flores M, et al: ACTH and CRF-producing bronchial carcinoid associated with Cushing's syndrome. *Clin Endocrinol* **24**:523, 1986.

107. Asa SL, Kovacs K, Tindall GT, et al: Cushing's disease associated with an intrasellar gangliocytoma producing corticotrophin-releasing factor. *Ann Intern Med* **101**:789, 1984.

108. Schteingart DE, Lloyd RV, Akil H, et al: Cushing's syndrome secondary to ectopic corticotropin-releasing hormone-adrenocorticotropin secretion. *J Clin Endocrinol Metab* **63**:770, 1986.

109. Howlett TA, Price J, Hale AC, et al: Pituitary ACTH dependent Cushing's syndrome due to ectopic production of a bombesin-like peptide by a medullary carcinoma of the thyroid. *Clin Endocrinol* **22**:91, 1985.

110. Neumann PE, Horoupian DS, Goldman JE, et al: Cytoplasmic filaments of Crooke's hyaline change belong to the cytokeratin class. *Am J Pathol* **116**:214, 1984.

111. Felix IA, Horvath E, Kovacs K: Crooke's hyalinization in corticotroph cell adenomas of the human pituitary: A histological, immunocytological and electron microscopic study of three cases. *Acta Neurochir* **58**:235, 1981.

112. Neville MA, MacKay AM: The structure of the human adrenal cortex in health and disease. *Clin Endocrinol Metab* **1**:361, 1976.

113. Flattet A, Heidinger C: La morphologic de la cortico-surrenale dans le syndrome de Cushing. *Schweiz Med Wochenschr* **110**:1300, 1980.

114. Bahu RM, Battifora H, Schambaugh G: Functional black adenoma of the adrenal gland. *Arch Intern Med* **98**:139, 1974.

115. Zaniewski M, Sheeler LR: Cushing's syndrome associated with functional black adenoma of the adrenal cortex. *South Med J* **73**:1410, 1980.

116. Visser JW, Boeijingar JK, Meer CV: A functioning black adenoma of the adrenal cortex: A clinico-pathological entity. *J Clin Pathol* **27**:955, 1974.

117. Komiya I, Takasu N, Aizawa T, et al: Black (or brown) adrenal cortical adenoma: Its characteristic features on computed tomography and endocrine data. *J Clin Endocrinol Metab* **61**:711, 1985.

118. Ross EJ, Marshall JP, Friedman M: Cushing's syndrome: Diagnostic criteria. *Q J Med* **35**:149, 1966.

119. Urbanic RC, George JM: Cushing's disease—18 years' experience. *Medicine* **60**:14, 1981.

120. Ross J, Linch DC: Cushing's syndrome—Killing disease: Discriminatory value of signs and symptoms aiding early diagnosis. *Lancet* **2**:646, 1982.

121. Biglieri EG, Hane S, Slaton PE Jr, et al: In vivo and in vitro studies of adrenal secretions in Cushing's syndrome and primary aldosteronism. *J Clin Invest* **42**:516, 1963.

122. Krakoff L, Nicolis G, Amsel B: Pathogenesis of hypertension in Cushing's syndrome. *Am J Med* **58**:216, 1975.

123. Mendlowitz M, Gitlow S, Noftchi N: Work of digital vasoconstriction produced by infused norepinephrine in Cushing's syndrome. *J Appl Physiol* **13**:252, 1958.

124. Ganguly A, Weinberger MH, Grim CE: The renin–angiotensin–aldosterone system in Cushing's syndrome and pheochromocytoma. *Hormone Res* **17**:1, 1983.

125. Saruta T, Suzuki H, Handa M, et al: Multiple factors contribute to the pathogenesis of hypertension in Cushing's syndrome. *J Clin Endocrinol Metab* **62**:275, 1986.

126. Gabrilove JL: Neurologic and psychiatric manifestations of the classic endocrine syndromes. *Res Pub ARNMD* **43**:419, 1966.

127. Spillane JD: Nervous and mental disorders in Cushing's syndrome. *Brain 74:*72, 1951.

128. Glaser GH: Psychotic reactions induced by corticotrophin (ACTH) and cortisone. *Psychosom Med* **15**:280, 1953.

129. Jeffcoate WJ, Silverstone JT, Edwards CRW, et al: Psychiatric manifestations of Cushing's syndrome: Response to lowering of plasma cortisol. *Q J Med* **48**:465, 1979.

130. Saad MF, Adams F, Mackay B, et al: Occult Cushing's disease presenting with acute psychosis. *Am J Med* **76**:759, 1984.

131. Starkman M, Schteingart DE: Neuropsychiatric manifestations of patients with Cushing's syndrome. *Arch Intern Med* **141**:215, 1981.

132. Terenius L, Wahlstrom A, et al: Increased CSF levels of endorphins in chronic psychosis. *Neurosci Lett* **3**:157, 1976.

133. Khaleeli AA, Edwards RHT, Gohil K, et al: Corticosteroid myopathy: A clinical and pathological study. *Clin Endocrinol* **18**:155, 1983.

134. Khaleeli AA, Betteridge DJ, Edwards RHT, et al: Effect of treatment of Cushing's syndrome on skeletal muscle structure and function. *Clin Endocrinol* **19**:547, 1983.

135. Cryer P, Ludmerer KM, Kissane JM, et al: Bruising and thin skin in a 54 year old woman. *Am J Med* **79**:101, 1985.

136. Ferguson JK, Donald RA, Weston TS, et al: Skin thickness in patients with acromegaly and Cushing's syndrome and response to treatment. *Clin Endocrinol* **18**:347, 1983.

137. Imura H, Matsukusa S, Yamamoto H, et al: Studies on ectopic ACTH-producing tumors II. Clinical, biochemical features of 30 cases. *Cancer* **35**:1430, 1975.

138. Gold EM: The Cushing's syndromes: Changing views of diagnosis and treatment. *Ann Intern Med* **90**:829, 1979.

139. Teates CD: Steroid-induced mediastinal lipomatosis. *Radiology* **96**:501, 1970.

140. Santini LC, Williams JL: Mediastinal widening (presumable lipomatosis) in Cushing's syndrome. *N Engl J Med* **284**:1357, 1971.

141. Van De Putte LB, Wagenaar JPM, San KH: Paracardiac lipomatosis in exogenous Cushing's syndrome. *Thorax* **28**:653, 1973.

142. Lipson SJ, Naheedy MII, Kaplan MM, et al: Spinal stenosis caused by epidural lipomatosis in Cushing's syndrome. *N Engl J Med* **302**:36, 1980.

143. Lee M, Lekias J, Gubby SS, et al: Spinal cord compression by extradural fat after renal transplantation. *Med J Aust* **1**:201, 1975.

144. George WE, Wilmott M, Greenhouse A, et al: Medical management of steroid-induced epidural lipomatosis. *N Engl J Med* **308**:316, 1983.

145. Sowerbutts, JG: Some uses for presacral oxygen insufflation. *J Faculty Radiol* **10**:201, 1959.

146. Christian CD Jr, Schneider RP: Fatty tumor of the liver in a patient with Cushing's syndrome. *Arch Intern Med* **143**:1605, 1983.

147. Chajek T, Romanoff H: Cushing's syndrome with cyclical edema and periodic secretion of corticosteroids. *Arch Intern Med* **136:**441, 1976.
148. Anthony LB, Greco FA: Pneumocystis carinii pneumonia: A complication of Cushing's syndrome. *Ann Intern Med* **94:**488, 1981.
149. Kramer M, Corrado ML, Bacci V, et al: Pulmonary cryptococcosis and Cushing's syndrome. *Arch Intern Med* **143:**2179, 1983.
150. Graham BS, Tucker WS Jr: Opportunistic infections in endogenous Cushing's syndrome. *Ann Intern Med* **101:**334, 1984.
151. Dal Bo Zanon R, Fornasiero L, Boscaro M, et al: Increased factor VIII associated activities in Cushing's syndrome: A probable hypercoaguable state. *Thrombosis Hemostasis* **47:**116, 1982.
152. Small M, Lowe GDO, Forbes CD, et al: Thromboembolic complications in Cushing's syndrome. *Clin Endocrinol* **19:**503, 1983.
153. Didolkar MS, Bescher AR, Elias EG, et al: Natural history of adrenal cortical carcinoma. *Cancer* **47:**2153, 1981.
154. Hutter AM, Kayhoe DE: Adrenal cortical carcinoma—Clinical features of 138 patients. *Am J Med* **41:**572, 1966.
155. Gabrilove JL, Seman AT, Sabet R, et al: Virilizing adrenal adenoma with studies on the steroid content of the adrenal venous effluent and a review of the literature. *Endocr Rev* **2:**462, 1981.
156. Mahesh VB, Greenblatt RB, Coniff RF: Urinary steroid excretion before and after dexamethasone administration and steroid content of adrenal tissue and venous blood in virilizing adrenal tumors. *Am J Obstet Gynecol* **100:**1043, 1968.
157. Burr JM, Grahan T, Sullivan J, et al: A testosterone secreting tomour of the adrenal producing virilization in a female infant. *Lancet* **2:**643, 1973.
158. Cahill GF, Robinson JN: Androgenic symptom tumors of adrenal cortex in children. A report of four cases. *J Urol* **61:**680, 1949.
159. Lloyd CW, Lobotsky J, Jones J, et al: Hormone studies in a case of adrenogenitalism due to neoplasm of the adrenal cortex. *J Clin Endocrinol Metab* **11:**857, 1951.
160. Lynch J: Functional tumor of adrenal cortex in male child. *JAMA* **144:**921, 1950.
161. Riedel HA: Adrenogenital syndrome in a male child due to adrenocortical tumor. *Pediatrics* **10:**19, 1952.
162. Mortimer JG, Rudd BT, Butt WB: A virilizing adrenal tumor in a prepuberal boy. *J Clin Endocrinol Metab* **24:**842, 1964.
163. Trost BN, Koenig MP, Zimmermann A, et al: Virilization of a post-menopausal woman by a testosterone-secreting Leydig cell type adrenal adenoma. *Acta Endocrinol* **98:**274, 1981.
164. Fuller PJ, Pettigrew IG, Pike JW, et al: An adrenal adenoma causing virilization of mother and infant. *Clin Endocrinol* **18:**143, 1983.
165. Comite F, Schiebinger RJ, Albertson BD, et al: Isosexual precocious pseudopuberty secondary to a feminizing adrenal tumor. *J Clin Endocrinol Metab* **58:**435, 1984.
166. Scully RE, Galdabini JJ, McNeely BU: Case 23-1979. *N Engl J Med* **300:**1322, 1979.
167. Leditschke J, Arden F: Feminizing adrenal adenoma in a five year old male. *Aust Paediatr J* **10:**217, 1974.
168. Mosier DH, Goodwin WE: Feminizing adrenal adenoma in a 7-year old boy. *Pediatrics* **27:**1016, 1961.
169. Fachnie JD, Zafar MS, Mellinger R, et al: Pituitary carcinoma mimics the ectopic adrenocorticotropin syndrome. *J Clin Endocrinol Metab* **50:**1062, 1980.
170. Liddle GW, Nicholson WE, Island DP, et al: Clinical and laboratory studies of ectopic humoral syndromes. *Recent Prog Horm Res* **25:**283, 1969.
171. Singer W, Kovacs K, Ryan N, et al: Ectopic ACTH syndrome: Clincopathological correlations. *J Clin Pathol* **31:**591, 1978.
172. Birkenhager JC, Upton GV. Meduallary thyroid carcinoma. Ectopic production of peptides with ACTH-like, CRF-like and prolactin production stimulating activity. *Acta Endocrinol* **83:**280, 1976.
173. Spark RF, Connolly PB: ACTH secretion from a functioning pheochromocytoma. *N Engl J Med* **301:**416, 1979.
174. Howland WJ Jr, Pugh DG, Sprague RG: Roentgenologic changes of the skeletal system in Cushing's syndrome. *Radiology* **71:**69, 1958.

175. Hahn TJ: Corticosteroid-induced osteopenia. *Arch Intern Med* **138**:882, 1978.
176. Hahn TJ, Halstead LR, Baran DT: Effects of short term glucocorticoid administration on intestinal calcium absorption and circulating vitamin D metabolite concentrations in man. *J Clin Endocrinol Metab* **52**:111, 1981.
177. Findling JW, Adams ND, Lemann J, et al: Vitamin D metabolites and parathyroid hormone in Cushing's syndrome: Relationship to calcium and phosphorus homeostasis. *J Clin Endocrinol Metab* **54**:1039, 1982.
178. Teitelbaum SL, Malong JD, Kahn AJ: Glucocorticoid enhancement of bone resorption by rat peritoneal macrophages in vitro. *Endocrinology* **108**:795, 1981.
179. Jee WSS, Park HZ, Roberts WE, et al: Corticosteroid and bone. *Am J Anat* **129**:477, 1970.
180. Nosadini R, Del Prato A, Valerio TA, et al: Insulin resistance in Cushing's syndrome. *J Clin Endocrinol Metab* **57**:529, 1983.
181. McArthur RG, Cloutier MD, Hayles AB, et al: Cushing's disease in children. *Mayo Clin Proc* **47**:318, 1972.
182. Frantz AG, Rabkin MT: Human growth hormone: Clinical measurement, response to hypoglycemia and suppression by corticosteroids. *N Engl J Med* **271**:1375, 1964.
183. Vazquez AM, Schutt-Aine JC, Kenny FM, et al: Effect of cortisone therapy on the diurnal pattern of growth hormone secretion in congenital adrenal hyperplasia. *J Pediatr* **80**:433, 1972.
184. Sturge RA, Beardwell C, Hartog M, et al: Cortisol and growth hormone secretion in relation to linear growth: Patients with Still's disease on different therapeutic regimens. *Br Med J* **3**:547, 1970.
185. Preece MA: The effect of administered corticosteroids on the growth of children. *Postgrad Med J* **52**:625, 1976.
186. Morris HG, Jorgensen JR, Elrick H, et al: Metabolic effects of human growth hormone in corticosteroid-treated children. *J Clin Invest* **47**:436, 1968.
187. Phillips LS, Belosky DC, Young HS, et al: Nutrition and somatomedin. VI. Somatomedin activity and somatomedin inhibitory activity in serum from normal and diabetic rats. *Endocrinology* **104**:1519, 1979.
188. Green OC, Winter RJ, Kawathara FS, et al: Pharmacokinetic studies of prednisolone in children: plasma levels, half-life values, and correlation with physiologic assays for growth and immunity. *J Pediatr* **93**:299, 1978.
189. Elders MJ, Wingfield BS, McNatt ML, et al: Glucocorticoid therapy in children: Effect on somatomedin secretion. *Am J Dis Child* **129**:1393, 1975.
190. Clark I, Umbreit W: Effect of cortisone and other steroids upon in vitro synthesis of chondrotin sulfate. *Proc Soc Exp Biol Med* **86**:558, 1954.
191. Phillips LS, Herington AC, Daughaday WH: Steroid hormone effects on somatomedin I. Somatomedin action in vitro. *Endocrinology* **97**:780, 1975.
192. Phillips LS, Fusco AC, Unterman TG, et al: Somatomedin inhibitor in uremia. *J Clin Endocrinol Metab* **59**:764, 1984.
193. Unterman TG, Phillips LS: Glucocorticoid effects on somatomedins and somatomedin inhibitors. *J Clin Endocrinol Metab* **61**:618, 1985.
194. Bessler W: Vertebral growth arrest lines after Cushing's syndrome. Case report. *Diag Imaging* **51**:311, 1982.
195. Krieger DT, Allen W, Rizzo F, et al: Characterization of the normal temporal pattern of plasma corticosteroid levels. *J Clin Endocrinol Metab* **32**:266, 1971.
196. Crapo L: Cushing's syndrome: A review of diagnostic tests. *Metabolism* **28**:955, 1979.
197. Nugent CA, Nichols T, Tyler FH: Diagnosis of Cushing's syndrome—Single dose dexamethasone suppression test. *Arch Intern Med* **116**:172, 1965.
198. McHardy-Young S, Harris PWR, Lessof MH, et al: Single-dose dexamethasone suppression test for Cushing's syndrome. *Br Med J* **2**:740, 1967.
199. Tucci JR, Jagger PI, Lauler DP, et al: Rapid dexamethasone suppression tests for Cushing's syndrome. *JAMA* **199**:379, 1967.
200. Seidensticker JF, Folk RL, Wieland RG, et al: Screening test for Cushing's syndrome with plasma 11-hydroxycorticosteroids. *JAMA* **202**:87, 1967.
201. Meikle AW: Dexamethasone suppression tests: Usefulness of simultaneous measurement of plasma cortisol and dexamethasone. *Clin Endocrinol* **16**:401, 1982.

202. Meikle AW, Lagerquist LG, Tyler FH: Apparently normal pituitary–adrenal suppressibility in Cushing's syndrome: Dexamethasone metabolism and plasma levels. *J Lab Clin Med* **86**:472, 1975.

203. Caro JF, Meikle AW, Check JH, et al: Normal suppression to dexamethasone in Cushing's disease: An expression of decreased metabolic clearance for dexamethasone. *J Clin Endocrinol Metab* **47**:667, 1978.

204. Murphy BEP: Clinical evaluation of urinary cortisol determinations by competitive protein-binding radioassay. *J Clin Endocrinol Metab* **28**:343, 1968.

205. Eddy RL, Jones AL, Gilland PF, et al: Cushing's syndrome: A prospective study of diagnostic methods. *Am J Med* **55**:621, 1973.

206. Daughaday WH, Mariz IK: Corticosteroid-binding globulin: Its properties and quantitation. *Metabolism* **10**:936, 1961.

207. Trecan GV, Laudat MH, Thomopoulos JP, et al: Urinary free corticoids: evaluation of their usefulness in diagnosis of Cushing's syndrome. *Acta Endocrinol (Copenh)* **103**:110, 1983.

208. Haigh SE, Tevaarwerk GJM: A rise in the glomerular filtration rate as the cause of a "paradoxical" increase in urinary free cortisol during dexamethasone suppression in a patient with an adrenal adenoma: A case report. *Clin Endocrinol* **15**:53, 1981.

209. Hellman L, Nakada F, Curti J, et al: Cortisol is secreted episodically by normal man. *J Clin Endocrinol Metab* **30**:411, 1970.

210. Hellman L, Weitzman ED, Roffwarg H, et al: Cortisol is secreted episodically in Cushing's syndrome. *J Clin Endocrinol Metab* **30**:686, 1970.

211. de Lacerda L, Kowarski AA, Migeon CJ: Integrated concentration and diurnal variation of plasma cortisol. *J Clin Endocrinol Metab* **36**:227, 1973.

212. Zadik Z, de Lacerda L, Kowarski A: Evaluation of the 6-hour integrated concentration of cortisol as a diagnostic procedure for Cushing's syndrome. *J Clin Endocrinol Metab* **54**:1072, 1982.

213. Halbreich U, Zumoff B, Kream J, et al: The mean 1300–1600 h plasma cortisol concentration as a diagnostic test for hypercortisolism. *J Clin Endocrinol Metab* **54**:1262, 1982.

214. Ernest I: Steroid excretion and plasma cortisol in 41 cases of Cushing's syndrome. *Acta Endocrinol* **51**:511, 1966.

215. Nichols T, Nugent CA, Tyler FH: Steroid laboratory tests in the diagnosis of Cushing's syndrome. *Am J Med* **45**:116, 1968.

216. Streeten DHP, Stevenson CT, Dalakos TG, et al: The diagnosis of hypercortisolism. Biochemical criteria differentiating patients from lean and obese normal subjects and from females on oral contraceptives. *J Clin Endocrinol Metab* **29**:1191, 1969.

217. Liddle GW: Cushing's syndrome. In Eisenstein AB (ed): *The Adrenal Cortex.* Little, Brown, Boston, 1967, pp. 523–551.

218. Doe RP, Vennes JA, Flink EB: Diurnal variation of 17-hydroxycorticosteroids, sodium, potassium, magnesium and creatinine in normal subjects and in cases of treated adrenal insufficiency and Cushing's syndrome. *J Clin Endocrinol Metab* **20**:253, 1960.

219. Glass AR, Zavadil AP III, Halberg F, et al: Circadian rhythm of serum cortisol in Cushing's disease. *J Clin Endocrinol Metab* **59**:161, 1984.

220. Van Cauter E, LeClerq R, Van Haelst L, et al: Simultaneous study of cortisol and TSH daily variations in normal subjects and patients with hyperadrenalcorticism. *J Clin Endocrinol Metab* **39**:645, 1974.

221. Olsen NJ, Fang VS, DeGroot LJ: Cushing's syndrome due to adrenal adenoma with persistent diurnal cortisol secretory rhythm. *Metabolism* **27**:695, 1978.

222. Mizutani S, Sonoda T, Seki T, et al: Excretion patterns of urinary 17-KS and 17-OHCS in patients with Cushing's syndrome. *Urol Int* **29**:341, 1974.

223. Moolenaar AJ, Van Seters AP: Gas chromatographic determination of steroids in the urine of patients with Cushing's syndrome. *Acta Endocrinol* **67**:303, 1971.

224. Phillipou G: Investigation of urinary steroid profiles as a diagnostic method in Cushing's syndrome. *Clin Endocrinol* **16**:433, 1982.

225. Liddle GW: Tests of pituitary-adrenal suppressibility in the diagnosis of Cushing's syndrome. *J Clin Endocrinol Metab* **20**:1539, 1960.

226. Lamberts SWJ, de Jong FH, Birkenhager JC: Evaluation of diagnostic and differential diagnostic tests in Cushing's sysdorme. *Netherlands J Med* **20**:267, 1977.

227. Findling JW, Aron DC, Tyrell JB, et al: Selective venous sampling for ACTH in Cushing's syndrome. Differentiation between Cushing's disease and ectopic ACTH syndrome. *Ann Intern Med* **94:**647, 1981.

228. Linn JE, Bowdoin B, Farmer A, et al: Observations and comments on failure of dexamethasone suppression. *N Engl J Med* **277:**403, 1967.

229. Carey RM: Suppression of ACTH by cortisol in dexamethasone-non suppressible Cushing's disease. *N Engl J Med* **302:**275, 1980.

230. Lieberman B, Wajchenberg BL, Tambascia MA, et al: Periodic remission in Cushing's disease with paradoxical dexamethasone response: An expression of periodic hormonogenesis. *J Clin Endocrinol Metab* **43:**913, 1976.

231. Brown RD, Van Loon GR, Orth DN, et al: Cushing's disease with periodic hormogenesis. *J Clin Endocrinol Metab* **36:**445, 1973.

232. Kapcala LP, Hamilton SM, Meikle AW: Cushing's disease with "normal suppression" due to decreased dexamethasone clearance. *Arch Intern Med* **144:**636, 1984.

233. Mason AMS, Ratcliffe JG, Buckle RM, et al: ACTH secretion by bronchial carcinoid tumors. *Clin Endocrinol J* 1:3, 1972.

234. Northrop G, Baldwin D, Faber LP, et al: Dexamethasone suppression of urinary 17-hydroxycorticoids in a patient with an ACTH-producing bronchial adenoma. *Presbyterian–St. Luke's Medical Bull* **9:**43, 1970.

235. Findling JW, Tyrrell JB: Occult ectopic secretion of corticotropin. *Arch Intern Med* **146:**929, 1986.

236. Mellinger RC: The conundrum of Cushing's syndrome. *Arch Intern Med* **146:**858, 1986.

237. Aron DC, Tyrell JB, Fitzgerald PA, et al: Cushing's syndrome: Problems in diagnosis. *Medicine* **60:**25, 1981.

238. Gruen PH: Endocrine changes in psychiatric diseases. *Med Clin North Am* **62:**285, 1978.

239. Kennedy L, Atkinson AB, Johnston H, et al: Serum cortisol concentrations during low dose dexamethasone suppression test to screen for Cushing's syndrome. *Br Med J* **289:**1188, 1984.

240. Tyrrell JB, Findling JW, Aron DC, et al: An overnight high-dose dexamethasone suppression test for rapid differential diagnosis of Cushing's syndrome. *Ann Intern Med* **104:**180, 1986.

241. Horrocks PM, London DR: Diagnostic value of 9 am plasma adrenocorticotrophic hormone concentrations in Cushing's disease. *Br Med J* **285:**1302, 1982.

242. Meikle AW, Jubiz W, Hutchings M, et al: Simplified metyrapone test with determination of plasma 11-deoxycortisol (metyrapone test with plasma S). *J Clin Endocrinol Metab* **23:**985, 1969.

243. Spiger N, Jubiz W, Meikle AW, et al: Single dose metyrapone test. Review of a four-year experience. *Arch Intern Med* **135:**698, 1975.

244. Sindler BH, Griffing GT, Melby JC: The superiority of the metyrapone test versus the high-dose dexamethasone test in the differential diagnosis of Cushing's syndrome. *Am J Med* **74:**657, 1983.

245. Carballeira A, Fishman LM, Jacobi GD: Dual sites of inhibition by metyrapone of human adrenal steroidogenesis: Correlation of in vivo and in vitro studies. *J Clin Endocrinol Metab* **42:**687, 1976.

246. Matthews JI, Fariss BB, Chertow BS, et al: Adrenal adenoma with variable response to dexamethasone. *J Clin Endocrinol Metab* **34:**902, 1972.

247. Vale W, Spiess J, Rivier C, et al: Characterization of a 41-residue ovine hypothalamic peptide that stimulates secretion of corticotropin and β-endorphin. *Science* **213:**1394, 1981.

248. Rivier C, Brownstein M, Spiess J, et al: In vivo corticotropin-releasing factor-induced secretion of adrenocorticotropin, β-endorphin, and corticosterone. *Endocrinology* **110:**272, 1982.

249. Orth DN, DeBold CR, DeCherney GS: Pituitary microadenomas causing Cushing's disease resond to corticotropin-releasing factor. *J Clin Endocrinol Metab* **55:**1017, 1982.

250. Muller OA, Stalla GK, Werder K: Corticotropin releasing factor: A new tool for the differential diagnosis of Cushing's syndrome. *J Clin Endocrinol Metab* **57:**227, 1983.

251. Pieters GFFM, Hermus ARMM, Smals AGH: Responsiveness of the hypophyseal-adrenocortical axis to corticotropin-releasing factor in pituitary dependent Cushing's disease. *J Clin Endocrinol Metab* **57:**513, 1983.

252. Suda T, Tomori N, Toxawa F: Effects of corticotropin-releasing factor and other materials on adrenocorticotropin secretion from pituitary glands of patients with Cushing's disease in vitro. *J Clin Endocrinol Metab* **59:**840, 1984.

253. Shibasaki T, Nakahara M, Shizume K, et al: Pituitary adenomas that caused Cushing's disease or Nelson's syndrome are not responsive to ovine corticotropin-releasing factor in vitro. *J Clin Endocrinol Metab* **56:**414, 1983.

254. Gold PW, Loriaux DL, Roy A, et al: Responses to corticotropin-releasing hormone in the hypercortisolism of depression and Cushing's disease. *N Engl J Med* **314:**1329, 1986.

255. Catania A, Cantalamessa L, Orsatti A, et al: Plasma ACTH-response to the corticotropin releasing factor in patients with Cushing's disease. Comparison with the lysine–vasopressin test. *Metabolism* **33:**478, 1984.

256. Abraham RR, Campbell EA, Gillham B, et al: The effect of ovine corticotrophin releasing factor (oCRF), bromocriptine and TRH on the secretion of ACTH and α-MSH in Nelson's syndrome and Cushing's disease. *Clin Endocrinol* **25:**75, 1986.

257. Nieman LK, Chrousos GP, Oldfield EH, et al: The ovine corticotropin-releasing hormone stimulation test and the dexamethasone suppression test in the differential diagnosis of Cushing's syndrome. *Ann Intern Med* **105:**862, 1986.

258. Hermus ADRMM, Pieters GFFM, Pesman GJ, et al: Responsivity of adrenocorticotropin to corticotropin-releasing hormone and lack of suppressibility by dexamethasone are related phenomena in Cushing's disease. *J Clin Endocrinol Metab* **62:**634, 1986.

259. Lytras N, Grossman A, Perry L, et al: Corticotrophin releasing factor: Responses in normal subjects and patients with disorders of the hypothalamus and pituitary. *Clin Endocrinol (Oxf)* **20:**71, 1984.

260. Orth DN: The old and new in Cushing's syndrome. *N Engl J Med* **310:**649, 1984.

261. Weyman PJ, Glazer HS: The adrenals. In Lee JKT, Sagel SS, Stanley RJ (eds): *Computed Body Tomography.* Raven Press, New York, 1983, p. 379.

262. Karstaedt N, Sagel SS, Stanley RJ, et al: Computed tomography of the adrenal gland. *Radiology* **129:**723, 1978.

263. Wilms G, Baert A, Marchal G, et al: Computed tomography of the normal adrenal glands: Correlative study with autopsy specimens. *J Comput Assist Tomogr* **3:**467, 1979.

264. White FE, White MC, Drury PL, et al: Value of computed tomography of the abdomen and chest in investigation of Cushing's syndrome. *Br Med J* **284:**771, 1982.

265. Dunnick NR, Loppman JL, Gill JR Jr, et al: Localization of functional adrenal tumors by computed tomography and venous sampling. *Radiology* **142:**429, 1982.

266. Goldman SM, Gatewood OMB, Walsh PC, et al: CT configuration of the enlarged adrenal gland. *J Comput Assist Tomogr* **6**(2):276, 1982.

267. Green HT: The venous drainage of the human hypophysis cerebri. *Am J Anat* **100:**435, 1957.

268. Oldfield EH, Girton ME, Doppman JL: Absence of intercavernous venous mixing: evidence supporting lateralization of pituitary microadenomas by venous sampling. *J Clin Endocrinol Metab* **61:**644, 1985.

269. Kley HK, Stolze T, Kruskemper HL: Jugular-vein sampling of ACTH. *N Engl J Med* **297:**731, 1977.

270. Corrigan DF, Schaaf M, Whaley RA, et al: Selective venous sampling to differentiate ectopic ACTH secretion from pituitary Cushing's syndrome. *N Engl J Med* **296:**861, 1977.

271. Oldfield EH, Chrousos GP, Schulte HM, et al: Preoperative localization of ACTH secreting microadenomas by bilateral and simultaneous inferior petrosal sinus sampling. *N Engl J Med* **312:**100, 1985.

272. Manni A, Latshaw RF, Page R, et al: Simultaneous bilateral venous sampling for adrenocorticotropin in pituitary-dependent Cushing's disease: Evidence for lateralization of pituitary venous drainage. *J Clin Endocrinol Metab* **57:**1070, 1983.

273. Beierwaltes WH, Lieberman LM, Ansari AN, et al: Visualization of human adrenal glands in vivo by scintillation scanning. *JAMA* **216:**275, 1971.

274. Anderson BG, Beierwaltes WH: Adrenal imaging with radiocholesterol in the diagnosis of adrenal disorders. *Adv Intern Med* **19:**327, 1974.

275. Wahner HW, Northcutt RC, Salassa RM: Adrenal scanning: Usefulness in adrenal hyperfunction. *Clin Nucl Med* **2:**253, 1977.

276. Schteingart DE, Seabold JE, Gross MD, et al: Iodocholesterol adrenal tissue uptake and imaging in adrenal neoplasms. *J Clin Endocrinol Metab* **52:**1156, 1981.

277. Nicolis GL, Mitty HA, Modlinger RS, et al: Percutaneous adrenal venography. A clinical study of 50 patients. *Ann Intern Med* **76:**899, 1972.

278. Mitty HA, Nicolis GL, Gabrilove JL: Adrenal venography: Clinical–roentgenographic correlation in 80 patients. *Am J Roentgenol* **119:**564, 1973.

279. Sutton D: The radiological diagnosis of adrenal tumours. *Br J Radiol* **48:**237, 1975.

280. Lecky JW, Wolfman NT, Modic CW: Current concepts of adrenal angiography. *Radiol Clin North Am* **14:**309, 1976.

281. Bayliss RS, Edwards OM, Starer F: Complications of adrenal venography. *Br J Radiol* **43:**531, 1970.

282. Fellerman H, Dalakos TG, Streeten DHP: Remission of Cushing's syndrome after unilateral adrenal phlebography. Apparent destruction of adrenal adenoma. *Ann Intern Med* **73:**585, 1970.

283. Meikle AW, Stanchfield JB, West CD, et al: Hydrocortisone suppression test for Cushing's syndrome. *Arch Intern Med* **134:**1068, 1974.

284. Sachar EJ, Hellman L, Roffwarg HP, et al: Disrupted 24-hour patterns of cortisol secretion in psychotic depression. *Arch Gen Psychiatry* **28:**19, 1973.

285. Carroll BJ, Curtis GC, Davies BM, et al: Urinary free cortisol excretion in depression. *Psychol Med* **6:**43, 1976.

286. Butler PWP, Besser GM: Pituitary adrenal function in severe depressive illness. *Lancet* **1:**1234, 1968.

287. Chrousos GP: Clinical application of corticotropin-releasing factor (NIH conference). *Ann Intern Med* **102:**344, 1985.

288. Gold PW, Loriaux DL, Roy A, et al: Responses to corticotropin-releasing hormone in the hypercortisolism of depression and Cushing's disease. *N Engl J Med* **314:**1329, 1986.

289. Frajria R, Angeli A: Alcohol-induced pseudo–Cushing's syndrome. *Lancet* **1:**1050, 1977.

290. Jenkins RM, Page M McB: An atypical case of alcohol-induced cushingoid syndrome. *Br Med J* **282:**1117, 1981.

291. Lamberts SWJ, Klinjn JGM, de Jong FH, et al: Hormone secretion in alcohol-induced pseudo–Cushing's syndrome. *JAMA* **242:**1640, 1979.

292. James VHT, Landon J, Wynn U, et al: A fundamental defect of adrenocortical control in Cushing's disease. *J Endocrinol* **40:**15, 1968.

293. Elias AN, Meshkinpour H, Valenta LJ, et al: Pseudo–Cushing's syndrome: The role of alcohol. *J Clin Gastroenterol* **4:**137, 1982.

294. Sharp NA, Devlin JT, Rimmer JM: Case report. Renal failure obfuscates the diagnosis of Cushing's disease. *JAMA* **256:**2564, 1986.

295. Ramirez G, Gomez-Sanchez C, Meikle WA, et al: Evaluation of the hypothalamic hypophyseal adrenal axis in patients receiving long-term hemodialysis. *Arch Intern Med* **142:**1448, 1982.

296. Nolan GE, Smith JB, Chavre VJ, et al: Spurious overestimation of plasma cortisol in patients with chronic renal failure. *J Clin Endocrinol Metab* **52:**1242, 1981.

297. Cook DM, Meikle AW: Factitious Cushing's syndrome. *J Clin Endocrinol Metab* **61:**385, 1985.

298. Yamaji T, Ishibashi M, Sekihara H, et al: Serum dehydroepiandrosterone sulfate in Cushing's syndrome. *J Clin Endocrinol Metab* **59:**1164, 1984.

299. Bailey RE: Periodic hormogenesis: A new phenomenon; periodicity in function of a hormone producing tumor in man. *J Clin Endocrinol Metab* **32:**317, 1971.

300. Jordan RM, Ramos-Gabatin A, Kendall JW, et al: Dynamics of adrenocorticotropin (ACTH) secretion in cyclic Cushing's syndrome: Evidence for more than one abnormal ACTH biorhythm. *J Clin Endocrinol Metab* **55:**531, 1982.

301. Atkinson AB, Chestnutt A, Crothers E, et al: Cyclical Cushing's disease: Two distinct rhythms in a patient with a basophil adenoma. *J Clin Endocrinol Metab* **60:**328, 1985.

302. Sakiyama R, Ashcraft MW, Van Herle AJ: Cyclic Cushing's syndrome. *Am J Med* **77:**944, 1984.

303. Vagnucci AH, Evans E: Cushing's Disease with intermittent hypercortisolism. *Am J Med* **80:**83, 1986.

304. Oates TW, McCourt JP, Friedman WA, et al: Cushing's disease with cyclic hormonogenesis and diabetes insipidus. *Neurosurgery* **5:**598, 1979.

305. Spark RF, Connolly PB: ACTH secretion from a functioning pheochromocytoma. *N Engl J Med* **301:**416, 1979.

306. Cook DM, Kendall JW: Cushing's syndrome—Current concepts of diagnosis and therapy. *West J Med* **132:**111, 1980.

307. Smith DJ, Kohler PC, Helminiak R, et al: Intermittent Cushing's syndrome with an empty sella turcica. *Arch Intern Med* **142:**2185, 1982.

308. Scott RS, Espiner EA, Donald RA: Intermittent Cushing's disease with spontaneous remission. *Clin Endocrinol* **11:**561, 1979.

309. Burch WM: Cushing's disease. *Arch Intern Med* **145:**1108, 1985.

310. Smyth HS: Pituitary basophilism. Microsurgical observations on the pathogenesis of Cushing's disease. Workshop on Pituitary Pathology, 2nd Meeting International Pituitary Pathology Club, Oaxtepec, Morelos, Mexico, June 19–22, 1984, pp. 3–4.

311. Styne DM, Grumbach MM, Kaplan SL, et al: Treatment of Cushing's disease in childhood and adolescence by transsphenoidal microadenomectomy. *N Engl J Med* **310:**889, 1984.

312. Pont A, Gutierrez-Hartman A: Cushing's disease: Recurrence after a surgically induced remission. *Arch Intern Med* **139:**938, 1979.

313. McNichol AM: Patterns of corticotropic cells in the adult human pituitary in Cushing's disease. *Diag Histopathol* **4:**335, 1981.

314. Schnall AM, Kovacs K, Brodkey JS, et al: Pituitary Cushing's disease without adenoma. *Acta Endocrinol* **94:**297, 1980.

315. Ludecke D, Kautzky R, Saeger W, et al: Selective removal of hypersecreting pituitary adenomas. *Acta Neurochir* **35:**27, 1976.

316. Nelson DH, Meakin JW, Thorn GW: ACTH-producing pituitary tumors following adrenalectomy for Cushing's syndrome. *Ann Intern Med* **52:**560, 1960.

317. Cook DM, Kendall JW, Allen JP, et al: Nycothemeral variation and suppressibility of plasma ACTH in various stages of Cushing's disease. *Clin Endocrinol* **5:**303, 1976.

318. Moore TJ, Dluhy RG, Williams GH, et al: Nelson's syndrome: Frequency, prognosis, and effect of prior irradiation. *Ann Intern Med* **85:**731, 1976.

319. Kasperlik-Zaluska AA, Nielubowicz J, Wislawski J, et al: Nelson's syndrome: Incidence and prognosis. *Clin Endocrinol* **19:**693, 1983.

320. Carpenter PC, Wahner HW, Salassa RM, Duick DS: Demonstration of steroid-producing gonadal tumors by external scanning with the use of NP-59. *Mayo Clin Proc* **54:**321, 1979.

321. Papapetrou PD, Jackson I: Cortisol secretion in Nelson's syndrome, persistence after "total" adrenalectomy for Cushing's syndrome. *JAMA* **234:**847, 1975.

322. Bonner RA, Mukai K, Oppenheimer JH: Two unusual variants of Nelson's syndrome. *J Clin Endocrinol Metab* **49:**23, 1979.

323. Burke EF, Gilbert E, Uehling DT: Adrenal rest tumors of the testes. *J Urol* **109:**649, 1973.

324. Earll JM, Newman SG, DiRaimondo VC: Bilateral testicular tumors in untreated congenital adrenocortical hyperplasia. *JAMA* **209:**937, 1969.

325. Verdonk C, Guerin C, Lufkin E, et al: Activation of virilizing adrenal rest tissues by excessive ACTH production. *Am J Med* **73:**455, 1982.

326. Baranetsky NG, Zipser RD, Goebelsmann U, et al: Adrenocorticotropin-dependent virilizing paraovarian tumors in Nelson's syndrome. *J Clin Endocrinol Metab* **49:**381, 1979.

327. Sharpe GF, Kendall-Taylor P, Prescott RWG, et al: Pituitary function following megavoltage therapy for Cushing's disease: Long term follow up. *Clin Endocrinol* **22:**169, 1985.

328. Jennings AS, Liddle GW, Orth DN: Results of treating childhood Cushing's disease with pituitary irradiation. *N Engl J Med* **297:**957, 1977.

329. Barnett AH, Livesey JH, Friday K, et al: Comparison of preoperative and postoperative ACTH concentrations after bilateral adrenalectomy in Cushing's disease. *Clin Endocrinol* **18:**301, 1983.

330. Krieger DT, Amorosa L, Linick F: Cyproheptadine-induced remission of Cushing's disease. *N Engl J Med* **293:**893, 1975.

331. Krieger DT: Cyproheptadine: Drug therapy for Cushing's disease. In Muller EE (ed): *Neuroactive Drugs in Endocrinology.* Elsevier North Holland, Amsterdam, 1980, p. 361.

332. Lamberts SWJ, Birkenhager JC: Bromocriptine in Nelson's syndrome and Cushing's disease. *Lancet* **2:**811, 1976.

333. Miura K, Aida M, Mihara A, et al: Treatment of Cushing's disease with reserpine and pituitary radiation. *J Clin Endocrinol Metab* **41:**511, 1975.

334. Elias AN, Gwinup G, Valenta LJ: Effects of valproic acid, naloxone and hydrocortisone in Nelson's syndrome and Cushing's syndrome. *Clin Endocrinol* **15:**151, 1981.

335. Dornhorst A, Jenkins JS, Lamberts SWJ, et al: The evaluation of sodium valproate in the treatment of Nelson's syndrome. *J Clin Endocrinol Metab* **56:**985, 1983.

336. Jones MT, Gillham B, Beckford U: Effect of treatment with sodium valproate and diazepam on plasma corticotropin in Nelson's syndrome. *Lancet* **1:**1179, 1981.
337. Gomi M, Iida S, Itoh Y, et al: Unaltered stimulation of pituitary adrenocorticotrophin secretion by corticotrophin releasing factor following sodium valproate administration in a patient with Nelson's syndrome. *Clin Endocrinol* **23:**123, 1985.
338. Sonino N, Boscaro M, Merola G, et al: Prolonged treatment of Cushing's disease by keto-conazole. *J Clin Endocrinol Metab* **61:**718, 1985.
339. Välimäki M, Pelkonen R, Porkka L, et al: Long-term results of adrenal surgery in patients with Cushing's syndrome due to adrenocortical adenoma. *Clin Endocrinol* **20:**229, 1984.
340. Luton JP, Mahoudeau JA, Boughard PH, et al: Treatment of Cushing's disease by o,p'DDD. Survey of 62 cases. *N Engl J Med* **300:**459, 1979.
341. Couzinet B, Thomopoulos P, Schaison G: Durable remission of Cushing's disease after o,p'DDD treatment. *Acta Endocrinol* **100:**63, 1982.
342. Beardwell CG, Adamson AR, Shalet SM: Prolonged remission in florid Cushing's syndrome following metyrapone treatment. *Clin Endocrinol* **14:**485, 1981.
343. Jeffcoate WJ, Rees LH, Tomlin S, et al: Metyrapone in long-term management of Cushing's disease. *Br Med J* **2:**215, 1977.
344. Dickstein G, Lahav M, Shen-Orr Z, et al: Primary therapy for Cushing's disease with metyrapone. *JAMA* **255:**1167, 1986.
345. Loli P, Berselli ME, Tagliaferri M: Use of ketoconazole in the treatment of Cushing's syndrome. *J Clin Endocrinol Metab* **63:**1365, 1986.
346. Dewis P, Anderson DC, Bullock DE, et al: Experience with trilostane in the treatment of Cushing's syndrome. *Clin Endocrinol* **18:**533, 1983.
347. Komanicky P, Spark RF, Melby JC: Treatment of Cushing's syndrome with trilostane (WIN) 24,540, and inhibitor of adrenal steroid biosynthesis. *J Clin Endocrinol Metab* **47**(5):1042, 1978.
348. Gaillard RC, Poffet D, Riondel AM, et al: RU 486 inhibits peripheral effects of glucocorticoids in humans. *J Clin Endocrinol Metab* **61:**1009, 1985.
349. Bertagna X, Bertagna C, Luton JP, et al: The new steroid analog RU 486 inhibits glucocorticoid action in man. *J Clin Endocrinol Metab* **59:**25, 1984.
350. Nieman LK, Chrousos GP, Kellner C, et al: Successful treatment of Cushing's syndrome with the glucocorticoid antagonist RU 486. *J Clin Endocrinol Metab* **61:**536, 1985.
351. Kannan C: *Essential Endocrinology.* Plenum, New York, p. 256.

4

Glucocorticoid Therapy

Introduction

It has been estimated that more than 5 million patients are treated with glucocorticoids annually.[1] The anti-inflammatory and immunosuppressive effects of steroids have been exploited in the treatment of such diverse diseases as bronchial asthma, collagen vascular disease, rheumatoid arthritis, chronic active hepatitis, certain forms of nephritides, and inflammatory bowel diseases. In many instances the use of glucocorticoids is for short-term purposes, allowing the host to recover from a self-limited disease process. But, unfortunately, in some patients glucocorticoid therapy becomes necessary on a chronic basis because of exacerbation of the underlying process when steroids are withdrawn. It is in these patients that the adverse effects of chronic steroid drug therapy assume a magnitude conducive for the development of clinical problems. A detailed understanding of steroid pharmacology and pharmacokinetics is essential for the physician who treats patients with glucocorticoids for any length of time. The mechanism of action, as well as the systemic effects, of steroids must be clearly understood to provide the patient with the utmost benefit of these drugs while minimizing their side effects. The effects of glucocorticoids in causing suppression of the hypothalamic–pituitary axis (HPA) are of paramount importance, if the facets of the steroid withdrawal syndrome are to be understood. This chapter focuses on the clinical perspectives that relate to steroid therapy.

Pharmacokinetics

All glucocorticoids are 21-carbon steroid molecules. Cortisol (hydrocortisone) is the major circulating glucocorticoid in humans. The term *glucocorticoids* refers to the ability of these steroid hormones to increase hepatic glucose output by the induction of hepatic gluconeogenesis. The primary event that facilitates this process is the remarkable and profound effect of these steroids in inhibiting incorporation of amino acids into the protein of skeletal muscles. It appears that the resultant hyperaminoacidemia plays a major role in providing substrate for hepatic gluconeogenesis. This, coupled with the steroid-induced increase in hepatic gluconeogenic enzymes,[2] as well as increased glucagon secretion by the pancreatic alpha cells,[3] leads to the "gluco" corticoid effect of these steroids. Minor differences in chemical structure result in remarkable differences in the potency and duration of action of these glucocorticoid hormones.

Chemistry

As indicated earlier, cortisol (hydrocortisone) is the main circulating natural glucocorticoid in humans. Alterations at different positions of the steroid molecule result in the formation of several synthetic analogs of cortisol. The five important synthetic analogs of cortisol are cortisone, prednisone, dexamethasone, prednisolone, and methylprednisolone. It is remarkable that minimal changes in the molecular structure of cortisol should yield steroid substances with widely disparate half-lives, potency, and duration of action. Cortisone is derived from cortisol by replacement of the hydroxyl group at C-11 by a keto-group at C-11. Prednisone is also an 11-keto compound, but contains a double bond between C-1 and C-2. Dexamethasone, on the other hand, is an 11-hydroxy compound with a double bond between C-1 and C-2, but has the important characteristics of fluorination of the B rings and methylation at C-17. Prednisolone is derived from cortisol by double bonding between 1 and 2 positions, methylation of which yields methylprednisolone. The clinical relevance of these changes is twofold. First, the presence of a hydroxy group at C-11 is crucial for glucocorticoid activity. Thus, 11-keto compounds such as cortisone and prednisone require conversion to their corresponding 11-hydroxy compound to exert glucocorticoid action. This reaction occurs in the liver.[4] Prednisolone, methylprednisolone, and dexamethasone are 11-hydroxy compounds and therefore possess inherent glucocorticoid activity. Second, the introduction of double bond at C-1, C-2 positions, as well as fluorination of the B ring, prolongs the half-lives of these compounds. The uniqueness of the glucocorticoid potency of dexamethasone lies in the fact that it contains an 11-hydroxy group at C-11, a double bond at C-1, C-2, and a fluorine atom in ring B. Less commonly used synthetic steroid derivatives include triamcinolone, betamethasone, flunisolide, and budesomide; the latter two analogs have been used more extensively in Europe than in the United States. To emphasize, the biological activity of glucocorticoids depends on the presence of a hydroxyl group at carbon number 11.

Metabolism

Cortisol, the major glucocorticoid of the adrenal cortex, circulates in the blood at a concentration of 5–25 µg/dl of plasma. Approximately 80% of the circulating cortisol is bound to transcortin (or corticosteroid-binding globulin), an alpha globulin. A smaller portion circulates in the plasma bound to albumin. Only the albumin-bound cortisol and the "free" unbound cortisol are able to enter the target cells to express the biological effects of cortisol. In contrast to the natural glucocorticoid cortisol, the synthetic analogs are less bound to plasma proteins and hence diffuse more readily into tissues.

The liver is the major site of glucocorticoid metabolism. Several metabolic transformations, at different sites of the glucocorticoid molecule, can result in inactivation of glucocorticoids. Metabolic transformations occur at three major sites. The first is reduction of the double bond at position 4–5. The second is reduction of the ketone group at the third carbon atom to a hydroxyl group with subsequent esterification with glucuronic acid or sulfate. The third reaction involves hydroxylation at the sixth carbon atom. Less prominent and slower mechanisms of transformation include oxidation of the hydroxyl group at C-17 and reduction of the 20-ketone group to a 20-hydroxyl group. The metabolic rate of glucocorticoid molecules can be decreased by the incorporation of a double bond at C-1, C-2, or by the fluorination of the B ring.

The importance of the liver in the metabolism of glucocorticoids has several clinical ramifications.

1. In patients with chronic liver disease, cortisol is metabolized at a decreased rate. However, the plasma cortisol levels are generally normal in cirrhotics. This is due to the fact that although the metabolism of cortisol is decreased, the synthesis of cortisol is appropriately decreased, due to signals from the intact hypothalamic pituitary adrenal axis.[5]

2. The circulating half-life of predinsone is prolonged in patients with cir-

rhosis, owing to the impairment in the hepatic reduction of the 11-keto group. Since prednisone needs to be converted to prednisolone for exertion of its glucocorticoid activity, it is probably preferable to use prednisolone rather than prednisone in patients with chronic liver disease. The conversion of cortisone to cortisol in liver disease has not been studied extensively.

3. The hypoalbuminemia seen in patients with chronic liver disease can further complicate matters. Since a higher percentage of prenisolone circulates unbound in the presence of hypoalbumineamia, the dosage of prednisolone may require reduction to avoid adverse side effects of steroid therapy.

Half-Life and Biological Potency

The half-lives and biological potency of the various glucocorticoids depend on four factors—absorption, plasma protein binding, the volume of distribution, and the rate of elimination. The plasma half-lives of several glucocorticoids has been extensively studied by several investigators.[4,6–11] Table 22 compares the plasma half-lives of several commonly used glucocorticoids. The plasma "biological half-life" of the corticosteroid represents the time elapsed before half of the concentration of the administered steroid disappears at a given point.[12,13] This reflects the rate of degradation of the steroid by hepatic enzymes, as well as the metabolic stability of the glucocorticoid at its receptor site. While it is generally true that glucocorticoids with a longer plasma half-life tend to also exhibit a longer plasma biological half-life, one cannot extrapolate the biological activity of a particular glucocorticoid from the plasma half-life data. This is best illustrated in the example of dexamethasone and prednisone. Although these two glucocorticoids possess nearly similar half-life data, dexamethasone demonstrates a much longer biological effect and is clearly the more potent of the two.

The potency and duration of action of a given glucocorticoid depends on numerous factors. The duration of anti-inflammatory activity of orally administered glucocorticoids roughly approximates the duration of HPA suppression induced by the particular glucocorticoid. This concept has permitted study of the

TABLE 22.
Half-Lives of Various Steroids

Glucocorticoid	Plasma half-life (min)	Biological half-life (hr)
Cortisol	90	8–12
Cortisone	90	8–12
Prednisolone	200	12–36
Methylprednisolone	200	12–36
Triamcinolone	200	12–36
Betamethasone	300	36–54
Dexamethasone	300	36–54

potency of the orally administered glucocorticoids. As early as 1966, Harter[14] studied this facet of glucocorticoid activity, i.e., ACTH suppression, by administering several glucocorticoids to normal subjects. Based on the duration of ACTH suppression (which was determined by performing the metyrapone test), glucocorticoids were divided into three groups—the "short-acting" (cortisone, hydrocortisone, prednisone, prednisolone, methylprednisolone), where ACTH activity returned to normal within 24–36 hr following a single dose of the glucocorticoid; an "intermediate" group (triamcinolone, paramethasone), where ACTH suppression persisted for 48 hr; and a third, "long-acting" group (dexamethasone, betamethasone), which caused ACTH suppression for more than 48 hr. These studies were performed using glucocorticoids with comparable anti-inflammatory activities. The classic concept notwithstanding, it should be realized that the duration of ACTH suppression by a particular glucocorticoid is not exclusively an index of its anti-inflammatory activity.

Two decades after the study by Harter,[14] data on relative glucocorticoid potency continue to be puzzling, often discrepant. There are several reasons for these discrepancies. First, the differences in the circulating half-lives of various glucocorticoids contrast sharply with the differences in the ability of these steroids to cause ACTH suppression. Further, the duration of ACTH suppression exceeds the half-life of these steroids by at least a five-fold factor, implying ongoing ACTH suppression, despite disappearance of the steroid from the circulation. Second, it is now clear that the relative intrinsic potency of a given glucocorticoid is affected by the interaction of the steroid with its receptor.[15–17] The steroid molecule passes through the target cell membrane, enters the cytoplasm, and binds to a cytoplasmic receptor protein. This complex, subsequently, enters the nucleus, where it modifies transcription and causes RNA synthesis from the nuclear DNA template. The effects of the "internalized" steroid may continue even after the hormone has disappeared from the circulation. Therefore, the plasma half-life, and the biological half-life alone, may not be altogether accurate indicators to define the intrinsic potency or duration of a given glucocorticoid. A third factor, also relatively unknown, is the amount of glucocorticoid absorbed and delivered to the peripheral tissues. Even more unquantifiable is the relative amount of steroid taken up by the target tissue, as well as the catabolism of the steroid hormone within the target tissues. Obviously, the absorption of glucocorticoid from the gastrointestinal (GI) tract, its metabolism by splanchnic tissues, and its delivery to peripheral tissues through the circulation must have a major impact on duration and potency of the orally administered hormone. However, there is a paucity of data in the literature regarding these aspects of steroid metabolism, since these phenomena are rather difficult to quantitate. Meikle et al.[18] studied the kinetics and interconversion of prednisone to prednisolone by radioimmunoassay and found that 70% of an orally administered dose of prednisone was converted to prednisolone in the plasma within 8 hr. There is a paucity of data for dexamethasone or hydrocortisone. Finally, the confusion is further compounded by the use of different immunoassay systems in estimation of relative glucocorticoid potencies.[13,17,19–23] In spite of the above, emerging data indicate that the two most important factors that determine the relative glucocorticoid potencies of orally administered glucocorticoids are the "intrinsic" biological potency of the hormone at the tissue level and the relative rates of disappearance from the plasma.[24]

Clinically, the various glucocorticoids are classified as short acting, intermediate acting, and long acting. These merely reflect the relative duration of the biological potency of the administered drug. Cortisone and hydrocortisone represent the short-acting glucocorticoids. Prednisone, prednisolone, and methylprednisolone are classified as intermediate-acting glucocorticoids, while dexamethasone, betamethasone, and paramethasone are prototypical for long-acting glucocorticoids.

Bioavailability

This refers to the percentage of biologically active glucocorticoid available to the target tissue. Bioavailability depends on the in vivo ability to convert 11-keto glucocorticoids (such as cortisone or prednisone) to their respective 11-hydroxysteroids, since the intrinsic glucocorticoid activity depends on the presence of a hydroxyl group at C-11. Bioavailability also depends on the amount absorbed and the amount bound to carrier proteins in the plasma. The synthetic analogs of cortisol are less bound to carrier proteins and hence are more freely available to tissues.

Choice of Agent for Therapy

In 1966, Thorn[25] emphasized that several aspects of theapy should be carefully considered prior to initiating glucocorticoid therapy. The passage of time has reaffirmed the wisdom of such deliberation. The criteria recommended by Thorn[25] require reinforcement if the utmost benefits are to be gained by the use of glucocorticoid therapy. The seven important considerations prior to the use of glucocorticoids as pharmacological agents are the following:

1. The seriousness of the underlying disorder for which steroids are prescribed.
2. The anticipated duration of therapy.
3. The anticipated effective dose.
4. The presence of underlying factors that predispose patients to the adverse side effects of steroids.
5. The type of glucocorticoid chosen.
6. The choice of other treatment modalities.
7. The choice of an alternate-day regimen.

Each of these considerations deserves brief mention.

The Seriousness of the Underlying Disorder

Clearly, the option to institute glucocorticoid therapy is not one that can be chosen lightly. The benefits of steroid therapy in treatment of the underlying disease should be worthy of risking the adverse side effects of glucocorticoid therapy. The prompt treatment with steroids can be lifesaving in certain potentially lethal situations, such as allergic (anaphylactic) emergencies, bronchial asthma with respiratory compromise, cerebral edema, and idiopathic thrombocytopenic purpura. Intensive short-term therapy with steroids may also produce dramatic responses in conditions such as necrotizing vasculitis, central hyperthermia, acute hypercalcemia of vitamin D intoxication, and hormone therapy for metastatic breast cancer. In most of the aforementioned situations the glucocorticoid therapy is for short-term use, with very few sequelae. Unfortunately, most disorders that require corticosteroid therapy are those that involve long-term, high-dose treatment. The prototypical examples in this category include bronchial asthma, chronic active hepatitis, ulcerative colitis, nephritides, and collagen vascular disease. The dose and duration of glucocorticoid therapy are conducive for the development of iatrogenic Cushing's syndrome and severe suppression of the hypothalamic pituitary axis. The third variety of disorders are conditions where treatment consists of low-dose, chronic, palliative therapy. Chronic, complicated rheumatoid arthritis and certain cases of asthma and lupus erythematosus belong in this category. Glucocorticoids reduce the morbidity and mortality of these conditions, where the inflammatory response has imperiled the host. Of course, the use of glucocorticoid therapy is lifesaving in patients with primary or secondary adrenal insufficiency, such therapy constituting physiological replacement glucocorticoid therapy.

Anticipated Duration of Therapy

Treatment with glucocorticoids is lifelong for patients with hypoadrenalism. With the exception of situations where short-term, intensive steroid therapy is used, for the most part glucocorticoid therapy becomes a protracted therapeutic affair. The duration of therapy in chronic diseases that require glucocorticoids is determined by the response to therapy and the occurrence of exacerbations when therapy is discontinued. Some patients may never be able to get off steroids either because the underlying disease requires chronic, low-dose palliation or because of permanent HPA suppression from protracted steroid therapy.

Anticipated Effective Dose

Most patients who require treatment with glucocorticoids need *at least* 15 mg of prednisolone daily. The dosage for individual patients is highly variable, ranging from 15 to as high as 120 mg daily. In most conditions where prolonged high-dose suppressive therapy is indicated, the initial starting dose is the equivalent of 60–80 mg of prednisone per day. Chronic, palliative, low-dose corticosteroid therapy involves the use of 5–10 mg of prednisolone daily. Ideally, the smallest possible dose should be used for the shortest period of time.[26,27]

The Presence of Underlying Risk Factors

Patients with certain underlying disorders are paticularly prone to develop side effects of chronic glucocorticoid therapy. Patients who are to be started on steroids should be screened for the presence of diabetes mellitus, hypertension, cardiovascular disease, peptic ulcer disease, and osteoporosis. In addition, tuberculosis, other chronic infections, and psychiatric illness should be excluded. Patients with the above-mentioned illnesses are particularly likely to reap the ravages of steroid hazard. The potential problems with glucocorticoid therapy become magnified when the cited underlying disorders are present. In such settings use of glucocorticoids is not absolutely contraindicated, if such therapy is deemed absolutely essential to control the disease process. However, greater scrutiny and care are required in these patients.

The Type of Glucocorticoid Chosen

The glucocorticoid chosen for treatment should contain minimal mineralocorticoid activity. Cortisone and hydrocortisone have the highest mineralocorticoid activity, while dexamethasone, methylprednisolone, betamethasone, and triamcinolone have little or no mineralocorticoid activity. Prednisone and prednisolone fall somewhere in between and contain some mineralocorticoid activity.

The glucocorticoid chosen for treatment should be of the intermediate-acting type, which can be given as a single dose in the morning, a strategy that minimizes the potential for hypothalamic pituitary adrenal suppression. Prednisone, prednisolone, and methylprednisolone are ideally suited for this purpose. Of course, when large doses of intermediate-acting glucocorticoids are given for more than a 2-week period, HPA suppression occurs regardless of the duration of action.

In addition to these two features, the glucocorticoid chosen must possess substantial anti-inflammatory activity. On a weight for weight basis the intermediate and long-acting glucocorticoids are more anti-inflammatory than cortisone or hydrocortisone (Table 23). Betamethasone and dexamethasone are the most potent anti-inflammatory glucocorticoids. However, the long duration of action of these steriods renders them more conducive to causing HPA suppression. Prednisone, prednisolone, and methylprednisolone are ideally suited for long-term pharmacological therapy owing to their intermediate duration of action, their minimal mineralocorticoid activity, and their fairly potent anti-inflammato-

TABLE 23.
Anti-inflammatory Potency

Steroid	Equivalent anti-inflammatory dose (mg)
Cortisone	25
Hydrocortisone	20
Prednisone	5
Prednisolone	5
Methylprednisolone	5
Triamcinolone	4
Dexamethasone	0.75
Betamethasone	0.6

ry effects. In the presence of liver disease prednisolone is preferable to prednisone since conversion of prednisone to its 11-hydroxy metabolite is impaired. Dexamethasone is seldom used for long-term anti-inflammatory effect, although clearly this drug is unparalleled in its anti-inflammatory potency while totally lacking mineralocorticoid potency. However, its long duration of action with the attendant HPA suppression makes it unsuitable for long-term, high-dose therapy. Cortisone and hydrocortisone, which are short-acting and "natural" steroids, are ideal for physiological replacement since these compounds possess some mineralocorticoid effect, and since their use in such a setting is not primarily for anti-inflammatory purposes.

Other Treatment Modalities

Every attempt should be made to exploit other, less hazardous, forms of therapy prior to embarking on glucocorticoid therapy. When possible, the use of other symptomatic remedies may facilitate reduction of the glucocorticoid dosage. The use of salicylates, other nonsteroidal anti-inflammatory agents, and other agents can be combined with steroid therapy to obtain the maximal benefit with the lowest possible dose of steroids.

Alternate-Day Regimen

In the early 1960s, several workers hypothesized that the anti-inflammatory effects of glucocorticoids persist longer than the undesirable metabolic effects.[28–31] The validity of this hypothesis is now well established. Alternate-day glucocorticoid therapy not only reduces the incidence of some of the adverse side effects of steroids, but also cause less suppression of the hypothalamic pituitary axis. In addition, such therapy is nearly as effective as daily therapy in controlling the underlying disease. The use of alternate-day steroid therapy has been found to be beneficial in the treatment of nephrotic syndrome,[32,33] rheumatoid arthritis,[34] lupus nephritis,[35] asthma,[36] myasthenia gravis,[37,38] sarcoidosis,[39,40] and pemphigus vulgaris.[41] In some disorders alternate-day drug therapy as the initial therapeutic program may fail to induce a satisfactory response. Also, patients who had been controlled with a daily regimen may fail to respond when switched abruptly to an alternate-day regimen. Occasionally, alternate-day therapy can be hazardous in patients who have already developed HPA suppression from prior use of chronic daily therapy, because the patient has no glucocorticoid coverage during the latter part of the 48-hr cycle.

Route of Administration

Corticosteroids can be administered orally or parenterally. In addition, topical, intralesional, or intra-articular forms of administration can be resorted to, in appropriate circumstances. Topical corticosteroids are effective agents in the treatment of several dermatological disorders, such as seborrheic dermatitis, atopic or neurodermatides, pruritus, contact or allergic dermatitis, and pemphigus. Chronic and indiscriminate use of topical steroids, especially the high-potency topical steroids such as betamethasone and fluocinonide, may produce iatrogenic Cushing's syndrome as well as HPA suppression. Prolonged topical application can also result in atrophy of skin and development of striae. Intralesional therapy for cutaneous lesions and intra-articular steroid therapy for localized inflammatory noninfectious articular lesions can provide immense symptomatic relief in selected cases. The duration of the beneficial effect is also variable. Triamcinolone is usually preferred for intraarticular injections. When glucocorticoids are chosen for local therapy, the preparation chosen should be potent and should contain a hydroxyl group in C-11. Thus, prednisone and cortisone, which are 11-keto compounds, are not useful for this purpose.

Aerosolized corticosteroids have gained an increasing popularity in the treatment (and perhaps prophylaxis) of asthma attacks. In some patients receiving oral steroids for bronchial asthma, reduction in oral dosage can be attained by the concomitant use of aerosolized steroids. Beclomethasone dipropionate and dexamethasone phosphate are usually employed for this purpose. Although systemic side effects are considerably less than with oral therapy, the potential for HPA suppression still does exist with aerosolized steroids. In addition, one particular complication, oropharyngeal candidiasis, occurs as a common side effect of aerosolized steroids. Dexamethasone has also been used in the form of nosedrops for the treatment of allergic rhinitis.

Although adverse metabolic side effects and HPA suppression are less common with topical and aerosolized steroids, the potential for such phenomena should never be ignored in these patients.

Mechanism of Action

Glucocorticosteroids play a pivotal role in the therapy of diseases mediated by inflammatory or immune phenomena. Glucocorticoids affect several populations of leukocyte cells. Some effects of glucocorticoids are exerted on the traffic of leukocyte cell population to various sites, while other effects are exerted directly on the function of these cells. The steroid-induced alterations in granulocyte function and kinetics are brought about in several ways.[42]

1. Glucocorticoids induce neutrophilic leukocytosis, which reaches a peak 4–6 hr following administration of the drug.[43] This effect is brought about by dual phenomena—increased mobilization of neutrophils from the bone marrow, and a reduced egress from the circulation to the site of inflammation.[44]

2. One of the most important effects of glucocorticoids lies in the ability of these drugs to prevent accumulation of neutrophils and monocytes at the site of inflammation. This effect is most pronounced when glucocorticoids are administered on a daily basis and attenuated when the drug is given on alternate days.[45]

3. Glucocorticoids, in some unknown manner, alter the granulocyte surface, leading to decreased adherence of these cells to the vascular endothelium.[46]

4. Glucocorticoids suppress phagocytosis and bactericidal activity of neu-

trophils in vitro. It has been shown that granulocytes obtained from patients receiving glucocorticoids demonstrate a decreased reduction of the chemical nitroblue tetrazolium.[47] The significance of this effect on granulocyte function in vivo is uncertain.

5. The characteristic eosinopenia induced by glucocorticoids may be due to lysis of these cells or alternatively could represent redistribution of eosinophils out of the circulating compartment.

6. Glucocorticoids cause significant reductions in the lymphocytic population. All subpopulations of the lymphocytes are decreased, but the degree of suppression is more profound on the T lymphocytes as compared to the B lymphocytes.[48,49] The effect is transient, following the administration of single dose of glucocorticoids, returning to normal within 24 hr. The decrease in lymphocytes is probably brought about by redistribution of the circulating cells to other compartments of the body. Fauci and Dale[50] have convincingly demonstrated that the depletion of intravascular lymphocytes predominantly affects the cells that belong to the recirculating pool.

7. The mechanism by which glucocorticoids affect the function of the mononuclear cells has been a matter of controversy. Several mechanisms have been proposed; glucocorticoids can affect the stability of these cells by causing alterations in the lysosomes; or alternatively, corticosteroids react with cytoplasmic receptors, following which the steroid–receptor complex migrates to the nucleus and affects nucleic acid transcription.

8. Glucocorticoids also affect several facets of cell-mediated immunity. The abolition of delayed hypersensitivity reaction by steroids is believed to result from decreased recruitment of macrophages necessary for the expression of cellular immunity.[51] In addition, glucocorticoids can affect a variety of soluble mediators of cell-mediated immunity. These include macrophage migration inhibitory factor, monocyte chemotactic factor, and macrophage aggregating factor. In addition, corticosteroids suppress the proliferation of lymphocytes in response to antigens and mitogens in vitro and in vivo.

As a result of these effects glucocorticoids exert their profound antiinflammatory actions. These unique effects also lead to immunosuppression of the host, resulting in a predisposition to a multitude of infections. The effects of chronic steroid therapy on the immune system have important clinical implications.

1. Prolonged glucocorticoid administration results in atrophy of lymphatic tissue, including the spleen and the thymic tissue. Lymphopenia is the hematological hallmark of such lymphosuppression.

2. Glucocorticoid therapy suppresses delayed hypersensitivity reactions. Reduction in tuberculin reactivity is an important reflection of this phenomenon. Salomon and Angel[52] have convincingly shown this reduction in tuberculin sensitivity which developed within 3 months of therapy with ACTH or steroids and returned to normal within a week of discontinuation of therapy.

3. Chronic glucocorticoid therapy renders the host immunocompromised.

This is the single most dangerous complication of protracted steroid therapy, predisposing the steroid-treated patient to several common as well as unusual pathogens.

Clinical Considerations

To the clinician caring for patients on glucocorticoid therapy, the problems inherent in such a setting are threefold. First, the patient may develop all the features of iatrogenic Cushing's syndrome. The situation is particularly frustrating because the patient's need for glucocorticoids renders the physician powerless to halt this progression. Second, the numerous side effects of steroid therapy can occur in the absence of overt Cushing's syndrome. These side effects encompass a multitude of disciplines, necessitating that the physician keep a broad and informed perspective. Third, the complicated and often puzzling facets of HPA suppression and steroid withdrawal syndrome behoove the physician to understand the complexities of endocrine feedback.

Iatrogenic Cushing's Syndrome

Perhaps nowhere are the plethoric features of Cushing's syndrome more impressively manifest as with the development of iatrogenic Cushing's syndrome. The characteristic "moon facies" and fat deposition (dorsocervical and supraclavicular fat pads), cutaneous atrophy, striae, osteoporosis, and diabetes may often develop with alarming and unanticipated rapidity in patients receiving large doses of glucocorticoids. The resultant syndrome, to a large extent, resembles natural (endogenous) Cushing's syndrome, with some notable exceptions. These differences, originally pointed out by Ragan,[53] have been highlighted by Axelrod.[54] The differences arguably might be due to the differences in ACTH secretion between with pituitary-dependent Cushing's syndrome and iatrogenic Cushing's syndrome. Thus, as expected, features of androgen excess (hirsutism, acne, menstrual disturbances) and mineralocorticoid excess (hypertension) are more frequent in patients with ACTH-mediated Cushing's syndrome. The features that are unique to iatrogenic Cushing's syndrome are the development of benign intracranial hypertension,[55] posterior subcapsular cataracts,[56] aseptic necrosis of the bone,[57] glaucoma, and pancreatitis.[54] These complications are likely to occur in patients receiving very large doses of glucocorticoids for protracted periods of time. The reason for the unique occurrence of these features in iatrogenic Cushing's syndrome is not clear. Perhaps the degree of hypercortisolism achieved with heavy-dose steroid therapy is seldom attained in natural Cushing's syndrome. The difficulty in accepting this hypothesis lies in the fact that there are no significant differences in the steroid levels between patients who develop the Cushingoid features while on steroid therapy and those who do not. It is generally believed that although patients on high-dose therapy are susceptible to this complication, the determining factor is the metabolic clearance of the administered glucocorticoid. Thus, patients with a decreased clearance rate of prednisolone are more likely to develop iatrogenic Cushing's syndrome than those with a normal or increased clearance.[58] Once florid Cushing's syndrome has developed, the stigmata are slow to resolve even after withdrawal of the drug. Of course, withdrawal of steroids in this setting has to be done with the utmost caution, since these patients have profound suppression of their HPA axis. The process of weaning is often further complicated by the exacerbation of symptoms of the underlying disease, as well as the development of symptoms of steroid withdrawal such as nausea, vomiting, lethargy, and myalgia. The process of weaning patients with iatrogenic Cushing's syndrome from steroids is an exercise in patience and is frought with untold frustration on the part of patient and physician.

Multisystem Effects of Steroid Therapy

The adverse effects of chronic glucocorticoid therapy can affect almost every organ system. In addition to complications that involve the cardiovascular, musculoskeletal, gastrointestinal, and central nervous systems, chronic steroid therapy affects the bone, somatic growth, and the eyes and exerts a profound influence on other endocrine and metabolic phenomena.

Cardiovascular

The acute administration of large doses of glucocorticoids to normal subjects results in an increased cardiac output with a decrease in the peripheral resistance.[59] While fluid retention is not a significant problem in patients receiving glucocorticoids with little or no mineralocorticoid activity, the use of large doses of glucocorticoids may result in mild fluid retention, especially in patients with congestive failure. The incidence of hypertension in patients treated with glucocorticoids ranges between 4 and 25% and appears to be related to the dose and duration of therapy, as well as the presence of underlying renal disease.[60] The blood pressure elevation seen in association with glucocorticoid therapy is mild to moderate, although occasionally malignant hypertension has been reported to occur.[61] Fluid retention or sodium retention plays little role, if any, in the development of glucocorticoid-mediated hypertension.

Musculoskeletal

Corticosteroid excess, whether exogenous or endogenous, causes untoward effects on the muscle tissue. In its extreme form this assumes the form of "steroid myopathy."[62–71] This is a slowly evolving, chronic, painless myopathy that predominantly involves the proximal muscles. Muscles of the pelvic girdle are more often, and more severely inolved than those of the pectoral girdle. Mandel[72] characterized the features of steroid myopathy and has pointed out several important features:

1. The process is insidious in onset and slowly progressive.
2. Steroid myopathy can occur at any age but demonstrates a predilection to affect patients most commonly between 20 and 50 years of age.
3. Proximal, pelvic girdle myopathy is the most common form of the disease. Weakness and fatigue are the most common symptoms. Obvious

muscle wasting is often present. Exercise may, paradoxically, improve the weakness.

4. Pain is characteristically absent.

5. While any steroid preparation can result in myopathy when used in large doses, steroid myopathy is particularly common with the use of fluorinated compounds such as triamcinolone or dexamethasone. The longer tissue half-life of these steroids might be responsible for this.

6. When steroid myopathy has developed, other facets of steroid excess are evident.

7. The muscle enzymes in the serum are usually normal or mildly elevated. Creatinuria is nearly always present.

8. Electromyographic findings usually demonstrate nonspecific abnormalities. Characteristically, fibrillations or fasciculations are absent.

9. The most consistent finding in the muscle biopsy is type 2B atrophy. Histochemically, skeletal muscle fibers can be divided into two types— type I fibers have a low glycolytic and high oxidative activity, while type II fibers have just the opposite. Patients with Cushing's syndrome demonstrate characteristic, selective atrophy of type II glycolytic fibers.[73] Electron microscopy may reveal large mitochondria and accumulation of glycogen, with eventual development of mitochondrial disarray, degenerative changes, and alterations in the Z-line pattern of the sarcomere.

The clinical severity of steroid myopathy is quite variable. Occasionally, the muscle weakness can become so profound and generalized as to render the patient bedridden. The problem can be further compounded by the fact that several diseases for which steroid therapy is used can be complicated by the development of myopathy as part of the disease process. This is particularly so in the case of polymyositis and other connective tissue disorders, such as systemic lupus erythematosus and rheumatoid arthritis. In such cases the distinction between steroid myopathy and disease flare can become difficult, even impossible. This problem has been addressed by Askari et al.,[74] who described the development of steroid myopathy in eight patients treated with steroids for a variety of connective tissue disorders. Three important observations were made by these workers. First, myalgia was rather common when steroid myopathy first appeared. This symptom had not previously been associated with steroid myopathy. These myalgias rapidly disappeared upon reduction of steroid dosage. Second, the pattern of development of the myopathy induced by steroid therapy was not different from that of polymyositis. The steroid myopathy, however, progressed in a systematic and predictable fashion, initially involving the proximal pelvic muscle, then involving the shoulder girdle muscles, and ultimately becoming generalized with involvement of the distal muscle groups as well. Third, the combination of creatinuria with persistently normal enzyme levels supported the diagnosis of steroid myopathy, while the combination of creatinuria and progressive increase in enzyme levels favored disease flare. Of course, this fact must be interpreted with caution because some patients with polymyositis on steroids may not show an enzyme elevation during the early phase of disease flare.[75,76]

The pathogenesis of steroid myopathy continues to remain elusive. Several mechanisms may underlie the development of steroid myopathy.[77–81]

1. Glucocorticoids inhibit the uptake of amino acids by the skeletal muscle. This action, a primal effect of glucocorticoids, leads to decreased incorporation of amino acids into muscle protein.
2. Glucocorticoids interfere with oxidative metabolism of muscle. This fact, coupled with enhanced glycogen synthesis and decreased protein synthesis, leads to catabolism of muscle protein and eventual muscle wasting.
3. Glucocorticoids may induce membrane abnormalities that lead to a depolarization block. Decreased intracellular potassium concentration and impaired calcium uptake as well as binding in the sarcoplasmic reticulum have been postulated.[72]

Regardless of the pathogenesis, steroid myopathy resolves within 1–4 months after cessation of steroid therapy. Immediate strategies for treatment include reduction of dose, switching to a different preparation, switching to an alternate-day regimen, or, arguably, the use of phenytoin.[82] Since most of the adverse effects of steroid therapy are attenuated by an alternate-day regimen, consideration should be given to this method of administration to hopefully prevent this complication. This appears particularly relevant in the treatment of disorders where myopathy can complicate the disease process per se.

Gastrointestinal

The gastrointestinal effects of chronic glucocorticoid therapy include the development of peptic ulcer disease, particularly gastric ulceration, gastric hemorrhage, intestinal perforation, and, occasionally, pancreatitis. Of these, the association between steroid therapy and peptic ulceration merits special consideration. The literature is replete with anectodal associations of peptic ulcer disease in patients receiving steroid therapy.[83–89] The clinical, radiological, and pathological features of peptic ulcers in patients receiving steroid therapy closely resemble the ulceration seen in association with other ulcerogenic drugs. However, Conn and Blitzer,[90] in 1976, combined the data from 42 randomized, controlled trials and concluded that peptic ulceration does not occur with steroid therapy unless these drugs are given for a period exceeding 30 days, and in a total dose exceeding 1 g of prednisone. This controversial report prompted reexamination of the issue by Messer et al.,[91] who pooled the data from 71 controlled clinical trials where patients had been randomized to systemic glucocorticoid (or ACTH) therapy or to nonsteroidal therapy. Their report strongly suggests that of 3064 steroid-treated patients evaluated for peptic ulcer, 1.8% had ulcers, as compared with 0.8% of 2897 controls. Also, of the 3135 steroid-treated patients evaluated for GI bleeding 2.5% had bleeding as compared to 1.6% of the 2976 controls. Their conclusion was that the risk of peptic ulcer disease and GI hemorrhage is significant, even in patients who receive steroid therapy for less than 30 days, even when the total dose does not exceed 1000 mg/day. Based on the belief that steroid-induced ulcerogenesis is not a myth, many physicians routinely resort to concomitant treatment with antacids and or cimetidine.

Rarely, esophagitis[92] and esophageal perforation[93] have been reported in association with glucocorticoid therapy. The incidence of acute pancreatitis is clearly higher in glucocorticoid-treated patients, particularly children.[94] In one autopsy series of children with nephrotic syndrome, evidence of pancreatitis was seen in 40%, compared with the 10% incidence in the non–steroid-treated group.[95] The mechanism for this occurrence remains unclear.

Central Nervous System

The effects of glucocorticoid treatment on the nervous system are three-fold: effects on the electrical activity of the brain, effects on the mood and behavior, and effects on conduction of impulses across nerve tissue.

Several studies have documented that glucocorticoid excess can result in definite alterations in the electroencephalogram.[96–98] These changes include an increase in theta waves, changes in the alpha activity, and bursts of increased activity. These changes assume significance in view of the fact that seizures have been reported to occur in patients with no prior history of convulsive disorders placed on steroid therapy.[99,100] These rare reports emphasize the need for caution when treating patients with convulsive disorders with steroids.

The psychiatric manifestations seen in patients with hypercortisolism are of a broad spectrum. Euphoria is often present during the early phase of initiating glucocorticoid therapy. The range of behavior changes varies from mild depression and sleep disturbances to paranoid ideation with suicidal tendencies. The psychiatric manifestations that accompany the hypercortisolemic state are outlined in chapter 3.

The development of pseudotumor cerebri in association with prolonged corticosteroid treatment is unique for exogenous hypercortisolism.[55,101] Peculiarly, the occurrence of this phenomenon has been known to coincide with a reduction in the dose of steroid or during a switch in the glucocorticoid preparation being used.

Finally, although glucocorticoids experimentally decrease nerve conduction velocity,[102] peripheral neuropathy is extremely unusual with chronic steroid therapy.

Steroid-Induced Bone Disease

Corticosteroid-induced bone loss continues to pose one of the most disabling problems of steroid therapy. It has been known for decades that prolonged elevations in the corticosteroid levels can result in severe bone loss.[103,104] Steroid-induced osteopenia can be viewed from several important perspectives; the frequency of the phenomenon, the distribution of bone loss, the relationship of steroid-induced bone loss to the underlying disease treated, the recognition of patients at high risk, the pathophysiology of the osteopenia, and the preventive aspects of this dreaded complication must be placed in perspective.

Frequency of Steroid Osteopenia.   When patients with Cushing's syndrome are evaluated for the presence of osteopenia, a significant reduction of bone mass is seen in 80–90% of patients.[104,105] There is every reason to believe that bone loss, if sought carefully, can also be demonstrated in a significant percentage of patients with iatrogenic Cushing's syndrome. It is noteworthy that a vast proportion of patients with steroid-induced osteopenia tend to remain asymptomatic for surprisingly long periods of time. Equally noteworthy is the observation that symptomatic osteopenia from steroid therapy appears to be highest in two groups of patients[106,107] children (who have a rapid bone turnover) and women over the age of 50 (who have a relatively low bone mass to begin with). Attempts in the literature to correlate the development of osteopenia with the type of the glucocorticoid used, the duration of therapy, and the dose of the drug have yielded inconsistent data. The inherent difficulties in separating the effect of steroids from the effect of other important variables such as age, sex, underlying disease, and the use of other medications are the reasons for the widely discrepant data in the literature. Hahn[108] was able to demonstrate a definite relationship between the duration of corticosteroid therapy and degree of reduction in the bone mass. Other workers[109] have not found such a correlation.

Distribution Pattern of Bone Loss.   The corticosteroid-induced bone loss occurs in a characteristic distribution. The process predominantly involves the trabecular bone. Therefore, more severe osteopenia occurs in the vertebrae and the ribs, which have a higher content of trabecular bone; the trabecular bone has a higher turnover rate than cortical bone. Severe steroid-induced bone can also involve the cortical bone, but these changes are less dramatic. The concept that trabecular bone is markedly affected by steroid use has resulted in evaluating trabecular bone density measured by single (or dual) energy proton absorptiometry, as a method for evaluating the effect of corticosteroids on bone.

Steroid Osteopenia and Underlying Disease.   It is well known that in many disease for which steroid therapy is used, bone disease can be an inherent part of the primary disease process. This is best exemplified by patients with rheumatoid arthritis or renal disease on steroids.[107,108,110–113] Understandably, this poses difficulties in separating the effects of steroids from the effects of the underlying disease on the bone mass. In fact, Mueller[114] measured bone mass at the distal radius and reported that glucocorticoid therapy caused bone loss only in patients with rheumatoid arthritis, but not in those with asthma. Thus, the prevalent opinion in the past was that the underlying disease (i.e., rheumatoid arthritis, renal disease) had an impact on the development of steroid-induced osteopenia. However, in a recent study, Adinoff and Hollister[109] retrospectively reviewed patients with asthma who had taken daily or alternate-day corticosteroids for at least a year and found a significant incidence of fractures of the vertebrae or ribs. They also prospectively studied 30 hospitalized asthmatic patients between 10 and 70 years of age and found rib or vertebral fractures in 8 of the 19 asthmatic patients receiving long-term steroid treatment. In addition, the trabecular, but not the cortical, bone mass was decreased in the group of asthmatics on long-term steroid therapy. Clearly, steroid therapy can cause osteopenia regardless of the nature of the underlying disease.

Recognition of Patients at High Risk.   As indicated earlier, steroid-induced osteopenia is particularly common in women over 50 years of age, children, and patients whose disease has limited their ambulation. Identification of patients at high risk for steroid osteopenia entails identification of the subset of patients losing bone at a rate higher than that seen as a result of the normal aging process. The following approach has been suggested[115] for this purpose. The bone mass of the distal radius is measured at the start of glucocorticoid therapy and every 3–6 months thereafter for approximately 2 years. With these data the rate of bone loss can be determined. Should the bone loss occur at a rate higher than that seen as a consequence of aging, preventive measures can be considered in patients with a relatively low basal bone mass.

Pathophysiology of Steroid Osteopenia. Histological studies of bone biopsy samples from patients who have been treated with glucocorticoids suggest that dual mechanisms underlie the bone loss of steroid-treated patients—decreased bone formation and increased resorption.[116–118] The dramatic and rapid loss of trabecular bone in steroid-treated patients is because of this dual effect.

The decrease in bone formation in patients treated with steroids is believed to be a direct inhibitory effect of glucocorticoids on osteoblastic function.[119] In vitro, addition of cortisone to bone cells maintained in culture dramatically decreases protein synthesis.[120]

A great deal of controversy revolves around the mechanisms that increase osteoclastic resorption. While it is generally accepted that glucocorticoid excess suppresses intestinal calcium absorption, the mechanism for such a phenomenon is far from clear. Most investigators have described a secondary increase in parathyroid hormone levels to counteract the decrease in intestinal absorption.[108] The effects of short-term administration of glucocorticoids on intestinal calcium absorption and its relationship to circulating parathyroid hormone (PTH) and vitamin D metabolite concentrations have been studied by Hahn and his co-workers.[121] They concluded that the reduced intestinal calcium absorption following glucocorticoid administration could not be attributed to changes in immunoreactive PTH or circulating concentrations of the major known metabolites of vitamin D. To help resolve the differences cited in the literature, Findling et al.[122] studied the relationship of vitamin D metabolites and PTH to calcium and phosphorus homeostasis in seven patients with spontaneous Cushing's syndrome. Their data suggest that endogenous hypercortisolism decreases tubular reabsorption of phosphorus and increases PTH, which in turn results in an increase in 1,25-(OH)2 D3. These effects may contribute to the bone loss caused by the direct action of cortisol on bone. Histological and biochemical studies have demonstrated that glucocorticoids are not only capable of inhibiting bone formation, but also can *directly* stimulate bone-resorbing cells.[123] The increased PTH secretion assumes importance as a contributory factor, especially in view of data that support the notion that parathyroidectomy abolishes the osteoclastic effect of steroids on bone in animals.[124] These facts may play an important role in the prevention and treatment of steroid-related osteopenia.

Prevention of Steroid Osteopenia. It is still unclear whether steroid-induced osteopenia is preventable. Once the patient at high risk for steroid-induced osteopenia has been identified, i.e., patients with a relative decrease in basal bone mass, patients with accelerated rate of bone loss, children, and women over the age of 50, consideration should be given to prophlactic programs. None of the prophylactic programs have been evaluated on a large scale. Simple measures include promoting ambulation, estrogen replacement for postmenopausal females, and androgen replacement for hypogonadal males. The combination of sodium fluoride, vitamin D, and calcium has been tried, without much success in preventing the progressive loss of trabecular bone in patients receiving glucocorticoids.[125] When hypercalciuria is present, hydrochlorothiazide can be administered to minimize calcium loss from the renal tubules.[126] In the absence of hypercalciuria, the patient on glucocorticoids can be tried on a regimen of vitamin D and calcium, a regimen that has not been widely evaluated. Baylink[115] has emphasized that even in the absence of proof that such prophylactic programs are effective, it is far easier to maintain bone mass than it is to replace it. These prophylactic programs, at the very least, may help maintain bone mass.

The treatment of overt osteoporosis with fractures is as disappointing as the treatment modalities for postmenopausal osteoporosis.

Ophthalmological

The ophthalmological complications of steroid therapy are rare and generally occur when high doses are used for protracted periods of time. The spectrum of ophthalmological complications seen in association with steroid therapy is outlined in Table 24.

The most important ocular complication of steroid therapy is the development of posterior subcapsular cataracts.[127–130] The incidence of developing this complication ranges between 11 and 38%.[130] It is believed that while adults develop this complication only with large doses of steroids, children may develop this complication with much lower doses of glucocorticoids. The other complications listed in Table 24 are extremely rare.

Skin and Integument

The association between the development of transparent skin and steroid therapy is a controversial one. This complication is particularly likely to occur in patients with rheumatoid arthritis who are on glucocorticoid therapy.[131] Green-

TABLE 24.
Ophthalmological Complications
of Glucocorticoid Therapy[a]

Part of eye	Complications
Lens	Posterior subcapsular cataract
	Acute myopia
	Minor refractive changes
Anterior chamber	Glaucoma
Lids	Swelling, chemosis
	Ptosis
Fundus	Papilledema
	Retinal hemorrhage
Extraocular muscles	Ocular muscle myopathy
Conjunctiva	Subconjunctival hemorrhage
Sclera	Thinning of the sclera

[a]Adapted from David et al.[60]

wood[132] evaluated 96 patients with rheumatoid arthritis on steroid therapy and concluded that the alteration in the dermal connective tissue induced by steroid therapy caused loss of skinfold thickness, transparency of skin, and the frequent occurrence of purpuric skin lesions. Perhaps for this reason the purplish, wide striae are more impressive in patients with iatrogenic Cushing's syndrome than in endogenous hypercortisolism.

Endocrine and Metabolic Effects

Glucocorticoids exert a profound influence on other endocrine systems. The most important of these are the effects of chronic glucocorticoid therapy on the pituitary, on the thyroid, and on glucose tolerance.

The Pituitary Gland

Glucocorticoid administration can result in striking perturbations in the dynamics of TSH, prolactin, and growth hormone.

Thyroid-Stimulating Hormone.   Wilber and Utiger,[133] in the late sixties, demonstrated that 24–48 hr following the administration of 2–8 mg of dexamethasone, the basal serum thyroid-stimulating hormone (TSH) levels fell by 18–47% in normal subjects and more dramatically (23–96%) in patients with primary hypothyroidism. Similar finding were reported by Nicoloff et al.,[134] who also found suppression of basal TSH levels in euthyroid subjects given 60 mg of prednisolone per day for 3 days. The glucocorticoid-induced TSH suppression appears to be exerted at the level of the thyrotropes. It is now abundantly evident that endogenous or exogenous hypercortisolism is associated with a blunted response of TSH to thyrotropin-releasing hormone (TRH) administration. Several studies have documented this impairment in the responsiveness of the thyrotropes to TRH in patients with various forms of Cushing's syndrome, including iatrogenic hypercortisolism.[135–140] This abnormality is usually reversible following the restoration of eucortisolemia. The glucocorticoid-induced functional blunting of the thyrotropes does not significantly affect thyroid function. Nevertheless, this phenomenon should be kept in mind while evaluating TSH dynamics in patients with any form of hypercortisolism.

Prolactin.   Although physiologically glucocorticoids play no discernible role in the regulation of prolactin secretion, perturbations in glucocorticoid dynamics are often associated with striking reciprocal changes in prolactin dynamics. Pharmacological doses of glucocorticoids suppress prolactin secretion in humans.[141–144] However, the effect is less striking and less consistent in comparison with the suppressive effect of glucocorticoids on TSH secretion. Thus, acute or chronic administration of glucocorticoids results in the blunting of prolactin responsiveness to TRH, insulin hypoglycemia,[143,145] and metoclopramide.[141] Although the effects of glucocorticoid excess on TSH and prolactin are analogous, these responses may be dissociated, i.e., abolition of TSH response to TRH, while the prolactin response to TRH is preserved. Glucocorticoids are believed to exert their suppressive action on prolactin secretion at the level of the pituitary. A large body of in vitro studies, as well as animal experiments performed with rats, support this tenet.[146–148] This intriguing effect of glucocorticoids on prolactin suppression has no clinical consequences.

Growth Hormone.   It has long been known that endogenous or exogenous hypercortisolemia impairs the response of growth hormone to a variety of provocative stimuli. Thus, the growth hormone responses to hypoglycemia,[149] arginine,[150] lysine, vasopressin,[150] and L-dopa[151] are impaired during hypercortisolemia, returning to normal once the hypercortisolemia has been corrected. The inhibitory effect of glucocorticoids on growth hormone release is believed to exerted at the pituitary level. This concept has been substantiated recently with the use of human pancreatic growth hormone–releasing hormone. Smals et al.[152] found that patients with endogenous hypercortisolism had an absent or severely blunted growth hormone response to intravenous administration of hp GRF 1-44.

The consequences of hypercortisolism on growth assume importance in youngsters. Although the dynamics of growth hormone in adults with hypercortisolism may demonstrate a blunting in provoked secretion, such is not the case in children; growth hormone dynamics in children with iatrogenic or spontaneous steroid excess is generally considered to be normal.[153,154] These observations, coupled with the fact that growth hormone therapy fails to induce linear growth in hypercortisolemic children,[155,156] suggest that glucocorticoids may interfere with somatomedin generation or action. Indeed, several studies have demonstrated low levels of net circulating somatomedin activity following glucocorticoid administration.[157–159] There is some controversy as to whether this reflects decreased production or indicates antagonism to the activity of these peptides. The results from studies using bioassay systems that test the direct effect of glucocorticoids on cartilage growth have yielded conflicting results. A recent concept gaining favor is that hypercortisolemia can lower somatomedin activity by inducing changes in somatomedin inhibitors. These inhibitors of somatomedin activity are present in normal subjects. Unterman and Phillips[160] have demonstrated that glucocorticoid administration is followed by an increase in circulating somatomedin inhibitors, which can account for the fall in the net somatomedin activity in the sera of patients with hypercortisolemia. Regardless of the mechanism, hypercortisolemia results in a significant impairment of linear growth, resulting in retarded skeletal maturation. The vertebral "growth arrest lines" (zones of increased density corresponding with the vertebral plates) may linger as residual radiological changes, years after cure, serving as a grim reminder of the childhood Cushing's.[161]

Thyroid Function

Although patients with iatrogenic Cushing's syndrome are mostly euthyroid, glucocorticoid therapy can alter several aspects of the convential thyroid function tests.

1. The serum T_4 concentration is frequently decreased in patients receiving prolonged glucocorticoid therapy.[162,163] This lowering is believed to be a result

of lowering of the thyroxin binding globulin (TBG) levels by glucocorticoids. While glucocorticoid-induced TSH suppression could also lead to lowering of serum T_4 concentrations, this appears less likely, since the free T_4 is normal in patients receiving glucocorticoids. The free T_4 is the unbound fraction of hormone available to the tissues and represents the biologically active fraction of the hormone. Therefore, hypothyroidism seldom results from glucocorticoid treatment. Glucocorticoids can cause a slight delay in thyroxine turnover.[164] This occurs as a result of steroid-induced redistribution of T_4. As a consequence, there is a reduction in the extravascular, hepatic T_4 pool and an increased reentry of T_4 into the plasma.

The ^{131}I uptake by the thyroid gland may be slightly decreased by steroid therapy. This effect is secondary to steroid-induced TSH suppression.

2. The serum T_3 concentration is also decreased, often strikingly, in patients receiving glucocorticoids. This is due to the well-known inhibitory effect of glucocorticoids on T_4-to-T_3 conversion. In normal subjects the major portion of circulating T_3 is derived by monodeiodination of T_4. This reaction, mediated by the enzyme "outer-ring deiodinase," takes place in the peripheral tissues, particularly the liver and the kidney. Short-term administration of glucocorticoids in pharmacological doses causes a prompt reduction in the circulating T_3 levels within 24 hr without inducing any change in the T_4 level.[165–167] This effect is particularly impressive with the use of dexamethasone.[165,168] The reduction in the circulating concentration of T_3 is accompanied by a contemperaneous and reciprocal increase in the levels of reverse T_3 (3,3',5'-tri-iodothyronine), the metabolically inactive isomer of T_3. The prompt and dramatic decreases in circulating T_3 levels brought about by dexamethasone can be exploited in emergent situations such as thyroid crisis, where immediate reductions in circulating T_3 concentrations may be desirable. However, the same effect is often achieved by other drugs (propylthiouracil, propranolol, and sodium ipodate) that are commonly employed in the treatment of thyroid crisis.

3. The effects of steroid therapy on basal and stimulated TSH secretion are detailed in Chapter 2, in the section dealing with extraadrenal dysfunction in Addison's disease.

The triple phenomena just mentioned i.e., lowering of T_4, lowering of T_3, and suppression of basal and stimulated TSH levels, assume importance when thyroid disease is suspected in patients receiving glucocorticoid therapy. For instance, measuring the TSH response to TRH for the purpose of confirming or excluding hyperthyroidism is of no value in patients receiving glucocorticoids. When patients on glucocorticoid therapy do develop hyperthyroidism, the hormonal elevations may not be striking. Further, the profile of circulating thyroidal hormones resembles that of "T_4 toxicosis" with elevated total and free T_4, in the presence of a normal or even low circulating T_3 concentrations. Finally, the development of primary hypothyroidism can be easily missed since the basal TSH levels in steroid-treated patients may be suppressed, and the abnormalities in T_4 and T_3 could be erroneously attributed to "steroid effect." Some guidelines may help prevent these pitfalls. First, since abnormalities in TBG are often present in glucocorticoid-treated patients, measurement of free hormone concentrations represents a superior method for assessment of thyroid status. Second, the diagnosis of hypothyroidism may occasionally pose difficulties in the steroid-treated patient. In addition to the conventional tests, measurement of reverse T_3 may assist in establishing the diagnosis of hypothyroidism. This isomer is usually elevated when the low T_3 is caused by steroids, but is lowered in presence of concomitant hypothyroidism. Finally, the TSH response to TRH must be interpreted with caution in patients receiving glucocorticoid therapy.

Glucocorticoid Excess and Glucose Intolerance

Glucocorticoid-induced glucose intolerance is usually characterized by hyperglycemia and hyperinsulinemia, a combination that suggests insulin resistance.[169–171] Several mechanisms have been proposed to explain the glucose intolerance seen in association with glucocorticoid administration.

1. Glucocorticoids increase hepatic gluconeogenesis. This primal effect of glucocorticoids results in increased production of glucose by the liver. In addition, glucocorticoids produce glycogenolysis, an effect that is also conducive to increased hepatic glucose output. The latter effect is probably mediated by the glucocorticoid-facilitated release of glucagon. Hepatic glucose production in patients with hypercortisolism can be inhibited by insulin. However, much higher doses of insulin are required, indicating a certain degree of resistance to the action of insulin at the hepatic level. Studies performed on isolated perfused rat liver have favored the concept that glucocorticoids support gluconeogenesis, as well as glycogen synethesis, and facilitate the actions of glucagon.[172]

2. In addition to the hepatic effects of glucocorticoids that lead to increased glucose production, steroids cause insulin resistance at the level of peripheral tissues. The insulin resistance seen in glucocorticoid-treated patients is reflected in the combination of hyperinsulinemia in the presence of abnormal glucose tolerance. In addition, decreased tissue sensitivity to insulin can be demonstrated by the use of euglycemic clamp procedures which permit calculation of the maximum glucose disposal following insulin infusions. Patients with endogenous and exogenous hypercortisolism demonstrate dose–response curves that are significantly shifted to the right, indicating the higher insulin requirements for maintenance of adequate glucose disposal. The exact mechanism that underlies the insulin resistance associated with glucocorticoid excess has not been completely elucidated. While decreased glucose transport[173] or interference in the intracellular metabolic pathways by glucocorticoids[174] remains a feasible mechanism, the focus of interest has clearly shifted to insulin receptors. Studies that evaluate the effect of glucocorticoid excess on insulin binding to receptors have yielded highly inconsistent results. Thus, insulin binding to receptors has been reported to be normal,[175] decreased,[176] or even increased[177] in hypercortisolemia. These conflicting data on insulin receptor studies have led to the search for other mechanisms. It has been proposed that glucocorticoids impair tissue glucose disposal through a postreceptor mechanism. Rizza et al.[178] and Nosadini et al.[179] have demonstrated that cortisol excess results in an insulin-resistant state without significant alterations in the insulin receptor binding to monocytes and erythrocytes.

3. In addition to increased hepatic glucose production and increased tissue resistance to insulin, glucocorticoids may have direct effects on the islets of Langerhans. Glucocorticoids stimulate glucagon secretion by the alpha islet cells, an action that is conducive for accentuation of glucose intolerance. The direct effects of glucocorticoids on beta cells are controversial.In rats glucocorticoid administration has been shown to induce striking histological alterations in the beta cells.[180] However, in humans, glucocorticoids are not believed to directly exert significant effects on the beta-cell population.

The clinical implications of the impact of glucocorticoids on glucose metabolism are several. First, although abnormal glucose tolerance may be encountered in a high proportion of steroid-treated patients, overt, severe, symptomatic diabetes develops in only a few. This group represents patients with a predisposition to the development of diabetes. Second, the "steroid diabetes" usually requires insulin for control. Occasionally, "clinical" insulin resistance develops in steroid-induced diabetes, necessitating extremely large amounts of insulin. Third, steroid-treated patients show a tendency to develop hyperosmolar diabetic dehydration. The risk, however, is quite small. Fourth, the dose and duration of steroid therapy, as well as the type of glucocorticoid employed, may have an influence on the steroid-induced glucose intolerance. Interestingly, the impairment in glucose tolerance seen in normal subjects placed on steroids paradoxically tends to improve during chronic glucocorticoid administration.[169,181] Finally, the alternate-day regimen may cause less alterations in glucose tolerance than daily treatment with glucocorticoids.

Opportunistic Infections

In 1952, Plotz et al.[182] warned that the recent introduction of cortisone and ACTH into the therapeutics of medicine might produce a heightened proclivity for infections in steroid-treated patients. The prophetic nature of this observation was evident 6 years later, when Kass and Finland[183] first documented the association between infections and corticosteroid therapy. Extensive laboratory studies[184,185] have established that glucocorticoid treatment enhances susceptibility to various infections. The steroid-treated host becomes easy prey to a variety of bacterial, fungal, viral, and parasitic organisms.

Bacterial Infections

Bacteria account for the majority of infections that complicate steroid therapy. Thus, pneumonia, pyelonephritis, peritonitis, liver abscess, endocarditis, and osteomyelitis have all been reported to occur in glucocorticoid-treated patients.[186–188] More important is the effect of chronic glucocorticoid therapy in reactivating tuberculosis. It is customary to recommend chemoprohylaxis with isoniazid in tuberculin-positive patients who are to be placed on chronic glucocorticoid therapy.[189,190] The incidence of steroid-treated patients contracting tuberculosis is low. However, this complication should be kept in mind while evaluating pulmonary symptoms in patients receiving steroid therapy. An association between steroid therapy and the development of infection with *Listeria monocytogenes* has been implied in the literature.[191]

Fungal Infections

The association between several fungal infections and chronic hypercortisolism is well documented. The association is particularly strong in the case of two fungal infections—candidiasis and aspergillosis.[192–195] In approximately 30% of patients with disseminated candidiasis, and in 50% of patients with disseminated aspergillosis, prior or present exposure to systemic glucocorticoid therapy can be encountered.[60] These infections sometimes can become rampant, with disastrous sequelae, culminating in fatalities.[192] In addition to candidiasis and aspergillosis, patients on steroid therapy are predisposed to cryptococcosis, nocardiosis, and even mucormycosis. The tendency for the predisposition to fungal infections is not limited to iatrogenic Cushing's syndrome. Graham and Tucker[196] described a series of six patients with endogenous Cushing's syndrome and reviewed 17 other cases described in the literature where opportunistic fungal infections complicated endogenous hypercortisolism. In addition to appropriate antifungal treatment, rapid reduction of hypercortisolism can be lifesaving.

Viral Infections

Glucocorticoid therapy can be complicated by the development of several viral infections, such as varicella, herpes zoster, and cytomegalovirus infections. Fatalities from varicella infection have been reported in patients on glucocorticoid therapy.[197]

Parasitic Infections

Infections with *Pneumocystis carinii* represent the most common parasitic infection associated with chronic glucocorticoid therapy. It is unclear whether the dose and duration of glucocorticoid therapy influence the susceptibility to opportunistic infections. It would seem so, if one extrapolates the data regarding the prevalence of opportunistic infections in patients with endogenous hypercortisolism. Graham and Tucker[196] made the observation that patients with endogenous hypercortisolism who developed opportunistic infections had higher morning cortisol levels.

The mortality and morbidity in patients with hypercortisolism are greatly influenced by the infections that complicate this disease. The devasting toll exacted by infections that complicate hypercortisolemia was appreciated more than three decades ago by Plotz et al.[182] In a review of data collected from 107 patients with Cushing's syndrome in whom autopsy findings were made available, the cause of death in almost 50% of patients was an infectious disease. Thus, hypercortisolism with its attendant effects of immunosuppression exacts a very heavy price.

Hypothalamic Pituitary Adrenal Suppression from Steroid Therapy

HPA suppression represents potentially the most hazardous complication of chronic steroid therapy. This effect compromises the ability of steroid-treated patients in terms of responding to and combating stress. This potential and dangerous complication of steroid therapy was recognized soon after the introduction of steroids for therapeutic purposes. Fraser et al.[198] described a patient on prolonged glucocorticoid therapy who developed fatal postoperative shock following arthroplasty of the hip. The next year, Salassa et al.[199] described two additional cases of the same surgical complication. The authors recommended, with prophetic wisdom, that "it seems safest to suppose that any patient who has received cortisone in significant quantities within three to six months should receive prophylactic therapy." Several aspects of HPA suppression are relevant to the clinician:

1. The minimum dose of steroid therapy that causes appreciable suppression of the HPA axis.
2. The minimum duration of steroid therapy that causes appreciable suppression of the HPA axis.
3. The correlation between the dose and duration of steroid therapy in causing HPA suppression.
4. The pattern of suppression and the pattern of recovery after withdrawal of the steroids.
5. The utilization of tests that predict the presence and degree of PHA suppression.
6. The effect of the method of administration, i.e., daily versus alternate-day regimens, in causing HPA suppression.
7. The general recommendations for steroid-treated patients planning to undergo elective procedures and the recommendations for therapy of such patients during stress.

The *minumum* dose of steroids that causes significant suppression of the HPA axis has not been established with certainty. Some of the difficulties in evaluating the prolific data in the literature regarding steroids and HPA suppression lie in the heterogeneity of tests performed on patients on widely different doses with significant variations in individual responses. Physiologically, it seems that any dose of steroid in excess of the replacement doses would constitute amounts capable of inducing suppression of the HPA axis. Christy et al.[200] have demonstrated that an impairment in the response of HPA axis occurs in patients receiving 20–30 mg of prednisone daily for 5–7 days. In another study, by Plager and Cushman,[201] abnormal ACTH responses were seen following the use of 100 mg of cortisol daily for 3 days. Thus, it is reasonable to assume that any dose of steroid in excess of physiological replacement has the potential to cause suppression of the HPA axis. Clearly, large doses of steroids, even for short periods of time, or small doses of steroids for long periods of time have the capability of suppressing the HPA axis.

The duration of steroid therapy required to cause suppression of the HPA axis is probably one of the major determinants in this phenomenon. While it is generally true that the degree of suppression has a direct correlation with the duration of exogenous steroid therapy, impressive evidence can be found in the literature to suggest that even short-term steroid therapy can result in significant, even dangerous sequelae. In 1953, Salassa et al.,[199] in a series of postmortem studies, demonstrated significant reductions in the adrenal weight within 5–10 days of initiating steroid therapy. From animal experiments in various species, it is generally accepted that adrenal atrophy occurs after 10 days of high-dose steroid therapy. In humans, several studies have convincingly demonstrated the occurence of abnormalities of HPA function following even brief administration of glucocorticoids.[200–204] The spectrum of changes that occur following the short-term administration of glucocorticosteroids includes reduction in the basal cortisol levels, loss of the diurnal rhythm of plasma cortisol, diminished adrenal response to ACTH, and a decreased ACTH response to the administration of metyrapone and insulin-induced hypoglycemia.[14,201] Streck and Lockwood[204] studied the effects of short-term administration of prednisone at the dosage of 25 mg twice a day on hypothalamic pituitary adrenal function in normal subjects. It was noticed that barely 2 days after prednisone therapy was begun, the peak cortisol response to exogenous ACTH and hypoglycemia were significantly impaired in comparison to the pretreatment status. Five days after conclusion of pednisone therapy, the peak cortisol response to hypoglycemia had returned to normal, although the peak responses to synthetic ACTH remained reduced. These data suggest that brief exposure to high-dose steroid therapy may limit the adrenal component of the HPA systems for up to 5 days. These studies are in keeping with the obserations by Salassa et al.[199] that fatal adrenal failure can occur even following short-term administration of glucocorticoids. However, these data differ from other studies that have shown no appreciable abnormalities in the HPA axis following short-term administration of steroids. For instance, in a study by Wilson et al.,[205] patients with myeloma and lymphoma who received short courses of prednisolone failed to show consistent suppression of the HPA. It has been reported that modest doses of prednisone (less than 40 mg/day) given in the morning for less than 5–7 days do not result in appreciable pituitary adrenal suppression.[200,206] Despite the conflicting data in the literature, it seems clear that the potential for HPA suppression exists even with short-term steroid therapy. It also seems clear that the shorter the course, the more rapid the recovery.

The degree of ACTH and adrenal suppression seems to be highly variable from patient to patient. Treadwell et al.[207] examined pituitary reserve in 41 patients on daily prednisolone therapy and found no abnormal responses to metyrapone before 15 months of treatment and no normal responses after 35 months of prednisolone treatment. These data differ from other studies where abnormal responses to metyrapone were seen much earlier in the course of treatment,[1] at times as early as 3 days after initiation of steroid therapy.[20] These variations may be related to the total dose of steroid, the time of administration, the methods of study, as well as the marked individual variations in patient responses to the drug. Regardless of the controversies that surround these data, one fact clearly stands out: all patients who have been on daily doses of steroid equivalent to or greater than 40 mg of prednisone for a period greater than 2–3 weeks should be considered as compromised in terms of pituitary adrenal function. In such a setting, adrenal failure can occur in response to stress as long as 1 year after the discontinuation of steroids. The abnormalities are most profound and the recovery slowest in patients placed on larger daily doses for longer than 6 months.

Pattern of Recovery

The course and time frame of recovery of the HPA axis from the suppressive effects of chronic glucocorticoid therapy have been the subject of extensive and intense investigation. Three studies in particular have been especially informative. Graber et al.[208] studied the time course and pattern of normalization of pituitary adrenal function in patients recovering from prolonged pituitary suppression. Several phases were identified during the process of recovery. The first phase, seen during the first month following withdrawal, was characterized by low plasma cortisol level, low plasma ACTH level, subnormal cortisol response to a standard dose of exogenous ACTH, and clinical symptoms of mild adrenal insufficiency. The second phase was characterized by a rising ACTH level in the plasma, suppressed plasma cortisol levels despite the elevated plasma ACTH level, and a suboptimal response of plasma cortisol to exogenous ACTH administration. This phase, lasting from 2 to 5 months, highlights the dormancy of the adrenal cortex despite stimulation by endogenous ACTH. In the third phase, the plasma cortisol levels had normalized, but the response of the adrenal to ACTH remained blunted. The final phase represents normalization of plasma cortisol levels, along with restoration of normal adrenal responsiveness to ACTH administration. This phase evolves 9–12 months after steroid withdrawal. This study, a hallmark study for its time, emphasized two major clinical concepts—that the adrenal recovery lags behind the recovery of the pituitary, and that a normal adrenal response to exogenous ACTH heralds total recovery of the entire HPA axis from the suppressive effects of glucocorticoid therapy. Livanou et al.[206] used the plasma cortisol response to insulin-induced hypoglycemia as a parameter and demonstrated a slow but steady response in patients weaned from glucocorticoid therapy. It appears that in most studies, regardless of the parameters used, the time taken for total recovery of the HPA axis approximates 1 year. The third important study that further sheds light on the various patterns of corticosteroid withdrawal is that of Dixon and Christy.[209] Their data on HPA function in patients studied during corticosteroid withdrawal demonstrated several interesting patterns. In one type, symptomatic and biochemical evidence of HPA suppression developed; in a second group of patients, steroid withdrawal was associated with significant recrudescence of the disease, but with evidence of physical or psychological dependence on steroids. The final group of patients were those with biochemical evidence of HPA suppression but who remained asymptomatic. Studies such as these point out the immense heterogeneity of patient responses and preclude predictions in a given case.

The recovery of adrenal function following removal of cortisol-secreting adenomas provides an excellent model for studying chronic HPA suppression and its recovery.[206,210–212] In contrast to iatrogenic hypercortisolism, where the steroid levels show a great deal of variability, the hypercortisolism caused by adenomas results in a sustained elevation in plasma cortisol levels. This may underlie the reason for the longer period of time taken for complete recovery of the HPA axis in patients with Cushing's syndrome following removal of the adrenal adenoma.

In addition to evaluating recovery of the HPA axis by conventional stimuli,

Melby[21,213] has studied the response of the recovering axis to stress. By use of the pyrogen test, he showed that the responsiveness to stress may be recovered much earlier than the usually quoted 12 months following steroid withdrawal. This was particularly evident in patients on low-dose or alternate-day regimens. This group of patients often showed preservation of the response to pyrogen even when adrenal responsiveness was impaired. These studies strongly support the concept that alternate-day therapy causes lesser and incomplete degress of suppression in the HPA axis.

Tests to Determine Recovery of the HPA Axis

The tests to determine that the hypothalamic pituitary adrenal axis has recovered from prolonged suppression can be divided into those that evaluate the pituitary ACTH recovery, those that evaluate restoration of adrenal responsiveness, and those that evaluate both these aspects together. The following general comments are applicable to the laboratory testing of HPA function in patients recovering from chronic glucocorticoid therapy.

1. Testing should be undertaken only during the period when the patient has been gradually weaned off from supraphysiological doses of glucocorticoids, and when on physiological replacement with the equivalent of 20 mg of hydrocortisone administered orally in the morning.

2. The basal cortisol level obtained 24 hr after the preceding dose provides limited but important information regarding the recovery process; a basal 8 A.M. cortisol below 10 µg% is clearly abnormal, indicating subnormal adrenal recovery. It serum cortisol values are in excess of 10 µg, it may be concluded that baseline adrenal pituitary function is adequate, but this does not provide information as to the ability of the adrenal cortex to respond to stress.

3. Since the pituitary ACTH usually recovers earlier than the adrenal, it can be assumed that a normal adrenal response to an exogenous pulse of ACTH implies recovery of both components of the axis.[214,215]

4. The insulin hypoglycemia challenge, when positive. i.e., a good cortisol response to hypoglycemia, also indicates complete recovery of the HPA axis. However, the test can be hazardous and requires close supervision of the patient, since profound symptomatic hypoglycemia can occur in patients with compromised adrenal function. If the hypoglycemia test is performed with measurements of both ACTH and cortisol, patients with complete restoration of the HPA axis to normal can be identified, but this is hardly necessary in clinical practice. ACTH recovery, without restoration of adrenal responsiveness to the endogenously released ACTH, still places the patient in a compromised setting. For the same reason testing ACTH responsiveness to lysine vasopressin, metyrapone, or ovine CRH is of limited value in clarifying the status of the adrenal cortex. Of course, a concomitantly normal cortisol response clearly indicates total recovery of the entire axis, but the same information can be directly, and more simply, obtained by measurement of the cortisol response to a bolus of synthetic ACTH.

5. Thus, for the most part, the rapid ACTH test, performed in a patient

whose basal 8 A.M. cortisol is greater than 10 μg%, serves as an excellent guide that reflects recovery of the entire axis. When the cortisol rise is suboptimal, adrenocortical dormancy is implied, even though the ACTH may have recovered. The criteria to define a "normal" response in this setting are not very well defined, but an increment in plasma cortisol greater than 6 μg over the baseline and to more than 20 μg per 100 ml is considered adequate.[214,215]

6. The concordance between demonstrable adrenal recovery and the ability to withstand the stress of anesthesia or surgery without any steroid coverage is a matter of great individual variability. Even in patients who have shown normalization of adrenal function, i.e., in terms of responsiveness of the plasma cortisol to exogeneous ACTH, it is wise to cover with steroids during stress up to a period of 1 year following withdrawal of steroid. After that period, it is a matter of physician judgment, the decision often resting with the nature and severity of the stress.

7. Finally, while the sequence of recovery of the HPA axis following glucocorticoid therapy is believed to be characterized by pituitary recovery first, followed by adrenal recovery later,[208,216,217] there have been a number of reports suggesting that the sequence may be reversed.[218-220] In this setting, normal cortisol response to exogenous ACTH may be seen, despite an impaired response of endogenous ACTH release to insulin hypoglycemia. In such patients, the dangers in assuming the integrity of the entire HPA axis, based on a normal ACTH stimulation test, are obvious. Such patients may fail to muster an ACTH response under stress, despite the demonstration of a normal cortisol response to exogenous ACTH administration. Hopefully, such patients are rare. This diagnostic pitfall can be avoided if both facets of the HPA axis are viewed together before declaring normalcy of the entire system.

Alternate-Day Glucocorticoid Regimen

Administration of glucocorticoids on an alternate-day basis causes less suppression of ACTH.[28] Since adrenal dormancy is a consequence of ACTH suppression, it follows that alternate-day regimens cause lesser degrees of adrenal responsiveness. Ackerman and Nolan[221] as well as well as others[32,222] have shown that the response of plasma 17-hydroxycorticoids to insulin-induced hypoglycemia is normal, near normal, or only mildly suppressed 24 hr after the last dose in patients who had received supraphysiological doses of prednisone on an alternate-day basis for as long as 50 months. The pituitary adrenal responsiveness to CRH has been studied by Schurmeyer et al.[223] These workers showed that patients maintained on alternate-day glucocorticoid regimens demonstrated preservation of adrenal responsiveness, even though the ACTH response to CRH was blunted—markedly so on the day of treatment and mildly on the day off treatment. This intriguing finding of responsive adrenals in patients on alternate-day regimen perhaps can be explained by the concept that even small amounts of endogenous ACTH may be sufficient to maintain the function of the adrenal glands. Regardless, as a group, patients on alternate-day regimen demonstrate suppression of the HPA axis to a smaller magnitude than those on the daily regimen. The periodicity in circulating levels of the administered steroids probably allows time for some recovery of the HPA axis from the prior day's therapy. Alternate-day therapy with long-acting glucocorticoids will impair HPA responsiveness.[224,225] Since alternate-day therapy also carries a lower incidence of adverse side effects, this method of administration should be considered in all candidates for chronic glucocorticoid therapy.

The general recommendations for steroid-treated patients planning to undergo elective procedures and the recommendations for therapy during stress constitute an important facet of glucocorticoid therapy. Patients who require elective surgery while on steroid treatment definitely need to be covered with "stress doses" of hydrocortisone, usually 100 mg t.i.d. parenterally. Patients who are being gradually weaned off steroid therapy, at any time during withdrawal, should be covered for stress or surgery with parenteral steroids. Patients who have managed to get off steroid therapy completely should be considered adrenally insufficient for a period of 1 year, and acute coverage for stress or surgery is indicated.

The stress associated with general anesthesia or surgery is not a major concern in patients who had received *replacement* doses of steroids. This represents a daily dosage equivalent to or less than 5 mg of prednisone, 20 mg of hydrocortisone, 25 mg of cortisone, or 0.75 mg of dexamethasone. In a study by Danowski et al.,[226] 117 patients who had received replacement doses of glucocorticoids passed through 80 episodes of pregnancy, surgery, acute illness, or stressful diagnostic procedures without features of adrenocortical failure. Suppression of HPA function need not be a concern in patients receiving less than the physiological replacement dosage.

HPA suppression represents potentially the most hazardous side effect of chronic glucocorticoid therapy. Efforts to minimize HPA suppression include

1. Use of the lowest possible dose that is effective in controlling the underlying disease process
2. Use of a single morning dose, with avoidance of the night dose, which has the greatest suppressive potential
3. Use of an alternate-day regimen
4. Avoidance of long-acting steroids such as dexamethasone, which can produce protracted HPA suppression even when administered on an alternate-day basis

Withdrawal from Glucocorticoid Therapy

Some aspects of the steroid withdrawal syndrome, particularly those that pertain to the suppression of HPA function, were outlined in the preceding section. The steroid withdrawal syndrome consists of several components. It represents a constellation of clinical and hormonal facets seen during withdrawal of steroids in patients who had been on these drugs for protracted periods of time. The four major aspects of this syndrome are

1. The development of several symptoms, most of which are unrelated to adrenocortical failure.

2. The recrudescence of symptoms of the underlying disease process for which steroid therapy was instituted.
3. Hormonal evidence for suppressed hypothalamic pituitary adrenal function.
4. Overt clinical and hormonal evidence of adrenal decompensation, usually precipitated by stress.

Symptoms of the Steroid Withdrawal Syndrome

Steroid therapy should never be withdrawn abruptly. Sudden withdrawal of prolonged or intensive steroid therapy is extremely dangerous and can result in death.[198,199,227,228] The only cases where steroids can be abruptly discontinued are instances where intensive therapy was given for less than 48 hr. When steroids are gradually reduced, patients may develop several symptoms. Amatruda et al.[229] described a syndrome characterized by anorexia, fever, nausea, lethargy, weakness, weight loss, desquamation, and arthralgia that developed after withdrawal of steroid therapy. These symptoms were unrelated to adrenal insufficiency since the steroid levels as well as the response to exogenous ACTH were often not significantly impaired. In a subsequent study Amatruda et al.[230] further noted that there were no significant differences in the response of plasma and urine steroids to metyrapone in the group of patients who experienced withdrawal symptoms when compared to those who did not. Since these symptoms are unrelated to true adrenocortical insufficiency, and since such symptoms improve with steroid administration, the possibility of physical or psychological dependence to steroids is a distinct consideration. Addiction or undue dependence on steroids has been reported in the literature.[231–236] The extent to which dependence on steroids is related to its mood-altering effects is unclear. Further, it is difficult to separate the effects of physical dependency on steroids from subtle alterations in the HPA axis. The various forms of corticosteroid withdrawal were outlined by Dixon and Christy.[209] Upon closely studying the clinical and hormonal course of five patients who had received corticosteroids for several years, these authors noted several subgroups of patients—those with symptomatic as well as biochemical evidence of HPA suppression; those with flareup of the disease; those with normal HPA function but with evidence of physical or psychological dependence; and those without any symptoms despite demonstrable biochemical evidence of HPA suppression. The single most important parameter that governs the rate and degree of corticoisteroid withdrawal is the status of the underlying disease. It is almost impossible to completely wean a patient from corticosteroid therapy when the disease flares up upon withdrawal of steroid. This is especially so when the disease responds only to steroids.

Recrudescence of Underlying Disease

When glucocorticoid withdrawal is contemplated, the process should begin with gradual and gentle reduction in the dosage. Reduction in prednisone dosage by no more than 2.5–5 mg every 3–7 days is a good place to start. Should the patient develop a reactivation of the underlying process, the glucocorticoid dose is increased again, until clinical benefit is accrued. When withdrawal is recontemplated, the dose reduction has to be even slower and more gradual than the preceding effort. Switching to alternate-day regimens may or may not be of help, depending on the underlying disease that is being treated and the individual variability of patient response. Consideration of alternate drug therapies— particularly other forms of immunosuppression—is valid at this juncture.

Hormonal Studies

Since eventually, with sufficient time, function of the HPA axis will indeed recover, the task of the physician is to employ tests to determine when the patient has satisfactorily recovered. Byyny[216] has suggested an excellent design to facilitate glucocorticoid withdrawal with minimum symptoms in the patient and relative certainty for the physician. The first step consists of gradual tapering of the dose of glucocorticoid, until the biological equivalent of 20 mg of hydrocortisone per day is reached. The duration of this phase will, naturally, depend on the flareup of the underlying disease, as well as the development of symptoms of steroid withdrawal. It is crucial to supplement the steroid dose during times of stress during this period. The second phase of this exercise begins 4 weeks after the patient has been maintained on the equivalent of 20 mg hydrocortisone per day. At this point an 8 A.M. plasma cortisol is obtained; if the plasma cortisol is less than 10 µg/100 ml, the patient is maintained on 20 mg of hydrocortisone, and an attempt is made to gradually taper the dose (by 2–5-mg decrements) to 10 mg/day, with supplementation during stress. If, on the other hand, the plasma cortisol is greater than 10 µg/100 ml, the hydrocortisone can be stopped, but supplemented during stress. The third phase consists of hormonal evaluation, usually 4 weeks after the basal cortisol is found to be greater than 10 µg/100 ml. During this phase, the adrenal responsiveness to ACTH administration is evaluated by performing the rapid ACTH stimulation test. If the response is suboptimal, i.e., the response in serum cortisol is less than 6 µg over the baseline, or the maximal cortisol response is under 20 µg (or both), steroid supplementation for stress is recommended. If, on the other hand, the cortisol response to ACTH is clearly normal, stress supplementation can be discontinued.

The above-mentioned design is a generalized approach, understanding fully well that a normal basal 8 A.M. cortisol together with a normal cortisol response to ACTH implies complete recovery in most but not all patients. Some workers have recommended stress coverage for as long as 5 years following steroid withdrawal.[220]

Overt Adrenal Failure

Overt adrenal failure seldom occurs when corticosteroids are gradually tapered. In situations where intensive short-term therapy (limited to a maximum of 48 hr) with steroids was instituted, steroid drugs can be withdrawn as abruptly as they were started, provided the underlying disease is not exacerbated. More often it is the activity of the underlying condition that determines the need for tapering, when steroids are given for a brief period of intensive therapy. When overt clinical or hormonally documented adrenal failure develops in the setting of steroid withdrawal, more often than not, an underlying stressful factor is present that triggers the event. The importance of supplementation during all phases of steroid withdrawal cannot be overemphasized.

References

1. Christy NP (ed): *The Human Adrenal Cortex.* Harper & Row, New York, 1971, p. 395.
2. Julian JA, Chytil F: A two-step mechanism for the regulation of tryptophan pyrrolase. *Biochem Biophys Res Commun* **34:**734, 1969.
3. Marco J, Calle C, Roman D, et al: Hyperglucagonism induced by glucocorticoid treatment in man. *N Engl J Med* 288:128, 1973.
4. Jenkins JS: The metabolism of cortisol by human extra-hepatic tissues. *J Endocrinol* **34:**51, 1966.
5. Peterson RE: Adrenocortical steroid metabolism and adrenal cortical function in liver disease. *J Clin Invest* **39:**320, 1960.
6. Ely RS, Done AK, Kelley VC: δ1-Hydrocortisone: Plasma 17-hydroxycorticosteroid concentrations following oral and i.v. administration. *Proc Soc Exp Biol Med* **91:**503, 1956.
7. Migeon CJ, Sandberg AA, Decker HA, et al: Metabolism of 4-C^{14}-cortisol in man: Body distribution and rates of conjugation. *J Clin Endocrinol Metab* **16:**1137, 1956.
8. Nugent CA, Eik-nes K, Tyler FH: A comparative study of the metabolism of hydrocortisone and prednisolone. *J Clin Endocrinol Metab* **19:**526, 1959.
9. Slaunwhite WR, Sandberg AA: Disposition of radioactive 17α hydroxyprogesterone, 6α methyl-17α acetoxyprogesterone and 6α methylprednisolone in human subjects. *J Clin Endocrinol Metab* **21:**753, 1961.
10. Sandberg AA, Slaunwhite WR: Differences in metabolism of prednisolone-C^{14} and cortisol-C^{14}. *J Clin Endocrinol Metab* **17:**1040, 1957.
11. Peterson RE: Metabolism of adrenocorticosteroids. *Ann NY Acad Sci* **82:**846, 1959.
12. Melby JC: Adrenocorticosteroids in medical emergencies. *Med Clin North Am* **45:**875, 1961.
13. Melby JC: Systemic corticosteroid therapy: Pharmacology and endocrinologic considerations. *Ann Intern Med* **81:**505, 1974.
14. Harter JG: Corticosteroids: Their physiologic use in allergic disease. *NY State J Med* **66:**827, 1966.
15. O'Malley BW: Mechanism of action of steroid hormones. *N Engl J Med* **284:**370, 1971.
16. Thompson EB, Lippman ME: Mechanism of action of glucocorticoids. *Metabolism* **23:**159, 1974.
17. Ballard PL, Carter JP, Graham BS, et al: A radioreceptor assay for evaluation of the plasma glucocorticoid activity of natural and synthetic steroids in man. *J Clin Endocrinol Metab* **41:**290, 1975.
18. Meikle AW, Weed JA, Tyler FH: Kinetics and interconversion of prednisolone and prednisone studied with new radioimmunoassay. *J Clin Endocrinol* **41:**717, 1975.
19. West KM: Relative eosinopenic and hyperglycemic potencies of glucocorticoids in man. *Metabolism* **7:**441, 1958.
20. Buus O, Munthe Fog CV, Moller PR: The plasma hydrocortisone level during prednisolone administration as a measure of pituitary function. *Danish Med Bull* **9:**210, 1962.
21. Melby JC: Assessment of adrenocorticotropic activity with bacterial pyrogen in hypopituitary states (Abstract). *J Clin Invest* **38:**1025, 1959.
22. Sayers GL, Trains RH: *Adrenocorticotrophic Hormone: Adrenocortical Steroids and Their Synthetic Analogs in the Pharmacological Basis of Therapeutics,* 4th ed. Macmillan, New York, 1971, p. 1627.

23. DeKloet ER, Van Der Vies J, DeWied D: The site of the suppressive action of dexamethasone on pituitary-adrenal activity. *Endocrinology* **94**:61, 1974.

24. Meikle AW, Tyler FH: Potency and duration of action of glucocorticoids. Effects of hydrocortisone, prednisone and dexamethasone on human pituitary-adrenal function. *Am J Med* **63**:200, 1977.

25. Thorn GW: Clinical considerations in the use of corticosteroids. *N Engl J Med* **274**:775, 1966.

26. Boston Collaborative Drug Surveillance Program: Acute adverse reactions to prednisone in relation to dosage. *Clin Pharmacol Ther* **13**:694, 1972.

27. Newman S: Hormone-induced diseases. In Moser RH (ed): *Disease of Medical Progress: A Study of Iatrogenic Disease*. Charles C Thomas, Springfield, IL, 1969, p. 361.

28. Harter JG, Reddy WJ, Thorn GW: Studies on an intermittent corticosteroid dosage regimen. *N Engl J Med* **269**:591, 1963.

29. Haugen HN, Reddy WJ, Harter JG: Intermittent steroid therapy in bronchial asthma. *Nord Med* **63**:15, 1960.

30. Reichling GH, Kligman AM: Alternate-date corticosteroid therapy. *Arch Dermatol* **83**:980, 1961.

31. Jacobson ME: The rationale of alternate-day corticosteroid therapy. *Postgrad Med* **49**:181, 1971.

32. Sadeghi-Nejad A, Senior B: Adrenal function, growth, and insulin in patients treated with corticoids on alternate days. *Pediatrics* **43**:277, 1969.

33. Soyka LF: Treatment of the nephrotic syndrome in childhood: Use of an alternate-day prednisone regimen. *Am J Dis Child* **113**:693, 1967.

34. Carter ME, James, VHT: Effects of alternate-day, single-dose, corticosteroid therapy on pituitary-adrenal function. *Ann Rheum Dis* **31**:379, 1972.

35. Ackerman GL: Alternate-day steroid therapy in lupus nephritis. *Ann Intern Med* **72**:511, 1970.

36. Kuzemko JA, Lines JG: Adrenal cortical function in asthmatic children on alternate day steroids. *Arch Dis Child* **46**:366, 1971.

37. Engel WK, Festoff BW, Patten BM, et al: Myasthenia gravis. *Ann Intern Med* **81**:225, 1974.

38. Seybold ME, Drachman DB: Gradually increasing doses of prednisone in myasthenia gravis: Reducing the hazards of treatment. *N Engl J Med* **290**:81, 1974.

39. Block AJ, Light RW: Alternate day steroid therapy in diffuse pulmonary sarcoidosis. *Chest* **63**:495, 1973.

40. Sheagren JN, Simon HB, Rich RR: Therapy of sarcoidosis initiated with alternate-day prednisone. *J Natl Med Assoc* **65**:391, 1973.

41. Haim S, Benderly A, Shafrir A, et al: Alternate-day corticosteroid regimen. *Dermatologica* **142**:171, 1971.

42. Fauci AS, Dale DC, Balow JE: Glucocorticosteroid therapy: Mechanisms of action and clinical considerations. *Ann Intern Med* **84**:304, 1976.

43. Dale DC, Fauci AS, Guerry D, et al: Comparison of agents producing a neutrophilic leukocytosis in man: Hydrocortisone, prednisone, endotoxin and etiocholanolone. *J Clin Invest* **56**:808, 1975.

44. Bishop CR, Athens JW, Boggs DR, et al: Leukokinetic studies: XIII. A now-steady-state kinetic evaluation of the mechanism of cortisone-induced granulocytosis. *J Clin Invest* **47**:249, 1968.

45. Dale DC, Fauci AS, Wolff SM: Alternate-day prednisone: Leukocyte kinetics and susceptibility to infections. *N Engl J Med* **291**:1154, 1974.

46. Ward PA: The chemosuppression of chemotaxis. *J Exp Med* **124**:209, 1966.

47. Chretien JH, Garagusi VF: Suppressed reduction of nitroblue tetrazolium by polymorphonuclear neutrophils from patients receiving steroids. *Experientia* **27**:1343, 1971.

48. Yu DTY, Clements PJ, Paulus HE, et al: Human lymphocyte subpopulations. Effect of corticosteroids. *J Clin Invest* **53**:565, 1974.

49. Fauci AS, Dale DC: Alternate-day prednisone therapy and human lymphocyte subpopulations. *J Clin Invest* **55**:22, 1975.

50. Fauci AS, Dale DC: The effect of hydrocortisone on the kinetics of normal human lymphocytes. *Blood* **46**:235, 1975.

51. Weston WL, Mandel MJ, Yeckley JA, et al: Mechanism of cortisol inhibition of adoptive transfer of tuberculin sensitivity. *J Lab Clin Med* **82**:366, 1973.

52. Salomon H, Angel JH: Corticotropin-induced changes in the tuberculin skin test. *Am Rev Respir Dis* **83**:235, 1961.

53. Ragan C: Corticotropin, cortisone and related steroids in clinical medicine: Practical considerations. *Bull NY Acad Med* **29:**355, 1953.

54. Axelrod L: Glucocorticoid therapy. *Medicine* **55:**39, 1976.

55. Walker AE, Adamkiewicz JJ: Pseudotumor cerebri associated with prolonged corticosteroid therapy. *JAMA* **188:**779, 1964.

56. David DS, Berkowitz JS: Ocular effects of topical and systemic corticosteroids. *Lancet* **2:**149, 1969.

57. Heimann WG, Freiberger RH: Avascular necrosis of the femoral and humeral heads after high-dosage corticosteroid therapy. *N Engl J Med* **263:**672, 1960.

58. Kozower M, Veatch L, Kaplan MM: Decreased clearance of prednisolone, a factor in the development of corticosteroid side effects. *J Clin Endocrinol Metab* **38:**407, 1974.

59. Sambhi MP, Weil MH, Udhoji VN: Acute pharmacodynamic effects of glucocorticoids. *Circulation* **31:**523, 1965.

60. David DS, Grieco MH, Cushman P: Adrenal glucocorticoids after twenty years: A review of their clinically relevant consequences. *J Chron Dis* **22:**637, 1970.

61. Good RA, Vernier RL, Smith RT: Serious untoward reactions to therapy with cortisone and adrenocorticotropin in pediatric practice. *Pediatrics* **19:**95, 1957.

62. Afifi AK, Bergman RA, Harvey JC: Steroid myopathy, clinical, histologic and cytologic observation. *Johns Hopkins Med J* **123:**158, 1968.

63. Williams RS: Triamcinolone myopathy. *Lancet* **1:**698, 1959.

64. MacLean K. Schurr PH: Reversible amyotrophy complicating treatment with fludrocortisone. *Lancet* **1:**701, 1959.

65. Perkoff GT, Silber R, Tyler FH, et al: Studies in disorders of muscles. XII. Myopathy due to the administration of therapeutic amounts of 17-hydroxycorticosteroids. *Am J Med* **26:**891, 1959.

66. Harman JB: Muscular wasting and corticosteroid therapy. *Lancet* **1:**887, 1959.

67. Syme JR: Muscle wasting complicating treatment with dexamethasone. *Med J Aust* **2:**240, 1960.

68. Hagstrom JWC, Roseman DM, Ellis JT: Debilitating muscular wasting and steroid therapy. *Arch Neurol* **5:**60, 1961.

69. Byers RK, Bergman AB, Joseph MC: Steroid myopathy: Report of five cases occurring during treatment of rheumatic fever. *Pediatrics* **29:**26, 1962.

70. Bauer FK, Dubois EL, Teleer N: Total exchangeable potassium in SLE with reference to triamcinolone myopathy. *Proc Soc Exp Biol Med* **105:**671, 1960.

71. Powell RJ: Steroid and hypokalemic myopathy after corticosteroids for ulcerative colitis. *Am J Gastroenterol* **52:**425, 1969.

72. Mandel S: Steroid myopathy: Insidious cause of muscle weakness. *Postgrad Med* **72:**207, 1982.

73. Pleasure DE, Walsh GO, Engel WK: Atrophy of skeletal muscle in patients with Cushing's syndrome. *Arch Neurol* **22:**118, 1970.

74. Askari A, Vignos PJ Jr, Moskowitz RW: Steroid myopathy in connective tissue disease. *Am J Med* **61:**485, 1976.

75. Vignos PJ Jr, Goldwyn J: Evaluation of laboratory test in diagnosis and management of polymyositis. *Am J Med Sci* **263:**291, 1972.

76. Vignos PJ Jr, Bowling GF, Watkins MP: Polymyositis, effect of corticosteroids on final results. *Arch Intern Med* **114:**263, 1964.

77. Bullock GR, Christian RA, Peters RF, et al: Rapid mitochondrial enlargement in muscle as a response to triamcinolone acetonide and its relationship to the ribosomal defect. *Biochem Pharmacol* **20:**943, 1971.

78. Comez-Puyow A, Pena-Dias A, Gwzman-Garcia J, et al: Effect of triamcinolone and other steroids on the oxidative phosphorylation reaction. *Biochem Pharmacol* **12:**331, 1963.

79. Kimberg DV, Loud AV, Wiener J: Cortisone-induced alterations in mitochondrial function and structure. *J Cell Biol* **37:**63, 1968.

80. Afifi AK, Bergman RA: Steroid myopathy. A study of evolution of the muscle lesion in rabbits. *Johns Hopkins Med J* **124:**66, 1969.

81. Smith B: Histological and histochemical changes in muscles of rabbits given the corticosteroid triamcinolone. *Neurology* **14:**857, 1964.

82. Stern R, Amundsen P: Diphenylhydantoin for steroid-induced muscle weakness. *JAMA* **223:**1287, 1973.

83. Smyth GA: Activation of peptic ulcer during pituitary adrenocorticotropic hormone therapy: Report of three cases. *JAMA* **145**:474, 1951.

84. Lubin RI, Misbach WD, Zemke EM, et al: Acute perforation of duodenal ulcer during ACTH and cortisone therapy. *Gastroenterology* **18**:308, 1951.

85. Meltzer LE, Bockman AA, Kanenson W, et al: The incidence of peptic ulcer among patients on long term prednisone therapy. *Gastroenterology* **35**:351, 1958.

86. Kern F Jr, Clark GM, Lukens JG: Peptic ulceration occurring during therapy for rheumatoid arthritis. *Gastroenterology* **33**:25, 1957.

87. Freiberger RH, Kammerer WH, Rivelis AL: Peptic ulcers in rheumatoid patients receiving corticosteroid therapy. *Radiology* **71**:542, 1958.

88. Gedda PO, Moritz U: Peptic ulcer during treatment of rheumatoid arthritis with cortisone derivatives. *Acta Rheum Scand* **4**:249, 1958.

89. Sherwood H, Epstein JI, Buckley WE: Peptic ulcer among allergic patients on long-term triamcinolone therapy. *J Allergy* **31**:21, 1960.

90. Conn HO, Blitzer BL: Nonassociation of adrenocorticosteroid therapy and peptic ulcer. *N Engl J Med* **294**:473, 1976.

91. Messer J, Reitman D, Sacks HS, et al: Association of adrenocorticosteroid therapy and peptic-ulcer disease. *N Engl J Med* **309**:21, 1983.

92. Sauer WG, Dearing WH, Wollaegher EE: Serious untoward gastrointestinal manifestations possibly related to administration of cortisone and corticotrophin. *Proc Mayo Clin* **28**:641, 1953.

93. Glenn F, Grafe WR: Surgical complications of adrenal steroid therapy. *Ann Surg* **165**:1023, 1967.

94. Riemenschneider TA, Wilson JF, Vernier RL: Corticosteroid-induced pancreatitis in children. *Pediatrics* **41**:428, 1968.

95. Oppenheimer EH, Boitnott JK: Pancreatitis in children following adrenal corticosteroid therapy. *Bull Johns Hopkins Hosp* **107**:297, 1960.

96. Glaser GH, Kornfeld DS, Knight RP: Intravenous hydrocortisone, corticotropin and the electroencephalogram. *Arch Neurol Psychiatry* **73**:338, 1955.

97. Feldman S, Todt JC, Porter RW: Effect of adrenocortical hormones on evoked potentials in the brain stem. *Neurology* **11**:109, 1961.

98. Nishitani H: Electroencephalogram in endocrine diseases. II. Adrenal diseases. *Jpn Arch Intern Med* **9**:413, 1962.

99. Dorfman A, Apter NS, Smull K, et al: Status epilepticus coincident with the use of pituitary adrenocorticotropic hormone: Report of three cases. *JAMA* **146**:25, 1951.

100. Glaser GH: On the relationship between adrenal cortical activity and the convulsive state. *Epilepsia* **2**:7, 1953.

101. Editorial: Intracranial hypertension and steroids. *Lancet* **2**:1052, 1964.

102. Woodbury DM: Relation between the adrenal cortex and the central nervous system. *Pharmacol Rev* **10**:275, 1958.

103. Sussman ML, Coplman B: The roentgenologic appearance of the bones in Cushing's syndrome. *Radiology* **39**:288, 1942.

104. Howland WJ, Pugh DG, Sprague RG: Roentgenologic changes of the skeletal system in Cushing's syndrome. *Radiology* **71**:69, 1968.

105. Soffer LJ, Iannaccone A, Gabrilove JL: Cushing's syndrome: A study of fifty patients. *Am J Med* **30**:129, 1961.

106. Bradley BW, Ansell BM: Fractures in Still's disease. *Ann Rheum Dis* **19**:135, 1960.

107. Saville PD, Kharmosh O: Osteoporosis of rheumatoid arthritis: Influence of age, sex and corticosteroids. *Arthritis Rheum* **10**:423, 1967.

108. Hahn TJ: Corticosteroid-induced osteopenia. *Arch Intern Med* **138**:882, 1978.

109. Adinoff AD, Hollister JR: Steroid-induced fractures and bone loss in patients with asthma. *N Engl J Med* **309**:265, 1983.

110. Curtiss PH Jr, Clark WS, Herndon CH: Vertebral fractures resulting from prolonged cortisone and cortricotropin therapy. *JAMA* **156**:467, 1954.

111. McConkey B, Fraser GM, Bligh AS: Osteoporosis and purpura in rheumatoid disease: Prevalence and relation to treatment with corticosteroids. *Q J Med* **31**:419, 1962.

112. Hajiroussou VJ, Webley M: Prolonged low-dose corticosteroid therapy and osteoporosis in rheumatoid arthritis. *Ann Rheum Dis* **43**:24, 1984.

113. Chesney RW, Mazess RB, Rose P: Effect of prednisone on growth and bone mineral content in childhood glomerular disease. *Am J Dis Child* **132:**768, 1978.

114. Mueller MN: Effects of corticosteroids on bone mineral in rheumatoid arthritis and asthma. *Am J Roentgenol* **126:**1300, 1976.

115. Baylink DJ: Glucocorticoid-induced osteoporosis. *N Engl J Med* **309:**306, 1983.

116. Frost HM, Villanueva AR: Human osteoblastic activity. III. The effect of cortisone on lamellar osteoblastic activity. *Henry Ford Hosp Med Bull* **9:**97, 1961.

117. Jowsey J, Riggs BL: Bone formation in hypercortisolism. *Acta Endocrinol* **63:**21, 1970.

118. Bressot C, Meunier PJ, Chapuy MC, et al: Histomorphometric profile, pathophysiology and reversibility of corticosteroid-induced osteoporosis. *Metab Bone Dis Relat Res* **1:**303, 1979.

119. Canalis E: Effect of glucocorticoids on type I collagen synthesis, alkaline phosphatase activity, and deoxyribonucleic acid content in cultured rat calvariae. *Endocrinology* **112:**931, 1983.

120. Peck WA, Brandt J, Miller I: Hydrocortisone-induced inhibition of protein synthesis and uridine incorporation in isolated bone cells in vitro. *Proc Natl Acad Sci USA* **57:**1599, 1967.

121. Hahn TJ, Halstead LR, Baran DT: Effects of short term glucocorticoid administration on intestinal calcium absorption and circulating vitamin D metabolite concentrations in man. *J Clin Endocrinol Metab* **52:**111, 1981.

122. Findling JW, Adams ND, Lemann J, et al: Vitamin D metabolites and parathyroid hormone in Cushing's syndrome: Relationship to calcium and phosphorus homeostasis. *J Clin Endocrinol Metab* **54:**1039, 1982.

123. Teitelbaum SL, Malong JD, Kahn AJ: Glucocorticoid enhancement of bone resorption by rat peritoneal macrophages in vitro. *Endocrinology* **108:**795, 1981.

124. Jee WSS, Park HZ, Roberts WE, et al: Corticosteroid and bone. *Am J Anat* **129:**477, 1970.

125. Rickers H, Deding A, Christiansen C, et al: Corticosteroid-induced osteopenia and vitamin D metabolism: Effect of vitamin D_2, calcium phosphate and sodium fluoride administration. *Clin Endocrinol* **16:**409, 1982.

126. Condon JR, Dent CE, Nassim JR, et al: Possible prevention and treatment of steroid-induced osteoporosis. *Postgrad Med J* **54:**249, 1978.

127. Braver DA, Richards RD, Good TA: Posterior subcapsular cataracts in steroid-treated children. *Arch Ophthalmol* **77:**161, 1967.

128. Toogood JH, Dyson C, Thompson CA, et al: Posterior subcapsular cataracts as a complication of adrenocortical steroid therapy. *Can Med Assoc J* **86:**52, 1962.

129. Oglesby RB, Black RL, Von Sallmann L, et al: Cataracts in patients with rheumatic diseases treated with corticosteroids—Further observations. *Arch Ophthalmol* **66:**625, 1961.

130. Giles CL, Mason GL, Duff IF, et al: The association of cataract formation and systemic corticosteroid therapy. *JAMA* **182:**719, 1962.

131. McConkey B, Fraser GM, Bligh AS: Transparent skin and osteoporosis: A study in patients with rheumatoid disease. *Ann Rheum Dis* **24:**219, 1965.

132. Greenwood BM: Capillary resistance and skin-fold thickness in patients with rheumatoid arthritis. Effect of corticosteroid therapy. *Ann Rheum Dis* **25:**272, 1966.

133. Wilber JF, Utiger RD: The effect of glucocorticoids in thyrotrophin secretion. *J Clin Invest* **48:**2096, 1969.

134. Nicoloff JT, Fisher DA, Appleman JD Jr: The role of glucocorticoids in the regulation of thyroid function in man. *J Clin Invest* **49:**1922, 1970.

135. Otsuki M, Dakoda M, Baba S: Influence of glucocorticoids on TRF-induced TSH response in man. *J Clin Endocrinol Metab* **36:**95, 1973.

136. Hall R, Ormston BJ, Besser GM, et al: Thyrotropin releasing hormone test in diseases of the pituitary and hypothalamus. *Lancet* **1:**759, 1973.

137. Kuku SF, Child DF, Nader S, et al: Thyrotrophin and prolactin responsiveness to thyrotrophin releasing hormone in Cushing's disease. *Clin Endocrinol* **4:**437, 1975.

138. Visser TJ, Lamberts SWJ: Regulation of TSH secretion and thyroid function in Cushing's disease. *Acta Endocrinol* **96:**480, 1981.

139. Duick DS, Wahner HW: Thyroid axis in patients with Cushing's syndrome. *Arch Intern Med* **139:**767, 1979.

140. Lamberts SWJ, Timmermans HAT, de Jong FH, et al: The role of dopaminergic depletion in the pathogenesis of Cushing's disease and the possible consequences for medical therapy. *Clin Endocrinol (Oxf)* **7:**185, 1977.

141. Sowers JR, Carlson HE, Brautbar N, et al: Effect of dexamethasone on prolactin and TSH responses to TRH and metoclopramide in man. *J Clin Endocrinol Metab* **44:**237, 1977.

142. Dussault JH: The effect of dexamethasone on TSH and prolactin secretion after TRH stimulation. *Can Med Assoc J* **111:**1195, 1974.

143. Copinschi G, L'Hermite M, Leclerq R, et al: Effects of glucocorticoids on pituitary hormonal response to hypoglycemia. Inhibition of prolactin release. *J Clin Endocrinol Metab* **40:**442, 1975.

144. Bratusch-Marrain P, Vierhapper H, Waldhausl W, et al: Acute suppressive effect of ACTH-induced cortisol secretion on serum prolactin levels in healthy man. *Acta Endocrinol* **99:**352, 1982.

145. Lantigua RA, Streck WF, Lockwood DH, et al: Glucocorticoid suppression of pancreatic and pituitary hormones: Pancreatic polypeptide, growth hormone, and prolactin. *J Clin Endocrinol Metab* **50:**298, 1980.

146. Dannies PS, Tashjian AH: Effect of thyrotropin-releasing hormone and hydrocortisone on synthesis and degradation of prolactin in a rat pituitary cell strain. *J Biol Chem* **248:**6174, 1973.

147. Adams EF, Brajkovich IE, Mashiter K: Growth hormone and prolactin secretion by dispersed cell cultures of human pituitary adenomas: Long term effects of hydrocortisone, estradiol, insulin, 3,5,3'-triodothyronine and thyroxine. *J Clin Endocrinol Metab* **53:**381, 1981.

148. Leung FC, Chen HT, Verkaik SJ, et al: Mechanism(s) by which adrenalectomy and corticosterone influences prolactin release in the rat. *J Endocrinol* **87:**131, 1980.

149. Krieger D: Lack of responsiveness to L-dopa in Cushing's disease. *J Clin Endocrinol Metab* **36:**277, 1973.

150. Demura R, Demura H, Nunokawa T, et al: Response of plasma ACTH, GH, LH and 11-hydroxycorticosteroids to various stimuli in patients with Cushing's syndrome. *J Clin Endocrinol Metab* **34:**852, 1972.

151. Krieger D, Luria M: Plasma ACTH and cortisol responses to TRF, vasopressin or hypoglycemia in Cushing's disease and Nelson's syndrome. *J Clin Endocrinol Metab* **44:**361, 1977.

152. Smals AEM, Pieters GFFM, Smals AGH, et al: Human pancreatic growth hormone releasing hormone fails to stimulate human growth hormone both in Cushing's disease and in Cushing's syndrome due to adrenocortical adenoma. *Clin Endocrinol* **24:**401, 1986.

153. Vazquez AM, Schutt-Aine JC, Kenny FM, et al: Effect of cortisone therapy on the diurnal pattern of growth hormone secretion in congenital adrenal hyperplasia. *J Pediatr* **80:**433, 1972.

154. Sturge RA, Beardwell C, Hartog M, et al: Cortisol and growth hormone secretion in relation to linear growth: patients with Sitll's disease on different therapeutic regimens. *Br Med J* **3:**547, 1970.

155. Preece MA: The effect of administered corticosteroids on the growth of children. *Postgrad Med J* **52:**625, 1976.

156. Morris HG, Jorgensen JR, Elrick H, et al: Metabolic effects of human growth in corticosteroid-treated children. *J Clin Invest* **47:**436, 1968.

157. Phillips LS, Belosky DC, Young HS, et al: Nutrition and somatomedin. VI. Somatomedin activity and somatomedin inhibitory activity in serum from normal and diabetic rats. *Endocrinology* **104:**1519, 1979.

158. Green OC, Winter RJ, Kawathara FS, et al: Pharmacokinetic studies of prednisolone in children: plasma levels, half-life values, and correlation with physiologic assays for growth and immunity. *J Pediatr* **93:**299, 1978.

159. Elders MJ, Wingfield BS, McNatt ML, et al: Glucocorticoid therapy in children: Effect on somatomedin secretion. *Am J Dis Child* **129:**1393, 1975.

160. Unterman TG, Phillips LS: Glucocorticoid effects on somatomedins and somatomedin inhibitors. *J Clin Endocrinol Metab* **61:**618, 1985.

161. Bessler W: Vertebral growth arrest lines after Cushing's syndrome. Case report. *Diag Imaging* **51:**311, 1982.

162. Oppenheimer JH, Werner SC: Effect of prednisone on thyroxine-binding proteins. *J Clin Endocrinol Metab* **26:**715, 1966.

163. Werner SC, Platman SR: Remission of hyperthyroidism (Graves' disease) and altered pattern of serum-thyroxine binding induced by prednisone. *Lancet* **2:**751, 1965.

164. Ingbar SH, Freinkel N: The influence of ACTH, cortisone and hydrocortisone on the distribution and peripheral metabolism of thyroxine. *J Clin Invest* **34:**1375, 1955.

165. Burr WA, Ramsden DB, Griffiths RS, et al: Effect of a single dose of dexamethasone on serum concentrations of thyroid hormones. *Lancet* **2:**58, 1976.

166. Croxson MS, Duick DS, Nicoloff JT: Effect of glucocorticoids on serum triiodothyronine (T₃) concentrations in man. *Excerpta Medica Int Congr Ser* **378:**266, 1976.

167. Duick DS, Warren DW, Nicoloff JT, et al: Effect of a single dose dexamethasone on the concentration of serum triiodothyronine in man. *J Clin Endocrinol Metab* **39:**1151, 1974.

168. DeGroot LJ, Hoye K: Dexamethasone suppression of serum T₃ and T₄. *J Clin Endocrinol Metab* **42:**976, 1976.

169. Olefsky J, Kimmerling G: Effects of glucocorticoids on carbohydrate metabolism. *Am J Med Sci* **271:**202, 1976.

170. Berger S, Downey J, Traisman H: Mechanism of the cortisone-modified glucose tolerance test. *N Engl J Med* **274:**1460, 1966.

171. Perley M, Kipnis D: Effect of glucocorticoids on plasma insulin. *N Engl J Med* **274:**1237, 1966.

172. Exton JH, Miller JB, Harper SC, et al: Carbohydrate metabolism in perfused livers of adrenalectomized and steroid replaced rats. *Am J Physiol* **230:**163, 1976.

173. Olefsky J: Effect of dexamethasone on insulin binding, glucose transport and glucose oxidation of isolated rat adipocytes. *J Clin Invest* **56:**1499, 1975.

174. Cahill G: Action of adrenal cortical steroids on carbohydrate metabolism. In Christy NP (ed): *The Human Adrenal Cortex.* Harper & Row, New York, 1971, p. 205.

175. Fantus I, Ryan J, Hiraka N, et al: The effect of glucocorticoids on the insulin receptor: An in vivo and in vitro study. *J Clin Endocrinol Metab* **52:**953, 1981.

176. De Pirro R, Green A, Yung-Chin Kao M, et al: Effects of prednisolone and dexamethasone in vivo and in vitro: Studies of insulin binding, deoxyglucose uptake, and glucose oxidation in rat adipocytes. *Diabetologia* **21:**149, 1981.

177. Beck-Nielsen H, De Pirro R, Pedersen O: Prednisone increases the number of insulin receptors on monocytes from normal subjects. *J Clin Endocrinol Metab* **50:**1, 1980.

178. Rizza R, Mandarino L, Gerich J: Cortisol-induced insulin resistance in man: Impaired suppression of glucose production and stimulation of glucose utilization due to a postreceptor defect of insulin action. *J Clin Endocrinol Metab* **54:**131, 1982.

179. Nosadini R, Del Prato S, Tiengo A, et al: Insulin resistance in Cushing's syndrome. *J Clin Endocrinol Metab* **57:**529, 1983.

180. Kinash B, Haist RE: Effect of ACTH and cortisone on islets of Langerhans and pancreas in intact and hypophysectomized rats. *Am J Physiol* **178:**441, 1954.

181. Bastenie P, Conrad V, Frankson J: Effect of cortisone on carbohydrate metabolism measured by the "glucose assimilation coefficient." *Diabetes* **3:**205, 1954.

182. Plotz CM, Knowlton AI, Ragan C: The natural history Cushing's syndrome. *Am J Med* **13:**597, 1952.

183. Kass EH, Finland M: Corticosteroids and infections. *Adv Intern Med* **9:**45, 1958.

184. Kass EH, Finland M: Adrenocortical hormones in infection and immunity. *Annu Rev Microbiol* **7:**361, 1953.

185. Beisel WR, Rapoport MI: Adrenocortical functions and infectious illness. *N Engl J Med* **280:**541, 1969.

186. Rokseth R: Agranulocytosis and sepsis associated with prednisolone. *Lancet* **1:**680, 1960.

187. Sparberg M, Gottschalk A, Kirsner JB: Liver abscess complicating regional enteritis: Report of two cases. *Gastroenterology* **49:**548, 1965.

188. Baker MA, Brain MC, Miller JK, et al: Osteomyelitis: An unusual sequel to neutropenia. *Br Med J* **1:**722, 1967.

189. Des Prez RM, Muschenheim C: The chemoprophylaxis of tuberculosis. *J Chronic Dis* **15:**599, 1963.

190. Statement by an Ad Hoc Committee of the American Thoracic Society: Chemoprophylaxis for the prevention of tuberculosis. *Amer Rev Respir Dis* **96:**559, 1967.

191. Louria DB, Hensle T, Amstrong D, et al: Listeriosis complicating malignant disease. *Ann Intern Med* **67:**261, 1967.

192. Rifkind D, Marchioro TL, Schneck SA, et al: Systemic fungal infections complicating renal transplantation and immunosuppressive therapy. *Am J Med* **43:**28, 1967.

193. Torack RM: Fungus infections associated with antibiotic and steroid therapy. *Am J Med* **22:**872, 1957.

194. Hurley R: Acute dissmeninated (septicemic) moniliasis in adults and children. *Postgrad Med* **40:**644, 1964.

195. Sidransky H, Pearl MA: Pulmonary fungus infections associated with steroid and antibiotic therapy. *Dis Chest* **39:**630, 1961.

196. Graham BS, Tucker WS Jr: Opportunistic infections in endogenous Cushing's syndrome. *Ann Intern Med* **101:**334, 1984.

197. Haggerty RJ, Eley RC: Varicella and cortisone. *Pediatrics* **18:**160, 1956.

198. Fraser CG, Preuss FS, Bigford WD: Adrenal atrophy and irreversible shock associated with cortisone therapy. *JAMA* **149:**1542, 1952.

199. Salassa RM, Bennett WA, Keating FR Jr: Postoperative adrenal cortical insufficiency: Occurrence in patients previously treated with cortisone. *JAMA* **152:**1509, 1953.

200. Christy NP, Wallace EZ, Jailer JW: Comparative effects of prednisone and of cortisone in suppressing the response of the adrenal cortex to exogenous adrenocorticotropin. *J Clin Endocrinol Metab* **16:**1059, 1956.

201. Plager JE, Cushman P Jr: Suppression of the pituitary–ACTH response in man by administration of ACTH or cortisol. *J Clin Endocrinol Metab* **22:**147, 1962.

202. Grant S, Forsham P, Diraimondo V: Suppression of 17-hydroxycorticosteroids in plasma and urine by single and divided doses of triamcinolone. *N Engl J Med* **273:**1115, 1965.

203. Nichols T, Nugent C, Tyler F: Diurnal variation of suppression of adrenal function by glucocorticoids. *J Clin Endocrinol Metab* **25:**343, 1965.

204. Streck WF, Lockwood DH: Pituitary adrenal recovery following short term suppression with corticosteroids. *Am J Med* **66:**910, 1979.

205. Wilson K, Gray C, Cameron E: Hypothalamic pituitary adrenal function in patients treated with intermittent high dose prednisone and cytotoxic chemotherapy. *Lancet* **1:**610, 1976.

206. Livanou T, Ferriman D, James VHT: Recovery of hypothalamopituitary–adrenal function after corticosteroid therapy. *Lancet* **2:**856, 1967.

207. Treadwell BLJ, Savage O, Sever ED, et al: Pituitary–adrenal function during corticosteroid therapy. *Lancet* **1:**355, 1963.

208. Graber AL, Ney RL, Nicholson WE, et al: Natural history of pituitary–adrenal recovery following long-term suppression with corticosteroids. *J Clin Endocrinol* **25:**11, 1965.

209. Dixon RB, Christy NP: On the various forms of corticosteroid withdrawal syndrome. *Am J Med* **68:**224, 1980.

210. Krieger DT, Gewirtz GP: Recovery of hypothalamic–pituitary–adrenal function, growth hormone responsiveness and sleep EEG pattern in a patient following removal of an adrenal cortical adenoma. *J Clin Endocrinol Metab* **38:**1075, 1974.

211. Kyle LH, Meyer RJ, Canary JJ: Mechanism of adrenal atrophy in Cushing's syndrome due to adrenal tumor. *N Engl J Med* **257:**57, 1957.

212. Tucci JR, Meloni CR, Carreon GG, et al: Pituitary–adrenal functional abnormalities in corticogenic adrenal atrophy. *J Clin Endocrinol Metab* **25:**823, 1965.

213. Melby JC: Assessment of adrenocorticotropic activity in man following steroid therapy. *Acta Endocrinol (Kbh)* **35**(Suppl 51):347, 1960.

214. Kehlet H, Binder C: Value of an ACTH test in assessing hypothalamic pituitary–adrenocorical function in glucocorticoid-treated patients. *Br Med J* **2:**147, 1973.

215. Wood JB, James VHT, Franckland AW, et al: A rapid test of adrenocortical function. *Lancet* **1:**243, 1965.

216. Byyny RL: Withdrawal from glucocorticoid therapy. *N Engl J Med* **295:**30, 1976.

217. Lessof MH: Reducing the problems of cortico-steroid therapy. *Br J Hosp Med* **18:**360, 1977.

218. Roscoe P, Choo-Kang YFJ, Horne NW: Betamethasone valerate in cortico-steroid-dependent asthmatics. The integrity of the hypothalamic–pituitary–adrenal axis. *Br J Dis Chest* **69:**240, 1975.

219. Spitzer SA, Kaufman H, Koplovitz A, et al: Beclomethasone dipropionate and chronic asthma. The effect of long-term aerosol administration on the hypothalamic–pituitary–adrenal axis after substituion for oral therapy with cortico-steroids. *Chest* **70:**38, 1976.

220. Harrison BDW, Rees LH, Cayton RM, et al: Recovery of hypothalamo–pituitary–adrenal function in asthmatics whose oral steroids have been stopped or reduced. *Clin Endocrinol* **17:**109, 1982.

221. Ackerman GL, Nolan CM: Adrenocortical responsiveness after alternate-day corticosteroid therapy. *N Engl J Med* **278:**405, 1968.

222. MacGregor RR, Sheagren JN, Lipsett MB, et al: Alternate-day prednisone therapy—Evaluation of delayed hypersensitivity responses, control of disease and steroid side effects. *N Engl J Med* **280:**1427, 1969.

223. Schurmeyer TH, Tsokos GC, Avgerinos PC, et al: Pituitary–adrenal responsiveness to corticotropin-releasing hormone in patients receiving chronic, alternate day glucocorticoid therapy. *J Clin Endocrinol Metab* **61:**22, 1985.

224. Easton JG, Busser RJ, Heimlich EM: Effect of alternate-day steroid administration on adrenal function in allergic children. *J Allergy Clin Immunol* **48:**355, 1971.

225. Rabhan NB: Pituitary–adrenal suppression and Cushing's syndrome after intermittent dexamethasone therapy. *Ann Intern Med* **69:**1141, 1968.

226. Danowski TS, Bonessi JV, Sabeh G, et al: Probabilities of pituitary–adrenal responsiveness after steroid therapy. *Ann Intern Med* **61:**11, 1964.

227. Sampson PA, Brooke BN, Winstone NE: Biochemical confirmation of collapse due to adrenal failure. *Lancet* **1:**1377, 1961.

228. Sampson PA, Winstone NE, Brooke BN: Adrenal function in surgical patients after steroid therapy. *Lancet* **2:**322, 1962.

229. Amatruda TT Jr, Hollingsworth DR, D'Esopo ND, et al: A study of the mechanisms of the steroid withdrawal syndrome: Evidence for integrity of the hypothalamic–pituitary–adrenal system. *J Clin Endocrinol Metab* **20:**339, 1960.

230. Amatruda TT Jr, Hurst MM, D'Esopo ND: Certain endocrine and metabolic facets of the steroid withdrawal syndrome. *J Clin Endocrinol* **25:**1207, 1965.

231. Cook DM, Meikle AW: Factitious Cushing's syndrome. *J Clin Endocrinol Metab* **61:**385, 1985.

232. Morgan HC, Boulnois J, Burns-Cox C: Addiction to prednisone. *Br Med J* **2:**93, 1973.

233. Kimball CP: Psychological dependency on steroids. *Ann Intern Med* **75:**111, 1971.

234. Champion PK: Cushing's syndrome secondary to abuse of dexamethasone nasal spray. *Arch Intern Med* **134:**750, 1974.

235. Ehrig U, Rankin JG: Dependence on ACTH. *Ann Intern Med* **77:**482, 1971.

236. Flavin DK, Fredrickson PA, Richardson JW, et al: Corticosteroid abuse—An unusual manifestation of drug dependence. *Mayo Clin Proc* **58:**764, 1983.

5

The Mineralocorticoid Hormones

Introduction

Aldosterone, the potent mineralocorticoid, was first identified in 1955. The role played by this hormone in the renal regulation of sodium and potassium is of life-sustaining importance. Aldosterone is an integral part of the control system that regulates electrolytes and controls blood pressure. Even in the early sixties the crucial role played by the renin–angiotensin–aldosterone system had become evident to investigators.[1,2] Understanding of the role of the renin–angiotensin–aldosterone system in the causation of clinical disorders has been slow to evolve in the three decades following the discovery of aldosterone. Perturbations in aldosterone secretion can underlie a common disorder such as essential hy-

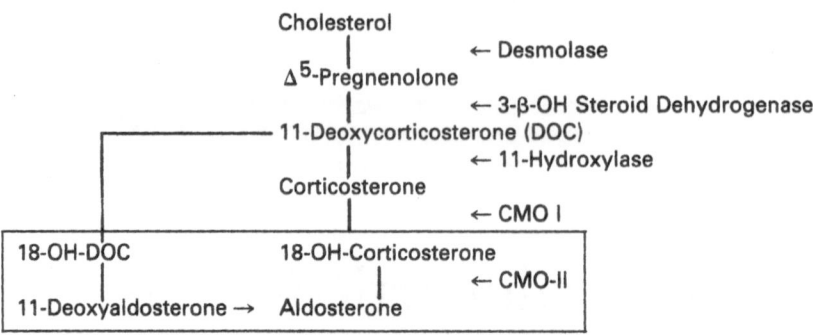

FIGURE 8. Mineralocorticoid synthesis by the adrenal cortex.

pertension, rare disorders such as primary hyperaldosteronism and Bartter's syndrome, or life-threatening alterations in potassium concentration in the body. The physiological principles that control the synthesis and release of aldosterone are briefly reviewed in this chapter.

Synthesis (Fig. 8)

Aldosterone, the prototypical mineralocorticoid secreted by the adrenal cortex, is synthesized almost exclusively by the cells of the zona glomerulosa. Aldosterone and its precursors are synthesized from cholesterol. There are several sequential steps in the synthetic process. First is the desmolase reaction that converts cholesterol into Δ^5-pregnenolone. Next comes the process of 3-β-hydroxyoxidation and double-bond shift to form progesterone. Progesterone, thus formed, undergoes a series of hydroxylations, first at position 21 to form 11-deoxycorticosterone (DOC) and next at position 11 to form corticosterone. Although these steps take place primarily in the zona glomerulosa, the zona fasciculata also has the enzymatic machinery to initiate and complete the steps. However, the subsequent steps that result in conversion of 11-deoxycorticosterone and corticosterone to aldosterone take place only in the zona glomerulosa. The enzyme corticosterone methyloxidase is unique to the glomerulosa cells, and hence the penultimate and ultimate steps in the biosynthesis of aldosterone deserves brief mention.

The Desmolase Reaction (Fig. 9)

The substrate for all steroidogenesis is cholesterol. This precursor is available to the adrenals from the dietary sources of animal fat, as well as endogenous cholesterol snythesized in the body. In addition to their ability to extract cholesterol from the low-density lipoprotein (LDL) in the circulation, the adrenal cortices are capable of de novo cholesterol synthesis.[3] However, this is only a backup mechanism, and for the most part the substrate for steroidogenesis is derived from the cholesterol associated with LDL.[4] The adrenocortical cells contain LDL receptors on the cell membrane. The first step in "trapping" cholesterol by the adrenal cortex involves the combination of LDL cholesterol with

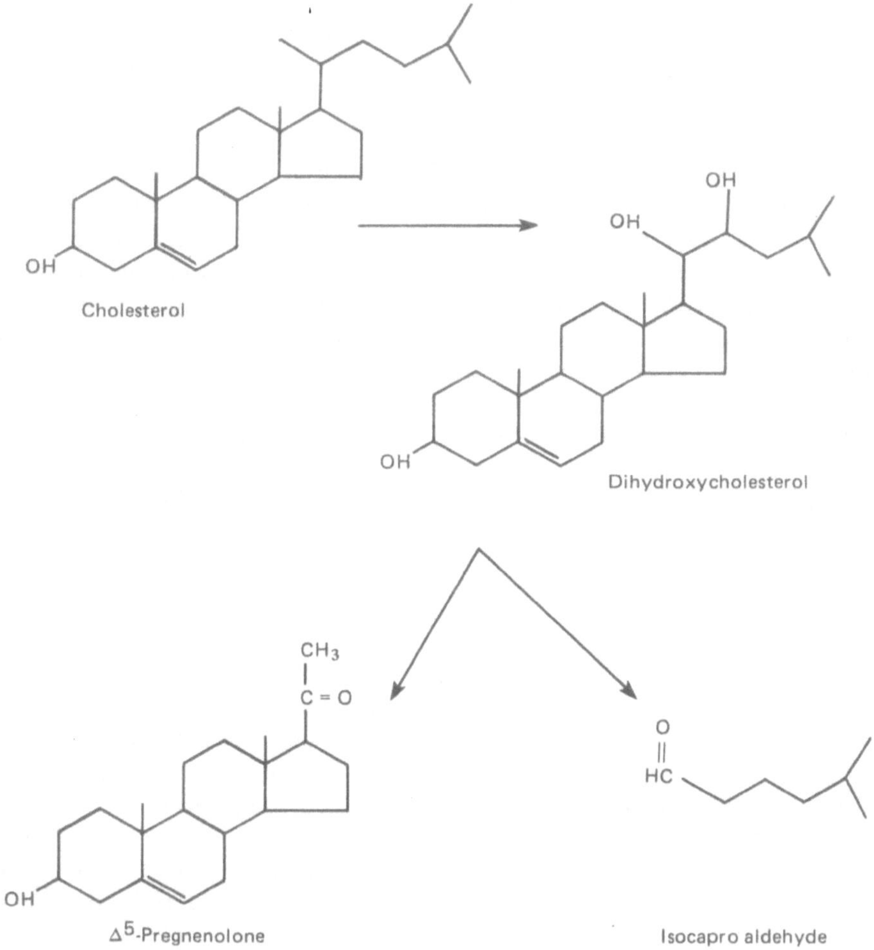

FIGURE 9. Cholesterol side-chain cleavage reaction.

the LDL receptors. The LDL–receptor complex is then internalized by receptor-mediated endocytosis. Once within the adrenocortical cells, the cholesterol is hydrolyzed by the lysosomes to release free cholesterol. Some of this is used for steroid biosynthesis while the rest is stored as cholesterol esters within the cells.[5] This depot form of cholesterol esters serves as an important reservoir of substrate, available to be released when the need arises. A constant process of reesterification and hydrolysis keep this storage pool in dynamic equilibrium with a smaller free pool of cholesterol.[6] As the depot pool becomes depleted, the adrenal cortex extracts more LDL cholesterol from the plasma to replenish its depleting stores. This process is stimulated by both ACTH and angiotensin II. The first alteration that is undergone by the cholesterol is hydroxylation at positions 20 and 22, followed by cleavage of the side chain of the newly formed dihydroxycholesterol. The main product formed by this cleavage is pregnenolone; the by-product of this reaction is isocaproic aldehyde. The formation of pregnenolone from cholesterol takes place in the mitochondria and requires the enzyme desmolase. This crucial, early, and rate-limiting step is stimulated by various adrenocorticotropic stimuli, most notably ACTH.[7] The reaction of side-chain cleavage requires molecular oxygen, as well as NADPH.

3-β-Hydroxyoxidation

The Δ^5-pregnenolone, formed in the mitochondria, is transported to the endoplasmic reticulum for further processing. The crucial event here is conversion of pregnenolone into progesterone. The adrenal cortex cannot make any steroids from cholesterol as long as the double bond exists between carbons 5 and 6. Therefore, the first reaction involves the shift of the double bond to carbons 4 and 5. In addition, the hydroxyl group derived from cholesterol in the A ring is oxidized. These two changes (shift in double bond and oxidation of the hydroxyl group) result in conversion of pregnenolone to progesterone. The enzymes involved are 3-β-hydroxysteroid dehydrogenase and 5-3-ketosteroid isomerase. The product formed is progesterone, which is ready to undergo the necessary hydroxylations that will convert it into DOC and corticosterone.

21- and 11-Hydroxylations

Hydroxylations at specific sites are carried out by specific hydroxylating enzymes. Substrate-specific cytochrome oxidases (cytochrome P-450) are involved in providing specificity and stereo-orientation of these hydroxylation reactions. The process of steroid hydroxylation requires molecular oxygen. Activation of molecular oxygen is mediated by the cytochrome P-450. This process requires two reducing equivalents for each hydroxylated product that is formed. This energy is supplied by electron transport within the adrenal cell.[8] In the mitochondrion the immediate source of the reducing equivalent is the nucleotide TPNH. The sequence of reactions that results in reduction of the cytochrome P-450 is as follows: TPNH — Fpt — ISP — P-450, where Fpt is a flavoprotein, and ISP is an iron–sulfur protein. TPNH is oxidized by Fpt, which is further oxidized by ISP (adrenodoxin). ISP, or adrenodoxin, eventually reduces cytochrome P-450. In the endoplasmic reticulum, there exist two electron transport systems that provide reducing equivalents to the oxygenases. The transport mechanism is represented by the reaction DPNH — Fpd -Cyt b5 — P-450, where DPN-linked flavoprotein dehydrogenase transports electrons via cytochrome b5 to cytochrome P-450. The second system is represented by the following formula: TPNH — Fpt — X — P-450, where TPN-linked flavoprotein dehydrogenase transports electons by an unidentified carrier (X) to cytochrome P-450.

Cytochrome P-450 binds steroid substrates and oxygen and catalyzes the reaction that yields one molecule of hydroxylated product and one molecule of water.

The exact mechanisms involved in hydroxylation have not been completely elucidated; however, several steps have been postulated.

1. Binding of steroid substrate with enzyme.
2. Alteration of redox properties of cytochrome P-450, resulting in electron transport and reduction of P-450 Fe_3 to P-450 Fe_2.

3. The reduced ferrous heme iron can combine with CO or O_2.
4. The heme iron–0_2 couple undergoes two electronic transitions, one before and one after donation of a second electron from the reduced adrenodoxin.
5. Formation of a highly reactive ferryl iron complex that rapidly hydroxylates substrate by a process of two-electron oxidation.
6. The resultant product is a hydroxylated steroid substrate. The ferric enzyme is regenerated during the process. Hydroxylation proceeds in a sequential process; i.e., 21-hydroxylation precedes 11-hydroxylation. It is believed that 21-hydroxylation takes place in the smooth endoplasmic reticulum, and 11-hydroxylation at the mitochondrion. 21-Hydroxylation of progesterone results in the formation of DOC, which is returned to the mitochondrion for 11-hydroxylation that results in the formation of corticosterone.

The Formation of Aldosterone (Fig. 10)

Synthesis of aldosterone is exclusively carried out in the cells of the zona glomerulosa. The major pathway involves the use of corticosterone as the substrate. The enzyme corticosterone methyloxidase type 1 (CMO-I) converts corticosterone to a labile, oxygenated steroid–metalloenzyme complex. This complex can either decompose to 18-hydroxycorticosterone or serve as substrate for corticosterone methyloxidase type II (CMO-II). CMO-II further hydroxylates the methyl group, following which spontaneous dehydration yields the terminal aldehyde aldosterone. Thus CMO-I and CMO-II are important methyloxidases that operate the penultimate and ultimate steps in the biosynthesis of aldosterone.[9,10] The compound 18-hydroxycorticosterone is believed to be formed as a by-product of the first action of CMO.[9] It is unclear whether 18-hydroxycorticosterone is also an immediate precursor of aldosterone. Regardless, 18-hydroxycorticosterone is exclusively formed in the zona glomerulosa. This contrasts with the other aldosterone precursors, such as DOC and corticosterone, which are also synthesized by the zona fasciculata. The assumption that 18-hydroxycorticosterone is produced predominantly, perhaps exclusively, by the zona glomerulosa is based on the observation that parallel increases in 18-hydroxycorticosterone and aldosterone occur in response to salt depletion, angiotensin II infusion, and ACTH administration.[11] 18-Hydroxycorticosterone is believed to be secreted in a ratio of 2.1:1 compared to aldosterone.[11] The terminal reactions in aldosterone biosynthesis are mediated by dehydrogenation (by CMO enzymes) and hydroxylation at position 18. Loss of water from the 18-hydroxy derivative results in the aldehyde aldosterone.[12–14]

A minor pathway for aldosterone formation exists in the adrenal zona glomerulosa. DOC serves as the substrate in this pathway. By a series of hydroxylations and dehydrogenation reactions, DOC is converted into 18-OH-DOC, 11-deoxyaldosterone, and finally aldosterone (Fig. 11).

Control of Aldosterone Secretion

Aldosterone secretion by the zona glomerulosa is controlled by several factors (see Chap. 6), including the renin–angiotensin system, potassium, ACTH,

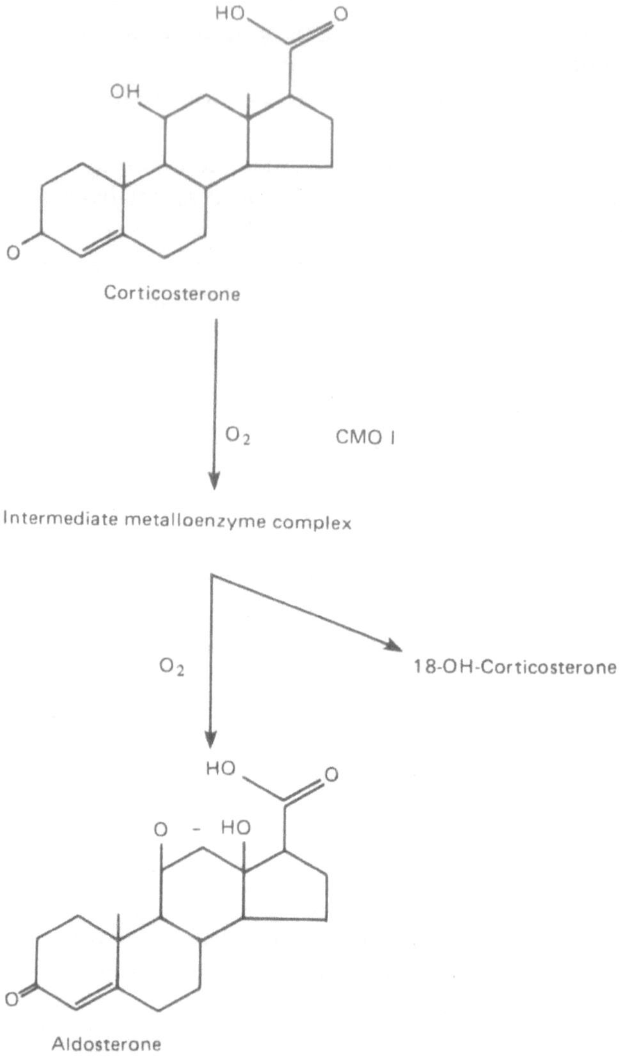

FIGURE 10.   Major pathway for aldosterone synthesis.

and other factors such as Na^+ and dopamine. Of these, the renin–angiotensin system dominates the control of aldosterone secretion.

The Renin–Angiotensin System

Renin is an enzyme synthesized by the juxtaglomerular cells of the kidney. Following release into the circulation, renin acts on an α-2-glycoprotein synthesized by the liver. This substrate is termed renin substrate or angiotensinogen. As a result the decapeptide angiotensin I is released from the renin substrate. Angiotensin I is biologically inactive, but is quickly converted into angiotensin II

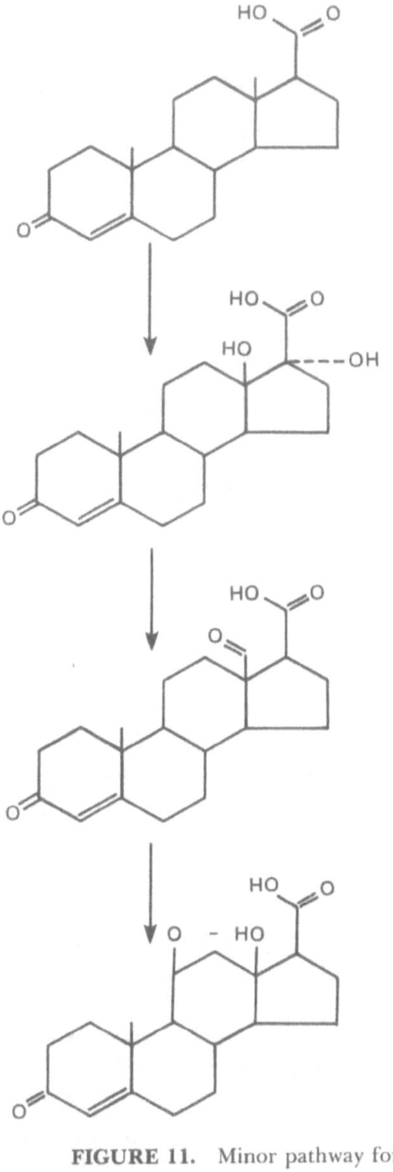

11-Deoxycorticosterone (DOC)

18-Hydroxy-DOC

11-Deoxyaldosterone

Aldosterone

FIGURE 11. Minor pathway for aldosterone synthesis.

by the angiotensin-converting enzyme. This conversion takes place predominantly in the lungs. As a result of this conversion, the biologically active octapeptide angiotensin II is derived. Angiotensin II is degraded by peptidases into numerous smaller peptides, including angiotensin III, which retains its biological activity. Angiotensin III is further degraded into inactive peptide fragments by angiotensinases present in the circulation.

Figure 12 illustrates the major components of the renin–angiotensin system. A consideration of the physiological principles that underlie the function-

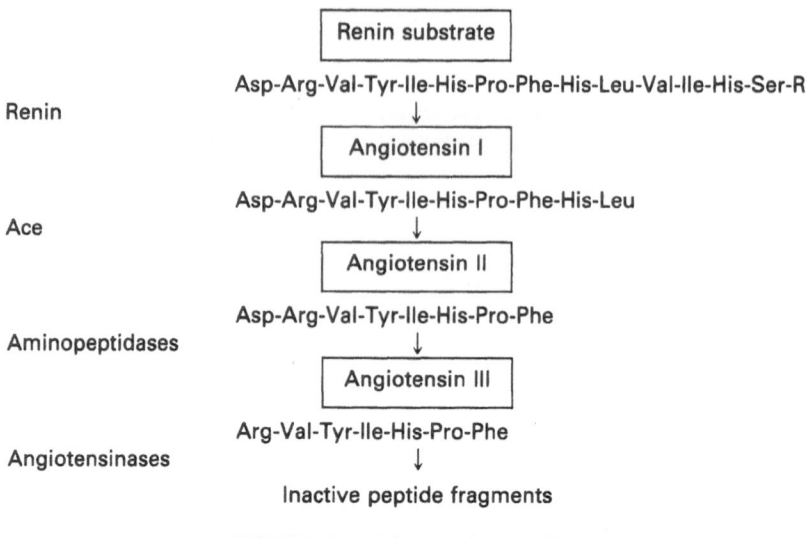

FIGURE 12.   The "renin cascade."

ing of each component is essential in order to understand disorders of the renin–angiotensin–aldosterone axis.

Renin

The first component of the renin–angiotensin system is the all-important enzyme renin. Renin is produced by the juxtaglomerular cells which are located within the afferent arteriole adjacent to the early part of the distal tubule, the macula densa.[15] The two important stimuli that regulate renin release are decreased pressure within the afferent arteriole (a baroreceptor, stretch mechanism) and decreased sodium concentrations in the tubular fluid at the macula densa. Hypotension can stimulate renin release through this baroreceptor mechanism, even in the absence of sodium alterations. To this end, Blaine et al.,[16] using an experimental model, demonstrated that hemorrhage and suprarenal aortic constriction caused a brisk release of renin in dogs without any tubular fluid flow to the macula densa (a nonfiltering kidney). Conversely, increased pressure within the afferent arteriole and renal arterial infusion of sodium chloride are profound inhibitors of renin release. While volume expansion from any cause can theoretically inhibit renin, the most profound inhibition is noticeable when volume expansion is secondary to sodium chloride, imparting importance to transport of NaCl. The relative importance of volume and chloride transport in causing renin release is best illustrated with the use of diuretics that inhibit chloride transport. When chloride transport in the ascending limb is inhibited, by use of furosemide, release of renin occurs even without volume contraction.[12]

In addition to these two major stimuli (decreased pressure within the afferent arteriole and decreased sodium) there are several other stimuli for release of renin (Table 25). Of these the role of the sympathethic system in the regula-

TABLE 25.
Factors That Affect Renin Release

Factors that stimulate renin release	Factors that inhibit renin release
Volume depletion	Volume expansion
Decreased renal arterial pressure	Increased renal arterial pressure
Decreased Na$^+$ at macula densa	Increased Na$^+$ at macula densa
Sympathetic system	
Prostaglandins	Prostaglandin synthethase inhibition
Hypokalemia	Hyperkalemia
	Angiotensin II
	Vasopressin
	Calcium

tion of renin release and the role of prostaglandins in the synthesis of renin merit consideration.

The Sympathethic System. Several lines of evidence suggest a significant role for the sympathethic system in the modulation of renin release:

1. Maneuvers that increase sympathethic activity, such as hypoglycemia and vagotomy are associated with accentuated release of renin.
2. Catecholamines can directly stimulate release of renin, an effect probably mediated by adrenergic receptor agonism.
3. β blockers such as propranolol inhibit release of renin, while alpha-adrengeric blockers do not.[17]
4. Decreased autonomic function, as exemplified in patients with idiopathic orthostatic hypotension, is associated with hyporeninemia.[18,19]
5. The sympathethic mediation of renin secretion assumes clinical importance in the syndrome of hyporeninemic hypoaldosteronism. The hyporeninemia seen in this syndrome may be related to the autonomic dysfunction that is often associated with diabetes, a disorder frequently encountered in patients with hyporeninemic hypoaldosteronism.

Prostaglandins and Renin Secretion. In the kidney, arachidonic acid metabolism through the cyclo-oxygenase pathway generates an array of compounds— prostaglandins (PGs) E$_2$, F$_{2\alpha}$, D$_2$, G, I$_2$, and thromboxane A$_2$. The metabolic pathway involved in the synthesis of these substances is referred to as the renal PG system. The type and amount of PG generated vary with the different parts of the nephron and the renal vasculature. The effect of these prostaglandins on important renal functions such as renin release, renal blood flow, and urinary concentration differs with the type of PG secreted.[20] In a landmark article Larsson and colleagues[21] first demonstrated the existence of a close link between renal PG synthesis and renin release. These workers demonstrated that arachidonic acid stimulates renin release in rabbits, while indomethacin had the opposite effect. Further research rapidly followed this discovery and several facts were established:

1. The renin-releasing effect of arachidonic acid is mediated by PG endo-peroxidases.[22]
2. The metabolite of arachidonic acid that mediates renin release is PGI_2 and not PGE_2.[23,24]
3. A substantial amount of renal PG is biosynthesized in the renal cortex,[25] a feature that may allow direct action of PG on the renin-producing cells. The ability of the renal cortex to synthesize PGI_2 has been convincingly demonstrated by several workers.[26-28]
4. The availability of synthetic PGI_2 has permitted study of the effectiveness of this PG in stimulating renin release. While numerous in vitro data have been compiled to document the dose-dependent increase of renin release following arachidonic acid,[22,29] the studies in human subjects are more interesting and convincing. Patrano et al.[30] demonstrated that PGI_2, when infused at a dosage that inhibited platelet aggregation, caused a dose-dependent rise in plasma renin activity (PRA). Similar findings have been reported by others.[31]
5. The renin-releasing effect of PGI_2 is not related to the vasodilatation caused by the PG.[30,32]
6. The stimulatory effects of arachidonic acid on renin release from cortical slices can be completely prevented by inhibitors of PG synthetase, such as indomethacin.

Clearly, PGs play a significant role in secretion and release of renin. It is possible that the renal PG system mediates a major portion of the noradrenergic release of renin. The clinical relevance of PG-mediated renin release asumes importance in two disorders—hyporeninemic hypoaldosteronism and Bartter's syndrome. The development of severe renin deficiency has been described with the use of indomethacin.[33] Further, deficient production of renal prostacyclin is emerging as one explanation for the development of the syndrome of hypo-reninemic hypoaldosteronism. As for Bartter's syndrome, the role of PG mediation in that disorder is suggested by the dramatic amelioration of some aspects of that syndrome with indomethacin therapy.[34-43]

Renin Substrate

Once renin has been released, it reacts with an α-2-globulin called "renin substrate." Following cleavage at the leucine–leucine' bond, the decapeptide angiotensin I is formed. The source of renin substrate is the liver, and therefore, its concentration in the plasma is increased by estrogens and decreased in chronic liver disease. The generation of angiotensin I from renin substrate by the action of renin is the method employed to measure renin activity in the plasma (PRA). In vivo, the amount of angiotensin I generated by renin is a delicate and dynamic process, determined by several variables—the amount of renin substrate, the half-life of renin in the circulation, the activity of renin, and, most interestingly, the presence of "renin activators" and "renin inhibitors" in the circulation. Renin has no physiological actions of its own, but its ability to generate angiotensin I sets up a cascade of events.

Angiotensin I

Angiotensin I is the biologically inactive precursor of angiotensin II. It must be converted to angiotensin II by the converting enzyme.

The Angiotensin-Converting Enzyme

Angiotensin I is cleaved at its carboxy terminal by angiotensin-converting enzyme (ACE) to form angiotensin II. Although the main source of ACE is in the pulmonary tissue, ACE activity is not limited to the lungs; such activity has been noted in the plasma and in the endothelial vasculature. ACE activity is decreased in hypoxic states and in chronic obstructive lung disease and increased in granulomatous diseases (particularly sarcoidosis) and hyperthyroidism. ACE is inactivated by angiotensinases, which are proteolytic enzymes. These enzymes also cleave angiotensin II at its carboxy terminal, freeing a heptapeptide fragment with considerable biological activity. This fragment is called Angiotensin III (Fig. 12).

Angiotensin-converting enzyme is also responsible for inactivating bradykinin, a vasodepressor substance. Thus, on one hand ACE allows the formation of the vasopressor angiotensin II and on the other hand inactivates the vasodepressor kinins. The beneficial effects on the blood pressure by inhibiting ACE (with captopril) are obvious.

Angiotensin II

Angiotensin II has powerful effects on the smooth muscle and the adrenal zona glomerulosa. This peptide is a strong pressor agent that effects smooth muscle contraction of the vascular bed and stimulates the synthesis and release of aldosterone by the zona glomerulosa. In addition to these two effects, angiotensin II has far-reaching effects on numerous systems. Thus, angiotensin II stimulates the release of catecholamines from the adrenal medulla, facilitates the release of norepinephrine from sympathethic nerve terminals, stimulates the thirst center, promotes redistribution of blood flow in the kidneys, affects renal handling by the tubules (retaining Na at low infusion rates, while being natriuretic at high infusion rates), and may even decrease cardiac output. Angiotensin III, the heptapeptide derived from angiotensin II, is viewed by some as the more active form of angiotensins. While the aldosterone-stimulating potency of angiotensin III is as good as, or even better than, that of its parent compound, its vasoconstrictor properties are far less pronounced.

Focusing on the effect of angiotensin II on the zona glomerulosa, this peptide stimulates the early steps of aldosterone synthesis. The angiotensins bind to specific receptors located in the cells of the zona glomerulosa. Infusions of angiotensin II provoke a brisk and robust release of aldosterone in normal subjects. The sodium balance of the individual has an important bearing in the aldosterone response to infusion of angiotensin II or III. It has been known for a long time that salt deprivation blunts the response of aldosterone to exogenous infusion of angiotensin II. It is believed that the adrenal sensitivity is altered in the salt-deprived individual, who had already attained maximal stimulation of the glomerulosa by heightened endogenous angiotensin II levels. In such a setting, further administration of exogenous angiotensin II may not evoke a response. This "end-organ sensitivity" can considerably modify the aldosterone response to angiotensin II infusion. An inverse relationship between the endogenous angiotension II levels and receptor sensitivity to exogenous infusion of angiotensin II may also help explain the blunted response of aldosterone in settings of salt or volume depletion.

Angiotensin III

While it is clear that angiotensin III is quite effective in stimulating aldosterone secretion and release from the zona glomerulosa, the exact site to which angiotensin III binds is controversial. The receptor for angiotensin II in the glomerulosa cells does not react with angiotensin III.[44] However, it has been shown that cells of the zona fasciculata bind both angiotensins at the same site.[45] Angiotensin III is far less potent than angiotensin II in terms of its vasopressor effects.

The renin–angiotensin system is the most physiological modulator of aldosterone secretion and release on a day-to-day basis. Volume or sodium depletion signals release of renin, which triggers the angiotensin–aldosterone system with eventual restoration of sodium and volume balance. Under conditions of severe sodium restriction (below 25 meq/day) the plasma aldosterone levels may increase to fivefold of that seen in normal subjects.

Potassium

Potassium concentrations in the plasma can directly affect aldosterone secretion, independent of the renin–angiotensin system. Several important lines of evidence ascribe as important role to potassium in regulation of aldosterone secretion.[46–48]

1. Administration of potassium augments aldosterone secretion. Potassium directly stimulates the early (desmolase system) as well as the terminal (corticosterone methyloxidase systems) steps in aldosterone biosynthesis.[49]
2. In the anephric state, the potassium-mediated aldosterone release constitutes the first line of defense against the dangerous development of hyperkalemia.
3. In animals, combined depletion of sodium and potassium results in attenuation of the aldosterone response to salt depletion.[50]
4. In humans, an increase in the serum K^+ concentrations by as little as 0.1 meq/liter is attended by a brisk increase in plasma aldosterone levels. Conversely, hypokalemia lowers plasma aldosterone. While the general consensus is that potassium directly stimulates aldosterone secretion by

the zona glomerulosa, recent data suggests some dependency on circulating angiotensin II levels. Pratt[51] has shown that the stimulatory effect of potassium on plasma aldosterone secretion can be markedly blunted following the administration of captopril, an ACE inhibitor that lowers circulating angiotensin II levels. In vitro studies[52] have also demonstrated an attenuation in the potassium-mediated aldosterone release in the absence of angiotensin II. Thus, it appears that aldosterone regulation by the renin–angiotensin system and by potassium are not as far apart and independent as originally thought.

In addition to its effects on aldosterone secretion, potassium also exerts direct and indirect influenzes on renin secretion. These effects are just the opposite of the effects of potassium on aldosterone secretion. Thus, hypokalemia stimulates, while hyperkalemia suppresses, renin. The hypokalemia-induced renin release is possibly mediated by PGs, since a variety of clinical disorders characterized by hypokalemia (e.g., Bartter's syndrome, habitual vomiting) are associated with increased PG excretion and hyperreninemia.

ACTH and Aldosterone Secretion

The role of ACTH in the physiological regulation of aldosterone secretion is not as important as the renin–angiotensin system or serum K^+ concentrations. A permissive role for ACTH in mineralocorticoid secretion has been suggested by observations in patients with hypopituitarism. As early as 1960, Lieberman and Luetscher[53] postulated that while aldosterone secretion in hypopituitarism may be nearly normal in the basal state on a normal sodium intake, the increase in aldosterone following sodium deprivation is clearly blunted. This reduced sensitivity of zona glomerulosa in the presence of chronic ACTH deficiency has been supported by other workers.[54] Physiologically, administration of an intravenous bolus of ACTH is associated with a brisk rise in serum aldosterone levels. The fact that such a response is preserved even in patients with chronic ACTH deficiency supports the long-held notion that the zona glomerulosa does not atrophy as a consequence of long-term ACTH deprivation. While the classic teaching has been that the reaction of aldosterone to a supraphysiological dose of ACTH is characterized by a brisk response followed by a suboptimal secretory responses with continued stimulation, recent studies have differed somewhat. Thus, Nicholls et al.,[55] as well as Kem et al.,[56] have studied the plasma aldosterone response to low-dose ACTH stimulation in normal, dexamethasone-suppressed, and sodium-depleted states. These studies support the notion that when physiologial concentrations of ACTH are maintained in plasma, the zona glomerulosa responds in a manner parallel to that of the zona fasciculata.

The role of other non-ACTH pituitary peptides in the regulation of aldosterone secretion is even more controversial. The melanotropic hormones of the pituitary play little role, if any, in the physiological regulation of aldosterone secretion. Several of these peptides, however, had been implicated in the development of "idiopathic hyperaldosteronism." Therefore, the role of the melanotropic hormones of the anterior pituitary as well as the "aldosterone-stimulating factor" is discussed in detail in Chapter 6.

Testing Mineralocorticoid Function

Studies of mineralocorticoid function are often resorted to in clinical practice. As in other endocrine disorders, when hyperfunction is suspected suppression maneuvers are attempted, while provocative tests are employed when hypofunction is suspected. Studies of mineralocorticoid function can be divided into basal and dynamic tests.

Basal Measurements of Aldosterone

Aldosterone is easily measured in the plasma and urine. Measurement of plasma aldosterone by radioimmunoassay is readily available. Basal aldosterone measurements are carried out in the morning in the supine position. These measurements are affected by several physiological factors such as dietary salt intake, posture, the volume status, and renal function. Since so many variables affect the diagnostic interpretation of a single plasma level of aldosterone, this measurement seldom confers diagnostic specificity. Suppression and stimulation maneuvers are usually carried out to confirm or exclude the diagnosis.

Urinary aldosterone measurements in a 24-hr urine collection are qualitatively subject to the same variables that affect measurements in the plasma, but to a lesser extent. In clinical practice measurement of 24-hr urinary aldosterone is performed as part of a suppressive maneuver.

Measurements of other mineralocorticoids such as DOC or 18-hydroxycorticosterone are seldom used in clinical practice. Measurement of DOC is a valuable test when congenital adrenal hyperplasia due to 11-β-hydroxylase deficiency is suspected. Measurement of 18-hydroxycorticosterone, the immediate precursor to aldosterone in the terminal steps of aldosterone biosynthesis, may be a valuable adjunct in differentiating hyperaldosteronism caused by adenoma from hyperplasia (Chap. 6).

Dynamic Studies of Aldosterone

The indication for these studies is to exclude or confirm hyper- or hypofunctional states involving aldosteronism.

Suppression Tests

The physiological suppressor of aldosterone is sodium. Conditions characterized by relative or absolute aldosterone hypersecretion are associated with partial or complete inability to suppress. The classic indication where evaluation of aldosterone response to salt loading is helpful is primary aldosteronism. In its most widely applied format, the salt-loading test is performed by measuring plasma aldosterone levels in the supine position before and after administering 2 liters of normal saline intravenously in 4 hr. The test can also be performed by administering an oral salt load (200 meq/day for 3 days) and measuring urinary aldosterone excretion before and after the salt load. The diagnostic value of these tests is discussed in Chapter 6.

Another form of suppression test is the mineralocorticoid suppression test. Here, the aldosterone excretion in urine is evaluated before and after administration of oral 9-α-fludrocortisone, a potent synthethic mineralocorticoid or intramuscular DOC. Physiologically, administration of these powerful mineralocorticoids results in volume expansion and suppression of PRA with resultant prompt suppression of aldosterone excretion in normal subjects and is attended by a prompt decline in urinary aldosterone excretion. The role of these tests in evaluating patients with primary aldosteronism is discussed in Chapter 6.

Stimulation Tests

Measurement of basal aldosterone level is not sensitive enough to detect hypomineralocorticism. When adrenocortical insufficiency is suspected, the mineralocorticoid response to exogenous ACTH administration is a useful screening as well as confirmatory test. In subjects with normal reserve of zona glomerulosa the aldosterone levels in plasma show a three- to fourfold increase following the intravenous administration of ACTH. The aldosterone response in Addisonian patients is characteristically flat, while being normal or minimally attenuated in patients with secondary adrenocortical insufficiency. The aldosterone response to ACTH is usually evaluated by measuring aldosterone in the plasma before and 30, 60, and 90 min after the intravenous administration of a bolus containing 2.5 mg of synthethic corticotropin. The screening value of this test as well as the ability to differentiate primary from secondary adrenal failure is discussed in Chapter 2.

A second type of stimulation test is evaluation of aldosterone response to administration of angiotensin II. The angiotensin infusion test is not widely employed, for several reasons. First, the potential for inducing hypertension from the infused angiotensin is a real and limiting factor; second, the infusion of angiotensin involves use of an infusion pump to deliver graded doses of the peptide; third, the variability in individual responses has precluded adequate standardization; finally, the availability of synthethic ACTH provides an easier, safer, and consistent stimulus for aldosterone release, with much better standardization. For all these reasons the angiotensin infusion test is restricted to research centers.

A third type of stimulation test involves the use of maneuvers to provoke release of renin and, therefore, indirectly aldosterone. The renin–aldosterone response to the combined stimuli of salt depletion (by diuretics) and erect posture is widely used in evaluating patients with hyporeninemic hypoaldosteronism. In this situation, the blunted or absent renin responses to salt depletion (and posture) result in failure of aldosterone to rise after these maneuvers. The PRA response to posture and diuretic is also an important diagnostic aid in evaluating patients with suspect primary hyperaldosteronism. In this situation the PRA suppression is a phenomenon secondary to autonomous hypersecretion of aldosterone. The role of PRA response to provocative maneuvers is discussed in Chapter 6.

The Renin–Angiotensin–Aldosterone System in Health and Disease

The renin–angiotensin–aldosterone system works in a concerted manner to ensure proper sodium and water balance. Thus, volume and salt depletion result in stimulation of the axis with consequent retention of sodium and water. Severe salt restriction or depletion is the most potent form of stimulus for aldosterone release. The aldosterone–renin axis is controlled by negative feedback indirectly in two ways. The retention of salt caused by aldosterone results in volume expansion which shuts off renin production. In addition, juxtaglomerular (JG) cells are also suppressed by angiotensin II levels. Thus, the physiologically activated renin–angiotensin system eventually returns to the basal state owing to these negative feedback mechanism.

The disorders that affect the renin angiotensin aldosterone axis can be of the following varieties:

1. Primary hyperreninism, resulting in activation of the entire axis.
2. Secondary hyperreninism, which also results in activating the angiotensin–aldosterone axis.
3. Primary hyperaldosteronism, where the secretion of aldosterone by the zona glomerulosa is autonomous and independent of the renin–angiotensin axis.
4. Hyporeninemia, which results in decreased activation of the angiotensin–aldosterone components of the axis.
5. Intrinsic failure of the zona glomerulosa to secrete aldosterone, which results in hypoaldosteronism, with secondary hyperreninemia due to negative feedback.

Primary Hyperreninism

Very rarely, hyperreninemia can result from autonomous secretion of renin by a tumor of the JG cells. This results in a constellation of features characterized by hypertension, hypokalemic alkalosis, and hyperaldosteronism with hyperreninemia. The JG cell tumors are discussed on page 297.

Secondary Hyperreninism

Secondary hyperaldosteronism is commonly seen in association with edematous states (such as cirrhosis with ascites and nephrotic syndrome) where the effective extracellular fluid volume is decreased. Less commonly secondary hyperreninemia is seen in association with renovascular hypertension, malignant hypertension, and Bartter's syndrome (Chapter 7).

Primary, autonomous secretion of aldosterone is encountered classically in primary hyperaldosteronism caused by adenoma, bilateral hyperplasia, or adrenal carcinoma. This entity is described in detail in Chapter 6.

Hyporeninism with secondary hypoaldosteronism is encountered in selective hypoaldosteronism (discussed in Chapter 8). The characteristic feature here is hyperkalemia with metabolic acidosis.

Finally, intrinsic defects in the zona glomerulosa can result in aldosterone deficiency. This is most classically seen in Addison's disease or enzymatic defects in aldosterone biosynthesis. The situation here is characterized by hyperreninemia, in an attempt by the JG cells to produce aldosterone from a failing zona glomerulosa.

The disorders of the renin–angiotensin–aldosterone system are discussed in Chapters 6, 7, and 8.

References

1. Laragh JH, Angers M, Kelly WG, et al: Hypotensive agents and pressor substances: The effect of epinephrine, norepinephrine, angiotensin II, and others on the secretory rate of aldosterone in man. *JAMA* **174**:234, 1960.
2. Ames RP, Borskowski AJ, Sicinski AM, et al: Prolonged infusions of angiotensin II and norepinephrine and blood pressure, electrolyte balance, and aldosterone and cortisol secretion in normal man and in cirrhosis with ascites. *J Clin Invest* **44**:1171, 1965.
3. Goodman AD: Studies on the effect of omega-methylpantothenic acid on corticosterone secretion in the rat. *Endocrinology* **66**:420, 1960.
4. Gwynne JT, Strauss JF III. The role of lipoproteins in steroidogenesis and cholesterol metabolism in steroidogenic glands. *Endocr Rev* **3**:299, 1982.
5. Szabo D, Szabon J: The role of liquid crystalline behaviour of adrenocortical lipids in the control of steroidogenesis. *Endokrinologie* **80**:275, 1982.
6. Moses HL, David WW, Rosenthal AS, et al: Adrenal cholesterol localization by electron-microscope autoradiography. *Science* **163**:1203, 1969.
7. Davis WW, Garren LD: Evidence for the stimulation by adrenocorticotropic hormone of the conversion of cholesterol esters to cholesterol in the adrenal in vivo. *Biochem Biophys Res Commun* **24**:805, 1966.
8. Estabrook RW, Mason IJ, Baron J, et al: Drugs, alcohol and sex hormones. A molecular perspective of the receptivity of cyto chrome P450. *Ann NY Acad Sci* **212**:27, 1973.
9. Ulick S: Diagnosis and nomenclature of the disorders of the terminal portion of the aldosterone biosynthetic pathway. *J Clin Endocrinol Metab* **43**:92, 1976.
10. Neher R: Aldosterone: chemical aspects and related enzymology. *J Endocrinol* **81**:25, 1979.
11. Biglieri EG, Schambelan M: The significance of elevated levels of plasma 18-hydroxycorticosterone in patients with primary aldosteronism. *J Clin Endocrinol Metab* **49**:87, 1979.
12. Nicolis GL, Ulick S: Role of 18-hydroxylation in the biosynthesis of aldosterone. *Endocrinology* **76**:514, 1965.
13. Pasqualini JR: Conversion of tritiated 18-hydroxycorticosterone to aldosterone by slices of human cortico-adrenal gland and adrenal tumor. *Nature* **201**:501, 1964.
14. Raman PB, Ertel RJ, Ungar F: Conversion of progesterone-^{14}C to 18-hydroxycorticosterone and aldosterone by mouse adrenals in vitro. *Endocrinology* **74**:865, 1964.
15. Davis JO, Freeman RH: Mechanism regulating renin release. *Physiol Rev* **56**:1, 1976.
16. Blaine EH, Davis JO, Witty RT: Renin release after hemorrhage and after suprarenal aortic constriction in dogs without sodium delivery to the macula densa. *Circ Res* **27**:1081, 1970.
17. Ganong WF: Sympathetic effects on renin secretion: Mechanisms and physiological role. *Adv Exp Med Biol* **17**:17, 1972.
18. Corder CN, Kanefsky TM, McDonald RH, et al: Postural hypotension: Adrenergic responsivity and levodopa therapy. *Neurology* **27**:921, 1977.
19. Love DR, Brown JJ, Chinn RH, et al: Plasma renin in idiopathic orthostatic hypotension: Differential response in subjects with probable afferent and efferent autonomic failure. *Clin Sci* **41**:289, 1971.
20. Dunn MJ, Hood VL: Prostaglandins and the kidney. *Am J Physiol* **233**:F169, 1977.
21. Larsson C, Weber P, Anggard E: Arachidonic acid increases and indomethacin decreases plasma renin activity in the rabbit *Eur J Pharmacol* **28**:391, 1974.
22. Weber PC, Larsson C, Anggard E, et al: Stimulation of renin release from rabbit renal cortex by arachidonic acid and prostaglandin endoperoxides. *Circ Res* **39**:868, 1976.
23. Whorton AR, Smigel M, Oates JA, et al: Regional differences in prostacyclin formation by the kidney. Prostacyclin is a major prostaglandin of renal cortex. *Biochim Biophys Acta* **529**:176, 1978.
24. Oates JA, Whorton AR, Gerkens JR, et al: The participation of prostaglandins in the control of renin release. *Fed Proc* **38**:72, 1979.
25. Larsson C, Anggard E: Mass spectrometric determination of prostaglandin E_2, F_{2a} and A_2 in the cortex and medulla of the rabbit kidney. *J Pharm Pharmacol* **28**:326, 1976.
26. Oliw E, Lunden I, Sjoquist B, et al: Determination of 6-ketoprostaglandin F_{1a} in rabbit kidney and urine and its relation to sodium balance. *Acta Physiol Scand* **105**:359, 1979.
27. Terragno NA, McGiff JC, Terragno A: Prostacyclin ($PG1_2$) production by renal blood vessels: relationship to an endogenous prostaglandin synthesis inhibitor (EPSI). *Clin Res* **26**:545A 1978 (Abstract).
28. Hassid A, Konieczkowski M, Dunn MJ: Prostaglandin synthesis in isolated rat kidney glomeruli.

Proc Natl Acad. Sci USA **76:**1155, 1979.

29. Whorton AR, Misono K, Hollifield J, et al: Prostaglandins and renin release. I. Stimulation of renin release from rabbit renal cortical slices by PGI_2. *Prostaglandins* **14:**1095, 1977.

30. Patrono C, Ciabattoni G, Cinotti GA, et al: Prostacyclin and renin release in man. *Clin Res* **27:**426A, 1979 (Abstract).

31. FitzGerald GA, Friedman LA, Miyamori I, et al: A double blind placebo controlled crossover study of prostacyclin in man. *Life Sci* **25:**665, 1979.

32. Gerber JG, Branch RA, Nies AS, et al: Prostaglandins and renin release: II. Assessment of renin secretion following infusion of PGI_2, E_2, and D_2 into the renal artery of anesthetized dogs. *Prostaglandins* **15:**81, 1978.

33. Tan SY, Shapiro R, Franco R, et al: Indomethacin-induced prostaglandin inhibition with hyper-kalemia. A reversible cause of hyporeninemic hypoaldosteronism. *Ann Intern Med* **90:**783, 1979.

34. Garin EH, Fennell RS III, Iravani A, et al: Treatment of Bartter's syndrome with indomethacin. *Am J Dis Child* **134:**258, 1980.

35. Bowden RE, Gill JR Jr, Radfar N, et al: Prostaglandin synthetase inhibitors in bartter's syn-drome. Effect on immunoreactive prostaglandin E excretion. *JAMA* **239:**117, 1978.

36. Dillon MJ, Shah V, Mitchell MD: Bartter's syndrome: 10 cases in childhood. Results of long-term indomethacin therapy. *Q J Med* **48:**429, 1979.

37. Delaney VB, Oliver JF, Simms M, et al: Bartter's syndrome: Physiological and pharmacological studies. *Q J Med* **50:**213, 1981.

38. Vierhapper H, Waldhausl W: Effect of indomethacin upon the renin-angiotensin system in patients with Bartter's syndrome. *Eur J Clin Invest* **10:**119, 1980.

39. Kornerup JH, Pedersen EB, Petersen VP: Bartter's syndrome without hyperplasia of the jux-taglomerular apparatus, treated with indomethacin. *Acta Med Scand* **204:**235, 1978.

40. Shimoyama R: Reversal of altered vascular responsiveness in Bartter's syndrome by indo-methacin treatment. *J Clin Endocrinol Metab* **51:**908, 1980.

41. Sasaki H, Kawasaki T, Okumura M, et al: Bartter's syndrome: Effect of indomethacin on prostaglandins, urinary kallikrein, renin-angiotensin-aldosterone system, and the response to angiotensin II antagonist. *Endocrinol Jpn* **27:**417, 1980.

42. Lechi A, Mantero F, Opocher G, et al: Effect of indomethacin on urinary kallikrein excretion in Bartter's syndrome of the adult. *J Endocrinol Invest* **4:**17, 1981.

43. Halushka PV, Wohltmann H, Privitera PJ, et al: Bartter's syndrome: Urinary prostaglandin E-like material and kallikrein; indomethacin effects. *Ann Intern Med* **87:**281, 1977.

44. Braley LM, Menachery AI, Underwood RH, et al: Is the adrenal angiotensin receptor angioten-sin II or angiotensin III like? *Acta Endocrinol* **102:**116, 1983.

45. Vallotton MB, Capponi AM, Grillet C, et al: Characterization of angiotensin receptors on bovine adrenal fasciculata cells. *Proc Natl Acad Sci USA* **78:**592, 1981.

46. Fraser R, Brown JJ, Lever AF, et al: Control of aldosterone secretion. *Clin Sci* **56:**389, 1979.

47. Chan JCM: Control of aldosterone secretion. *Nephron* **23:**79, 1979.

48. Himathongkam T, Dluhy RJ, Williams JH: Potassium aldosterone renin interrelationships. *J Clin Endocrinol Metab* **41:**153, 1975.

49. McKenna TJ, Island DP, Nicholson WP, et al: The effects of potassium on early and late steps in aldosterone biosynthesis in cells of the zona glomerulosa. *Endocrinology* **103:**1411, 1978.

50. Campbell WB, Schmitz JM: Effect of alterations in dietary potassium on the pressor and steroidogenic effects of angiotensins-II and III. *Endocrinology* **103:**2098, 1978.

51. Pratt JH: Role of angiotensin II in potassium mediated stimulation of aldosterone secretion in the dog. *J Clin Invest* **70:**667, 1982.

52. Fredlund P, Saltman S, Kondo T, et al: Aldosterone production by isolated glomerulosa cells. Modulation of sensitivity to angiotensin II and ACTH by extracellular potassium concentration. *Endocrinology* **100:**481, 1977.

53. Lieberman AH, Leutscher JA: Some effects of abnormalities of pituitary adrenal or thyroid function on secretion of aldosterone and the response to corticotropin or sodium deprivation. *J Clin Endocrinol Metab* **20:**1004, 1960.

54. Fraser R, Brown JJ, Lever AF, et al: Control of aldosterone secretion. *Clin Sci* **56:**389, 1979.

55. Nicholls MG, Espiner EA, Donald RA: Plasma aldosterone response to low dose ACTH stimula-tion. *J Clin Endocrinol Metab* **41:**186, 1975.

56. Kem DC, Gomez-Sanchez C, Kramer NJ, et al: Plasma aldosterone and renin activity in response to ACTH infusion in dexamethasone suppressed normal and sodium depleted man. *J Clin Endocrinol Metab* **40:**116, 1975.

6

Primary Hyperaldosteronism

Introduction

In 1955 Jerome Conn described a patient with hypertension, hypokalemia, and aldosterone-secreting adrenocortical adenoma.[1] This event was largely responsible for the "crossover" of aldosterone from the realm of pure biochemistry and physiology into mainstream clinical medicine. The excitement generated by the notion of "curable hypertension" led to an intense search for diagnostic methods that would permit identification of hypertensive patients who might be suffering from "Conn's syndrome" and hence would be candidates for adrenal surgery. The laboratory and clinical investigations of Davis, Genest, Laragh, and their respective coworkers[2–4] had by the early sixties clearly established that the renal pressor system is one of the major regulators of aldosterone secretion by the adrenal zona glomerulosa. When Boucher et al.[5] developed assays that permitted the measurement of plasma angiotensin and renin activity in human plasma, time was ripe for what would be referred to in the future as a "classic example of technology transfer." In a landmark article resonating with clarity Conn et al.[6] reported that suppression of plasma renin activity (PRA) was the hallmark of primary aldosteronism and that measurement of this activity would delineate primary from secondary aldosteronism in hypertensive states. The authors believed that "security and confidence in the preoperative diagnosis of primary aldosteronism is now afforded by this measurement."

The reader who follows the evolution of this syndrome from that first description of an aldosterone-producing adenoma in 1955 to the present time cannot fail to be impressed by an evolutionary process that provides compelling reading. Regardless of whether that literature is viewed through the tinted glasses of the endocrinologist, or through the biochemist's glasses darkly, the information revealed in the past 3 decades have assumed a kaleidoscopic quality. The entire subject of aldosterone excess is an area of clinical and investigational flux, where even well-backed hypotheses of the day have quickly become yesterday's fancy. A brief look at the developments following the discovery of PRA methodology will make the point.

The concept held by Conn and associates[7] that an estimated 20% of patients with essential hypertension may indeed harbor an aldosterone-producing adenoma turned out to be incorrect and misleading. The notion that primary aldosteronism is not a single disease entity took hold when it was discovered that a substantial percentage of patients with hypertension, hypokalemia, low PRA, and aldosterone excess had no tumor on exploration; such patients were instead found to have nodular or simple hyperplasia of the zona glomerulosa or very occasionally had normal histology.[8] The observation that surgery held no cure for the group of patients with hyperplasia emphasized the importance of pre-operative delineation between the adenomatous versus idiopathic hyperplastic forms of the syndrome. In the following years data accumulated supporting the notion that PRA measurements, while helpful, do not have the requisite sensitivity and specificity needed for reliable separation of patients with hyperaldosteronism from those with just essential hypertension. The description of "low-renin essential hypertension" added confusion to the clinical and hormonal spectrum of mineralocorticoid hypertension, as was the elusive search for the offending mineralocorticoid responsible for causing that syndrome. The prevailing confusion, at one point, led to the speculation that idiopathic hyperaldosteronism is part of a continuum with essential hypertension. Prevailing dogma was disputed when workers from Glasgow suggested that idiopathic hyperaldosteronism is "at the upper end of a wider-than-normal distribution of aldosterone in essential hypertension, from which it has been separated wrongly."[9] The subsequent years saw new additions to an already enlarging spectrum; thus, the entities of dexamethasone-suppressible hyperaldosteronism, the syndrome of unilateral adrenal hyperplasia, aldosterone hypersecretion by adrenal carcinoma, and ectopic secretion of aldosterone by malignant neoplasms were all discovered in the seventies. The term *indeterminate hyperaldosteronism* was coined to denote an entity that was in the process of evolution into adenomatous hyperaldosteronism.

The literature of late seventies and early eighties is peppered and punctuated by "yet another new test" that provides a hormonal means to distinguish the aldosterone-producing adenoma from idiopathic hyperplasia. An impressive array of hormonal tests have been devised, all generating a rising interest, which in many instances fell into disrepute after a brief reign. The list of such tests forms an impressive catalogue. Nonetheless, the bulk of data suggests that no single test consistently separates patients with adenoma from those with hyperplasia. Thus, combinations of response patterns are likely to yield more information than any single test.

The notion that a non-ACTH pituitary secretagogue may be involved in the development of at least one form of hyperaldosteronism, i.e., idiopathic hyperplasia, has gained momentum since the early eighties. As in other spheres involving aldosterone pathophysiology, the contender for such a role is riddled with myth, mystery, and minutiae. Thus, γ-melanotropin, β-lipotropin, α-melanotropin, pro-opiomelanocortin, and a glycoprotein termed *aldosterone-stimulating factor* have all, at one time or another, vied for that title. That a neurotransmitter connection may exist in regulation of aldosterone secretion is not longer just theoretical, and dopamine is rapidly becoming accepted as a physiological inhibitor of aldosterone secretion—a relationship analogous to that between dopamine and prolactin. The concept that dopamine dysregulation of a non-ACTH pituitary secretagogue may underlie the pathophysiology of idiopathic hyperaldosteronism awaits elucidation.

While the hormonal developments in the diagnosis of various subtypes of hyperaldosteronism were abounding, new radiological procedures to permit localization of tumors were also keeping pace. Iodocholesterol scintigraphy, high-resolution computerized scans, and invasive procedures such as adrenal venous sampling emerged as important techniques to demonstrate the presence and site of tumors. Thus, the apparently simple syndrome that Jerome Conn described in 1955 can be approached and viewed from several vantage points. The perspectives provided below are predominantly those seen through a clinician's eyes.

Pathophysiology

Primary hyperaldosteronism results from autonomous secretion of aldosterone by the zona glomerulosa, secondary to either a tumor or hyperplasia. The condition can be viewed as a situation where the renin–angiotensin–aldosterone nexus is no longer in control of aldosterone secretion by the zona glomerulosa. To understand the numerous dynamic phenomena seen in association with primary hyperaldosteronism it is essential to briefly review the physiology of aldosterone secretion.

Physiology of Aldosterone Secretion

The major control mechanisms involved in the secretion and release of aldosterone are

1. The renin–angiotensin–aldosterone system
2. Serum potassium
3. ACTH
4. Non-ACTH pituitary secretagogues

The influence of each of these mechanisms on aldosterone secretion merits brief consideration.

The major physiological factor that regulates aldosterone secretion on a day-to-day basis is renin, via angiotensin. Renin is an enzyme produced by the juxtaglomerular cells which are located within the afferent arteriole adjacent to the early part of the distal tubule, the macula densa.[10] The two important stimuli that regulate renin release are decreased pressure within the afferent arteriole (a baroreceptor stretch mechanism) and decreased sodium concentrations in the tubular fluid at the macula densa. Hypotension (the baroreceptor mechanism) can stimulate renin release, even in the absence of sodium alterations. Blaine et al.,[11] using an experimental model, demonstrated that hemorrhage and suprarenal aortic constriction caused a brisk release of renin in dogs without any tubular fluid flow to the macula densa (a non-filtering kidney). Similarly, increased pressure within the afferent arteriole and renal arterial infusion of sodium chloride are profound inhibitors of renin release. While volume expansion from any cause can theoretically inhibit renin, the most profound inhibition is noticeable when volume expansion is secondary to sodium chloride, imparting importance to transport of NaCl. The relative importance of volume and chloride transport in causing renin release is best illustrated with the use of diuretics that inhibit chloride transport. When chloride transport in the ascending limb is inhibited, by use of furosemide, release of renin occurs even without volume contraction.[12]

In addition to baroreceptor and chemomediated mechanisms of release, renin is also regulated by a neural, predominantly sympathetic mechanism. Several maneuvers that increase sympathetic activity, most prototypical of which are hypoglycemia and vagotomy, are associated with release of renin. Catecholamines can directly stimulate renin release, an effect probably mediated by β-adrenergic-receptor agonism. The decreased renin secretion that is encountered in autonomic neuropathic states as well as following beta-blocker therapy is a clinical correlate based on the sympathetic mediation involved in release of renin.

Besides the three factors already mentioned—i.e., pressure, sodium chloride, and agonist sympathetic control—several other factors, albeit minor, have been implicated in the regulation of renin release (Fig. 13). Thus, calcium,[13]

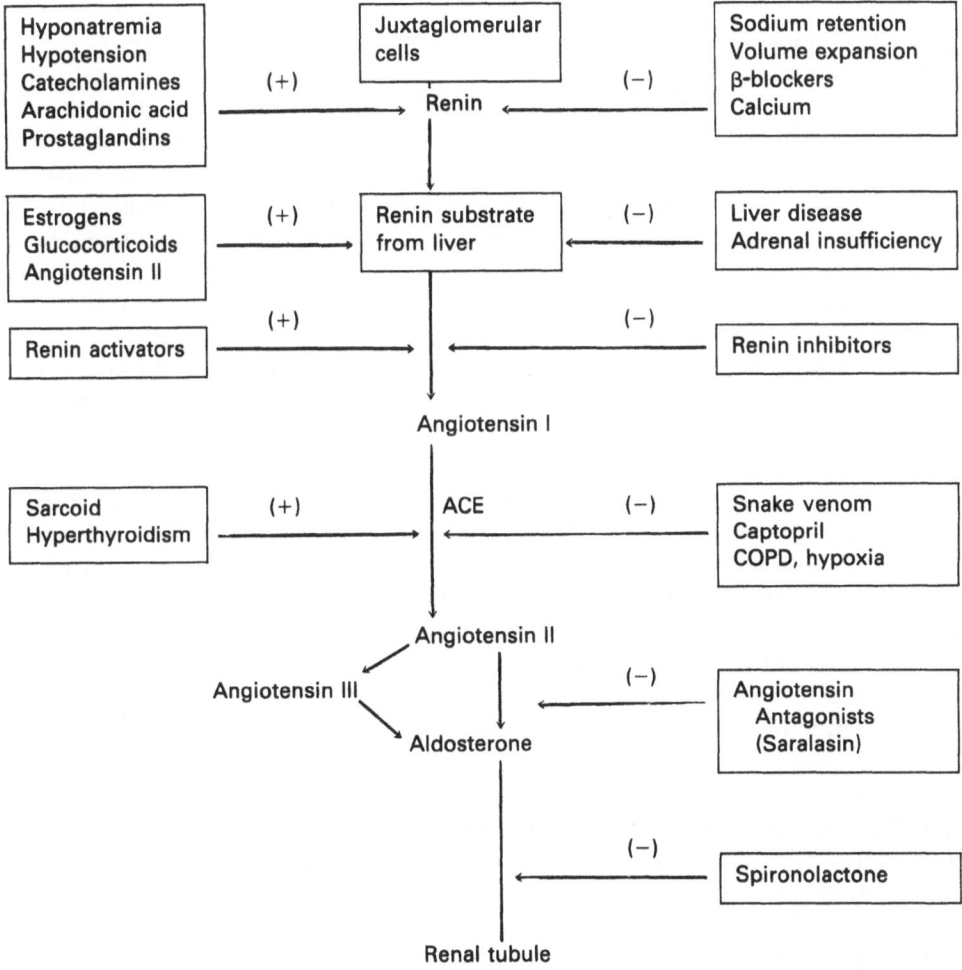

FIGURE 13. Factors that affect the renin–angiotensin–aldosterone system.

arachidonic acid[14] ("the renal eicosanoid system"), and prostaglandins[15] variably participate in regulation of renin release. Of these, calcium exerts an inhibitory effect, while arachidonic acid and prostaglandins have a trophic effect on release of renin.

Recent evidence suggests that renin is synthesized as a 50,000-dalton single-chain polypeptide.[16–18] This prorenin ("big renin," "inactive renin") has generated considerable controversy—according to some workers, the inactive renin may represent renin–inhibitor protein complexes, which could reflect an in vivo storage form of the enzyme, or alternatively might represent an in vitro artifact.[19,20]

Once renin has been released, it reacts with an α_2 globulin called "renin substrate." Following cleavage at the leucine–leucine bond, the decapeptide angiotensin I is formed. The source of renin substrate is the liver, and therefore, its concentration in the plasma is increased by estrogens and decreased in chronic liver disease. The generation of angiotensin I from renin substrate by the action of renin is the method employed to measure renin activity in the plasma (PRA). In vivo, the amount of angiotensin I generated by renin is a delicate and dynamic process, determined by several variables—the amount of renin substrate, the half-life of renin in the circulation, the activity of renin, and, most interestingly, the presence of "renin activators" and "renin inhibitors" in the circulation.[21] Renin has no physiological actions on its own, but its ability to generate angiotensin I sets up a cascade of events. Angiotensin I, which is biologically inactive, is cleaved at its carboxy terminal by angiotensin-converting enzyme (ACE), to form angiotensin II. Although the main source of ACE is the pulmonary tissue, ACE activity is not limited to the lungs; such activity has been noted in the plasma and in the endothelial vasculature. ACE activity is decreased in hypoxic states and chronic obstructive lung disease and increased in granulomatous diseases (particularly sarcoidosis) and hyperthyroidism. ACE is inactivated by angiotensinases, which are proteolytic enzymes. These enzymes cleave angiotensin II at its carboxy terminal, freeing a fragment, a heptapeptide with considerable biological activity. This fragment is called Angiotensin III.

Angiotensin II and angiotensin III have powerful effects on the smooth muscle and the adrenal zona glomerulosa. These peptides are strong pressor agents that effect smooth muscle contraction of the vascular bed, as well as stimulate the synthesis and release of aldosterone by the zona glomerulosa. In addition to these two effects, angiotensin II has far-reaching effects on numerous systems. Thus, angiotensin II stimulates the release of catecholamines from the adrenal medulla, facilitates the release of norepinephrine from sympathetic nerve terminals, stimulates the thirst center, promotes redistribution of blood flow in the kidneys, affects renal handling of ions by the tubules (retaining Na at low infusion rates, while being natriuretic at high infusion rates), and may even decrease cardiac output. Angiotensin III, the heptapeptide derived by from angiotensin II, is viewed by some as the more active form of angiotensins. While the aldosterone-stimulating potency of angiotensin III is as good as, or even better than, that of its parent compound, its vasoconstrictor properties are far less pronounced.

Focusing on the effect of angiotensin II and III on the zona glomerulosa, these peptides stimulate the early steps of aldosterone synthesis. The angiotensins bind to specific receptors located in the cells of the zona glomerulosa. Infusions of angiotensin II provoke a brisk and robust release of aldosterone in normal subjects. The sodium balance of the individual has an important bearing in the aldosterone response to infusion of angiotensin II. It has been known for a long time that salt deprivation blunts the response of aldosterone to exogenous infusion of angiotensin II.[22,23] It is believed that the adrenal sensitivity is altered in the salt-deprived individual, who had already attained maximal stimulation of the glomerulosa by heightened endogenous angiotensin II levels. In such a setting, further administration of exogenous angiotensin II may not evoke a response. This "end-organ sensitivity" can considerably modify the aldosterone response to angiotensin II infusion.[24] An inverse relationship between the endogenous angiotensin II levels and receptor sensitivity to exogenous infusion of angiotensin II may also help explain the blunted response of aldosterone in settings of salt or volume depletion.

The renin–angiotensin–aldosterone axis works in harmony to maintain sodium balance and maintain volume of the extracellular space and the blood pressure. The angiotensin II–dependent vasoconstrictor mechanism and the aldosterone-mediated sodium—and volume-dependent mechanism are responsible for maintaining the blood pressure of the normal and most hypertensive states.[25] Physiologically, factors that stimulate the renin–angiotensin system provoke aldosterone release, while the converse occurs when the renin–angiotensin system is inhibited (Fig. 13).

Pathophysiology of Hyperaldosteronism

The aforementioned principles that underlie normal physiology undergo considerable alteration in patients with hyperaldosteronism. The most important alterations are the following:

1. Suppression of plasma renin activity is a biochemical hallmark seen in a vast majority of patients with hyperaldosteronism, regardless of etiology. The prevalence of such a phenomenon in patients with primary hyperaldosteronism, the incidence of such a finding in patients with essential hypertension ("low-renin type"), and the screening value of measurement of PRA are discussed under the section on laboratory diagnosis. The most important physiological principle to note is that PRA suppression delineates the primary from the secondary varieties of hyperaldosteronism. The reasons for the suppressed PRA activity in patients with primary aldosteronism are the increased plasma volume and expansion of the exchangeable sodium space. In the classic patient with primary hyperaldosteronism the PRA is low with failure to rise following standing and furosemide administration.

2. The sustained and autonomous secretion of aldosterone by the zona glomerulosa results in kaliuresis and retention of sodium. The initial response to aldosterone excess is positive sodium balance followed by a gradual expansion of the exchangeable sodium space. The hypertension develops gradually, as does the hypokalemia. After these initial responses to aldosterone, sodium balance is reestablished since the renal tubules "escape" from the progressive salt-retaining effect of chronic hyperaldosteronism. However, the potassium-losing effect continues, with progressive kaliuresis. With passage of time the intravascular vol-

ume readjusts downward, albeit at a higher level than normal. While the aldosterone excess is clearly the cause of hypokalemia in the syndrome, the link between the hormonal excess and hypertension is less certain. The hypertension in primary aldosteronism is related more to other probable factors, such as increased sodium content of arterioles, hypersensitivity of vascular smooth muscle to pressor substances, and chronic expansion of exchangeable sodium space. The results of surgery in "curing" the hypertension of hyperaldosteronism provide further insight into the aldosterone–hypertension connection. Patients with hyperaldosteronism due to hyperplasia seldom are cured of their hypertension following surgery, although the hypokalemia substantially improves. Even in patients with adenomas, the cure rate of the hypertension following surgery is indeed not universal.

3. The autonomous secretion of aldosterone in patients with hyperaldosteronism serves as the phenomenon that establishes the diagnosis of the syndrome. Several procedures have been used to demonstrate nonsuppressible aldosterone secretion by the zona glomerulosa—e.g., oral salt loading, intravenous saline infusion, administration of mineralocorticoids such as desoxycorticosterone (DOC) or fludrocortisone. All these maneuvers normally suppress aldosterone secretion, but fail to do so in patients with hyperaldosteronism. In the normal person, salt loading is associated with a prompt suppression of aldosterone secretion and hense there are no changes in potassium homeostasis. In contrast, patients with primary aldosteronism will demonstrate a tendency toward potassium losses in the urine, resulting in hypokalemia. The reason for this is readily apparent if it is remembered that the delivery of an increased sodium load to the distal tubule, coupled with the presence of high, nonsuppressible aldosterone levels, provides a most favorable setting to exchange K^+ for Na^+. This results in increased reabsorption of sodium with increased excretion of K^+ in the urine.

4. Aldosterone-producing adenoma tissue, while autonomously secreting aldosterone, is not entirely autonomous. In the classic setting, aldosterone-producing adenomas demonstrate a remarkable paradox—on one hand, adenomas are estranged from renin–angiotensin-mediated stimuli, but on the other, they tend to demonstrate exquisite sensitivity to ACTH. Thus, minor diurnal fluxes in ACTH levels seem to affect the aldosterone output by the adenoma. Exogenous ACTH administration seems to activate release as well as all steps in the synthesis of aldosterone by the adenoma. The fact that adenomas show a tendency to respond to physiological cues is demonstrable by noting that profound hypokalemia can lower aldosterone secretion by the adenoma.

Potassium and Aldosterone Secretion

Potassium concentrations in the plasma regulate aldosterone secretion independent of the renin–angiotensin system or sodium balance. Hyperkalemia is a powerful stimulus for secretion and release of aldosterone, an effect that is noticeable even when the renin–angiotensin system is suppressed. Thus, physiologically, aldosterone secretion is stimulated by hyperkalemia and inhibited by hypokalemia. This homeostatic mechanism prevents dangerous rises in serum potassium concentrations and protects inappropriate kaliuresis in the presence of hypokalemia. Potassium stimulates both the proximal and the distal steps in aldosterone synthesis.[26] The independence from renin mediation is supported by observations that hyperkalemia stimulates aldosterone secretion in the anephric state.[27] The relative importance of the renin–angiotensin system versus the serum K^+ concentrations in stimulating aldosterone release has been studied in uremic patients on hemodialysis. Aldosterone release in such a setting can be provoked by acute salt depletion despite low serum potassium concentrations, implying the dominance of the renin–angiotensin system in that milieu.

The importance of potassium-mediated aldosterone regulation has direct clinical relevance while evaluating patients with suspect hyperaldosteronism. First, it is essential to correct the hypokalemia prior to studying aldosterone dynamics, because marked hypokalemia can lower the aldosterone secretory rate of even the "autonomous" adenomas. Second, the mere demonstration of kaliuresis in the presence of hypokalemia suggests inappropriate renal losses and thus, indirectly, abnormal aldosterone dynamics, provided that intrinsic tubular disease has been excluded. Unlike K^+, serum Na^+ concentrations play only a minor role in *directly* stimulating aldosterone synthesis and release.

ACTH and Aldosterone Secretion

The role of ACTH in the physiological regulation of aldosterone secretion is not as important as the renin–angiotensin system or serum K^+ concentrations. A permissive role for ACTH in mineralocorticoid secretion has been suggested by observations in patients with hypopituitarism. As early as 1960, Lieberman and Luetscher[28] postulated that while aldosterone secretion in hypopituitarism may be near normal in the basal state on a normal sodium intake, the increase in aldosterone following sodium deprivation is clearly blunted. The concept of reduced sensitivity of the zona glomerulosa in the presence of chronic ACTH deficiency has been supported by other workers.[29] Physiologically, administration of an intravenous bolus of ACTH is associated with a brisk rise in serum aldosterone levels. The fact that such a response is preserved even in patients with chronic ACTH deficiency supports the long-held notion that the zona glomerulosa does not atrophy as a consequence of long-term ACTH deprivation. While the classic teaching has been that the response of aldosterone to a supraphysiological doses of ACTH is characterized by a brisk response followed by a suboptimal secretory response to continuous stimulation, recent studies have challenged that notion. Thus, Nicholls et al.,[30] as well as Kem et al.,[31] have studied the plasma aldosterone response to low-dose ACTH stimulation in normal, dexamethasone-suppressed, and sodium-depleted states. Their studies support the notion that when physiological concentrations of ACTH are maintained in the plasma, the zona glomerulosa responds in a manner parallel to that of the zona fasciculata.

At least one variety of hyperaldosteronism, glucocorticoid-suppressible hyperaldosteronism, appears to be related to ACTH. The entire syndrome (the hypokalemia, hypertension, and aldosterone excess) can be promptly reversed by administration of small doses of dexamethasone, which effectively suppress ACTH. In another variety of hyperaldosteronism caused by an aldosterone-producing adenoma, the tumor tends to demonstrate exquisite sensitivity to endogenous ACTH. This is reflected in several phenomena that are associated with aldosterone-producing adenomas; these include parallel declines in plasma aldosterone level with the circadian fluxes in plasma ACTH, the robust increase in aldosterone levels following ACTH infusion, and so forth.

In summary, while ACTH plays a minimal role in the day-to-day regulation of aldosterone secretion, in some pathological states it may assume a significant role. It is important to point out that ACTH is not the only pituitary hormone involved in aldosterone dynamics. The role of non-ACTH pituitary secretagogues in controlling aldosterone secretion in general, and in idiopathic hyperplasia in particular, is discussed next.

Non-ACTH Pituitary Peptides and Aldosterone

In addition to the intact ACTH molecule, it is becoming increasingly apparent that other fragments of pro-opiomelanocortin (POMC) may have "adrenoglomerulotropic" activity. Several POMC peptides have contended for such a role in physiological as well as pathological states, particularly in patients with idiopathic hyperaldosteronism. Thus, γ-melanotropin, β-MSH, β-endorphin, and an "aldosterone-stimulating factor" have all at various times been considered *the* non-ACTH adrenoglomerulotropin (Table 26).

γ-Melanotropin

The data that confer a glomerulotropic action to γ-melanotropin are mostly derived from studies that have evaluated the concentration of this peptide in the plasma of patients with idiopathic hyperaldosteronism (IH). Griffing et al.[32] estimated circulating levels of γ-melanotropin in the plasma of nine patients with IH and contrasted them with plasma levels of this peptide in five patients with aldosterone-producing adenoma (APA) as well as in nine patients with essential hypertension. Patients with IH demonstrated clearly elevated levels of gamma melanocyte-stimulating hormone (γ-MSH) in their sera. Further, the authors were able to demonstrate that dispersed adenomatous adrenal tissue from one such patient with high serum γ-MSH demonstrated a dose-dependent increment in aldosterone secretion in response to infusion with human Lys-γ-3-MSH in vitro. The authors speculated that pro-γ-MSH may play a causal role in some forms of primary hyperaldosteronism, especially the IH variety. These data of Griffing et al.[32] support the earlier observations by Berelowitz et al.,[33] who also reported elevated plasma IR-γ-MSH in idiopathic hyperaldosteronism. While

TABLE 26.
Non-ACTH Pituitary Peptides with
Aldosterone-Stimulating Properties

γ-Melanotropin
β-Melanotropin
β-Endorphin
"Aldosterone-stimulating factor"

mounting evidence to impart a causal role to γ-MSH in the evolution of IH continues, there have been reports in the literature to temper this tenet. Thus, Güllner et al.[34] measured immunoreactive levels of the POMC-derived peptides (ACTH, β-endorphin–lipotropin, and γ-3-MSH) in patients with APA, IH, and glucocorticoid-suppressible hyperaldosteronism (GSH) and essentially found no differences in these groups when compared to normal controls. Their data suggest that POMC-derived peptides are not overproduced in any form of hyperaldosteronism.

β-Melanotropin

In 1981, Matsuoka et al.[35] were among the earliest workers who ascribed an aldosterone-stimulating role to the peptide β-MSH. Working with rat adrenal capsular cells, these workers showed a β-MSH–dependent increase in the production of aldosterone by these cells. This effect was similar in magnitude to that observed by infusion of angiotensin II, a conventional glomerulotropin. The glomerulotropic effect of β-MSH was abolished by the use of a synthetic analog that normally blocks the effects of β-MSH.

In addition to β-MSH, at least two reports[36,37] have characterized the aldosterone-stimulating effect of alpha melanocyte-stimulating hormone (α-MSH) in animals. It is however, unclear whether these findings are clinically applicable in terms of conferring a causal role to these melanotropic hormones in the development of any particular variety of hyperaldosteronism.

β-Endorphin

In addition to the melanotropins, β-endorphin is being increasingly recognized as a physiological stimulator of aldosterone release. Güllner and Gill,[38] using the hypophysectomized, nephrectomized dog as an experimental model, showed that infusion of β-endorphin causes a selective rise in plasma aldosterone levels without a concomitant rise in cortisol. In contrast, Szalag and Stack,[39] using dispersed rat adrenal cells, were unable to demonstrate an impressive increase in aldosterone secretion following β-endorphin. More recently, the effect of synthetic human β-endorphin on the entire renin–angiotensin–aldosterone axis of normal subjects was studied by Rabinowe et al.[40] The administration of graded doses of β-endorphin intravenously was associated with a significant increase in both PRA and plasma aldosterone levels, with a simultaneous decrease in plasma cortisol levels. The authors concluded that the endorphin-induced aldosterone release is complex and in part, especially during the early phase, is mediated by an increase in renin release.

Aldosterone-Stimulating Factor

The concept that the anterior pituitary may in fact secrete an "aldosterone-stimulating factor" (ASF) unrelated to ACTH and the POMC-derived peptides is hardly new.[41] However, until recently, this factor defied characterization. Two sets of happenings have provided compelling proof for its existence as well as postulates for its role in causation of disease in humans. Sen et al.[42,43] isolated a glycoprotein from urine and demonstrated that this factor was capable of stimulating the rat adrenal glomerulosa cells to produce aldosterone. It appears that this ASF is a glycoprotein, with a molecular weight of 26,000 daltons. Further characterization of this glycoprotein using high-pressure liquid chromatography revealed a retention time that was quite different from that of ACTH and angiotensin II. More important, Saito et al.,[44] while demonstrating the steroidogenic properties of ASF, were able to demonstrate that specific antagonists of angiotensin II and ACTH failed to abolish the aldosterone-stimulating effect of this glycoprotein. The final piece of laboratory evidence was provided by Sen et al.,[45] who developed a new assay for measuring ASF and demonstrated its presence in the human anterior pituitary gland. The second important event in the evolution of the ASF story is the clinical revelation that this glycoprotein may play a role in the development of IH. Carey et al.[46] measured circulating levels of ASF in seven patients with idiopathic hyperaldosteronism and four with surgically proven aldosterone-producing adenomas and compared these levels with those in 15 normal subjects. All patients with IH demonstrated significantly high levels of ASF in the supine position, when compared to normals and patients with APA. Further, these workers were able to demonstrate that the plasma levels of ASF in patients with IH showed an increase in response to posture pari pasu with a parallel rise in aldosterone; this rise occurred in the absence of a rise in renin levels. Also, the administration of dexamethasone, while suppressing the ACTH and cortisol levels to undetectable levels, failed to suppress the circulating ASF levels in the plasma and urine of patients with IH. The assay employed to detect ASF in their study was specific and did not cross-react with ACTH (1-24 as well as 1-39), β-LPH, or γ-3-MSH. This is especially significant since, as discussed earlier, all these peptides have aldosterone-stimulating properties and have been at one time or another, implicated in the causation of IH. One of the most compelling concepts to emerge from the study of Carey et al.[46] is the postulate that in patients with idiopathic hyperaldosteronism, the aldosterone-stimulating factor can stimulate posture-mediated aldosterone responses without mediation of the renin–angiotensin system, or alternatively ASF might facilitate the aldosterone response to posture by interacting with angiotensin II.

While these data have indeed narrowed the search for an etiological agent involved in the causation of idiopathic hyperaldosteronism, the role of ASF remains conjunctural at the time of writing. Even if one accepts that the reason for increased aldosterone secretion is in fact the ASF, an intriguing question that remains unanswered is whether this factor is the cause of hypertension in patients with idiopathic hyperaldosteronism.

Dopaminergic Regulation of Aldosterone

The emerging data that implicate neurotransmitter mediation in the production of aldosterone do support the notion that central mechanisms underlie the hyperaldosteronism seen in association with idiopathic hyperaldosteronism. Especially relevant is the role of dopamine in the regulation of aldosterone secretion in patients with idiopathic hyperaldosteronism. Two reports in the late seventies by Norbiato et al.[47] and by Carey et al.[48] documented that metoclopramide, a dopamine antagonist, caused doubling of plasma aldosterone levels following i.v. administration to normal subjects. Noth et al.[49] provided further insight into this phenomenon by demonstrating that prior administration of dopamine blunted the integrated incremental change in aldosterone levels following metoclopramide, suggesting that dopamine exerted a tonic inhibitory influence on aldosterone secretion, a situation analogous to the effect of dopamine on prolactin secretion. In the early eighties several investigators[50–53] conclusively demonstrated the provocative effect of dopamine antagonism on the release of aldosterone secretion in normal subjects. The mechanism of metoclopramide-induced aldosterone release was believed to be exerted by antagonism of central and peripheral dopamine receptors. Nishida et al.[54] evaluated the effects of L-dopa and dexamethasone on metoclopramide-induced aldosterone release in normal subjects and concluded that the dopamine antagonist increases plasma aldosterone concentration through dual mechanisms—one involving its inherent antidopaminergic effect on dopamine receptors, and another involving ACTH-dependent, centrally mediated (perhaps stress-related) mechanism(s). It was also shown that the dopaminergic modulation of aldosterone secretion was unaffected by glucocorticoids and angiotensin blockade.[55]

Further proof for a neurotransmitter connection in patients with idiopathic hyperplasia comes from a study that evaluated the effect of cyproheptadine in patients with hyperaldosteronism caused by adenomas and hyperplasia. Gross et al.[56] noted that cyproheptadine, a serotonin antagonist, caused an impressive decline in the plasma aldosterone levels in patients with idiopathic hyperplasia but not in patients with adenoma. The concept that serotonin is involved in the secretion of the ASF is attractive, although a direct effect on the adrenal cortex has not been ruled out.

In summary, several factors control aldosterone secretion. In the physiological state the four important ones are the renin–angiotensin system, serum K^+, sodium balance, and ACTH. Under pathological states other putative substances may play a role. Understanding the physiological regulation of aldosterone secretion will clarify some of the problems surrounding the classification of the various states of hyperaldosteronism.

Etiology and Classification

Primary aldosteronism has been traditionally classified into four subtypes:

1. Aldosterone-producing Adenoma (APA)
2. Bilateral hyperplasia, micro- as well as macronodular, also referred to in the literature as "idiopathic hyperplasia" (IH) and "pseudoaldosteronism"
3. Glucocorticoid-suppressible hyperaldosteronism (GSH)
4. Aldosterone-secreting carcinoma (CA)

In addition, four rare variants of the syndrome have been described: a loosely defined entity referred to as "indeterminate hyperaldosteronism"; uni-

TABLE 27.
Classification of Hyperaldosteronism

1. Aldosterone-producing adenoma (APA)
2. Idiopathic hyperplasia (IH)
3. Glucocorticoid-suppressible hyperaldosteronism (GSH)
4. Rare variants
   Aldosterone-producing carcinoma
   Unilateral hyperplasia
   Ectopic secretion of aldosterone
   "Indeterminate" hyperaldosteronism

lateral hyperplasia; multiple adenomas; and ectopic secretion of aldosterone by nonadrenal neoplasms, particularly ovarian carcinoma (Table 27).

Aldosterone-Producing Adenoma

Adrenocortical adenomas that secrete excessive quantities of aldosterone ("Conn's syndrome") are the most common form of hyperaldosteronism, accounting for 60–70% of cases of primary hyperaldosteronism. Aldosterone-producing adenomas are characterized by their small size, unilaterality, and benign nature. In vitro studies of excised tumor tissue have demonstrated that adenomas can produce aldosterone, a feature notably absent in the case of the nodular tissue removed from patients with idiopathic hyperplasia.[57] The aldosterone secretion by adrenal adenomas is autonomous and unresponsive to changes in the renin–angiotensin system, while retaining varying degrees of sensitivity to ACTH. Excision of the adenoma substantially normalizes blood pressure in 50–60% of patients.[58–60] This contrasts with the uniform lack of surgical response encountered in patients with bilateral hyperplasia; therefore, the preoperative delineation between these two entities is of crucial importance.

Idiopathic Hyperplasia

Idiopathic hyperplasia is characterized by micro- (or rarely macro-) nodular hyperplasia of the zona glomerulosa bilaterally. Occasionally, the histology of the adrenal cortices has been found to be entirely normal.[61,62] The existence of such a syndrome was postulated when patients with established criteria for hyperaldosteronism (elevated aldosterone with suppressed PRA) were operated on in anticipation of finding a tumor, but proved to have none upon exploration. The terms *pseudo* and *idiopathic hyperaldosteronism*, coined in the early seventies, indicate the uncertainties that surrounded the classification of this entity. More important, the demonstration that such patients failed to respond to surgery, despite lowering of aldosterone levels by bilateral adrenalectomy, cast doubt on the causal relationship between aldosterone excess and the hypertension associated with this syndrome.

To gain some perspective on the nature of idiopathic hyperaldosteronism the disorder must be viewed from several vantage points. First, it is essential to consider the possible etiologies that underlie this syndrome. Second, an overview of the dynamic hormonal milieu that characterizes idiopathic hyperplasia requires elucidation. Finally, the probability that idiopathic hyperplasia is a diagnostic artifact and a victim of misclassification by overzealous application of rigid but arbitrary criteria needs to be addressed.

Etiology

The reasons for the development of bilateral hyperplasia are far from clear. In contrast to the bilateral hyperplasia involving the zona fasciculata (Cushing's disease) or the zona reticularis (adrenogenital syndrome), ACTH mediation does not seem to be involved in the development of this disorder. Although the hormonal similarities between APA and IH can at times be striking, the adrenal tissue excised during surgery does not demonstrate identical behavior patterns. As early as 1975, the contrasting nature of adenoma and hyperplasia tissue was studied in vitro by Nicholls et al.[63] These workers studied two patients with primary hyperaldosteronism, one with a solitary adenoma and the other with bilateral adrenal hyperplasia. Although the biochemical and hormonal characteristics of both patients were identical, the biochemistry of the excised adrenal tissue showed impressive differences. The adenoma tissue demonstrated enhanced incorporation of tritiated pregnenolone into aldosterone, while hyperplastic tissue did not. Thus, enhanced 18-hydroxylase activity was noted in adenoma tissue, but not in hyperplastic tissue. The authors also showed that injection of the plasma from the patient with hyperplasia into the sheep's transplanted adrenal gland caused a definite aldosterone response; such a response was not elicited when plasma from the patient with adenoma was injected into the sheep. These studies, in a sense, were landmark studies for two reasons: first, the concept that IH may have fundamental differences from adenoma was raised, and second, the notion that IH may result from a non-ACTH trophic substance was proven in the laboratory.

Since these studies, several substances have been cited as the possible trophic substances that underlie the etiology of bilateral hyperplasia. Aldosterone-stimulating factor, a glycoprotein, as well as several melanotropic hormones of the POMC molecule, such as γ-melanotropin, β-lipotropin, and even α-melanotropin, have all been implicated as causal factors in the development of IH.[32–37, 41–46] Regardless of the trophic factor involved, the hormonal milieu of hyperplasia shares several features with that of APA. Despite these similarities, there are significant differences that separate this entity from adenoma.

The hormonal and biochemical milieu of IH, when juxtaposed between low-renin essential hypertension and APA, shares several similarities with these entities. The characteristics of IH are

1. Increased plasma concentrations of aldosterone.
2. Decreased plasma renin activity.
3. The nonsuppression of plasma aldosterone to salt loading is more pronounced than in APA.
4. The suppression of PRA is also less pronounced than in adenoma; thus, the PRA and aldosterone in idiopathic hyperplasia can be shown to rise, at least partially, upon assumption of erect posture.

5. The plasma aldosterone of patients with IH can be stimulated further by angiotensin infusion.
6. The exchangeable sodium pool of patients with IH is less impressively and less consistently expanded in comparison with adenoma.

IH in the Spectrum of Low-Renin Essential Hypertension

For several years workers from Glasgow have been proposing that IH has more in common with low-renin essential hypertension than with primary hyperaldosteronism.[9,64] Davies et al.[64] have carefully compared renin–angiotensin–aldosterone dynamics in patients with adenoma, hyperplasia, and low-renin essential hypertension and have shown fundamental differences between these two entities, which have traditionally been considered subtypes of a single disorder. In normals, sodium is negatively related to renin and angiotensin II, renin is positively related to angiotensin II, and angiotensin II is positively related to aldosterone. In addition, normal subjects exhibit a strong positive correlation between systolic blood pressure and the exchangeable sodium pool. Davies et al.[64] studied these relationships in 28 patients with APA, and 17 with "nontumorous aldosterone excess" ("hyperplasia" by traditional classification), and compared these data with those obtained in 37 normal subjects, 72 with untreated hypertension, and 43 with renal artery stenosis (a classic example of secondary hyperaldosteronism). The study is one of the largest of its kind, permitting accurate and valid statistical analyses. Upon examination of several correlates in pathological states, the authors made several interesting observations. Patients with aldosterone tumors showed characteristic autonomy—the abnormally elevated aldosterone was associated with low levels of renin and angiotensin II. The normal negative correlation between sodium and renin–angiotensin was maintained in adenoma patients. The data in patients with essential hypertension and nontumorous aldosteronism showed several similarities. Most important, the correlations of aldosterone with sodium and that of sodium with renin were found to be insignificant. Another striking observation was that patients with nontumorous aldosteronism, like those with essential hypertension, showed a positive correlation between angiotensin II and aldosterone, contrasting with the negative correlation between these two measurements in patients with adenoma.

When compared together, the data sharply bring to focus the differences in the milieu of idiopathic hyperplasia and aldosterone-producing tumors (Table 28). Thus, the exchangeable Na^+ in patients with adenoma was clearly increased, and the exchangeable K^+ in such patients was decreased. These abnormalities were milder and sometimes even absent in patients with idiopathic hyperplasia. The correlation between angiotensin II and aldosterone, as befitting an autonomous tumor, was strongly negative in adenoma, while being positive in hyperplasia. In patients with adenoma the plasma aldosterone correlated negatively with exchangeable K^+ pool and positively with the exchangeable Na^+ pool. Patients with hyperplasia did not show significant correlations with either. And finally, the patients with adenoma showed blunted, even absent, plasma aldosterone responses to an infusion of angiotensin II, while patients with hyperplasia briskly responded.

TABLE 28.
Differences between Aldosterone Adenoma and Hyperplasia

	Aldosterone adenoma	Hyperplasia
Exchangeable Na^+	Strongly increased	N or increased
Exchangeable K^+	Strongly decreased	N or decreased
Correlation between angiotensin II and aldosterone	Strong negative	Quite positive
Plasma aldosterone to plasma K^+ and exchangeable K^+	Strong negative	Insignificant
Plasma aldosterone to exchangeable Na^+	Strong positive	Insignificant
Aldosterone response to angiotensin infusion	Blunted	Increased

Where Does IH Belong?

The similarities between IH and essential hypertension are as striking as the differences between IH and APA. "Essential" hypertension is also characterized by an abnormal relation between aldosterone and renin activity. This abnormality is particularly marked in essential hypertension with low renin. Collins et al.,[65] as well as Re et al.,[66] have shown that plasma aldosterone is "relatively" high compared to the PRA in patients with "low-renin hypertension." Padfield et al.[67,68] have argued that low-renin hypertension is not a distinct disorder separate from normal-renin essential hypertension. This is supported by the observation that when frequency-distribution curves of PRA and angiotensin II of patients with essential hypertension are plotted, these indices extend into the subnormal range without interruption.[69,70] Thus, low-renin hypertension reflects a group of essential hypertensives in the same spectrum of essential hypertension. Data from the study of Davies et al.[64] seem to imply that patients with IH may represent a disorder at the upper end of a "wider than normal distribution of aldosterone in essential hypertension."

The arguments in favor of this hypothesis are several, and will be outlined below, drawing the similarities between IH and essential hypertension on one hand and the dissimilarities between hyperplasia and adenoma on the other.

1. The measurement of exchangeable Na^+ and K^+ pools in the body is the most direct method to indicate effects of excessive aldosterone production. The measurements are unequivocally and classically abnormal in patients with adenoma, who show an increase in the exchangeable sodium and a decrease in the exchangeable K^+.[64] These changes are absent, or at best minimal, in patients with hyperplasia, as well as in patients with essential hypertension. It is generally agreed that increased exchangeable Na^+ is not a feature of low-renin hypertension.[71]

2. The hypokalemia, the PRA, and aldosterone abnormalities are more pronounced in APA. Hypokalemia, profound PRA suppression, and striking elevations in plasma aldosterone concentrations are more often encountered in patients with adenoma.[8,72] The abnormalities in K^+, PRA, and aldosterone,

when subjected to quadric analysis, clearly separate adenoma from hyperplasia, which forms a continuum with data from patients with essential hypertension.

3. Aldosterone hypersecretion in patients with adenoma is clearly nonsuppressible. Maneuvers that normally suppress aldosterone, such as saline infusion or mineralocorticoid administration, fail to suppress aldosterone levels to normal in patients with adenoma. In contrast, patients with essential hypertension usually suppress, while patients with hyperplasia fall somewhere in between.

4. The PRA of patients with adenoma is clearly suppressed and nonresponsive to posture-mediated mechanisms. In contrast, the posture-mediated rise in PRA, angiotensin, and aldosterone are often preserved in nonadenomatous aldosteronism.[60,73,74] In this regard, hyperplasia behaves more like essential hypertension than like autonomous secretion.

5. The nondependency of aldosterone secretion upon angiotensin in patients with adenoma is further illustrated by studying the response of plasma aldosterone to angiotensin infusion. Patients with adenoma generally show a blunted, even absent, plasma aldosterone response to angiotensin infusion.[64,75] This contrasts with the brisk response of aldosterone to angiotensin in patients with hyperplasia.[64,76] Patients with essential hypertension, including those with the low-renin variety, also demonstrate an enhanced, brisk rise in plasma aldosterone following angiotensin infusion.[77,78]

6. The relationship between plasma aldosterone and plasma angiotensin II concentrations, when plotted, separates adenoma from nonadenomatous cases at a glance. The relationship is significantly negative in adenoma, while being significantly positive in hyperplasia and insignificantly positive in essential hypertension. This abnormal relationship between angiotensin II and aldosterone is a common thread that characterizes the milieu in the hypertension of IH and essential hypertension. Although the mechanism for this abnormality is uncertain, this finding clearly seems to impart a significant role for aldosterone in the pathogenesis of essential hypertension, at least on some patients.

7. When renin, angiotensin II, and aldosterone concentrations in the plasma are pooled and analyzed, the data from patients with idiopathic hyperplasia and essential hypertension are continuously distributed, while data from patients with the adenoma are discontinuous from the other groups.

8. The significance of age on the renin data in patients with adenoma, hyperplasia, and essential hypertension is also interesting. Padfield et al.[68] and Ferris et al.[8] have shown that age is inversely related to PRA in essential hypertension and in IH, while bearing no relationship to one another in patients with adenoma.[8]

9. The ACTH influence on aldosterone secretion by adenomatous issue is evident in several different expressions—e.g., circadian fluxes in aldosterone levels in patients with adenoma, the anomalous decline in aldosterone levels between 8 A.M. and noon with ambulation, the aldosterone response to exogenous ACTH. These phenomena are seldom noted in patients with hyperplasia, and almost never in essential hypertension.

10. Anatomically, Conn's syndrome is characterized by a unilateral adenoma, whereas bilaterial hyperplasia of the zona glomerulosa, often nodular, is encountered in hyperplasia. The observation that nodular hyperplasia may be

encountered in essential hypertension[9] draws the similarities between hyperplasia and essential hypertension one step closer.

11. The failure of surgery to normalize blood pressure in patients with IH is a feature that sharply contrasts with adenoma; this brings the similarities between essential hypertension and IH even closer.

12. Even medical therapy with spironolactone or amiloride is less effective in IH when compared to that in patients with adenoma.[79] This appears paradoxical since the aldosterone secretory rate is considerably greater in the latter. This supports the notion that Conn's syndrome and IH may be fundamentally different disorders with nothing in common.

The Glasgow workers have suggested that Conn's syndrome is a distinct entity, while IH, low-renin hypertension, and essential hypertension comprise a continuum. The reason for misclassification may have to do with rigid applications of arbitrarily defined norms. In an important article, Padfield et al.[9] have historically traced the reason for such misclassification and conclude by quoting a passage from *Analysis of the Phenomenon of the Human Mind* by John Stuart Mill[80]—"The tendency has always been strong to believe that whatever receives a name must be an entity or being, having an independent existence of its own; and if no real entity answering to the name could be found man did not for that reason suppose that none existed, but imagined that it was something peculiarly abstruse and mysterious, too high to be an object of sense": indeed an apt quote to demystify a presumed myth.

Glucocorticoid-Suppressible Hyperaldosteronism

Sutherland et al.[81] in 1966 described an unusual variety of hyperaldosteronism where all the hormonal abnormalities could be reversed by long-term administration of dexamethasone. The condition is particularly important to consider in young patients with hyperaldosteronism, and in those with a family history of aldosterone excess. New et al.[82] have demonstrated an autosomal-dominant type of inheritance pattern in a kindred with dexamethasone-suppressible hyperaldosteronism. The uniqueness of dexamethasone-suppressible hyperaldosteronism lies in the fact that following dexamethasone, the previously elevated plasma aldosterone levels decline to almost unmeasurable levels. Thus, the triad that characteristics this particular form of hyperaldosteronism is the young age onset, autosomal familial tendency, and predominant regulatory control exerted by ACTH on aldosterone hypersecretion.

This predominant regulation by ACTH in GSH is illustrated by several features demonstrable in this disorder.

1. The aldosterone levels impressively decline following dexamethasone administration. This characteristic is reminiscent of the ACTH response to dexamethasone in congenital adrenal hyperplasia. Indeed, in the early years of its discovery, some workers even considered the possibility that this disorder might be a new form of congenital adrenal hyperplasia.[83] However, it became apparent subsequently that there were no enzymatic defects of steroidogenesis in this syndrome, and that it represented true hypersecretion of aldosterone. In addi-

tion to aldosterone, there is prompt decline in circulating concentrations of other adrenal steroids such as cortisol, DOC, compound B, and urinary aldosterone as well as 18-OH-DOC excretion following dexamethasone.[84]

2. In addition to lowering aldosterone levels, dexamethasone administration for varying durations normalizes the blood pressure of patients with dexamethasone-suppressible hyperaldosteronism.[81, 85–87]

3. GSH is characterized by an enhanced sensitivity of aldosterone to ACTH administration. Several studies have amply documented this point.[88–90] The aldosterone response to exogenous ACTH administration in patients with GSH is typical and is characterized by a brisk rise in aldosterone, the response being sustained. This contrasts with the response in normal subjects where, after an initial rise of aldosterone following ACTH, the response become gradually attenuated, even obliterated. This continuous aldosterone response to ACTH, amply documented by Oberfield et al.[84] in a series of 10 patients with GSH, mimics the cortisol response to continuous ACTH infusion.

4. The PRA and aldosterone of patients with GSH in general fail to respond to maneuvers such as salt depletion or ambulation prior to treatment with dexamethasone. Once ACTH suppression has been attained with dexamethasone therapy, the PRA and even plasma aldosterone respond to salt depletion.[84] This indicates that upon removal of the influence of ACTH, the renin–angiotensin–aldosterone system of patients with GSH is capable of responding to this physiological axis.

5. Finally, it has been shown by Ganguly et al.[91] that patients with GSH often demonstrate an anomalous decline in plasma aldosterone following assumption of erect posture. This phenomenon, a characteristic of APA, has been cited as a test for delineating patients with unilateral aldosterone-secreting tumors.[60,73,74,92–94] The explanation for this phenomenon is based on the enhanced responsiveness of tumors to circadian change in the plasma ACTH concentration, resulting in a decline, despite the ongoing effect of posture-mediated stimuli. That GSH patients also respond in an identical manner is not surprising, if one views the disorder as being ACTH dependent.

Thus, it appears that ACTH (or a related peptide) is involved in the pathogenesis of GSH. The reasons for the development of this syndrome, however, are unclear. One suggestion has been that the aldosterone hypersecretion in GSH may in fact be arising from the zona fasciculata. Miura et al.[95] have described the histological finding of hyperplastic areas of the zona fasciculata in patients with GSH. Oberfield et al.[84] have pointed out that parallel changes in aldosterone and cortisol occur when patients with GSH are placed on dexamethasone. The demonstration that aldosterone responses to renin–angiotensin-mediated stimuli are restored when ACTH is suppressed prompted the speculation that the zona glomerulosa can normally respond to the renin–angiotensin regulation, once the ACTH has been suppressed by dexamethasone. The reason for the failure of aldosterone to respond to such maneuvers before treatment may reflect dormancy (or atrophy) of the zona glomerulosa as a consequence of hypersecretion of aldosterone from the zona fasciculata. More recently, Fallo et al.[96] extended these observations to provide further proof of the notion that GSH may be a disorder of the zona fasciculata. These workers[96] evaluated aldosterone responses to infusions of angiotensin II and demonstrated impairment in responses to angiotensin II before and even up to 2 weeks after dexamethasone treatment. Their data are consistent with the concept that functional impairment of the zona glomerulosa may be present in patients with GSH, and that hypersecretion of the mineralocorticoid may indeed originate from the zona fasciculata.

The repeatedly documented enhanced sensitivity of aldosterone secretion to ACTH in GSH[97] has fostered the notion that perhaps the zona glomerulosa (or zona fasciculata) may be unduly sensitive to ACTH. Alternately, Mulrow[98] has suggested that the pituitary gland of patients with GSH may elaborate a "more potent form of β-lipotropin" that stimulates aldosterone secretion by both the adrenal cortices; clearly, the secretion of ACTH or related peptides by the pituitary in GSH is quite suppressible by glucocorticoid, indicating preservation of physiological cues.

Although the etiology of GSH is far from clear, its recognition has important therapeutic implications. The clinical and hormonal behavior of GSH can be identical to APA. Thus, hypertension, hypokalemia, increased aldosterone secretion nonsuppressible with saline infusion, suppressed PRA, and other factors are similar in both disorders. The similarity is carried steps further when it is considered that both conditions may be characterized by anomalous decline in plasma aldosterone concentrations with posture and absent aldosterone responses to angiotensin infusion. The aldosterone response to dexamethasone will help delineate GSH from the computerized tomography–negative adenoma and constitutes an important step in the diagnosis prior to embarking on selective adrenal vein catheterization studies.

The other relevant aspect to the clinician is the heredofamilial nature of GSH. Ganguly et al.[99] have described the disorder in three successive generations. Owing to the relatively small number of cases in the literature, the precise manner in which the condition is transmitted remains unclear. In the kindred described by Ganguly et al.,[99] the affected members were a 7-year-old boy, his mother, and his grandmother. Father-to-son transmission of GSH has been documented by others.[81,87,100] It is generally agreed that GSH in transmitted as an autosomal dominant trait. An absence of HLA linkage is suggested by the studies of New et al.[82] These and other studies[96] emphasize the need for screening family members when GSH is diagnosed in an index case.

Grim and Weinberger[87] have extended their observations on the familial nature of GSH to demonstrate the existence of "normokalemic" variants of this disorder in several members of affected families. These workers have shown that suppressed PRA and nonsuppressible aldosterone excretion can be encountered in the presence of persistently normal potassium levels in hypertensive relatives of children with classical GSH. Thus, the absence of hypokalemia should not deter the physician from excluding GSH in the setting of a family history.

Clinical Features

The dual clinical expressions of primary hyperaldosteronism, regardless of the underlying etiology, are hypertension and hypokalemia.

Hypertension

The hypertension associated with hyperaldosteronism is usually mild. The clinical features of the hypertension associated with the syndrome are often indistinguishable from those of benign essential hypertension. Headaches are often the only complaint experienced by the patient. While the classic teaching is that patients with primary hyperaldosteronism tend to have mild hypertension, often with a benign course, there are reports emphasizing the contrary in some patients with the syndrome. Clarke et al.[59] reported 11 patients with hyperaldosteronism caused by adenoma who presented with severe hypertension. In all, the blood pressure response following surgery was excellent. Malignant hypertension has also been described in association with primary hyperaldosteronism.[101,102] The percentage of such patients with severe, malignant hypertension was thought to constitute only a small proportion of patients with primary aldosteronism. However, Bravo et al.[103] reported in a prospective study of 80 patients with primary aldosteronism that several had moderate to severe hypertension; further, in 30% of patients the blood pressure could not be controlled with the use of conventional antihypertensive medications.

Occasionally, patients with hyperaldosteronism may have a perfectly normal blood pressure. Normotensive hyperaldosteronism is an extremely rare phenomenon. Snow et al.[104] first described a patient with a typical APA without hypertension. Zipser and Speckart[105] also reported a 45-year-old woman with hypokalemia, inappropriate kaliuresis, a perfectly normal blood pressure, and an elevated aldosterone level that failed to suppress with fludrocortisone. The best-studied case of normotensive hyperaldosteronism was that reported by Shiroto et al.[106] The patient, a 25-year-old Japanese woman, presented with tetany caused by severe hypokalemia, normotension, low plasma renin activity, elevated aldosterone levels, and an APA at surgery. Notably, the patient demonstrated an attenuated pressor response to angiotensin II administration. The authors postulated that the existence of a hypotensive mechanism in this patient counteracted the pressor effect of aldosterone excess. Other mechanisms for normotensive hyperaldosteronism are early phase of the disease and a state of severe sodium restriction.

Hypokalemia

Symptoms of hypokalemia dominate the symptomatology of primary aldosteronism. Thus, paresthesia (tingling, numbness), cramps, muscle weakness, and even tetany may be encountered. It should be remembered that in nearly 20–25% of patients with primary aldosteronism the serum potassium levels are in the normal range, and therefore, these patients would not be expected to suffer from symptoms of hypokalemia. In an equal number of patients the hypokalemia may be very mild. In general, patients with adenomas tend to be more hypokalemic than those with IH. The hypokalemia is more likely to be unmasked when diuretics are used. In a prospective series by Bravo et al.,[103] it is notable that in 54% of patients with primary hyperaldosteronism, diuretic therapy produced moderately severe hypokalemia (serum potassium < 3.0 mEq/ liter).

Both the hypertension and the hypokalemia can be masked by severe dietary sodium restriction. Since most patients with essential hypertension are placed on a salt-restricted diet, this factor may be responsible for underdiagnosing primary aldosteronism. However, since most hypertensives are also placed on some form of diuretic therapy, the hypokalemia is likely to manifest sooner or later. The pitfall to be avoided is the mistake of attributing the hypokalemia to diuretic therapy. Rather, such a phenomenon should signal the search for underlying primary hyperaldosteronism. There are no physical findings unique for primary aldosteronism. Edema is usually absent, and mild orthostatic changes may be found. A surprising finding reported by Bravo et al.[103] was the demonstration that hypovolemia was present in 25% of their 80 patients with hyperaldosteronism. A hyperkinetic circulation was noticed in some patients in this series.

The clinical settings that should prompt a suspicion of primary aldosteronism are usually three:

1. The occurrence of spontaneous or diuretic-induced hypokalemia in a hypertensive patient is one indicator. The incidence of normokalemia in patients with primary aldosteronism has been variably reported to range from 7 to 38%.[103,107–109] A more representative figure is perhaps in the range of 25–30%. It has been aptly pointed out that sole reliance on serum potassium concentrations could be conducive to missing the diagnosis of primary aldosteronism in one-fourth of patients with the syndrome. In the series reported by Bravo et al.,[103] 27.5% of patients were completely normokalemic while on a normal sodium diet, and more important, nearly half of this group (12.5%) remained so despite salt loading for 3 days. Yet, the presence of spontaneous or provoked hypokalemia continues to remain the most important marker in clinical practice to suspect this disorder.
2. The presence of unexplained polyuria or nocturia in a hypertensive patient may reflect the vasopressin resistance seen in association with chronic hypokalemia.
3. Hypertension refractory to conventional antihypertensive therapy also constitutes an important indicator to suspect primary hyperaldosteronism, along with other causes of refractory hypertension.

Laboratory Diagnosis

The diagnostic approach to hyperaldosteronism consists of several phases. The first phase is screening for hyperaldosteronism. The second step is confirmation of aldosterone excess. The third step involves tenuous exercises that attempt to differentiate adenoma from hyperplasia. The fourth phase involves the localizational procedures that would accurately predict the site of the tumor. The fifth phase is formulation of tests that would exclude carcinoma and dexamethasone-suppressible disease. The final phase, in the event an adenoma is found, is the predictability of response to surgery (Table 29).

TABLE 29.
Diagnostic Approach to
Hyperaldosteronism

Step 1: Screening of hypertensives for hyper-
          aldosteronism
Step 2: Confirmation of aldosterone excess
Step 3: Maneuvers to differentiate adenoma
          from hyperplasia
Step 4: Localizational procedures
Step 5: Exclusion of GSH
Step 6: Predicting the response to surgery if an
          adenoma is found

Step 1: Screening

In the past, screening for primary hyperaldosteronism consisted of methods that focused on indirect effects of chronic hypersecretion of aldosterone. Thus, measurement of serum potassium level and evaluation of basal PRA used to be considered as initial screening tests to separate patients with hyperaldosteronism from those with essential hypertension. These two methods are clearly and notoriously unreliable as screening tests for hyperaldosteronism. Even the measurement of stimulated PRA, i.e., PRA after standing, diuretic therapy, and salt restriction, may not have the desirable sensitivity, much less the specificity, required of a screening test. Since hyperaldosteronism represents autonomous production of aldosterone, it is not surprising that tests that directly evaluate aldosterone dynamics offer greater diagnostic yield than tests that measure the biological consequences (hypokalemia or PRA suppression) of aldosterone excess. The value of the following five screening procedures is discussed in this section:

1. Serum potassium concentrations
2. Basal PRA
3. Stimulated PRA
4. Aldosterone dynamic studies
5. Aldosterone response to captopril

Serum Potassium Concentration

While spontaneous or provoked hypokalemia is the most frequent clue that identifies the patient with hyperaldosteronism, it is now abundantly clear that as a screening test, measurement of serum K^+ concentrations is grossly inadequate. Normal serum potassium levels are encountered in patients with hyperaldosteronism with a frequency that ranges between 7 and 38%.[106–109] In one recent report by Hiramatsu et al.,[110] an astounding 67% of patients with hyperaldosteronism were normokalemic at the time of screening. A more representative sampling is provided by the study of Bravo et al.,[103] who found that 27.5% of patients with primary hyperaldosteronism were normokalemic when evaluated on a normal dietary sodium intake. These results are in keeping with two other major studies[60,108] that also suggest that nearly one-fourth of patients with hyperaldosteronism are normokalemic, underscoring the unreliability of the serum K^+ concentrations as a screening test. Yet, spontaneous or provoked hypokalemia continues to remain the most important initial clue that leads the physician to the diagnosis of hyperaldosteronism in hypertensive patients.

Basal PRA

Measurement of *basal* PRA in hypertensive patients does not help in identifying patients with hyperaldosteronism with any degree of consistency or accuracy. In the 80 patients with primary aldosteronism prospectively evaluated by Bravo et al.,[103] the basal PRA measured in the resting supine state was suppressed in only 72.5% of patients, while being normal in 12.5% and elevated in 15%. Interpretation of renin levels must take into account several factors:

1. The *methodology:* PRA is generally measured by the ability of this enzyme to generate angiotensin from renin substrate. Almost all assays for PRA rely on measuring the angiotensin I generated by renin activity. The development of specific antirenin antibodies may permit *direct* measurement of plasma renin levels,[111,112] but these methods have not, as yet, found wide application. Measured by the conventional methodology, a crucial factor that affects PRA is the pH during in vitro incubation. The optimum pH is 5.5. Failure to control pH between 5.5 and 6 and failure to use diisopropylfluorophosphate (or phenylmethylsulfonyl fluoride) to inhibit angiotensinases during incubation are the most common reasons for laboratory errors in measuring PRA. Several assays have been developed for measuring the plasma renin activity employing simple and reliable kits.[113–115]

2. In vivo conditions may affect PRA activity. Since the liver is the major site of renin metabolism,[116] acute and chronic liver disease can affect the PRA. The presence of "circulating inhibitors" and "renin activators" in the plasma can affect enzymatic activity of renin.

3. Different forms of renin have been identified in plasma. It is now well known that acidification enhances renin activity ("acid-activated renin") which results from conversion into a smaller-molecular-weight renin.[117,118]

4. The most important factor that undermines the utility of a single basal PRA measurement is the fact that release of renin is controlled by a multitude of factors (neural and sodium-sensitive factors). Unless these determinants are taken into consideration, interpretation of PRA in the basal state has no meaning in relation to the hypertension. Thus, "renin-profiling" in hypertensive patients must be typed on the basis of PRA measured in terms of sodium status and posture, and not in the basal state. Esler et al.[119] have suggested that even the measurement of stimulated renin activity must take into account the abnormality in sympathetic tone that may be present in hypertensive patients.

5. The influence of age, sex, and race on PRA measurement has not been clearly defined. Until the effects of these correlates on basal (and stimulated) PRA are better understood, even the definition of normal limits for purposes of classification may be incorrect.

6. Finally, the effect of drugs such as diuretics, oral contraceptive agents, and beta blockers should be taken into consideration when interpreting PRA.

With so many variables affecting PRA, measurement of *basal* PRA has little to offer in terms of "classifying" hypertensives, and certainly has no place in diagnostic or prognostic speculations in the hypertensive patient. Therefore, when PRA is measured, it must be done *both* in the basal states as well as following maneuvers that normally stimulate PRA.

Stimulated PRA

The concept that patients with hyperaldosteronism will demonstrate suppressed PRA, unresponsive to salt deprivation and ambulation, dates back to the landmark article by Conn et al.[6] in 1964. In fact, this physiological principle has indeed led to the widespread use of measurement of stimulated PRA as the first-line test in screening for hyperaldosteronism. A natural extension of this principle was the practice of "renin profiling" of *all* hypertensive patients. It soon became apparent that all hypertensive patients could be "renin-typed" into "low," "normal," or "high" renin hypertensives.[120–122] The literature is replete with claims that such profiling has relevance in terms of prognostication as well as treatment.[120,122–125] While dividing hypertensives into groups based on PRA profiling may indeed be of some help in screening for renovascular hypertension and adrenocortical hypertension, it is becoming more apparent that renin levels in hypertension are influenced by several factors.[119] Attempts to subdivide hypertensives into renin subgroups, for the most part, are arbitrary; the consensus of opinion favors the notion that measurement of renin in essential hypertension does not significantly assist in predicting the blood pressure response to diuretics, beta blockers, or a combination of the two.[126]

How valuable is the measurement of stimulated PRA in the screening for hyperaldosteronism? This must be viewed in terms of specificity and sensitivity. As for specificity, it has been known for a long time that a subset of essential hypertensives have low PRA (low-renin essential hypertension). In fact, as many as 10–15% of essential hypertensives may demonstrate an abnormally low PRA in response to provocative stimulation. Thus, based on merely the stimulated PRA it is not possible to delineate low-renin essential hypertension from primary hyperaldosteronism. In terms of the sensitivity of the test, discrepancies in the literature abound. Weinberger[127] has reported that 1 day of a salt-restricted diet (10 mEq of Nacl for 1 day) with 40 mg of furosemide 3 times a day, on that day, followed by 2 hr of ambulation provided clear separation of patients with hyperaldosteronism from normals. A poststimulated PRA below 2 ng/ml per hr points to the need for further diagnostic studies to confirm or exclude hyperaldosteronism. In another study[60] the usefulness of PRA measurements (following sodium and volume depletion as well as erect posture) in screening for hyperaldosteronism has been emphasized. However, several reports have highlighted the limitations of stimulated PRA as a sensitive screening test for the diagnosis of primary hyperaldosteronism. The most important of these is by Bravo et al.[103] In a prospective series of 80 patients with primary aldosteronism, these workers noted that as many as 35% of patients with primary aldosteronism increased their PRA above 2.0 ng/ml per hr after salt and water depletion. They also noted that 17% of essential hypertensives had suppressed PRA levels identical to those of patients with primary hyperaldosteronism. Other workers have also observed normal PRA levels in patients with hyperaldosteronism.[8,128] The high percentage of false positives, and the impressive number of false negatives (at least in some studies), have led to the following conclusions regarding the value of stimulated PRA in screening for primary hyperaldosteronism:

1. The presence of a suppressed PRA, i.e., failure to rise following salt restriction, diuretic, and ambulation, is strong corroborative evidence for the presence of hyperaldosteronism.
2. The absence of a suppressed PRA, i.e., an elevation > 2 ng/ml per hr following salt restriction, diuretic, and ambulation, does not preclude the diagnosis of primary hyperaldosteronism.
3. The most common cause of suppressed PRA in a hypertensive patient is low-renin essential hypertension. Distinction between this entity and primary hyperaldosteronism cannot be made on the basis of PRA levels alone.[129,130]
4. The most common cause of *nonsuppressed* PRA in primary hyperaldosteronism is chronic diuretic use. This underscores the fact that, even though PRA suppression is secondary to hyperaldosteronism, this consequence can be overridden by other regulatory mechanisms involved in the secretion of renin.
5. PRA dynamics alone should not be used as the *sole* screening test to detect primary hyperaldosteronism.[130]

Aldosterone Dynamic Studies

Since primary hyperaldosteronism represents autonomous production of aldosterone, logically the tests designed to discover its presence should aim at attempting to suppress aldosterone production. Thus, the response of aldosterone to salt loading, volume expansion, and administration of mineralocorticoid (DOC or 9-α-fludrocortisone) is a straightforward physiological attempt to test suppressibility of aldosterone secretion. These tests carry the highest sensitivity to detect the disorder and hence serve as high-yield screening tests as well as a confirmatory tests. They are discussed in the next section.

Captopril Test

A recent addition to the aforementioned studies is the evaluation of aldosterone response to oral captopril. Lyons et al.[131] compared the plasma aldosterone response to a single dose of captopril in patients with essential hypertension and in those with primary hyperaldosteronism. Captopril is a drug that blocks the conversion of angiotensin I to angiotensin II by inhibiting the ACE. Thus, patients with an intact renin–angiotensin–aldosterone axis would be expected to lower plasma aldosterone concentrations following administration of captopril, while those with already suppressed angiotensin II (such as those with autonomous aldosterone secretion) would not. These authors[131] reported a marked fall in plasma aldosterone concentration, exceeding 50% of control values, 2 hr after the administration of a single dose of 25 mg of captopril in normal subjects as well as in essential hypertensives, but not in those with primary hyperaldosteronism. An absolute postcaptopril plasma aldosterone level exceeding 15 ng/dl was felt to be diagnostic of primary aldosteronism. More important, a comparison of the postcaptopril renin:aldosterone ratio with the control ratio provided clear separation of patients with primary aldosteronism from those with essential hypertension. The ratio remained high in autonomous aldosterone hypersecretion, since ACE inhibition would not be expected to inhibit aldosterone release or to affect the chronically suppressed renin.

The captopril test is a simple, and safe screening test. However, its utility as an effective test in screening for hyperaldosteronism has not been established in large numbers of patients.

Step 2: Diagnostic Confirmation

The confirmation of primary hyperaldosteronism depends on the demonstration that the secretion of aldosterone is autonomous and nonsuppressible to salt loading and volume expansion. This can be attempted in one of several ways:

1. The intravenous saline infusion test.
2. The oral salt-loading test.
3. The mineralocorticoid suppression test performed with DOC or fludrocortisone.

Of these, a simple, reasonably reliable screening test is the saline infusion test.

Saline Infusion Test

In 1971, Kem et al.[132] described a simple test that can be performed in outpatients to detect excessive aldosterone production. The test is performed by infusing 2 liters of normal saline over a 4-hr period between 8 A.M. and noon. Aldosterone is measured in the plasma before and after infusion of normal saline. Patients with primary aldosteronism failed to decrease aldosterone levels in the circulation to below 10 ng/dl following the saline infusion. While a small number of patients with essential hypertension may also behave in a nonsuppressible fashion, the diagnosis of primary aldosteronism in such patients could be excluded by subsequent studies. The saline infusion test should not be performed in patients with severe hypertension, as well as in those with grade III or IV retinopathy or with a history of congestive failure, myocardial infarction, or stroke. Two other major studies have corroborated the diagnostic utility of the saline infusion test. Streeten et al.,[129] using a cutoff number of 8.5 ng/dl following the saline infusion, reported that the test yielded an encouragingly high sensitivity of 0.77%. In what must be one of the largest populations of hypertensives screened, these workers evaluated the utility of the saline test in 1036 consecutive hypertensives and found that the test accurately identified the presence of primary hyperaldosteronism in 17 of 22 patients proven to have the disorder. On the false positive side, the test was positive, i.e., nonsuppressible, in only 20 of 767 patients with essential hypertension, most of whom were proven not to have primary hyperaldosteronism, based on other tests. Thus, the false positive rate of the test was below 20%. The authors felt that the sensitivity of saline infusion test approached 0.95%, when it was combined with other studies such as the suppressed PRA response to salt depletion, diuretic, and posture; the serum K concentrations; and the lack of depressor response to saralasin. In another study, Weinberger et al.[60] demonstrated nonsuppressible plasma aldosterone levels after saline infusion in all the 48 patients with primary hyperaldosteronism. The saline suppression test has clearly become a simple screening procedure, the results of which can direct the subsequent workup. In many centers, the saline suppression test is performed before PRA response to provocative maneuvers are attempted.[133,134] The saline suppression test will not, however, provide clear separation between adenoma and hyperplasia causing hyperaldosteronism. As already indicated, the minority of patients with essential hypertension who demonstrate "nonsuppression" to saline can be weeded out by performing additional studies such as PRA response to provocation, as well as plasma and urinary aldosterone response to extended oral salt loading and mineralocorticoid administration.

Oral Salt Load Test

The oral salt load test involves studying the suppressibility of aldosterone secretion to volume expansion in response to extended salt loading. This is usually done by placing the patient on a constant diet of 110 meq of sodium per day for 3–5 days and comparing the urinary aldosterone excretion before and after salt loading. Bravo et al.[103] have noted that an aldosterone excretion rate greater than 14.0 µg/24 hr following 3 days of salt loading provided the highest sensitivity and specificity in identifying patients with primary aldosteronism. When this finding was combined with suppressed PRA activity, the diagnosis of primary hyperaldosteronism was "virtually assured" in all 80 patients studied by these workers. Similar results have been reported by Vaughan et al.,[128] who report that the simple method of extracellular fluid volume expansion correctly identified underlying hyperaldosteronism in every instance; all 29 patients with hyperaldosteronism in this series demonstrated inadequate suppression of urinary aldosterone, in contrast to 25 patients with essential hypertension, who demonstrated complete suppression of urinary aldosterone excretion following a salt load. Further, when data from patients with adenoma, hyperplasia, and benign essential hypertension were analyzed in this study, the authors were able to derive clear separation between the three groups simply based on the response to salt loading. The group of patients with hyperplasia demonstrated partial lowering of aldosterone in contrast to patients with adenoma, who showed impressive nonsuppression, and patients with essential hypertension, who showed impressive and total suppression. Thus, as a confirmatory test for primary hyperaldosteronism, the formal 3-day salt-loading test serves as the "gold standard test." Vaughan et al.[128] have further shown that such a maneuver also improves the diagnostic sensitivity of other tests aimed at separating adenoma from hyperplasia. For instance, when the plasma aldosterone response to posture is used to differentiate adenoma from hyperplasia, the test works at its best when performed in the volume-expanded state. This study, as well as other maneuvers used to differentiate adenoma from hyperplasia, is discussed in a subsequent section.

Response to DOC or Fludrocortisone

The response of urinary aldosterone to exogenous administration of DOC or 9-α-fludrocortisone (Florinef) for 2–3 days is based on the following principle—when these mineralocorticoids are administered, the resultant expansion of extracellular fluid volume should suppress the endogenous renin–angiotensin–aldosterone system in normal individuals. In contrast, patients with autonomous secretion of aldosterone, who already have maximal renin suppression, would not be expected to further suppress the renin–angiotensin–aldosterone system and consequently will not reduce their aldosterone secretion following mineralocorticoid administration. The principle is analogous to the use of T_3 in hyperthyroidism or dexamethasone in hypercortisolism. The administered DOC (intramuscular) or 9-α-fludrocortisone (oral) does not interfere with measurement of aldosterone in the urine.[135–138]

In principle the use of oral salt, intramuscular DOC, or oral 9-α-fludrocortisone all attempt to achieve the same end; all three maneuvers evaluate suppressibility of aldosterone secretion in response to the physiological effect of expansion of extracellular fluid volume. There are three important aspects to remember with regard to extended salt loading, as well as mineralocorticoid administration: First, the reliability of these tests, to some extent, depends on the serum K^+ concentration. In the presence of hypokalemia, patients with primary hyperaldosteronism may show some suppression to salt loading or mineralocorticoid administration, providing false negative responses. Second, these maneuvers may accentuate potassium loss in the urine of patients with primary aldosteronism. As a consequence, profound hypokalemia and cardiac arrythmias may ensue, with occasionally fatal consequences. Third, on a banal level, the difficulties of urine collection limit the practical application of these tests.

During the initial workup of primary aldosteronism, the mineralocorticoid suppression tests provide little additional diagnostic benefit over the standard 3-day salt-loading test. Therefore, most experts reserve the mineralocorticoid test for the occasional patient with essential hypertension who fails to suppress with salt loading. The criteria for "suppression" with DOC or 9-α-fludrocortisone are almost the same as those for oral salt loading.

Step 3: Adenoma versus Hyperplasia

After the presence of primary hyperaldosteronism has been confirmed based on the demonstration of nonsuppressibility of aldosterone to salt loading, coupled with a nonresponsive PRA (to posture, salt restriction, and diuretic), the next step is the delineation of the cause of aldosterone excess. The distinction between APA and IH has important therapeutic implications. While the ultimate distinction between these two entities rests on measurement of aldosterone gradients in the venous effluents from both adrenals, there are several hormonal maneuvers that can be attempted prior to invasive venous cathetherization studies. In some instances the use of hormonal studies can be convincing enough to totally preclude the need for venous catheterization studies. This section deals with the numerous dynamic studies that attempt to distinguish between APA and IH. The basis for most of these studies revolves around two principles—the APA is more sensitive to ACTH and shows very little dependence on the renin–angiotensin system, while hyperplasia retains responsiveness to renin-mediated maneuvers and possibly to central stimuli that originate at the pituitary level. Understanding these two principles permits classification of the numerous tests that have mushroomed in the literature as "a new test to differentiate APA from IH."

The tests that attempt to distinguish between adenoma and hyperplasia can be classified into the following categories (Table 30).

1. Tests that evaluate ACTH control of aldosterone secretion. Important tests in this category include the response of plasma aldosterone to erect posture between 8 A.M. and noon and the response to exogenous ACTH administration.
2. Tests that evaluate dependency of aldosterone secretion on the renin–angiotensin system; included in this category are the evaluation of aldosterone response to angiotensin II (or III) and saralasin; measurement of circulating angiotensin II levels; and evaluation of the effects of salt depletion and posture on aldosterone secretion.
3. Tests that evaluate the dependency of aldosterone secretion on "central" mechanisms. Included in this category are the responses of aldosterone to administration of cyproheptadine and metoclopramide; and the measurement of ASF in urine.
4. Tests that evaluate serum concentration of nonaldosterone mineralocorticoids. The measurement of circulating levels of 18-hydroxycorticosterone is an important test in this category.

TABLE 30.
Hormonal Methods for Differentiating Adenoma
from Hyperplasia

I. Tests that evaluate ACTH dominance
    Aldosterone response to posture
    Aldosterone response to ACTH
II. Tests that evaluate renin–angiotensin dominance
    Aldosterone response to angiotensin II
    Aldosterone response to saralasin
    Plasma levels of circulating angiotensin II
    Response to salt depletion and posture
III. Tests that evaluate "central" control
    Response to cyproheptadine
    Response to metoclopramide
    Assay for "ASF" in urine
IV. Other steroid assays
    Recumbent basal 18-OH-corticosterone level

Tests That Evaluate Dependency of Secretion on ACTH

The important studies in this category are (1) evaluation of aldosterone response to standing between 8 A.M. and noon, and (2) evaluation of aldosterone response to ACTH infusion.

Evaluation of Aldosterone Response to Standing between 8 A.M. and Noon. In 1973 Ganguly et al.[73,92] described an anomalous decline in aldosterone concentration when patients with APA assumed an erect posture between 8 A.M. and noon. This contrasted with the response in patients with hyperplasia, who generally showed an increase in plasma aldosterone levels following erect posture.[73] This phenomenon of anomalous decline in aldosterone levels in adenomatous patients who stand between 8 A.M. and noon highlights the overriding effect of fluxes in ACTH on aldosterone concentrations. When ACTH levels fall between 8 A.M. and noon, there is a contemperaneous fall in aldosterone levels despite the erect posture, which is a normal stimulus for aldosterone release. Hence, the term *anomalous decline with posture.* This circadian influence on aldosterone levels of patients with APA has been confirmed by other workers.[74,109]

The test is positive in an impressive number of patients with adenoma. In the series reported by Weinberger et al.,[60] 23 of 32 patients with unilateral adenomas demonstrated an anomalous decline in aldosterone concentrations while ambulating between 8 A.M. and noon. In the series reported by Bravo et al.[103] of 33 patients with proven adenoma, an anomalous postural decline in aldosterone concentrations was seen in 20. The response in patients with hyperplasia is generally an increase in aldosterone level following standing. However, occasionally there may be no change, or even an anomalous decline following standing.

The utility of evaluating the plasma aldosterone response to standing between 8 A.M. and noon has been scrutinized in the past decade. Seven studies,[60,103,109,128,129,139,140] including more than 200 patients with surgically proven adenomas, have evaluated the false negative responses with this test in patients with APA and false positive responses in patients with hyperplasia. The mean rate of false negative responses in APA patients is approximately 35% and the rate of false positive responses (i.e., an anomolous postural decline in aldosterone levels in patients with hyperplasia), is approximately 15%. It is reasonable to assume that when an anomolous postural decline of aldosterone is encountered in a patient with primary hyperaldosteronism, it provides strong presumptive evidence for presence of an unilateral adenoma. Absence of an anomalous response, however, does not exclude an adenoma. Numerous studies have pointed out that a significant proportion of patients with adenoma failed to demonstrate this anomalous fall. Vaughan et al.[128] have proposed a method to improve the diagnostic yield of this test. These workers showed that the exceptions disappear when the test is performed in a volume-expanded state. In their study, when the test was performed on a normal sodium intake, 4 of 21 patients with adenoma showed an increase in aldosterone concentrations upon standing between 8 A.M. and noon. Curiously, two patients with essential hypertension showed an anomolous fall in plasma aldosterone upon standing. These exceptions disappeared when the study was done after volume expansion with 3 days of salt loading; all patients with adenoma demonstrated an anomalous decline in plasma aldosterone with standing, while none with idiopathic hyperplasia or essential hypertension demonstrated such a response. Thus, it appears that the diagnostic yield of the test can be greatly enhanced if it is performed in a volume-expanded setting. Salt loading apparently removes any influence the renin–angiotensin system may have on adenomas, permitting these tumors to be exclusively subject to the modulating influence of changes in the circulating ACTH concentrations.

Before this section on anomalous aldosterone response to posture is concluded, mention must be made of the observation that patients with GSH may also behave in a manner analogous to patients with adenoma. Ganguly et al.[91] studied six patients with GSH and showed that in all but one the plasma aldosterone levels declined with 2–4 hr of standing. This fall in aldosterone levels was accompanied by a simultaneous fall in cortisol levels, implying the powerful modulating influence of ACTH on aldosterone secretion in these patients. Also, with dexamethasone treatment there was restoration of the postural aldosterone response to normal, presumably because of a rise in PRA.[84]

Thus, it appears that in a patient with hyperaldosteronism, the demonstration of an anomalous fall in plasma aldosterone concentrations with standing implies the presence of unilateral adenoma, or the rare variant, GSH.

Evaluation of Aldosterone Response to Graded ACTH Infusion. Aldosterone secretion by APA is exquisitely sensitive to endogenous ACTH concentrations. This is illustrated by several observations: the circadian rhythm of plasma aldosterone concentrations in patients with primary aldosteronism caused by adenomas[141]; the episodic secretion of aldosterone in patients with primary hyperaldosteronism in general[142] and those with adenoma in particular[143]; and the anomalous fall in plasma aldosterone concentrations while standing between 8 A.M. and noon.[73,92] Despite these observations, plasma aldosterone responses to ACTH administration have not permitted diagnostic sensitivity either in diagnosing primary hyperaldosteronism or in making the distinction between adenoma and hyperplasia. An earlier study by Kem et al.[144] found that the aldosterone response to ACTH failed to distinguish adenoma from hyperplasia. It appears that the mere measurement of aldosterone response to supraphysiological doses of exogenous ACTH fails to distinguish between the two entities. However, the diagnostic yield can be improved if several mineralocorticoids are measured in response to *small* amounts of ACTH. Guthrie[145] evaluated plasma levels of steroids distal to progesterone, i.e., DOC, 18-OH-B, and aldosterone, in response to graded infusions of ACTH. The results indicated that patients with APA demonstrate a rise in DOC, 18-OH-B, and aldosterone with minute quantities of ACTH that failed to evoke any response in normal subjects or those with hyperplasia. This study highlighted the fact that precursor steroids (e.g., DOC, 18-OH-B) perhaps tend to emanate from tumor tissue in response to ACTH. This hypothesis is consistent with in vitro observations that corticosterone and DOC are found in higher quantities within adenoma tissue in comparison to normal or hyperplastic adrenal cortical tissue.[57,146,147]

Measurement of aldosterone and its precursor steroids in response to ACTH has not found wide application in differentiating adenoma from hyperplasia. Although the study by Guthrie[145] involved only a small number of patients, the implications are significant, postulating fundamental differences in the steroidogenic properties between adenoma and hyperplasia. This hypothesis is indeed in keeping with in vitro observations by Nicholls et al.[63] and by Dahl et al.[148] that adenoma tissue possesses an increased capacity for 18-OH-B secretion. The measurement of precursor steroid response to ACTH for differentiating adenoma from hyperplasia does have practical limitations. First, the test requires eliminating the effect of posture and endogenous ACTH (by dexamethasone administration). Second, the infusion of ACTH has to be graded, involving incremental doses of ACTH, a cumbersome procedure. Third, measurement of DOC and 18-OH-B may not be routine in many places. Finally, the heterogeneity of responses in primary aldosteronism precludes making predictions based on a small, albeit well-done study.

Tests That Evaluate Dependency of Aldosterone Secretion on the Renin–Angiotensin Axis

Several dynamic tests are based on the tenet that aldosterone production by hyperplasia is, at least partially, modulated by changes in the renin–angiotensin system. In contrast, aldosterone secretion by adenoma tends to be unaffected by maneuvers that involve changes in the renin–angiotensin system. The following dynamic studies belong to this category:

1. PRA and aldosterone response to salt depletion and posture.
2. Adrenal sensitivity to angiotensin infusion.
3. Response to angiotensin blockade with saralasin.

PRA and Aldosterone Response to Salt Depletion and Posture. It is generally agreed that patients with adenoma tend to demonstrate more profound suppression of renin than those with hyperplasia. This is expressed by several phenomena: the increase in plasma aldosterone in hyperplasia following prolonged standing; the basal PRA values in patients with hyperplasia on a normal sodium intake are slightly higher when compared to patients with adenoma; patients with hyperplasia, when salt depleted, show a tendency to increase urinary aldosterone excretion rates following salt depletion in comparison to patients with adenoma; finally, patients with hyperplasia tend to demonstrate lower urinary aldosterone excretion in comparison to adenoma when volume-expanded with 3 days of salt loading.

The increase in plasma aldosterone level in patients with hyperplasia following standing sharply contrasts with the anomalous postural decline in patients with adenoma. The increase in plasma aldosterone following standing is a reflection of the renin–angiotensin modulation of aldosterone secretion in patients with hyperplasia. Of course, such a response is not univeral in patients with hyperplasia since some patients with hyperplasia may fail to show any rise following standing, while occasionally the plasma aldosterone may actually decline following erect posture. The plasma aldosterone response to posture, in conjunction with other tests, is one of the main studies to help delineate adenoma from hyperplasia.

The basal PRA levels in patients with hyperplasia, although low, are not nearly as low as in patients with adenoma. Bravo et al.,[103] in a prospective study of 80 patients with hyperaldosteronism (70 with adenoma and 10 with hyperplasia), were able to demonstrate a significantly higher basal PRA in patients with hyperplasia on a normal Na^+ diet when compared to patients with APA. Also, these workers noted that during sodium deprivation, patients with hyperplasia had comparatively higher urinary aldosterone excretion rates, while in patients with adenoma the rate of aldosterone excretion was not appreciably affected by salt deprivation.

Finally, upon volume expansion a phenomenon opposite to that described above is noticeable. While the aldosterone secretion in patients with hyperplasia, as well as adenoma, is nonsuppressible to salt loading, the degree of nonsuppression in both groups is somewhat different. Vaughan et al.[128] evaluated the effect of oral salt loading on urinary excretion of aldosterone in 29 patients with hyperaldosteronism and were impressed by the fact that patients with hyperplasia lowered aldosterone excretion to a greater degree when compared with patients with APA. Thus, volume expansion was able to lower aldosterone excretion in patients with hyperplasia, although not to normal levels. The degree of suppression of urinary aldosterone in response to volume expansion clearly differentiated patients with adenoma, hyperplasia, and normals. Patients with hyperplasia suppressed to volume expansion better than patients with adenoma, but not as well as normals.

All these phenomena reflect the fact that patients with hyperplasia retain aldosterone modulation through the renin–angiotensin system. For the same reason, the aldosterone secretion in patients with hyperplasia also responds to exogenous angiotensin infusion and angiotensin agonism/antagonism, while adenomas do not.

Adrenal Sensitivity to Angiotensin Infusion. The postulate that aldosterone secretion in patients with hyperplasia is modulated by changes in the renin–angiotensin system supposes that hyperplastic adrenal tissue is exquisitely sensitive to even minor changes in circulating angiotensin II brought about by maneuvers such as orthostasis and salt depletion. This supposition was tested by studying the aldosterone responses to angiotensin infusion in patients with hyperaldosteronism caused by hyperplasia and comparing these responses with APA. Two studies, one by Wisgerhof et al.[76] and another by Hollifield et al.,[149] were supportive of the fact that patients with IH indeed showed larger-than-normal increases in plasma aldosterone concentrations during the infusion of angiotensin II. These observations were extended in an attempt to evaluate the utility of angiotensin infusion in separating patients with hyperplasia from APA. Wisgerhof et al.[94] evaluated the plasma aldosterone response to intravenous angiotensin II infused at a rate of 0.5 ng/kg per min to eight patients with surgically proven APA and nine with IH. The two significant observations that emerged were that patients with APA showed a significantly reduced aldosterone response to angiotensin II when compared with idiopathic hyperplasia, and that the threshold for response to angiotensin infusion was much higher in those with adenoma when compared to patients with hyperplasia. These results were consistent with similar response patterns described earlier in patients with

APA.[135,136] It was proposed that the response of aldosterone to angiotensin II might be used as a test to distinguish patients with idiopathic hyper-aldosteronism from those with adenoma.

Notwithstanding these reports, it is becoming increasingly apparent that patients with adenoma represent a heterogenous population, and indeed some with APA may, in fact, demonstrate preservation of aldosterone responsiveness to angiotensin infusion. This was illustrated in an early study reported by Spark et al.,[75] who noted that aldosterone secretion by adenoma can be activated by infusion of subpressor doses of angiotensin II. More recently, Carey et al.[150] studied aldosterone responses to infusion of the C-terminal heptapeptide fragment of angiotensin II. This peptide, also known as des-Aspartyl angiotensin II (des-Asp-A II) or angiotensin III (A III) possesses 25–100% of aldosterone-stimulating potency but has only less than 25% of the pressor potency of angiotensin II.[151] Thus, it is safe to use to study the effect of infused angiotensin on aldosterone secretion. As predicted, all three patients with IH showed responsiveness to A III infusion, supporting the tenet that aldosterone secretion by hyperplastic tissue is sensitive to angiotensin, a physiological stimulus. The unanticipated finding was that patients with APA behaved in an unpredictable fashion; of the seven patients with adenoma, five showed no aldosterone response, but two patients demonstrated aldosterone stimulation with des-Asp-A II. This observation documents the existence of a subpopulation of patients with adenomatous hyperaldosteronism, with preservation of responsiveness to angiotensin stimulation.

While these studies have thrown some light on the behavior pattern of adenomatous and hyperplastic adrenocortical tissue, there are no "universal responses" that permit clear separation of these two groups based on the angiotensin infusion test. Unless larger numbers of patients are studied, and the multitude of other factors that regulate aldosterone secretion are controlled, it is difficult to extrapolate the usefulness of this test in differentiating hyperplasia from adenoma. To make matters even more confusing, Wisgerhof et al.[78] have demonstrated an increased adrenal sensitivity to angiotensin II in patients with low-renin essential hypertension. The controversies surrounding the usage of this test in clinical practice led to yet another mode of evaluating aldosterone responsiveness to angiotensin, i.e., the use of saralasin in differentiating hyperplasia from adenoma.

The Use of Saralasin (1-Sar, 8-Ala Angiotensin II) in Differentiating Adenoma from Hyperplasia. Saralasin is a competitive partial antagonist of angiotensin II. Its effect, in terms of causing "angiotensin blockade" and lowering the blood pressure, as well as lowering aldosterone concentrations, revolves around the activity of the renin–angiotensin system; thus, in situations where the renin–angiotensin system is active (secondary hyperaldosteronism, high-renin hypertension) administration of saralasin results in lowering the blood pressure and reducing circulating aldosterone secretion.[152,153] In contrast, when saralasin is administered in low-renin situations (low-renin hypertension, for instance), the drug acts as an angiotensin II agonist and causes an increase in the blood pressure and in circulating aldosterone concentrations.[154] Since it has been postulated that patients with IH are exquisitely sensitive to angiotensin II and have very low circulating angiotensin II levels, it was anticipated that administration of sara-lasin to patients with IH would be attended by an increase in circulating al-dosterone concentrations, while patients with APA would show no such re-sponse. Brown et al.[155] infused saralasin in eight patients with hyperplasia and six with a solitary APA. It was noted that the plasma aldosterone concentration increased in all eight patients with hyperplasia in response to saralasin, while none of the patients with adenoma responded. It is noteworthy that the plasma aldosterone responded to saralasin even in patients with hyperplasia who failed to show the expected rise in aldosterone with posture. It is possible that pos-turally induced increases in angiotensin may be insufficient in some patients with IH to cause a rise in aldosterone, whereas controlled amounts of saralasin achieve the effect of stimulating aldosterone secretion by hyperplastic tissue. The authors concluded that a distinct increase in plasma concentrations of al-dosterone following saralasin in a patient with hyperaldosteronism strongly points to IH, while lack of such a response favors the diagnosis of APA.

While the data reported by Brown et al.[155] are impressive and indeed con-vincing, the test has not found widespread application. The need for graded infusion of saralasin, the potential for modest increases in the blood pressure, the lack of familiarity with saralasin, the availability of other tests to make the distinction between APA and IH, and above all, the inability of the test to distinguish patients with IH from those with low-renin hypertension[78,154] have all been conducive in limiting widespread application of this test.

Tests That Evaluate the Dependency of Aldosterone Secretion on "Central Mechanisms"

Data that support a neurotransmitter connection in the regulation of al-dosterone secretion have provided other tools to differentiate hyperplasia from adenoma. This, coupled with the fact that central aldosterone-stimulating secre-tagogues were being added to regulatory mechanisms that govern aldosterone secretion, led to measurement of several ASF in IH. The tests in this category include

1. Aldosterone response to cyproheptadine
2. Aldosterone response to metoclopramide
3. Assay for the glycoprotein ASF in the urine of patients with hyperaldos-teronism

Aldosterone Response to Cyproheptadine. Idiopathic hyperaldosteronism may be associated with overproduction of an unknown secretagogue, possibly origi-nating in the pituitary. Since neurotransmitter regulation is intimately linked with secretion of several adenohypophyseal hormones, it was postulated that serotonin may be involved in the regulation of the non-ACTH pituitary secre-tagogue that controls aldosterone secretion in IH. Gross et al.[56] studied the role of serotonin in regulating the release of aldosterone by administering a single oral dose of cyproheptadine to 14 patients with hyperaldosteronism. The influ-ence of salt restriction and ACTH was abolished by placing these patients on a high-salt diet and oral dexamethasone. It was noted that serum aldosterone levels declined significantly in patients with hyperplasia, while no fall was observed in patients with adenoma as well as in normal subjects with raised aldosterone levels due to salt restriction. The authors postulated that a serotonin-mediated aldosterone-stimulating system might be hyperactive in patients with IH.

The analogy between the neurotransmitter involvement in the hypercortisolism of pituitary-dependent Cushing's disease with bilateral hyperplasia of the zona fasciculata and hyperaldosteronism due to bilateral hyperplasia of the zona glomerulosa is inevitable. The effectiveness of cyproheptadine in suppressing pituitary ACTH in some patients with pituitary-dependent hypercortisolism[156] renders the analogy even more compelling. However, a direct effect of cyproheptadine on the zona glomerulosa has not been excluded. In vitro studies have demonstrated that serotonin may directly stimulate steroidogenesis by the isolated rat adrenal glomerulosa cells.[157] If confirmed in larger numbers of patients, the cyproheptadine suppression test may find an important place in the differentiation of hyperplasia from APA. More important, it may even find a therapeutic role in reducing aldosterone secretion in patients with hyperaldosteronism secondary to IH.

Aldosterone Response to Metoclopramide.   Dopamine is a physiological inhibitor of renin and aldosterone secretion. Administration of metoclopramide is associated with a rise in plasma aldosterone concentrations. The possibility that idiopathic hyperaldosteronism may result from central dopaminergic dysregulation is an attractive speculation. If one carries this speculation through, a decreased dopaminergic tone at the pituitary level may result in hypersecretion of a non-ACTH pituitary secretagogue, which may result in bilateral hyperplasia of the zona glomerulosa with consequent hyperaldosteronism. After all, such a situation has been described as a variant of pituitary-dependent Cushing's disease caused by intermediate lobe tumors[158] which are characterized by suppressibility to dopaminergic drugs. The possibility that a similar situation may exist in the case of idiopathic hyperaldosteronism is suggested by a unique case report. Franco-Saenz et al.[159] reported the case of a 37-year-old woman with documented primary aldosteronism who died of a cerebral hemorrhage. At autopsy both adrenal glands were enlarged due to hyperplasia; the pituitary gland was also enlarged, with nodular basophilic hyperplasia of the anterior and intermediate lobe. The authors postulated possible intermediate-lobe disease with consequent hypersecretion of non-ACTH, POMC peptides (β-lipotropin or γ-melanotropin). These findings, as well as the demonstration of increased levels of POMC-related peptides in patients with IH,[32–37] strongly suggest that a pituitary factor is probably involved in the development of IH. However, a connection between pituitary hypersecretion of an aldosterone secretagogue and disturbed dopamine regulation has not been established.

Very few studies have evaluated the role of dopamine antagonism in the differentiation of subtypes of hyperaldosteronism. Ganguly et al.[160] evaluated the effects of metoclopramide on plasma aldosterone in various forms of hyperaldosteronism. Their data indicate that dopaminergic influences may have a putative role in hyperaldosteronism. Following metoclopramide there was a greater-than-normal rise in aldosterone levels in patients with adenoma, hyperplasia, and dexamethasone-suppressible hyperaldosteronism. It is unclear whether this reflects a greater dopaminergic influence in primary hyperaldosteronism of all types or merely represents the output of hormone by increased numbers of aldosterone-producing cells. Other workers[161,162] have also reported that patients with hyperaldosteronism may show an exaggerated rise in plasma aldosterone following metoclopramide. The noteworthy observation in the study reported by Ganguly et al.[160] is that the metoclopramide-induced aldosterone rise could be abolished by dexamethasone only in patients with GSH. Patients with hyperaldosteronism caused by adenoma and IH continue to demonstrate an exaggerated plasma aldosterone response to metoclopramide when placed on dexamethasone. This differential response pattern may be related to the ACTH dependency of aldosterone secretion in GSH, a phenomenon that may be subject to greater modulation by dopaminergic influences.

The metoclopramide test does not help in clinical practice as a tool to separate the various subtypes of hyperaldosteronism. At best it provides interesting insights into the pathogenesis of one variety of primary hyperaldosteronism, the glucocorticoid suppressible variety.

Other Tests to Differentiate between APA and IH

Measurement of 18-OH-B. The measurement of circulating levels of 18-hydroxycorticosterone (18-OH-B) in the recumbent state is emerging as an important marker of primary aldosteronism caused by an APA. 18-OH-B is the penultimate steroid precursor of aldosterone biosynthesis and is believed to be exclusively secreted by the zona glomerulosa.[163] 18-OH-B either is formed as an immediate precursor of aldosterone or may represent a by-product of the action of corticosterone methyloxidase on corticosterone.[164] Early studies by Biglieri et al.[165] had suggested that aldosterone precursors such as DOC and 18-OH-B may be preferentially hypersecreted in patients with adenoma. Further support for this postulate was derived when it was shown that ACTH stimulation of adenomatous tissue resulted in enhanced secretion of precursor mineralocorticoids.[145,166] The first large study that systematically evaluated the utility of measuring plasma levels of 18-OH-B in patients with hyperaldosteronism was that reported by Biglieri and Schambelan[167] in 1979. These workers measured plasma 18-OH-B levels in 23 patients with primary hyperaldosteronism after an overnight recumbency and found that the levels were six times higher in patients with adenoma. Using a value of greater than 100 ng/dl, measurement of plasma 18-OH-B provided clear separation of the nine patients with adenoma from the 14 with idiopathic hyperplasia. Their results led the authors to conclude that the 18-OH-B level in plasma was an effective discriminator between APA and IH and may indeed be a "marker of the events in later aldosterone biosynthesis." The authors also noted a significant negative correlation between the ratio of 18-OH-B:aldosterone and the serum K concentrations, particularly in patients with IH. The utility of measuring plasma 18-OH-B concentrations in differentiating adenoma from hyperplasia has been confirmed by Kem et al.,[168] who observed that 22 of 23 patients with adenoma had basal 18-OH-B concentrations greater than 100 ng/dl, while all nine patients with idiopathic hyperplasia had values below 100 ng/dl. Thus, the measurement of a basal recumbent (or sitting) 18-

OH-B level in the morning provides a reasonably effective method for determining the etiology of hyperaldosteronism.

The reason for elevated plasma 18-OH-B levels in the plasma of patients with APA may be related to several factors; three in particular deserve mention—the effect of ACTH, the effect of K^+ balance, and the steroidogenic capabilities of tumor tissue. The effect of ACTH, which has its peak concentrations in the morning, may have an influence on the synthesis and release of 18-OH-B by adenoma tissue. It has been well documented that administration of graded doses of ACTH to patients with APA results in increased levels of precursor steroids (DOC, 18-OH-B) in the circulation. Hence, it is reasonable to assume that the elevated basal 18-OH-B concentrations drawn after overnight recumbency may be a reflection of the exquisite sensitivity of adenomas to ACTH. It would be of interest to study the 18-OH-B level in patients with dexamethasone-suppressible aldosteronism, a disorder in which aldosterone secretion is clearly dependent on circulating ACTH concentrations.

The role of K^+ balance in relation to the elevated 18-OH-B concentrations is conjectural. Potassium concentrations may significantly influence the conversion of 18-OH-B to aldosterone. Thus, hypokalemia may inhibit the 18-dehydrogenation activity of the enzyme methyloxidase type 2, resulting in a higher ratio of 18-OH-B to aldosterone in the plasma. Biglieri and Schambelan[167] evaluated this ratio in 14 patients with hyperplasia and showed that improvement of the hypokalemia was associated with decrease in the 18-OH-B:aldosterone ratio. This negative correlation is more apparent in patients with hyperplasia than in those with adenoma. The reason for this discrepancy is not very clear, since patients in both groups can have hypokalemia of an identical magnitude. In vitro studies[169] have demonstrated that incorporation of tritiated corticosterone into aldosterone and 18-OH-B is reduced in the presence of potassium depletion. If potassium depletion was the sole cause of an elevated plasma level of 18-OH-B, one would expect to find an elevated level in all subgroups of hyperaldosteronism, since hypokalemia is encountered in all variants of the disorder. Such is not the case, since basal levels in excess of 100 ng/dl are usually encountered only in patients with adenoma. Obviously, other factors must contribute to the elevated 18-OH-B levels in patients with APA.

The third possibility, that tumor tissue has different steroidogenic properties than hyperplastic tissue, is also an attractive hypothesis. The sharp difference in the in vitro behavior of adenomatous and hyperplastic tissue studied by tissue culture has been pointed out by Nicholls et al.[63] The incorporation of cholesterol into precursor steroids (such as DOC and 18-OH-B) is impressively enhanced in adenomatous tissue. Thus, it is possible that the elevated 18-OH-B concentrations in the plasma of patients with APA may be a reflection of the steroidogenic properties of the tumor.

The test for measurement of 18-OH-B concentrations in the plasma is not a commonly performed assay, and interpretation of the basal levels can be affected by several factors, including assay specificity, time of day, dietary sodium intake, and drugs that affect aldosterone secretion.[168] When a specific assay is used, measurement of basal 18-OH-B concentrations can be a valuable tool in differentiating patients with adenoma from those with hyperplasia.

Summary

From the foregoing it should be apparent that several tests have been devised and described in differentiating adenoma from hyperplasia on hormonal grounds. The multiplicity of available tests is an attestation to the difficulties involved in such separation. The inability to consistently differentiate unilateral from bilateral lesions by various routine clinical and hormonal data has been illustrated by several workers.[60,129,130,139,170] The ability to predict the underlying lesion improves when the entire picture is reviewed in conjunction with a combination of tests.[130]

In general, patients with APA demonstrate more profound hypokalemia and more pronounced abnormalities in plasma aldosterone and PRA, i.e., more marked elevations in plasma aldosterone concentrations and more profound suppression of renin, in comparison to patients with bilateral hyperplasia. The plasma aldosterone is more resistant to suppression with volume expansion, and the PRA is more resistant to stimulation by salt depletion and posture in comparison to hyperplasia. Aldosterone-secreting adenomas are more sensitive to ACTH, often demonstrate an anomalous decline in plasma aldosterone concentrations with posture, and are more frequently associated with a basal plasma 18-OH-B level in excess of 100 ng/dl in the supine state.

In general, patients with bilateral hyperplasia tend to have milder abnormalities in potassium, aldosterone, and PRA, when compared to those with adenoma. The plasma and urinary aldosterone in patients with hyperplasia, while not suppressing to normal with volume expansion, demonstrates some degree of partial suppression. The PRA of patients with hyperplasia can be stimulated, albeit to a minor degree, with protracted salt depletion and prolonged standing. The plasma aldosterone will often demonstrate a rise in response to erect posture in most patients with hyperplasia, and the basal plasma 18-OH-B levels in the supine state are often below 100 ng/dl. The glycoprotein ASF can be demonstrated in some, but not all, patients with IH.

In clinical practice the two most important hormonal tests that aid in differentiating APA from hyperplasia are evaluation of the aldosterone response to posture and measurement of plasma 18-OH-B levels in the basal state. The diagnostic utility of the plasma aldosterone response to posture in differentiating APA from IH can be enhanced by two additional maneuvers during the test. First, the volume-expanded patient with primary hyperaldosteronism is more likely to exhibit the predictable aldosterone response to posture in APA and hyperplasia. Vaughan et al.[128] have impressively shown that the maneuver of salt loading for 3 days eliminates false positives and false negatives when the aldosterone response to posture is used to differentiate adenoma from hyperplasia; i.e., volume-expanded patients with APA demonstrated the characteristic fall in aldosterone levels, while patients with hyperplasia demonstrated the characteristic rise in aldosterone concentrations upon assuming an erect posture. Another simple method that can enhance the utility of the posture test is to measure cortisol in the plasma simultaneously with aldosterone. Since the decline in plasma aldosterone in patients with APA is a reflection of the declining ACTH concentrations in the plasma of the patients standing between 8 A.M. and noon, the cortisol measurements can provide an indirect clue to the flux in ACTH concentrations. For instance, one of the factors that can mask the anomalous decline in aldosterone levels with posture in APA is the occurrence of an ACTH pulse, which naturally will elevate the aldosterone concentrations of patients with adenoma, and the response may, therefore, mimic hyperplasia. The presence of this interfering phenomenon can be detected by measuring cortisol, which would also be expected to be elevated contemporaneously. Thus, if the plasma cortisol did not show a decline at the end of the posture test, the procedure needs to be repeated to avoid misinterpretation. The diagnostic utility of measuring recumbent basal 18-OH-B level in differentiating between adenoma and hyperplasia has been alluded to already.

At our institution, Cook County Hospital, the first-line investigation in the hypertensive patient with suspected primary hyperaldosteronism is the saline infusion test, performed on a normal dietary sodium intake and in the normokalemic state. If the plasma aldosterone level following saline infusion is greater than 8.5 ng/ml, further investigations are undertaken. This means evaluating the PRA response to posture and salt depletion performed in the following manner. The patient is placed on a salt restricted diet for 1 day, along with oral furosemide, 40 mg three times a day. On the subsequent morning PRA is measured after an hour's recumbency. An extra tube of blood is drawn and stored for future measurement of aldosterone and cortisol, if indicated. The patient is ambulated for 4 hr, usually between 8 A.M. and noon, at the end of which the blood is assayed for PRA. Again, an extra tube of blood is drawn and saved for future determination of aldosterone and cortisol if necessary. Should the PRA levels after ambulation fail to show a rise above 2 ng/ml per hr, the diagnosis of primary hyperaldosteronism is made. Nonsuppressible aldosterone and suppressed, nonstimulatable PRA are the two criteria required to be satisfied. The subsequent line of workup focuses on the distinction between APA and IH. The extra test tube of blood saved during the posture test is used for this purpose. To avoid misinterpretation, the cortisol measurements are first performed in the stored samples obtained during the basal and postambulatory states. If the cortisol level in the noon sample shows a decline in comparison to the 8 A.M. sample, this can be assumed to reflect a decline in the ACTH levels between 8 A.M. and noon, and the plasma aldosterone measurements can be carried out in the samples. If, in contrast, the cortisol level does not show a decline, aldosterone measurements are not performed and the whole study is repeated. This cost-saving method avoids misinterpretation of the aldosterone response to posture. Depending on the results, further localizational procedures are carried out. These procedures are discussed in the next section.

Step 4: Localization of Adenoma

The three procedures that are most often used for localization of the etiology of hyperaldosteronism are computed tomography (CT) of the adrenals, iodocholesterol scintigraphy of the adrenals, and bilateral catheterization of adrenal veins for aldosterone measurement. Of these, CT has the advantages of easy availability, lower cost, and noninvasiveness. Iodocholesterol scintigraphy is considered by its advocates to be the only noninvasive test that evaluates the functional activity of the tumor, and when performed by experts, it has an excellent yield. The selective venous catheterization study, with its newer modifications, consistently identifies the etiology of the hyperaldosteronism. The procedure, however, is invasive.

In this section, the relative role of each of these major localizational tests is reviewed. Adrenal venography, another invasive test, is not discussed since it has been abandoned by most radiologists owing to hazards associated with the test and since other safer procedures are available. Adrenal venography can result in numerous complications, including adrenal hemorrhage and even adrenal insufficiency.[171-173] Remission of primary aldosteronism following venography is also a reported occurrence.[173,174] Although the risk of complications with venography is small, the availability of better and safer procedures has justifiably resulted in abandonment of venography as a localizational procedure. This section, therefore, focuses on the usefulness of CT, iodocholesterol scintigraphy, and adrenal venous sampling for aldosterone measurements in defining the etiology of primary hyperaldosteronism.

Computerized Tomography

Owing to its noninvasive nature, CT of the adrenals is the first-line localizational procedure to confirm the presence of an APA. Since APA are often the smallest of adrenal tumors, it is necessary to employ CT equipment with high-contrast resolving power, approaching 1 mm. In a prospective study of 22 consecutive patients with primary hyperaldosteronism, White et al.[175] evaluated the usefulness of CT in localizing APA. An adrenal mass was demonstrable in 12 of the 16 patients presumed to harbor adenomas, based on hormonal data. Adenomas were surgically confirmed in 11 of these patients. In the remaining four patients with biochemical and hormonal features suggestive of adenoma, the CT findings were not specific enough to permit localization; the presence of unilateral adenomas in those four patients was documented either by exploration or by adrenal vein catheterization studies. The demonstration of one normal gland with "indeterminate changes" in the contralateral gland merits mention. When such findings are seen in patients with powerful biochemical and hormonal data indicative of adenoma (such as severe hypokalemia, anomalous decline in aldosterone with posture, or marked elevations in 18-OH-B), the probability of adenoma on the side showing "indeterminate" changes is high. The CT appearance of hyperplasia is characterized by diffuse bilateral enlargement or by a normal configuration of both adrenals. The results of White et al.[175] are consistent with other reports that provide an 86% sensitivity and 85% accuracy of CT in localizing APA.[176,177]

The most important reason for the CT-negative adenomas is the small size of thse tumors. The difficulties in detecting adenomas smaller than 1 cm in size have been outlined by Linde et al.[178] In their study of nine patients with APA proven by exploration, the CT scan correctly identified the tumor preoperatively in only four patients. Further, two patients identified as having bilateral tumors by CT proved to have unilateral adenomas on exploration. In the remaining three, the CT appearance was negative, but unilateral adenomas were found during surgery. The authors were able to demonstrate a correlation between the size of the tumor and preoperative detection by CT, tumors below the size of 1.2 cm being undetectable by CT. Detection of small adenomas can be further hampered by difficulties on the part of patients to suspend respiration at precisely the same level for successive scans. Technical improvements in CT, such as dynamic scans and thin-slice CT studies, should improve the diagnostic yield of the procedure in detecting APA.

The use of CT as the first-line localizational procedure has several obvious advantages. First, its availability is almost universal, in contrast to other procedures such as iodocholesterol scans (which are limited to only a few centers) and bilateral adrenal venous catheterization (which requires great skill and expertise). Second, the procedure can be done on an outpatient basis, whereas selective venous catheterization requires hospitalization. Third, the radiation exposure is comparatively lower than in scintigraphy or catheterization studies. Fourth, the immediate availability of results provides diagnostic gratification. The only limitation is the rate of false negatives in localization, since approximately 20–30% of adenomas will not be visualized by CT owing to their small size. In these patients more elaborate localizing procedures can be undertaken.

The advantages of CT notwithstanding, caution should be exercized in diagnosing hyperplasia on the basis of a negative CT. Compared to the other hypersecretory tumors of the adrenal cortex (such as pheochromocytoma or cortisol-secreting adenoma), aldosterone-secreting tumors are generally less than 1.5 cm in diameter.[179] Some authorities believe that the resolution power of the currently available generation of CT equipment is such that the average-sized adenoma will be detected by the procedure. When adenoma is suspected on the basis of biochemical and hormonal data in the presence of a negative or equivocal CT, the localization options available to the physician are iodocholesteral scanning or bilateral adrenal venous sampling for aldosterone levels.

Iodocholesterol Scans

[^{131}I]19-iodocholesterol is an isotope used for adrenal imaging. The isotope is handled by the adrenal cortex in a manner analogous to cholesterol and is hence esterified by the adrenal cortex. The isotope is also concentrated by other organs, particularly the liver, spleen, gallbladder, and colon. The disparity in gland size (the asymmetry in appearance) lateralizes the site of tumor. The development of this technique in 1971 by Conn and co-workers[180] allowed photoscanning of aldosterone-producing tumors of the adrenal cortex. The prior administration of dexamethasone to suppress uptake by normal tissue enhances the sensitivity of this localizing procedure. The utility of this localizing procedure has been evaluated in several studies.[181–183] Hogan et at.[181] reported that iodocholesterol photoscanning documented the presence of an adenoma with 83% accuracy. In 10 of 12 patients with biochemical and hormonal data suggestive of an adenoma, the procedure was helpful in localizing the site of tumor. In contrast only 1 of 13 patients with hormonal evidence of hyperplasia or low-renin essential hypertension showed an asymmetrical adrenal uptake of isotope. Encouraging results with this procedure have been reported by Conn and associates,[184] who were consistantly able to separate tumor from hyperplasia by using the dexamethasone-modified adrenal photoscan using iodocholesterol.

A similar enthusiasm has been generated by reports of Lieberman et al.[182] and Seabold et al.[183] Despite the encouraging reports from Ann Arbor, where the procedure was developed and antedated the arrival of high-resolution CT on the diagnostic scene, several workers have reported less satisfactory results with iodocholesterol scans in localizing APA. Weinberger et al.[60] could correctly identify the presence of an adenoma in only 47% of their patients with tumors. In addition to the false negatives, which constituted a rather substantial (53%) number of patients in their series, the authors caution that bilateral uptake was encountered ("hyperplasia pattern") in 12 patients who subsequently were found to have adenomas. Thus, the Indiana University experience with iodocholesterol scanning is less encouraging than the pioneering reports from the University of Michigan. In retrospect, the limitations of iodocholesterol scanning may have been due to the small size of these tumors, which elude the resolution of adrenal photoscan. The other practical problem with iodocholesterol scanning is the need for repeated imaging to allow for the disappearance of localized extraneous activity of the isotope, which can be concentrated in several organs besides the adrenals. Since the interpretation of a positive scan relies heavily on asymmetry between the glands, minor inequalities, when present, need to be confirmed by repeat studies.

Another closely related isotope, [131I]6-β-iodomethyl norcholesterol (NP-59), has also been used extensively by the investigators at the University of Michigan for localization of APA. When scintiscans are performed with strict attention to the dose and duration of dexamethasone suppression as well as to the imaging intervals, adrenal imaging with NP-59 provides 90% sensitivity and accuracy.[185] In a large study of 87 patients with primary aldosteronism (50 with adenoma and 37 with hyperplasia), Gross et al.[186] using dexamethasone-suppressed adrenal scintiscan by NP-59, were able to accurately identify the underlying lesion in 82 of the 87 patients. In the same study, CT correctly identified the presence of adenoma in only 14 of 23 patients with adenoma and 2 of 10 patients with bilateral hyperplasia. The authors concluded that a correctly performed adrenal scintiscan with NP-59 is superior to CT in identifying the underlying lesion of primary hyperaldosteronism. Other workers[185,187] have also reported encouraging results with NP-59 as a useful agent for localization. Thus, the two factors of importance that decided the usefulness of iodocholesterol agents are expertise in performing and experience in interpreting the test. Advocates of the procedures dispute the degree of radiation caused by the procedure. Carey et al.[188] estimate that the absorbed radiation dose from NP-59 in normal adrenal tissue averages 25 rads/mCi (range 12–39 rads), with a total body exposure of 1.4 rads/mCi. With dexamethasone administration a further reduction in absorbed radiation can be anticipated. Gross et al.[189] estimate that the average radiation exposure to the adrenal by the use of dexamethasone-suppressed NP-59 scanning is no greater than the radiation exposure with CT. However, this does not take into account the total body radiation dose, a factor that is negligible with CT.

The availability of iodocholesterol scanning is limited to few centers. The procedure requires elaborate preparation and multiple examinations. The expertise required for interpretation is also limited to few centers. Clearly, most authorities use CT as the initial localization study. If it is negative or unsatisfactory, the patient can be referred to centers that perform the dexamethasone-suppressed iodocholesterol photoscanning with either [^{131}I]19-iodocholesterol or NP-59. However, if an expert radiologist trained in performing venous catheterization of adrenals is more readily available, this study would be the choice for localization.

Adrenal Vein Catheterization

Aldosterone measurement in the adrenal venous effluent is considered the most effective ("gold standard") method for localizing an APA. This technique has accurately located the source of aldosterone production in 80–100% of patients with primary hyperaldosteronism.[60,93,128,190,191] Comparison of several localizing procedures, such as CT, adrenal scintigraphy, venography, and adrenal venous sampling, for aldosterone concentrations reveals that measurement of aldosterone from venous effluents provides 100% accuracy in experienced hands.[60,103] The procedure, which is invasive, consists of cannulating both adrenal veins and obtaining blood samples from both adrenal veins as well as the low inferior vena cava. The vena caval sample reflects peripheral hormonal levels. Comparison of aldosterone levels in the different samples and calculation of the aldosterone gradient in each will readily reveal the unilateral source of hormone excess when an adenoma is present.

Despite its sensitivity and specificity, several areas of concern exist regarding bilateral adrenal vein catheterization.

1. First, the procedure is invasive and entails some degree of discomfort to the patient.
2. The right adrenal vein is difficult to catheterize. The technical difficulties involved in catheterizing the right adrenal vein have been outlined by Scoggins et al.[190]
3. There is a small, but significant, risk of trauma, even with the use of the advanced "cobra-shaped," tapered polyethylene catheters with reinforced wall. The risk, of course, is much less than with adrenal venography.
4. The potential problem of dilution of the venous effluent by blood from nonadrenal sources can lead to misinterpretation of results.
5. The episodic secretion of aldosterone may "mask" the true identity of the underlying lesion.

Several developments in the recent past have enhanced the diagnostic yield from bilateral venous catheterization of adrenal veins.[178,192,193] Thus, fluoroscopy with contrast injection can be used to document proper catheter placement; in addition, it is recommended that collection of blood by gravity flow tends to minimize errors in interpretation of the source of aldosterone.[191] The major problem in interpretation is the possibility of dilution of adrenal venous effluent by nonadrenal blood. Measurement of cortisol in the adrenal venous and venacaval blood has been recommended as a safeguard to minimize errors in interpretation.[192] The finding of a large cortisol gradient between adrenal and peripheral blood helps to confirm the relatively undiluted nature of the adrenal venous effluent. However, this is not foolproof, because, not uncommonly, there may be an overlap between the cortisol content of the adrenal venous effluent and the cortisol level in the peripheral veins. For instance, Spark et al.,[194] while studying cortisol dynamics in the adrenal venous effluent, noted that in 11 patients of the 27 studied, the cortisol levels in the adrenal veins were similar to the peripheral levels. Similar observations have been reported by Weinberger et al.[60] and more recently, by Levinson et al.[195] This prompted the introduction of measurement of epinephrine levels in the venous effluent to document the "relative purity" of the adrenal venous effluent. Levinson et al.[195] evaluated the usefulness of measuring the adrenal vein epinephrine levels in seven patients with primary hyperaldosteronism who underwent bilateral adrenal vein sampling studies. In six of the seven patients, they noted that epinephrine levels greater than 1500 pg/ml in the venous effluent confirmed the relatively undiluted nature of the adrenal venous efflux. Thus, measurement of epinephrine levels in the adrenal veins is a marker of the adrenal origin and appears to be superior to cortisol measurement because in normal subjects the adrenal vein epinephrine is always severalfold higher than the peripheral or inferior vena caval blood.[196–198] This phenomenal elevation of epinephrine content in the adrenal venous effluent contrasts with the cortisol levels in the adrenal venous effluent, which can overlap with peripheral blood levels. Therefore, in the absence of a pheochromocytoma, an adrenal venous epinephrine content greater than 1500 pg/ml clearly indicates to the radiologist that the source of the blood was the adrenal gland. When the epinephrine content of the adrenal vein ranges between 250 and 1500 pg/ml, it can be presumed that the sample did not come directly from the adrenal vein, or that there was significant dilution of adrenal venous effluent by blood from nonadrenal sources. The authors concluded that measurement of adrenal venous epinephrine concentration offers a sensitive and convenient method of detecting adrenal venous dilution and, therefore, should improve the accuracy of venous sampling in primary aldosteronism.

Another dimension has been added to enhance the sensitivity of the adrenal venous catheterization study by Weinberger et al.[60] They have minimized the artifact of episodic aldosterone secretion by measuring aldosterone levels from each adrenal and inferior vena cava during stimulation with ACTH. Exploiting the well-known sensitivity of APA to ACTH, Weinberger et al.[60] showed that the aldosterone gradient in the venous effluent on the side of the tumor was magnified after ACTH stimulation, permitting clear separation between APA and IH with 100% accuracy. This procedure helps to bring out the differences between the two sides in a manner that convincingly establishes the unilateral source of disease.

The remarkable localizational value of adrenal venous sampling has been illustrated in 11 different studies[60,93,103,128,190,193,195,199–202] and is considered by many as the only definitive localizing procedure. With improved technical placement of catheters and measurement of epinephrine (and cortisol), coupled with performing the study during ACTH stimulation, the procedure will provide a universal yield in localizing the adenoma. At present, in many institutions including ours, when an adenoma is suspected on hormonal grounds, but the CT study fails to demonstrate the adenoma, the venous sampling study is the next procedure of choice. It does involve hospitalization and obviously requires the expertise of well-trained and experienced angiographers.

Step 5: Exclusion of GSH

This variant of primary hyperaldosteronism needs to be considered under the following circumstances:

1. Hyperaldosteronism occurring in a young age group.
2. Hyperaldosteronism with a familial tendency.
3. Hyperaldosteronism with a negative CT study.

The hormonal similarities between APA and GSH are due to the fact that aldosterone secretion in both entities is characterized by exquisite dependency on ACTH. Thus, circadian fluxes in aldosterone levels and an anomalous fall in plasma aldosterone upon standing between 8 A.M. and noon may also be encountered in GSH. It has been proposed that young patients with primary hyperaldosteronism with a positive family history, who demonstrate an anomalous decline in aldosterone with posture and a normal CT, should be placed on dexamethasone for 4–6 weeks before invasive localizational studies are performed.[91] Patients with GSH, when treated with dexamethasone, demonstrate remarkable improvement in the hypertension and hypokalemia, with normalization of plasma aldosterone and PRA levels and even restoration of a normal renin responsiveness to posture. This typical response, when present, not only establishes the diagnosis of GSH, but also prevents performance of an unnecessary invasive localizational procedure. It has been argued that administration of dexamethasone for 4–6 weeks to a patient with CT-negative APA may cause hazardous increases in the blood pressure.[203] While this is indeed a valid concern, careful selection of patients for a 4–6-week course of dexamethasone would minimize, even obviate, undesirable consequences. Of course, close monitoring is mandatory during this period.

Dexamethasone may occasionally alter the aldosterone dynamics of adenomas, leading to confusion in terms of separating this response from that of GSH. The effect of short-term administration of dexamethasone on aldosterone secretion in patients with adenoma has been studied by several investigators. Kem et al.[142] have shown that dexamethasone reduces aldosterone levels in some patients with adenoma. Vetter et al.[143,204] have shown that the secretory spikes in plasma aldosterone can be attentuated, even eliminated, by dexamethasone in patients with APA. The transient nature of these phenomena has been emphasized by Ganguly et al.,[205] who noticed that while the plasma aldosterone level declined initially in patients with adenoma, these levels returned to their original elevated state with continued administration of dexamethasone. Therefore, the trial of dexamethasone for 4–6 weeks can clearly delineate patients with GSH, who would continue to have low plasma aldosterone levels with protracted dexamethasone treatment.

Step 6: Prediction of Surgical Response

Once the diagnosis of APA has been established, the effect of spironolactone therapy provides some index of predictability[206] in terms of a surgical cure. Patients with adenoma usually respond very well to treatment with spironolactone. In fact, the degree of responsiveness is better than in those with hyperplasia. In patients with adenoma showing a satisfactory lowering of blood pressure and normalization of serum K^+ levels, the prospects of reversing the abnormalities of hyperaldosteronism by excision of the adenoma are optimistic and high. Patients with adenoma who fail to respond to a trial of spironolactone would not be expected to respond too well to surgery either. The generally quoted figure for the surgical "cure" rate for adenoma is in the 60–70% range. It is unclear from the literature whether the "nonresponders" to surgery could have been satisfactorily delineated before surgery. Several recent reviews on primary hyperaldosteronism[60,93,103,134] have underplayed the importance of the predictive value of a trial with spironolactone before surgery. Even among those who respond to the drug, there is a population of patients who fail to be "cured" by surgery. Perhaps for this reason, the predictive accuracy of the test may be in question. Regarding the rare patient with adenoma unsatisfactorily responding to spironolactone, opinions vary as to whether or not surgery is recommended.

Special Variants of Hyperaldosteronism

In addition to the three major varieties of hyperaldosteronism—APA, IH, and GSH—there are some rare, but important variants of the disease. These include unilateral hyperplasia, indeterminant hyperaldosteronism, adrenocortical carcinoma with hypersecretion of aldosterone, and ectopic secretion of aldosterone by nonadrenal neoplasms.

Unilateral Hyperplasia

Although adrenal hyperplasia involves both glands, occasionally adrenal hyperplasia can be limited to just one gland. This form of "unilateral hyperplasia" is amenable to surgical cure with unilateral adrenalectomy. Ross[207] described a patient in whom the mineralocorticoid excess was alleviated following excision of a single gland, which histologically revealed hyperplasia. Similarly, Kawasaki et al.[208] described a patient who experienced complete remission following excision of a unilateral hyperplastic gland. The best-studied case is that reported by Ganguly et al.[209] These workers described a 45-year-old man with hypertension, hypokalemia, low PRA, and nonsuppressible aldosterone levels even on a 300-meq sodium intake. The aldosterone secretion in this patient exhibited marked circadian activity, temporarily suppressed with dexamethasone, and was stimulated by exogenous ACTH. All these features were similar to those of an adenoma. The plasma aldosterone response to posture was variable on eight separate occasions. An iodocholesterol scan performed without dexamethasone revealed a right adrenal larger than the left, and aldosterone concentration in the right adrenal vein was 50–150 times greater than in the left. Venography failed to demonstrate a tumor on either side. Based on the catheterization data, the right adrenal gland was excised and found to contain many microscopic subcapsular nests of clear-looking cells. Following surgery blood pressure, hypokalemia, PRA, and aldosterone dynamics were restored to normal. The patient had remained normotensive without any drugs for as long as 3 years after surgery. This rare variant behaves like adenoma in the dynamic tests and fails to visualize as a tumor by adrenal scintigraphy or CT. The diagnostic hallmark is the marked elevation in aldosterone concentration of the adrenal effluent in contrast to the opposite gland. Of course, care must be taken to avoid misinterpretation of data caused by dilution of the venous effluent from the opposite side. Proper diagnosis is essential since the unilateral hyperplasia, unlike bilateral hyperplasia, is amenable to surgical excision.

Indeterminate Hyperaldosteronism[210]

The term *indeterminate hyperaldosteronism* has been used in the past to denote a variant of idiopathic hyperaldosteronism. In retrospect, it may well be that this entity is a variant of essential hypertension. The characteristics of indeterminate hyperaldosteronism include

1. Mild to moderate hypertension
2. Minimal or no potassium depletion
3. Minimal to severe suppression of PRA
4. Elevated aldosterone levels in the urine
5. Suppressibility of the elevated urinary aldosterone level by DOC
6. Little or no response to surgery

There has been no unanimity of opinion as to the place of indeterminate hyperaldosteronism in the spectrum of hypertension. In contrast to patients with other forms of hyperaldosteronism and low-renin hypertension, patients with indeterminate hyperaldosteronism show little or no potassium depletion. In this sense these patients resemble those with the normokalemic varieties of hyperaldosteronism. Yet, the features of indeterminate hyperaldosteronism are closer to essential hypertension than to IH. The dynamics in indeterminate hyperaldosteronism, especially the excellent preservation of suppressibility to DOC, is analogous to that in patients with essential hypertension with low renin. Collins et al.[211] described a group of patients with essential hypertension and low-renin activity who had normal, but disproportionately high and partially suppressible, urinary aldosterone levels. Biglieri et al.[210] have suggested that the condition may be an early phase of APA. Yet, the ease with which aldosterone can be suppressed in this group, coupled with the mildness of the hypertension and lack of potassium depletion, is more reminiscent of an early phase of idiopathic hyperplasia.

Thus, it is unclear whether indeterminate hyperaldosteronism is an early phase in the evolution of APA, a milder version of IH, or a variant of essential hypertension with disproportionate but suppressible hypersecretion of aldosterone. The term *indeterminate* could not be more appropriate while discussing this entity.

Aldosterone-Producing Carcinoma

Adrenocortical carcinoma is a rare cause of hyperaldosteronism. Estimated to represent 1–2% of all cases of hyperaldosteronism, the mineralocorticoid excess that results from hypersecretion by adrenal carcinoma resembles that of adenoma. The rarity of aldosterone-secreting adrenal carcinoma has been emphasized in the literature.[212] As of 1980, only 29 cases of adrenal carcinoma with predominant aldosterone hypersecretion had been described.[213] There appear to be no differences in the age, sex, or clinical presentation between adrenal adenoma and adrenal carcinoma. Thus, the initial presentation is characterized, like APA, by hypertension, hypokalemic alkalosis, and mineralocorticoid excess. The clue that adrenocortical carcinoma underlies the hyperaldosteronism is generally derived by one of four features—the appearance by CT, the concomitant hypersecretion of other hormones, the histopathological evidence of cancer in the excised tissue, and presence of dissemination to the liver and lung.

The hormonal features of aldosterone-secreting carcinoma are several and include the following:

1. The elevated plasma aldosterone levels which show no suppression, as well as the low PRA activity in patients with carcinoma, resemble the hormonal milieu of aldosterone-producing adenoma. There is considerable overlap in the basal plasma aldosterone level in both groups, although in general, plasma aldosterone tends to be higher and hypokalemia more profound in carcinoma.

2. Patients with carcinoma often tend to show elevated basal levels of aldosterone precursors such as DOC and 18-OH-B. While these steroids are also elevated in patients with adenoma, the degree of elevation is higher with carcinoma.

3. Adrenocortical carcinomas that secrete aldosterone fail to show ACTH dependency, a feature that is characteristic of adenoma. Thus, the plasma aldosterone levels in patients with aldosterone-secreting carcinoma often demonstrate lack of circadian periodicity and an absence of response to exogenous ACTH infusion. The ACTH independence of aldosterone secretion by adrenocortical carcinoma was impressively documented in the report of Arteaga et al.[214] These workers performed several circadian studies in three patients with adrenocortical carcinoma and were able to demonstrate a correlation between progression of disease and loss of circadian variation in aldosterone. In one patient, the loss in circadian periodicity of the elevated aldosterone and desoxycorticosterone was observed before surgery and when the disease recurred 8 months later. Cortisol and corticosterone, on the other hand, remained normal and maintained circadian rhythmicity. As the patient's disease progressed there was impressive loss of circadian periodicity in aldosterone, DOC, cortisol, and corticosterone. In addition, the aldosterone secretion by carcinoma is unaffected by the administration of exogenous ACTH. This finding contrasts sharply with the aldosterone secretion by adenomas, where enhanced sensitivity to even small doses of ACTH is usually observed. Finally, when patients with aldosterone-secreting carcinomas assume erect posture between 8 A.M. and noon, there is no anomalous fall in aldosterone level, in contrast to that in aldosterone-producing adenoma.

4. "Pure" mineralocorticoid hypersecretion by adrenocortical carcinoma has been described,[212,215–217] but it is exceedingly rare. Usually there is concomitant hypersecretion of other adrenocortical steroids by the cancer. Even with tumors that secrete only mineralocorticoids, as the cancer progresses, autonomous secretion of cortisol and androgens by the tumor becomes noticeable. The hypersecretion of cortisol may be clinically silent or manifest as Cushing's syn-

drome. The different spectrum of hormone secretion by cancer is consistent with the behavior pattern of adrenal cancer in terms of indiscriminately secreting steroid hormones.[218]

5. Rarely, adenal carcinomas can hypersecrete DOC without excessive aldosterone secretion. While such a phenomenon has been described,[219,220] it must be exceptional. Kelly et al.[221] described a 51-year-old woman with hypertension, hypokalemia, low PRA, normal aldosterone levels, and extremely high 11-DOC due to ineffective 11-β-steroid hydroxylation caused by a large adrenal carcinoma metastatic to the liver. In tissue culture, it was shown that the tumor predominantly produced DOC from tritiated pregnenolone without detectable aldosterone secretion, indicating absence of 11-β-steroid hyroxylation. Isolated DOC hypersecretion is a common observation in patients with ectopic ACTH syndrome.[222] However, isolated DOC hypersecretion is an exceptionally uncommon feature of adrenocortical carcinoma.

In summary, aldosterone hypersecretion by adrenocortical carcinoma is characterized by the following:

1. Its rarity.
2. A close resemblence to APA in its clinical presentation.
3. The underlying tumor is large, often with locally invasive propensities.
4. Usually associated with subtle or overt abnormalities in glucocorticoid and/or androgen hypersecretion.
5. The aldosterone secretion by adrenocortical carcinoma is clearly independent of ACTH and, therefore, demonstrates no circadian periodicity or response to exogenous administration of ACTH.
6. Marked elevation in precursor steroids, such as 11-DOC, may be encountered in adrenal carcinoma. Isolated hypersecretion of 11-DOC (in the absence of aldosterone excess) is exceedingly rare.

Ectopic Secretion of Aldosterone

Ectopic secretion of aldosterone by nonadrenal neoplasms is exceedingly rare. The first description of such an occurrence dates back to 1963, when Ehrlich et al.[223] reported a 9-year-old girl with ovarian tumor presenting with precocious puberty with the metabolic and clinical picture of hyperaldosteronism. The removal of the ovarian tumor resulted in amelioration of symptomatology. When the neoplastic tissue was incubated with C-14-DOC, these authors observed the presence of 11-dehydrocorticosterone and small quantities of a steroid that migrated like aldosterone. One of the well-studied reports of ectopic aldosterone hypersecretion by a malignant ovarian tumor is that reported by Todesco et al.[224] These workers described a 31-year-old woman who presented with hypertension (230/140 mm Hg), hypokalemic alkalosis (serum K^+ of 2.5 meq/liter), weight loss, bilateral pleural effusions, normal renal arteries, and normal radioisotopic renography. Hormonal studies revealed markedly elevated urinary aldosterone levels, marked suppression of PRA, and elevated 17-β-estradiol levels. The poor general condition of the patient with neoplastic cachexia precluded surgery. Medical treatment with spironolactone helped ameliorate the hypokalemia, but not the hypertension. The patient died after a rapidly downhill course. At autopsy, the adrenal glands revealed an atrophic zona glomerulosa with no evidence of tumor or hyperplasia in either gland. A malignant ovarian neoplasm was found, which consisted of solid cellular agglomerates with circumscribed necrotic zones. Tumor extracts revealed a significant aldosterone content.

Thus, malignant ovarian neoplasms represent the commonest example of the uncommon phenomenon of ectopic aldosterone secretion. Rare as it is, the clinical correlate is that negative adrenal localization studies in the presence of hyperaldosteronism may be occasionally encountered when the underlying disorder is ectopic secretion of aldosterone by ovarian tumors. Thus, tomographic studies of the pelvis and hormonal measurement of sex steroids may be indicated to arrive at this rare diagnosis.

Differential Diagnosis

The combination of mild to moderate hypertension with hypokalemia, suppressed PRA, elevated nonsuppressible plasma or urinary aldosterone to salt loading, and/or mineralocorticoid administration is virtually diagnostic of primary hyperaldosteronism. However, several conditions can mimic the individual facets of primary hyperaldosteronism. The conditions that enter the differential diagnosis of primary hyperaldosteronism fall under four categories (Table 31):

1. Use or abuse of substances with mineralocorticoid activity. The use of licorice, carbenexolone, and mineralocorticoid-containing nasal spray or topical ointments belongs in this category.
2. Endogenous mineralocorticoid excess. In this category belong the various blocks in adrenal steroidogenesis, such as 11-β-hydroxylase deficiency, 17-α-hydroxylase deficiency, or 11-β-ketoreductase deficiency. Also included in this category is isolated secretion of DOC by benign (adenoma) or malignant (carcinoma) neoplasms, as well as the mineralocorticoid excess secondary to ectopic ACTH-secreting neoplasms.
3. Secondary hyperaldosteronism. This group includes conventional and classic examples of secondary hyperaldosteronism caused by decreased effective extracellular volume space (congestive failure and other edematous states). Also included in this category are patients with Bartter's syndrome and those with the rare primary renin-secreting tumors of the juxtaglomerular cells.
4. Essential hypertension. This important group includes patients with essential low-renin hypertension, accelerated or malignant hypertension, as well as essential hypertensives rendered hypokalemic with aggressive diuretic therapy.

Use or Abuse of Substances with Mineralocorticoid Activity

In patients presenting with the clinical and biochemical picture of primary hyperaldosteronism, a careful history should be obtained to exclude abuse of substances containing aldosteronelike activity. The important substances are lic-

TABLE 31.
Differential diagnosis of Hyperaldosteronism

Use or abuse of substances with mineralocorticoid activity
  1. Natural licorice
  2. Antacids containing carbenexolone
  3. Nasal sprays or topical ointments containing 9-α-flud-rocortisone

Endogenous mineralocorticoid excess
  1. Adrenogenital syndromes
      a. 11-β-Hydroxylase deficiency
      b. 17-α-Hydroxylase deficiency
      c. 11-β-Ketoreductase deficiency
  2. Exclusive hypersecretion of DOC by adrenal neo-plasms
      a. Adenoma
      b. Carcinoma
  3. Ectopic ACTH secretion

Secondary hyperaldosteronism
  1. In association with edematous states
      a. Nephrotic syndrome
      b. Ascites with cirrhosis
      c. Congestive heart failure
  2. In association with nonedematous, nonhypertensive states
      a. Renal tubular acidosis
      b. Salt-losing nephropathies
      c. Bartter's syndrome
  3. In association with hypertensive states
      a. Renovascular hypertension
      b. Renin-secreting tumors
      c. Essential hypertension

orice, carbenexolone, and nasal sprays or topical ointments containing 9-α-fluroprednisolone.

Natural licorice, which is an extract of glycyrrhiza glabra root, contains glycyrrhizinic acid. This compound has been documented to contain mineralocorticoid activity.[225,226] Cotterill and Cunliffe[227] described an interesting case of an Addisonian patient who successfully treated herself with licorice containing sweets for almost 1 year, in the absence of mineralocorticoid or glucocorticoid replacement. Licorice, and its active derivative glycyrrhizinic acid, have deoxycortonelike activity. Excessive ingestion of licorice and its derivatives may result in a syndrome that mimics primary aldosteronism (pseudoprimary hyperaldosteronism.[228] This syndrome is characterized by hypertension, hypokalemia, with suppressed PRA and aldosterone due to the mineralocorticoid properties of glycyrrhizinic acid.[229,230] Pseudoprimary hyperaldosteronism is rare in the United States, since most commercial products use artificial licorice flavoring. However, natural or true licorice can be found in certain chewing tobaccos. Blachley and Knochel[231] described an 85-year-old man who developed the classic features of mineralocorticoid excess, presenting with hypertension, hypokalemia, renal potassium wasting, metabolic alkalosis, and a suppressed

PRA. The aldosterone concentration in plasma was decreased, thus excluding primary hyperaldosteronism. The patient related a 50-year history of chewing 8–12 3-oz bags of chewing tobacco daily and swallowing the saliva produced. The authors estimated that he had consumed 0.88–1.33 g of glycyrrhizinic acid per day for 50 years, an amount that is capable of causing the syndrome of mineralocorticoid excess while suppressing endogenous renin and aldosterone. The patient rapidly recovered following withdrawal of chewing tobacco. The importance of a careful history to exclude "tobacco chewer's hypokalemia" is illustrated by this case.

Carbenexolone is contained in some antiulcer drugs. The side effects of carbenexolone in terms of causing mineralocorticoid excess have been described by Turpie and Thompson.[232] The underlying mechanisms are unclear, but this matters little since the drug is no longer used in the treatment of peptic ulcer disease.

Mineralocorticoid hypertension resulting from the use of nasal spray containing 9-α-fluroprednisolone is emerging as an important cause of iatrogenic disease. Mantero et al.[233] described in detail 10 patients using nasal sprays containing 9-α-fluroprednisolone for endonasal pathology, such as allergic rhinitis, chronic rhinitis, sinusitis, or prior nasal operations. This nasal spray is apparently used widely in Europe. All patients presented with severe hypertension, hypokalemic alkalosis, and suppressed PRA, mimicking primary hyperaldosteronism. Some had undergone in-depth diagnostic evaluation including numerous invasive studies such as arteriography to exclude mineralocorticoid-secreting tumors. One patient even underwent adrenalectomy with no improvement. The iatrogenic nature of the disorder was evident by low levels of aldosterone in the plasma, along with low to normal levels of precursors such as DOC and 18-OH-DOC. The potent effect of the 9-α-fluro derivatives in causing the syndrome and the long duration taken for recovery of the suppressed aldosterone levels upon withdrawal of the offending nasal spray were illustrated by these workers.[233] Again, the importance of a carefully taken history to exclude factitious disease cannot be overemphasized.

Finally, the development of pseudoprimary aldosteronism from the topical use of 9-α-fluroprednisolone has been described by Montoliu et al.[234] These workers described a 56-year-old man who presented with hypertension, severe hypokalemia (requiring intravenous potassium therapy), suppressed PRA, and normal plasma aldosterone concentrations. The CT of the adrenals was normal, but the [131]iodocholesterol revealed apparent uptake on the right side. The patient underwent a right adrenalectomy; the removed adrenal revealed an acid fast bucillus (AFB) positive caseating granuloma, but no evidence of an adenoma. The discontinuation of a powerful skin cream that the patient had used for 4 years resulted in complete normalization of the serum K^+. The case illustrates the unnecessary procedures that patients with pseudoprimary aldosteronism may undergo, when a proper history is either not obtained or overlooked. Many topical ointments used for dermatological disorders contain glucocorticoids, and occasionally mineralocorticoids. The potential for systemic absorption is obviously greater when applied on inflamed or denuded skin. With continual use for protracted periods of time, the cumulative effect can result in absorption of mineralocorticoid to toxic levels.

Endogenous Mineralocorticoid Excess

In addition to aldosterone secretion by adenoma or hyperplasia of the adrenal cortices, there are several situations characterized by a mineralocorticoid excess.

1. Enzymatic defects in steroidogenesis can result in mineralocorticoid excess. The three representative examples of adrenogenital syndrome characterized by mineralocorticoid excess are 11-β-hydroxylase deficiency[235] 17-α-hydroxylase deficiency,[236] and 11-β-ketoreductase deficiency.[237] The aldosterone levels in these circumstances can be high, normal, or low, depending on the type and severity of the enzymatic block. When the clinical picture is consistent with partial (or complete) adrenogenital syndrome, measurement of adrenal androgens and other precursor steroids (DOC, 18-OH-B, 11-deoxycortisol, and 17-α-OH progesterone) is appropriate in the evaluation of patients presenting with mineralocorticoid excess. The androgenital syndromes are discussed in Chapter 11.
2. The isolated secretion of DOC or corticosterone in the absence of hypersecretion of aldosterone is a rare but important cause of mineralocorticoid excess. Such a syndrome can occur in the absence of a tumor, in association with an adenoma or carcinoma of the adrenal cortex, and in the ectopic ACTH syndrome.

Brown et al.[238] described persistent elevation of plasma DOC in six patients with hypertension, suppressed PRA, and normal plasma aldosterone concentrations. None of these patients suffered from enzymatic defects in steroidogenesis as determined by steroid studies. The authors speculated regarding the existence of a subset of patients with essential hypertension in whom the only mineralocorticoid secreted in excess was DOC.

The isolated secretion of DOC or corticosterone by adrenal tumors is rare. Fraser et al.[239] described a patient with adrenal tumor characterized by only corticosterone hypersecretion. More recently, Kondo et al.[240] described a 35-year-old woman with a benign tumor of the left adrenal that caused hypertension, hypokalemia, suppressed PRA, and low normal aldosterone and cortisol coupled with an extremely high DOC level in the circulation. Iodocholesterol scans localized the tumor to the left adrenal. The excised tumor was a benign adenoma originating from the zona glomerulosa and was rich in DOC. The hypertension and hypokalemia reverted to normal following surgery. Isolated hypersecretion of DOC by *benign* adrenal neoplasms is extremely rare.

Adrenal carcinomas often hypersecrete DOC (and to a lesser extent corticosterone) along with aldosterone.[214] Rarely, however, isolated DOC secretion by adrenocortical carcinoma has been described.[221]

Finally, mineralocorticoid hypersecretion predominantly involving DOC is encountered in the ectopic ACTH syndrome.[222] The frequency and the severity of the hypokalemia encountered in the ectopic ACTH syndrome are partly attributed to mineralocorticoid hypersecretion, particularly of DOC and corticosterone. This contrasts with the mild and infrequent occurrence of hypokalemia in pituitary-dependent hypercortisolism. It is unclear whether the molecular differences between pituitary ACTH and ectopic ACTH determine the hypersecretion of DOC by the adrenal cortices in the ectopic ACTH syndrome.

Secondary Hyperaldosteronism

The common denominator for this group of disorders is hyperreninemia. The conditions that fall into this category are

1. Edematous states, such as ascites, congestive failure, and nephrotic syndrome, which are characterized by a decrease in the effective extracellular volume and thus serve as a potent stimulus for activating the renin–angiotensin–aldosterone axis.
2. Nonedematous states, such as salt-losing nephritis and renal tubular acidosis. Also included in this category are patients with Bartter's syndrome, a disorder of tubular transport that results in stimulating renin release by the juxtaglomerular (JG) apparatus, thus causing secondary hyperaldosteronism.
3. Renovascular hypertension, exemplified in its most dramatic form by renal artery stenosis affecting one or both sides.
4. Primary renin-secreting tumors, where the production of renin, in contrast to the other entities, is autonomous.

Edematous States

The prototypical situations characterized by edema and secondary hyperaldosteronism are nephrotic syndrome, cirrhosis with ascites, and congestive heart failure. All three situations are characterized by reduction in the effective extracellular fluid volume (intravascular volume) despite the presence of edema. In nephrotic syndrome the proteinuria and the resultant hypoalbuminemia lead to a dramatic fall in the oncotic pressure, with consequent movement of extracellular fluid from the intravascular space into the interstial compartment. In cirrhosis with ascites, the combination of increased portal pressure and decreased oncotic pressure (due to the profound hypoalbuminemia) results in translocation of fluid in the peritoneal cavity.

In congestive cardiac failure, the decreased activity of the cardiac pump results in persistence of increased venous blood volume which is not transmitted to the arterial system, with consequent lowering of the intraarterial volume. In all three situations the JG cells sense the decrease in the effective extracellular fluid volume and respond to this physiological stimulus by putting out more renin, which results in increased generation of angiotensin and aldosterone. Unlike nephrotic syndrome and ascites, where continued secretion of aldosterone plays a vital role in perpetuating the edema by progressive increases in sodium balance, aldosterone hypersecretion is believed to have little or no role in maintaining the edema of congestive failure.[241]

The recognition of secondary aldosteronism in edematous states has an important therapeutic impact. The administration of aldosterone antagonists to patients with ascites and nephrotic syndrome results in improvement in the edema as well as in potassium conservation. This is especially relevant in patients with cirrhotic ascites, where the use of diuretic alone to reduce edema will result in profound potassium losses in an already dangerously potassium-depleted setting. The concomitant use of spironolactone would prevent this disastrous occurrence.

Nonedematous, Nonhypertensive States

This category includes patients with renal tubular acidosis salt-losing nephritis and Bartter's syndrome.

1. Renal tubular acidosis can be caused by impaired reabsorption of bicarbonate at the proximal tubule or by impaired acidification of urine at the distal tubule. Both forms are characterized by renal losses of Na^+ and K^+. The constellation of metabolic features in both entities includes hyperreninemia, hyperaldosteronism, and hypokalemia in the presence of metabolic acidosis, a feature that distinguishes renal tubular acidosis from Bartter's syndrome.

2. Salt-losing nephropathies can occur during the course of chronic renal failure. The loss of sodium is a consequence of osmotic diuresis that results from decreased nephron population. The sodium-depleted state serves as a powerful stimulus for activation of the renin–angiotensin–aldosterone system. The condition can be recognized by the presence of underlying renal failure, decreased creatinine clearance, and enormous losses of sodium from the urine. The degree of K^+ losses in the urine is limited owing to the decreased population of the distal tubules and collecting ducts.

3. Bartter's syndrome[242] is a unique example of secondary hyperaldosteronism. The triad of hyperreninemia, hyperaldosteronism, and hypokalemic alkalosis in Bartter's syndrome is further characterized by resistance to the pressor effect of angiotensin. Thus, the clinical hallmark of Bartter's syndrome is normotension despite marked acceleration of the renin–angiotensin–aldosterone axis. While the underlying mechanism for Bartter's syndrome has long been elusive, it now appears that the disorder is a result of impairment in the active chloride transport in the thick ascending limb of the loop of Henle. As a consequence a cascade of events develop, including decreased medullary tonicity and increased distal tubular flow rate, which leads to increased distal tubular secretion of potassium. As a result of hypokalemia, prostaglandin secretion is increased, which sets off activation of the renin–angiotensin–aldosterone system.[243–247]

The mechanism(s), clinical features, hormonal dynamics, and treatment of Bartter's syndrome are discussed in Chapter 7. The salient features of this unique syndrome are outlined in Table 32.

Renovascular Hypertension

Renal artery constriction is a powerful stimulus for the release of renin. The chronic hyperreninemic state results in the development of a sustained increase in angiotensin II and aldosterone levels.

Two types of anatomical lesions are responsible for causing renal artery

TABLE 32.
Features of Bartter's Syndrome

Clinical
  Muscle weakness or cramps
  Polyuria
  Vomiting
  Salt craving
  Failure to thrive
  Tetany
Biochemical
  Hypokalemic alkalosis
  Hypomagnesemia
  Hypocalcemia
  Hyperuricemia
  Glucose intolerance
Endocrine
  Hyperreninemia
  Hyperaldosteronism
  Angiotensin insensitivity
  Vasopressin-resistant diabetes insipidus
Hematological
  Abnormal RBC sodium
  Abnormal platelet aggregation
  Polycythemia

stenosis: fibromuscular dysplastic lesions that can involve single or multiple layers of the renal artery or atheromatous plaques that constrict the renal artery at its origin from the aorta. Dysplastic lesions occur in the younger patient, often in females, and are found in the middle or distal third of the renal artery. This is the variety that appears as a "series of beads" in the arteriogram; it is more amenable to a successful outcome following surgical correction. Atheromatous plaques are more commonly encountered in elderly males and are less amenable to surgical cure.

Pathophysiologically, the decreased blood flow and reduced renal perfusion to the kidney affect unilateral renal function. When the renal function of the affected kidney is compared to that of its normal counterpart by performing split renal function studies (the "Howard test" or the "Stamey test"), distinct differences emerge if marked constriction of the renal artery is present. The affected kidney shows a smaller renal flow, a lower glomerular filtration rate, and a disproportionately large tubular reabsorption of sodium and water. The resultant urine from the affected side, therefore, will be of a smaller volume, with lower concentrations of sodium and higher concentration of nonreabsorbable solutes such as inulin and creatine. Further, when the renal veins are catheterized bilaterally and the renin values in the renal venous effluent are compared, the renin concentrations on the affected side are severalfold elevated in comparison to the normal side. While these studies are clearly indicative of unilateral abnormal hemodynamics, the relationship between hypertension and renal artery stenosis has been a topic of some controversy. It is now apparent that "incidental" renal artery stenosis can be a frequent finding in normotensive individuals. Holley et al.[248] analyzed 256 unselected autopsy subjects and found moderate to severe renal artery stenosis in 49%. This incidence rose to 64% when the analysis was limited to patients 50 years and older. Angiographic evaluation of the renal arteries in normotensive individuals by Eyler et al.[249] and Dustan et al.[250] has revealed the presence of impressive stenotic abnormalities in as many as a third of patients. Thus, the mere angiographic demonstration of renal artery stenosis has little meaning unless hemodynamic and physiological sequelae are demonstrable. The functional impairment demonstrable by an abnormal split renal function study performed by bilateral ureteral catheterization and abnormally high renin levels determined by bilateral renal vein catheterization are useful in predicting the degree of functional impairment caused by the stenosis and, therefore, the expected success rate with surgical relief of the stenosis.

Clinically, the blood pressure elevation in patients with functionally significant renal artery stenosis can resemble benign, labile, accelerated, or even malignant hypertension. Hypokalemia is relatively uncommon. The single, most useful, clinical sign to suspect renal stenosis is the presence of a bruit in the upper part of the abdomen. This sign is present in 50% of patients with renal artery stenosis, even when they are normotensive.[251]

Since patients with renal artery stenosis can be offered a chance for cure of hypertension, it is essential that all hypertensives, especially younger patients, be screened for the presence of this disorder. The most simple and useful screening procedure is the rapid-sequence intravenous pyelogram, which is positive in 85% of instances. The findings of a decrease in the renal size, a delay in the appearance of dye in early films, and hyperconcentration of the dye in the late films is the diagnostic triad that indicates moderate constriction of a magnitude approaching 60% or more. The abnormalities in the renal arterial tree can be demonstrated by renal arteriography, a procedure that yeilds anatomical information in virtually all cases.[252]

The special study that is crucial in terms of the predictive value of surgical correction is evaluation of renin dynamics. The peripheral renin activity in patients with renal artery stenosis may be normal in 50% of patients.[253] Therefore, this test is not useful in the screening or in the diagnosis of renal artery stenosis. On the other hand, the ratio of renin activity in the renal venous effluent of the affected kidney to that in the contralateral normal kidney has significant prognostic value. When an abnormally high ratio is present, the surgical results in terms of a cure or improvement in the hypertension are excellent.[254] A normal ratio is predictive of little or no improvement in the hypertension following surgery. Attempts to enhance the sensitivity of renin data obtained during catheterization after maneuvers such as controlled hypotension,[255] use of hydralazine,[256] or sodium depletion[257] have not met with consistent success. Rarely, normal PRA activity may be encountered in the venous effluent from the affected kidney.[254] The reason for this unexpected phenomenon is not clear. Captopril administration increases the PRA.

In summary, renal artery stenosis, at least in part, causes hypertension by stimulating the renin–angiotensin system. A similar, but more impressive, phenomenon develops in the case of primary renin-secreting JG cell tumors.

Renin-Secreting Tumors

Renin-secreting tumors are rare. The first description of a renin-secreting tumor dates back to 1967, when Robertson et al.[258] described a 16-year-old boy with a 2-year history of hypertension, hypokalemia, and a tumor in the left kidney. The renin content of the excised JG cell tumor was 50- to 100-fold greater than normal. In 1968, Kihara et al.[259] described a 23-year-old woman with recent-onset hypertension, mild hypokalemia, markedly increased urinary aldosterone excretion, elevated PRA activity, and a JG cell tumor in the left kidney. In both cases, excision of tumor was followed by a prompt and dramatic improvement in the clinical and biochemical abnormalities. Two other cases[260,261] were reported in the early seventies, with more complete preoperative studies of PRA and aldosterone measurements. The term *primary hyperreninism* was appropriately coined by Conn et al.[262] to describe the autonomous nature of renin secretion in these cases. These workers[262] reviewed the literature and added to it an exceptionally well-studied case of renin-secreting JG cell tumor. Their case was that of an 18-year-old patient with severe hypertension for 5 years, mild hypokalemia, and markedly increased urinary aldosterone excretion. The "secondary" nature of the hyperaldosteronism was evident when the plasma renin activity was found to be markedly elevated. Bilateral renal arteriograms revealed completely normal calibers of the renal arterial trees on both sides. It was at this point that a renin-secreting tumor was suspected, and bilateral catheterization studies of the renal veins were carried out. The renin activity and angiotensin I level in the right renal vein were double those of the left. Careful arteriographic studies localized the small, 1.5-cm tumor to the lower pole of the right kidney. On surgical exploration the tumor was located at the location identified by arteriography. Histologically the tumor consisted of cells containing granules that stained positive for renin antibodies, and the content of renin in the tumor was 27 times higher when compared to the adjacent kidney tissue.

The clinical dictum illustrated by the aforementioned cases is that the triad of high aldosterone levels, high renin levels, and normal renal arteries in the hypertensive patient should invoke the suspicion of renin-secreting tumors. The autonomous nature of renin secretion by these tumors is illustrated by the absence of changes in the PRA in response to changes in sodium intake. This contrasts sharply with renal artery stenosis, another disorder characterized by hypertension, high renin, and high aldosterone levels. Awareness of this entity has resulted in the recognition of primary reninism as yet another remediable, aldosterone-related cause of hypertension and hypokalemia.[263,264]

Essential Hypertension

Essential hypertension can be associated with abnormal minerlocorticoid dynamics in the following situations:

1. Malignant or accelerated hypertension can be associated with hyper-reninemic hyperaldosteronism.
2. Essential hypertensives on chronic diuretic therapy may demonstrate hyperreninemia and secondary hyperaldosteronism.

3. A subset of patients with essential hypertension may demonstrate suppressed PRA (low-renin hypertension).

Patients with *malignant* or *accelerated* hypertension, especially with renal disease, can develop a state of secondary hyperaldosteronism. The vasculitis (ar-
teriolitis) that complicates the hypertensive process leads to increased renin secretion and secondary aldosteronism. Thus, patients with severe hypertension,
retinopathy, and renal disease pass through a phase characterized by hyperreninemic hyperaldosteronism and hypokalemia. In addition to this group, patients with stable hypertension may have high basal PRA (high-renin essential
hypertension). The significance of renin levels in terms of predisposing to vascular complications of hypertension has been a controversial subject. Brunner et
al.[265] screened 219 patients with otherwise straightforward essential hypertension with "renin profiling" and compared the frequency of vascular complications in the groups with low, normal, and high renin activity. While there was a
comparable frequency of left ventricular hypertrophy in all three groups, the
incidence of cerebrovascular accidents and acute myocardial infarction was
clearly lower in the patients with low plasma renin. The concept that high renin
is a risk factor for the development of vascular complications of hypertension
has been disputed in the literature.[266–269]

Hypertensive patients receiving diuretic therapy can develop the syndrome
of hyperreninemic hyperaldosteronism. The sodium-depleted and volume-con-
tracted setting results in stimulation of the renin–angiotension–aldosterone
axis. It is recommended that diuretic therapy be withheld for at least 3 weeks
prior to evaluation of the renin and aldosterone dynamics in patients with
hypertension.

The subset of patients with low-renin "essential" hypertension has represented an enigma for the past two decades. The triad of features that characterize low-renin hypertension are suppressed PRA, normal aldosterone secretory rate, and exquisite sensitivity to aldosterone antagonist therapy. Several
characteristics of low-renin essential hypertension have been documented in the
literature:

1. The estimated incidence of low-renin hypertension in patients with essential hypertension is approximately 20%.
2. The plasma renin activity in patients with low-renin essential hypertension demonstrates little or no response to provocative maneuvers, such as
   sodium depletion, that normally raise PRA.[270,271]
3. Some patients with low-renin essential hypertension demonstrate an increase in exchangeable sodium, extracellular fluid volume, and plasma
   volume. In this regard, these patients resemble those with true mineralocorticoid excess. Hypokalemia of a mild nature can be encountered in
   patients with low-renin hypertension.
4. In contrast to patients with primary hyperaldosteronism, aldosterone
   secretion is normal in patients with low-renin hypertension.
5. The search for a nonaldosterone mineralocorticoid to explain the genesis
   of low-renin hypertension has not yielded consistent results. Corticosterone and 11-DOC are normal in most patients with low-renin essential hypertension.[272] A small number of patients with low-renin hy-

pertension have been found to have elevated 11-DOC levels in their plasma.[238] Increased plasma levels of 18-OH-DOC have been demonstrated in some patients with low-renin essential hypertension.[273,274] Despite these observations, no mineralocorticoid has been consitently found to be elevated in these patients.

6. Patients with low-renin essential hypertension respond excellently to the administration of spironolactone.[275]

7. The hypertension in patients with suppressed PRA also responds well to the use of the adrenocortical inhibitor aminoglutethimide.[272] The inference is that the hypertension in this subset of patients may be related to mineralocorticoid production by the adrenal cortex.

8. Finally, the excised adrenal glands from patients with low-renin essential hypertension exhibit structural abnormalities including micronodular and macronodular hyperplasia.[276]

Low-renin hypertension continues to remain a disorder in search of a mechanism. The concept that low-renin essential hypertension and hyperaldosteronism caused by bilateral hyperplasia may represent the same disorder in different ends of a spectrum has been postulated and discussed earlier.

Treatment

The therapy for primary hyperaldosteronism is simplistic and straightforward. After having persued a long and winding diagnostic avenue through numerous complicated byways of physiology, hormone dynamics, and radiology, it seems paradoxical that the treatment for hyperaldosteronism should be so simple. When a unilateral adenoma is localized, the therapy of choice is removal of tumor often coupled with unilateral adrenalectomy. When hyperaldosteronism is caused by bilateral hyperplasia ("idiopathic"), medical treatment with spironolactone or amiloride is the option of choice. The therapy for the rare GSH is chronic administration of small doses of dexamethasone. Thus, the three modalities of therapy for hyperaldosteronism, depending on the type, are surgery, spironolactone, and dexamethasone.

Surgery

Excision of the tumor by unilateral adrenalectomy is the treatment for unilateral APA. Preparation for surgery includes a 4–6-week preoperative course of spironolactone for normalization of blood pressure and restoration of normokalemia. The attainment of an excellent response to aldosterone antagonism portends a favorable outcome in terms of a surgical cure. In the immediate postoperative period, a mild transient phase of hypoaldosteronism may be encountered due to suppression of aldosterone secretion by the normal zona glomerulosa. The success rate of curing or improving the hypertension following surgery for APA ranges between 68 and 80%. Recurrence of hyperaldosteronism following unilateral adrenalectomy is unusual. When hypertension develops after years of normotension following surgery, the development should be viewed as a consequence of essential hypertension. Unilateral adrenalectomy is also the treatment of choice for patients with the rare condition of unilateral hyperplasia.

Spironolactone

Treatment for patients with idiopathic hyperaldosteronism is clearly not surgical, owing to the disappointing results of surgery in this group. Mineralocorticoid antagonism with spironolactone is the preferred choice of therapy for idiopathic hyperaldosteronism. Spironolactone is a specific pharmacological antagonist of aldosterone and exerts its action by competitive binding to receptors. These receptors are located at the aldosterone-dependent sodium–potassium exchange sites in the distal convoluted tubule. As a consequence, there is increased excretion of sodium and water, with a simultaneous retention of potassium. Following oral administration, spironolactone is rapidly absorbed, with attainment of maximal plasma levels within 30–60 min. The drug is excreted in the urine and by the biliary tract. The important metabolic products of spironolactone are canrenone and potassium canrenoate.[277,278]

The effectiveness of spironolactone in lowering, even normalizing, blood pressure, as well as in restoring normokalemia, is a matter of record.[279,280] The dose required in patients with hyperplasia varies from patient to patient, ranging between 100 and 200 mg/day in divided doses. Patients with hyperaldosteronism caused by adenoma respond better, and to lower doses of spironolactone, than patients with hyperplasia, who may require doses as high as 400 mg/day. The reason for the discrepancy is not clear; perhaps aldosterone antagonism is exerted best in the presence of higher aldosterone levels.

The limitations of spironolactone relate to its side effects, which are mostly due to the antiandrogenic effects of spironolactone. Decreased libido, impotence, and gynecomastia are the most distressing side effects in males, while menstrual irregularities and painful breast enlargement are noted in females.[281–283]

The antiandrogenic effects of spironolactone are due to its ability to block biosynthesis of testosterone at the 17-hydroxylase step. This enzyme is dependent on cytochrome P-450, which is destroyed by spironolactone. The effects of drug on plasma concentrations of selected adrenal, gonadal, and pituitary hormones in healthy male volunteers were evaluated by Stripp et al.[284] When spironolactone was given in a dose of 400 mg daily for 5 days, significant increases in plasma progesterone and 17-hydroxyprogesterone were noted. No changes in testosterone or estradiol were observed with short-term usage. When the drug was administered on a protracted basis, Taylor et al.[285] reported that therapeutic doses of the drug inhibit 17-hydroxylase and desmolase activity, which promptly reversed upon withdrawal of the drug. The effects on enzymatic blockade, however, could not entirely account for the estrogenic effects of the drug, since no changes in plasma testosterone or estradiol were seen at a time when patients experienced clinical side effects such as gynecomastia. Other mechanisms, therefore, are necessary to explain the antiandrogenic effects of spironolactone. In an interesting in vitro study, Biffignandi et al.[286] demonstrated an increase in free estradiol with a concomitant decrease of free testosterone in plasma from male subjects after incubation of plasma with spironolactone. This finding may, in part, account for the androgen/estrogen imbalance seen in patients who use spironolactone. In addition, spironolactone can cause alterations in sex steroid binding to plasma proteins, resulting in increased conversion of testosterone into estradiol.[287,288]

Finally, a peripheral blockade of androgen receptors by spironolactone has also been implicated in explaining the antiandrogenic effects of the drug.[289] Studies in rats[290] suggest that spironolactone may block the effects of androgen at the level of target tissue. Santen et al.[291] and Kulin et al.[292] have reported that spironolactone inhibits binding of dihydrotestosterone to target cytosol protein. Thus, it appears that the antiandrogenic effect of spironolactone is multifactorial and is exerted at several levels.

Dexamethasone

Small doses of dexamethasone (0.5–1 mg) are extremely effective in normalizing the hyperaldostronism of GSH. Normalization of blood pressure and serum potassium levels, as well as reversal of the abnormal renin and aldosterone dynamics, is observed with chronic administration of dexamethasone to patients with GSH. When and if side effects of chronic glucocorticoid therapy occur, the blood pressure can be controlled with other antihypertensive medications, while hypokalemia can be managed with spironolactone or amiloride.

References

1. Conn JW: Primary aldosteronism, a new clinical syndrome. *J Lab Clin Med* **45:**6, 1955.
2. Davis JO, Binnion PF, Brown TC, et al: Mechanisms involved in the hypersecretion of aldosterone during sodium depletion. *Circ Res* **18/19**(Suppl 1):1143, 1966.
3. Genest J, de Champlain J, Veyrat R, et al: Role of the renin–angiotensin system in various physiological and pathological states. In *Hypertension* XIII. American Heart Association, New York, 1965, pp. 97–116.
4. Laragh JH, Sealey JE, Sommers SC: Patterns of adrenal secretion and urinary excretion of aldosterone and plasma renin activity in normal and hypertensive subjects. *Circ Res* **18/19**(Suppl 1):1158, 1966.
5. Boucher R, Veyrat R, De Champlain J, et al: New procedures measurement of human plasma angiotensin and renin activity levels. *Can Med Assoc J* **90:**194, 1964.
6. Conn JW, Cohen EL, Rovner DR: Suppressed plasma renin activity in primary aldosteronism, distinguishing primary from secondary aldosteronism in hypertensive disease. *JAMA* **190:**213, 1964.
7. Conn JW, Cohen EL, Rovner DR, et al: Normokalemic primary aldosteronism: A detectable cause of curable "essential" hypertension. *JAMA* **193:**200, 1965.
8. Ferris JB, Beevers DG, Brown JJ, et al: Clinical, biochemical and pathological features of low-renin ("primary") hyperaldosteronism. *Am Heart J* **95:**375, 1978.
9. Padfield PL, Davis D, Lever AF, et al: The myth of idiopathic hyperaldosteronism. *Lancet* **2:**83, 1981.
10. Davis JO, Freeman RH: Mechanisms regulating renin release. *Physiol Rev* **56:**1, 1976.
11. Blaine EH, Davis JO, Witty RT: Renin release after hemorrhage and after suprarenal aortic constriction in dogs without sodium delivery to the macula densa. *Circ Res* **27:**1081, 1970.
12. Meyer P, Menard J, Papanicolaou N, et al: Mechanism of renin release following furosemide diuresis in rabbit. *Am J Physiol* **215:**908, 1968.
13. Kotchen TA, Maull KI, Luke RG, et al: Effect of acute and chronic calcium administration on plasma renin. *J Clin Invest* **54:**1279, 1974.
14. Patrono C, Pugliese F: The involvement of arachidonic acid metatolism in the control of renin release. *J Endocrinol Invest* **3:**193, 1980.

15. Oates JA, Whorton AR, Gerkens JF, et al: The participation of prostaglandins in the control of renin release. *Fed Proc* **38:**72, 1979.

16. Poulsen K, Vuust J, Lykkegaard S, et al: Renin is synthesized as a 50000 dalton single-chain polypeptide in cell-free translation systems. *FEBS Lett* **98:**135, 1979.

17. Nielsen AH, Lykkegaard S, Poulsen K: Renin in the mouse submaxillary gland has a molecular weight of 40000. *Biochim Biophys Acta* **576:**305, 1979.

18. Inagami T, Hirose S, Murakami K, et al: Native form of renin in the kidney. *J Biol Chem* **252:**7733, 1977.

19. Leckie BJ, McConnell A: A renin inhibitor from rabbit kidney: Conversion of a large inactive renin to a smaller active enzyme. *Circ Res* **36:**513, 1975.

20. Slater EE, Haber E: Inactive renin—"Through a glass darkly." *N Engl J Med* **301:**429, 1979.

21. Kotchen TA, Guthrie P Jr: Renin–angiotensin–aldosterone and hypertension. *Endocr Rev* **1:**78, 1980.

22. Blair-West JR, Coghlan JP, Denton DA, et al: The control of aldosterone secretion. *Recent Prog Horm Res* **19:**311, 1963.

23. Coghlan JP, Blair-West JR, Denton DA, et al: Control of aldosterone secretion. *J Endocrinol* **81:**55, 1979.

24. Laragh JH, Sealey JE: The renin–angiotensin–aldosterone hormonal system and regulation of sodium, potassium, and blood pressure homeostasis. In Orloff J, Berliner RW (eds): *Handbook of Physiology.* American Physiological Society, Washington, DC, 1973, sect. 8, p. 831.

25. Laragh JH: Vasconstriction–volume analysis for understanding and treating hypertension: The use of renin and aldosterone profiles. *Am J Med* **55:**261, 1973.

26. McKenna TJ, Island DP, Nicholson WP, et al: The effects of potassium on early and late steps in aldosterone biosynthesis in cells of the zona glomerulosa. *Endocrinology* **103:**1411, 1978.

27. McCaa RE, McCaa CS, Guyton AC: Role of angiotensin-II and potassium in the long term regulation of aldosterone secretion in intact conscious dogs. *Circ Res* **36**(Suppl 1):37, 1975.

28. Lieberman AH, Luetscher JA: Some effects of abnormalities of pituitary adrenal or thyroid function on excretion of aldosterone and the response to corticotropin or sodium deprivation. *J Clin Endocrinol Metab* **20:**1004, 1960.

29. Fraser R, Brown JJ, Lever AF, et al: Control of aldosterone secretion. *Clin Sci* **56:**389, 1979.

30. Nicholls MG, Espiner EA, Donald RA: Plasma aldosterone response to low dose ACTH stimulation. *J Clin Endocrinol Metab* **41:**186, 1975.

31. Kem DC, Gomez-Sanchez C, Kramer NJ, et al: Plasma aldosterone and renin activity in response to ACTH infusion in dexamethasone suppressed normal and sodium depleted man. *J Clin Endocrinol Metab* **40:**116, 1975.

32. Griffing GT, Berelowitz B, Hudson M, et al: Plasma immunoreactive gamma melanotropin in patients with idiopathic hyperaldosteronism, aldosterone-producing adenomas, and essential hypertension. *J Clin Invest* **76:**163, 1985.

33. Berelowitz B, Hudson M, Griffing GT, et al: Elevated plasma immunoreactive γ-melanotropin-stimulating hormone (IR-γ-MSH) in idiopathic hyperaldosteronism. *Clin Res* **32:**216A, 1984 (Abstract).

34. Güllner H-G, Nicholson WE, GIll JR Jr, et al: Plasma immunoreactive proopiolipomelanocortin-derived peptides in patients with primary hyperaldosteronism, idiopathic hyperaldosteronism with bilateral adrenal hyperplasia, and dexamethasone-suppressible hyperaldosteronism. *J Clin Endocrinol Metab* **56:**853, 1983.

35. Matsuoka H, Mulrow PJ, Franco-Saenz R: Stimulation of aldosterone production by β-melanotropin. *Nature* **291;**155, 1981.

36. Page RB, Boyd JE, Mulrow PJ: The effect of alpha-melanocyte stimulating hormone on adlosterone production in the rat. *Endocr Res Commun* **1:**53, 1974.

37. Vinson GP, Whitehouse BJ, Dell A, et al: Characterization of an adrenal zona glomerulosa-stimulating component of posterior pituitary extracts as α-MSH. *Nature (Lond)* **284:**464, 1980.

38. Güllner H, Gill JR Jr: Beta-endorphin selectively stimulates aldosterone secretion in hypophysectomized nephrectomized dogs. *J Clin Invest* **71:**124, 1983.

39. Szalag K, Stack E: Effect of beta-endorphin in the steroid production of isolated zona glomerulosa and zona fasciculata cells. *Life Sci* **29:**1355, 1981.

40. Rabinowe SL, Taylor T, Dluhy RG, et al: β-Endorphin stimulates plasma renin and aldosterone release in normal human subjects. *J Clin Endocrinol Metab* **60:**485, 1985.

41. Farrell G: Steroidogenic properties of extracts of beef diencephalon. *Endocrinology* **65**:29, 1959.
42. Sen S, Valenzuela R, Smeby R, et al: Localization, purification, and biological activity of a new aldosterone-stimulating factor. *Hypertension* **3**(3)(Suppl I):81, 1981.
43. Sen S, Shainoff JR, Bravo EL, et al: Isolation of aldosterone-stimulating factor and its effect on rat adrenal glomerulosa cells in vitro. *Hypertension* **3**:4, 1981.
44. Saito I, Bravo EL, Zanella T, et al: Steroidogenic characteristics of a new aldosterone-stimulating factor isolated from normal human urine. *Hypertension* **3**:300, 1981.
45. Sen S, Bumpus FM, Oberfield S, et al: Development and preliminary application of a new assay for aldosterone stimulating factor. *Hypertension* **5**(Suppl 1):27, 1983.
46. Carey RM, Sen S, Dolan LM, et al: Idiopathic hyperaldosteronism. A possible role for aldosterone-stimulating factor. *N Engl J Med* **311**:94, 1984.
47. Norbiato G, Bevilacqua M, Raggi U, et al: Metoclopramide increases plasma aldosterone concentration in man. *J Clin Endocrinol Metab* **45**:1313, 1977.
48. Carey RM, Thorner MO, Ortt EM: Effects of metoclopramide and bromocriptine on the renin–angiotensin–aldosterone system in man. *J Clin Invest* **63**:727, 1979.
49. Noth RH, McCallum RW, Contino C, et al: Tonic dopaminergic suppression of plasma aldosterone. *J Clin Endocrinol Metab* **51**:64, 1980.
50. Bevilacqua M, Norbiato G, Raggi U, et al: Dopaminergic control of serum potassium. *Metabolism* **29**:306, 1980.
51. Mantero F, Opocher G, Boscaro M, et al: Effect of metoclopramide on plasma aldosterone in normal subjects, primary aldosteronism and hypopituitarism. *Horm Metab Res* **13**:464, 1981.
52. Carey RM, Thorner MO, Ortt EM: Dopaminergic inhibition of metoclopramide-induced aldosterone secretion in man. *J Clin Invest* **66**:10, 1980.
53. Sowers JR, Brickman AS, Sowers DK, et al: Dopaminergic modulation of aldosterone secretion in man is unaffected by glucocorticoids and angiotensin blockade. *J Clin Endocrinol Metab* **52**:1078, 1981.
54. Nishida S, Matsuki M, Nagase Y, et al: Adrenocorticotropin-mediated effect of metoclopramide on plasma aldosterone in man. *J Clin Endocrinol Metab* **57**:981, 1983.
55. Sowers JR, Brickman AS, Sowers DK, et al: Dopaminergic modulation of aldosterone secretion in man is unaffected by glucocorticoids and angiotensin blockade. *J Clin Endocrinol Metab* **52**:1078, 1981.
56. Gross MD, Grekin RJ, Gniadek TC, et al: Suppression of aldosterone by cyproheptadine in idiopathic aldosteronism. *N Engl J Med* **305**:181, 1981.
57. Kaplan NM: The steroid content of adrenal adenomas and measurements of aldosterone production in patients with essential hypertension and primary aldosteronism. *J Clin Invest* **45**:728, 1967.
58. Biglieri EG, Schambelan M, Slaton PE, et al: The intercurrent hypertension of primary aldosteronism. *Circ Res* **26/27**(Suppl. 1):195, 1970.
59. Clarke D, Wilkinson R, Johnston IDA, et al: Severe hypertension in primary aldosteronism and good response to surgery. *Lancet* **1**:482, 1979.
60. Weinberger MH, Grim CE, Hollifield JW, et al: Primary aldosteronism: Diagnosis, localization and treatment. *Ann Intern Med* **90**:386, 1979.
61. Baer L, Brunner HR, Buhler F, et al: Pseudo-primary aldosteronism, a variant of low-renin essential hypertension? In Genest J, Loiw E (eds): *Hypertension '72.* Springer-Verlag, Berlin, 1972, p. 459.
62. Neville AM: The nodular adrenal. *Invest Cell Pathol* **1**:99, 1978.
63. Nicholls MG, Espiner EA, Hughes H, et al: Primary aldosteronism: A study in contrasts. *Am J Med* **59**:334, 1975.
64. Davies DL, Beevers DG, Brown JJ, et al: Aldosterone and its stimuli in normal and hypertensive man: Are essential hypertension and primary hyperaldosteronism without tumour the same condition? *J Endocrinol* **81**:79, 1979.
65. Collins RD, Weinberger MH, Dowdy AJ, et al: Abnormally sustained aldosterone secretion during salt loading in patients with various forms of benign hypertension; relation to plasma renin activity. *J Clin Invest* **49**:1415, 1970.
66. Re RN, Sancho J, Kliman B, et al: The characterization of low renin hypertension by plasma activity and plasma aldosterone concentration. *J Clin Endocrinol Metab* **46**:189, 1978.
67. Padfield PL, Beevers DG, Brown JJ, et al: Low renin hypertension: A diagnostic entity at-

tributable to mineralocorticoid excess? In Burley DM (ed): *Hypertension, Its Nature and Treatment*. CIBA, Horsham, England, 1975, pp. 135–146.

68. Padfield PL, Beever DG, Brown JJ, et al: Is low renin hypertension a stage in the development of essential hypertension or a diagnostic entity? *Lancet* **1**:548, 1975.

69. Beevers DG, Morton JJ, Nelson CS, et al: Angiotensin II in essential hypertension. *Br Med J* **1**:415, 1977.

70. Thomas GW, Ledingham JGG, Beilin LJ, et al: Reduced renin activity in essential hypertension. A reappraisal. *Kidney Int* **13**:513, 1978.

71. Lebel M, Schalekamp MA, Beevers DG, et al: Sodium and the renin–angiotensin system in essential hypertension and mineralocorticoid excess. *Lancet* **2**:308, 1974.

72. Baer L, Sommers SC, Krakoff LR, et al: Pseudoprimary aldosteronism: An entity distinct from true primary aldosteronism. *Circ Res* **26/27**(Suppl 1):203, 1970.

73. Ganguly A, Melada GA, Luetscher JA, et al: Control of plasma aldosterone in primary aldosteronism: Distinction between adenoma and hyperplasia. *J Clin Endocrinol Metab* **37**:765, 1973.

74. Schambelan M, Brust NL, Chang BCF, et al: Circadian rhythm and effect of posture on plasma aldosterone concentration in primary aldosteronism. *J Clin Endocrinol Metab* **43**:115, 1976.

75. Spark RF, Dale SL, Kahn PC, et al: Activation of aldosterone secretion in primary aldosteronism. *J Clin Invest* **48**:96, 1969.

76. Wisgerhof M, Carpenter PC, Brown RD: Increased adrenal sensitivity to angiotensin II in idiopathic hyperaldosteronism. *J Clin Endocrinol Metab* **47**:938, 1978.

77. Kisch ES, Dluhy RG, Williams GH: Enhanced aldosterone response to angiotensin II in human hypertension. *Circ Res* **38**:502, 1976.

78. Wisgerhof M, Brown RD: Increased adrenal sensitivity to angiotensin II in low renin hypertension. *J Clin Invest* **64**:1456, 1979.

79. Ferriss JB, Beevers DG, Boddy K, et al: The treatment of low-renin ("primary") hyperaldosteronism. *Am Heart J* **96**:97, 1978.

80. Mill JS: *Analysis of the Phenomenon of the Human Mind*, Vol. 2. London, 1869.

81. Sutherland DJA, Ruse JL, Laidlaw JC: Hypertension, increased aldosterone secretion and low plasma renin activity relieved by dexamethasone. *Can Med Assoc J* **95**:1109, 1966.

82. New MI, Oberfield SE, Levine LS, et al: Demonstration of autosomal dominant transmission and absence of HLA linkage in dexamethasone suppressible hyperaldosteronism. *Lancet* **1**:550, 1980.

83. New MI, Peterson RE: A new form of congenital adrenal hyperplasia. *J Clin Endocrinol Metab* **27**:300, 1967.

84. Oberfield SE, Levine LS, Stoner E, et al: Adrenal glomerulosa function in patients with dexamethasone-suppressible hyperaldosteronism. *J Clin Endocrinol Metab* **53**:158, 1981.

85. New MI, Siegal E, Peterson RE: Dexamethasone-suppressible hyperaldosteronism. *J Clin Endocrinol Metab* **37**:93, 1973.

86. Giebink GS, Gotlin RW, Biglieri EG, et al: A kindred with familial glucocorticoid-suppressible aldosteronism. *J Clin Endocrinol Metab* **36**:715, 1973.

87. Grim CE, Weinberger MH: Familial dexamethasone-suppressible normokalemic hyperaldosteronism. *Pediatrics* **65**:597, 1980.

88. Lee SM, Lightner E, Witte M, et al: Dexamethasone suppressible hyperaldosternism in a child with nephrosclerosis. *Acta Endocrinol* **99**:251, 1982.

89. Oberfield SE, Levine LS, Stoner E, et al: Adrenal glomerulosa and fasciculata response to low dose graded ACTH in DSH. In *Proceedings of 7th International Congress of Endocrinology*, July 1–7, Quebec City, Canada, Abstract, 1942. Excerpta Medica, Amsterdam, 1984.

90. Ganguly A: New insights and questions about glucocorticoid suppressible hyperaldosteronism. *Am J Med* **72**:851, 1982.

91. Ganguly A, Grim CE, Weinberger MH: Anomalous postural aldosterone response in glucocorticoid-suppressible hyperaldosteronism. *N Engl J Med* **305**:991, 1981.

92. Ganguly A, Dowdy AJ, Luetscher JA, et al: Anomalous postural response of plasma aldosterone concentration in patients with aldosterone-producing adrenal adenoma. *J Clin Endocrinol Metab* **36**:401, 1973.

93. Herf SM, Teates DC, Tegtmeyer CJ, et al: Identification and differentiation of surgically correctable hypertension due to primary aldosteronism. *Am J Med* **67**:397, 1979.

94. Wisgerhof M, Brown RD, Hogan MJ, et al: The plasma aldosterone response to angiotensin II infusion in aldosterone-producing adenoma and idiopathic hyperaldosteronism. *J Clin Endocrinol Metab* **52**:195, 1981.

95. Miura K, Yoshinaga K, Goto K, et al: A case of glucocorticoid-responsive hyperaldosteronism. *J Clin Endocrinol Metab* **28**:1807, 1968.

96. Fallo F, Sonino N, Armanini D, et al: A new family with dexamethasone-suppressible hyperaldosteronism: aldosterone unresponsiveness to angiotensin II. *Clin Endocrinol (Oxf.)* **22**:777, 1985.

97. Gill JR Jr, Bartter FC: Overproduction of sodium-retaining steroids by zona glomerulosa is adrenocorticotropin dependent and mediates hypertension in dexamethasone-suppressible aldosteronism. *J Clin Endocrinol Metab* **53**:331, 1981.

98. Mulrow P: Glucocorticoid-suppressible hyperaldosteronism: A clue to the missing hormone? *N Engl J Med* **305**:1012, 1981.

99. Ganguly A, Grim CE, Bergstein J, et al: Genetic and pathophysiologic studies of a new kindred with glucocorticoid-suppressible hyperaldosteronism manifest in three generations. *J Clin Endocrinol Metab* **53**:1040, 1981.

100. Grim CE, Weinberger MH, Anand SK: Dexamethasone-suppressible, normokalemic hyperaldosteronism. In New MI, Levine ES (eds): *Juvenile Hypertension.* Raven Press, New York, 1977, p. 109.

101. Baxter RH, Wang I: Malignant hypertension in a patient with Conn's syndrome. *Scott Med J* **19**:161, 1974.

102. Aloia JF, Beutow G: Malignant hypertension with aldosterone producing adenoma. *Am J Med Sci* **268**:241, 1974.

103. Bravo EL, Tarazi RC, Dustan HP, et al: The changing clinical spectrum of primary aldosteronism. *Am J Med* **74**:641, 1983.

104. Snow MH, Nicol P, Wilkinson R, et al: Normotensive primary aldosteronism. *Br Med J* **1**:1125, 1976.

105. Zipser RD, Speckart PF: "Normotensive" primary aldosteronism. *Ann Intern Med* **88**:655, 1978.

106. Shiroto H, Ando H, Ebitani I, et al: Normotensive primary aldosteronism. *Am J Med* **69**:603, 1980.

107. George JM, Wright L, Bell NH, et al: The syndrome of primary aldosteronism. *Am J Med* **48**:343, 1970.

108. Cain JP, Tuck ML, Williams GH, et al: The regulation of aldosterone secretion in primary aldosteronism. *Am J Med* **53**:627, 1972.

109. Biglieri EG, Schambelan M, Brust N, et al: Plasma aldosterone concentration: Further characterization of aldosterone-producing adenomas. *Circ Res* **34**(Suppl 1):1182, 1974.

110. Hiramatsu K, Yamada T, Yukimura Y, et al: A screening test to identify aldosterone-producing adenoma by measuring plasma renin activity. Results in hypertensive patients. *Arch Intern Med* **141**:1589, 1981.

111. Galen FT, Guyenne TT, Devaux C, et al: Direct radioimmunoassay of human renin. *J Clin Endocrinol Metab* **48**:1041, 1979.

112. Slater EE, Cohn RC, Dzau VJ, et al: Complete purification of human renin. *Circulation* **57/58**(Suppl 2):249, 1978 (Abstract).

113. Sealey JE, Laragh JF: How to do a plasma renin assay. *Cardiovasc Med* **2**:1079, 1977.

114. Delorme A, Guyene PT, Corvol P, et al: Methodologic problems in plasma renin activity measurements. *Am J Med* **61**:725, 1976.

115. Oparil S, Haber E: The renin-angiotensin system. *N Engl J Med* **291**:389, 1974.

116. Heacox R, Harvey AM, Vander AJ: Hepatic inactivation of renin. *Circ Res* **21**:149, 1967.

117. Boyd GW: An inactive higher-molecular-weight renin in normal subjects and hypertensive patients. *Lancet* **1**:215, 1977.

118. Leckie BJ, Brown JJ, Lever AF, et al: Inactive renin in human plasma. *Lancet* **2**:748, 1976.

119. Esler M, Zweifler A, Randall O, et al: The determinants of plasma-renin activity in essential hypertension. *Ann Intern Med* **88**:746, 1978.

120. Brunner HR, Laragh JH, Baer L, et al: Essential hypertension: Renin and aldosterone, heart attack, and stroke. *N Engl J Med* **286**:441, 1972.

121. Molzahn M, Dissmann T, Halim S, et al: Orthostatic changes of haemodynamics, renal func-

tion, plasma catecholamines and plasma renin concentration in normal and hypertensive man. *Clin Sci* **42:**209, 1972.

122. Adlin EV, Marks AD, Channick BJ: Spironolactone and hydrochlorothiazide in essential hypertension. *Arch Intern Med* **130:**855, 1972.

123. Laragh JH, Letcher RL, Pickering TG, et al: Renin profiling for diagnosis and treatment of hypertension. *JAMA* **241:**151, 1979.

124. Woods JW, Pittman AW, Pulliam CC, et al: Renin profiling in hypertension and its use in treatment with propranolol and chlorthalidone. *N Engl J Med* **294:**1137, 1976.

125. Case DB, Wallace JM, Keim HJ, et al: Possible role of renin in hypertension as suggested by renin-sodium profiling and inhibition of converting enzyme. *N Engl J Med* **296:**641, 1977.

126. Thurston H, Bing RF, Pohl JEF, et al: Renin subgroups in essential hypertension: an analysis and critique. *Q J Med* **47:**325, 1978.

127. Weinberger MH: Primary aldosteronism. In Genest J, Kuchel O Hamet P (eds): *Hypertension: Physiopathology and Treatment,* 2nd ed. McGraw-Hill, New York, 1983, pp. 922–947.

128. Vaughan NJA, Slater JD, Lightman SL, et al: The diagnosis of primary hyperaldosteronism. *Lancet* **1:**120, 1981.

129. Streeten DH, Tomycz N, Anderson GH: Reliability of screening methods for the diagnosis of primary aldosteronism. *Am J Med* **67:**403, 1979.

130. Streeten DHP, Anderson GH Jr: Simplified screening procedures for primary aldosteronism. Studies on the mechanism of the hyper-responsiveness to furosemide and standing. *Clin Exp Hypertension Theory Pract* **A4**(9,10):1663, 1982.

131. Lyons DF, Kem DC, Brown RD, et al: Single dose captopril as a diagnostic test for primary aldosteronism. *J Clin Endocrinol Metab* **57:**892, 1983.

132. Kem DC, Weinberger MH, Mayes DM, et al: Saline suppression of plasma aldosterone in hypertension. *Arch Intern Med* **128:**380, 1971.

133. Grim CE, Weinberger MH, Higgins JT, et al: Diagnosis of secondary forms of hypertension: A comprehensive protocol. *JAMA* **237:**1331, 1977.

134. Ganguly A, Grim CE, Weinberger MH: Primary aldosteronism. The etiologic spectrum of disorders and their clinical differentiation. *Arch Intern Med* **142:**813, 1982.

135. Slaton PE Jr, Schambelan M, Biglieri EG: Stimulation and suppression of aldosterone secretion in patients with an aldosterone-producing adenoma. *J Clin Endocrinol Metab* **29:**239, 1969.

136. Horton R: Stimulation and suppression of aldosterone in plasma of normal men and in primary aldosteronism. *J Clin Invest* **48:**1230, 1969.

137. Biglieri EG, Slaton PE Jr, Kronfield SJ, et al: Diagnosis of an aldosterone-producing adenoma in primary aldosteronism. An evaluative manuever. *JAMA* **201:**510, 1967.

138. Biglieri EG, Stockigt JR, Schambelan M: A preliminary evaluation for primary aldosteronism. *Arch Intern Med* **126:**1004, 1970.

139. Vetter H, Siebenschein R, Studer A, et al: Primary aldosteronism: Inability to differentiate unilateral from bilateral adrenal lesions by various routine clinical and laboratory data and by peripheral plasma aldosterone. *Acta Endocrinol (Copenh)* **89:**710, 1978.

140. Lund JO, Kamkjaer-Nielson M, Giese J: Prevalence of primary aldosteronism. *Acta Med Scand* **646**(Suppl):54, 1981.

141. Kem DC, Weinberger MH, Gomez-Sanchez C, et al: Circadian rhythm of plasma aldosterone concentration in patients with primary aldosteronism. *J Clin Invest* **42:**2272, 1973.

142. Kem DC, Weinberger MH, Gomez-Sanchez C, et al: The role of ACTH on the episodic release of aldosterone in patients with idiopathic adrenal hyperplasia, hypertension and hyperaldosteronism. *J Lab Clin Med* **88:**261, 1970.

143. Vetter H, Berger M, Armbruster H, et al: Episodic secretion of aldosterone in primary aldosteronism: Relationship to cortisol. *Clin Endocrinol* **3:**41, 1974.

144. Kem DC, Weinberger MH, Higgins JR, et al: Plasma aldosterone response to ACTH in primary aldosteronism and in patients with low renin hypertension. *J Clin Endocrinol Metab* **46:**552, 1978.

145. Guthrie GP Jr: Multiple plasma steroid responses to graded ACTH infusions in patients with primary aldosteronism. *J Lab Clin Med* **98:**364, 1981.

146. Louis LH, Conn JW: Primary aldosteornism: content of adrenocortical steroids in adrenal tissue. *Rec Prog Horm Res* **17:**415, 1961.

147. Biglieri EG, Hane S, Slaton PE Jr: In vivo and in vitro studies of adrenal secretions in Cushing's syndrome and primary aldosteronism. *J Clin Invest* **42**:516, 1963.

148. Dahl V, Scattini CM, Lantos CP: Comparative biosynthetic studies in a case of primary aldosteronism. *J Steroid Biochem* **7**:715, 1976.

149. Hollifield JW, Slaton PE Jr, Winn S, et al: Altered adrenal sensitivity in primary aldosteornism (PA), essential hypertension (EHT) and renovascular hypertension (RVH). *Clin Res* **26**:23A, 1978 (Abstract).

150. Carey RM, Ayers CR, Vaughan ED Jr: Activity of [des-aspartyl] angiotensin II in primary aldosteronism. *J Clin Invest* **63**:718, 1979.

151. Carey RM, Vaughan ED Jr, Peach MJ, et al: Activity of [des-aspartyl]-angiotensin II and angiotensin II in man. *J Clin Invest* **61**:20, 1978.

152. Johnson JA, Davis JO: Effects of a specific competitive antagonist of angiotensin II on arterial pressure and adrenal steroid secretion in dogs. *Circ Res* **32/33**(Suppl 1):1–159, 1973.

153. Hollenberg NK, Williams GH, Burger B, et al: Blockade and stimulation of renal, adrenal, and vascular angiotensin II receptors with I-Sar, 8-Ala angiotensin II in normal man. *J Clin Invest* **57**:39, 1976.

154. Brown RD, Tucker R, Tue K, et al: Effect of saralasin on plasma aldosterone in hypertensive man. *J Lab Clin Med* **91**:473, 1978.

155. Brown RD, Kem DC, Hogan MJ, et al: Evaluation of a test using saralasin to differentiate primary aldosteronism due to an aldosterone-producing adenoma from idiopathic hyperaldosteronism. *Metabolism* **33**:734, 1984.

156. Krieger DT, Amorosa L, Linick F; Cyproheptadine-induced remission of Cushing's disease. *N Engl J Med* **293**:893, 1975.

157. Bing RF, Schulster D: Steroidogenesis in isolated rat adrenal glomerulosa cells: response to physiological concentrations of angiotensin II and effects of potassium, serotonin and [Sar^1,Ala^8]-angiotensin II. *J Endocrinol* **74**:261, 1977.

158. Lamberts SWJ, de Lange SA, Stefanko SZ: Adrenocorticotropin-secreting pituitary adenomas originate from the anterior or the intermediate lobe in Cushing's disease: Differences in the regulation of hormone secretion. *J Clin Endocrinol Metab***286**:91, 1982.

159. Franco-Saenz R, Mulrow PJ, Kim K: Idiopathic aldosteronism: Possible disease of the intermediate lobe of the pituitary. *JAMA* **251**:2555, 1984.

160. Ganguly A, Pratt JH, Weinberger MH, et al: Differing effects of metoclopramide and adrenocorticotropin on plasma aldosterone levels in glucocorticoid-suppressible hyperaldosteronism and other forms of hyperaldosteronism. *J Clin Endocrinol Metab* **57**:388, 1983.

161. Brown RD, Hegstad R, Hogan MJ: Effect of metoclopramide, a dopamine antagonist, on aldosterone in primary aldosteronism. *Clin Res* **27**:678A, 1979 (Abstract).

162. Gniadek TC, Grekin RJ, Gross MD, et al: Hyperresponsiveness of aldosterone to metoclopramide in aldosteronism. *Clin Endocrinol (Oxf)* **16**:475, 1982.

163. Fraser R, Lantos CP: 18-Hydroxycorticosterone: A review. *Steroid Biochem* 9:273, 1978.

164. Ulick S: Diagnosis and nomenclature of the disorders of the terminal portion of the aldosterone biosynthetic pathway. *J Clin Endocrinol Metab* **43**:92, 1976.

165. Biglieri EG, Slaton PE, Schambelan M, et al: Hypermineralocorticoidism. *Am J Med* **45**:170, 1968.

166. Kem DC, Brown RD, Painton RP, et al: Differentiation of mineralocorticoid-induced hypertension. In Kaufmann W, Wamback G, Helber A, Meurer KA (eds): *Mineralocorticoids and Hypertension.* Springer-Verlag, Berlin, Heidelberg, 1983, p. 155.

167. Biglieri EG, Schambelan M: The significance of elevated levels of plasma 18-hydroxycorticosterone in patients with primary aldosteronism. *J Clin Endocrinol Metab* **49**:87, 1979.

168. Kem DC, Tang K, Hanson CS, et al: The prediction of anatomical morphology of primary aldosteronism using serum 18-hydroxycorticosterone levels. *J Clin Endocrinol Metab* **60**:67, 1985.

169. Baumann K, Müller J: Effect of potassium intake on the final steps of aldosterone biosynthesis in the rat. *Acta Endocrinol (Kbh)* **69**:701, 1972.

170. Espiner EA, Donald RA: Aldosterone regulation in primary aldosteronism: Influence of salt balance, posture and ACTH. *Clin Endocrinol* **12**:277, 1980.

171. Bayliss RIS, Edwards OM, Storer F: Complications of adrenal venography. *Br J Radiol* **43**:531, 1970.

172. Eagan RT, Page MI: Adrenal insufficiency following bilateral adrenal venography. *JAMA* **215:**115, 1971.

173. Bravo E: Primary aldosteronism: remission and development of adrenal insufficiency after adrenal venography. *Ann Intern Med* **85:**207, 1976.

174. Fisher EE, Turner FA, Horton R: Remission of primary hyperaldosteronism after adrenal venography. *N Engl J Med* **285:**334, 1971.

175. White EA, Schambelan M, Rost CR, et al: Use of computed tomography in diagnosing the cause of primary aldosteronism. *N Engl J Med* **303:**1503, 1980.

176. Dunnich NR, Scharer EG, Doppman JL, et al: Computed tomography in adrenal tumors. *Am J Roentgenol* **132:**43, 1979.

177. Prosser PR, Sutherland CM, Scullin DR: Localization of adrenal aldosterone adenoma by computerized tomography. *N Engl J Med* **300:**1278, 1979.

178. Linde R, Coulam C, Battino R, et al: Localization of aldosterone-producing adenoma by computed tomography. *J Clin Endocrinol Metab* **49:**642, 1979.

179. Smith DC, Mackett MCT, Billmoria PE: Computed tomography to determine the cause of primary aldosteronism. *N Engl J Med* **304:**1016, 1981.

180. Conn JW, Morita R, Cohen EL, et al: Primary aldosteronism. Photoscanning of tumors after administration of [131]I-19-idocholesterol. *Arch Intern Med* **129:**417, 1971.

181. Hogan MJ, McRae J, Schambelan M, et al: Location of aldosterone-producing adenomas with [131]I-19-Iodocholesterol. *N Engl J Med* **294:**410, 1976.

182. Lieberman LM, Beierwaltes WH, Conn JW, et al: Diagnosis of adrenal disease by visualization of human adrenal glands with [131]I-19-idocholesterol. *N Engl J Med* **285:**1387, 1971.

183. Seabold JE, Cohen EL, Beierwaltes WH, et al: Adrenal imaging with [131]I-19-iodocholesterol in the diagnostic evaluation of patients with aldosteronism. *J Clin Endocrinol Metab* **42:**41, 1976.

184. Conn JW, Cohen EL, Herwig KR: The dexamethasone-modified adrenal scintiscan in hyporeninemic aldosteronism (tumor versus hyperplasia). A comparison with adrenal venography and adrenal venous aldosterone. *J Lab Clin Med* **88:**841, 1976.

185. Freitas JE, Grekin RJ, Thrall JH, et al: Adrenal imaging with iodomethylnorcholesterol (I-131) in primary aldosteronism. *J Nucl Med* **20:**7, 1979.

186. Gross MD, Shapiro B, Grekin RJ, et al: Scintigraphic localization of adrenal lesions in primary aldosteronism. *Am J Med* **77:**830, 1984.

187. Miles JM, Wahner HW, Carpenter PC, et al: Adrenal scintiscanning with NP-59: A new radioiodinated cholesterol agent. *Mayo Clin Proc* **54:**321, 1979.

188. Carey JE, Thrall JH, Freitas JE, et al: Absorbed dose to the human adrenals from idomethylnorcholesterol (I-131) "NP-59." *J Nucl Med* **20:**60, 1979.

189. Gross MD, Freitas JE, Grekin RJ: Computed tomography to determine the cause of primary aldosteronism. *N Engl J Med* **304:**1016, 1981.

190. Scoggins BA, Oddis CJ, Hare WSC, et al: Preoperative lateralization of aldosterone producing tumors in primary aldosteronism. *Ann Intern Med* **76:**891, 1972.

191. Horton R, Finck E: Diagnosis and localization in primary aldosteronism. *Ann Intern Med* **76:**885, 1972.

192. Dunnick NR, Doppman JL, Mills SR, et al: Preoperative diagnosis and localization of aldosteronomas by measurement of corticosteroids in adrenal venous blood. *Radiology* **1:**331, 1979.

193. Yune HY, Klatte EC, Grim CE, et al: Radiology in primary hyperaldosteronism. *J Roentgenol* **127:**761, 1976.

194. Spark RF, Kettyle WR, Einsenberg H: Cortisol dynamics in the adrenal venous effluent. *J Clin Endocrinol Metab* **39:**305, 1974.

195. Levinson PD, Zadik Z, Hamilton BPM, et al: Adrenal vein epinephrine levels: A useful aid in venous sampling for primary aldosteronism. *Ann Intern Med* **97:**690, 1982.

196. Cryer PE: Isotope-derivative measurements of plasma norepinephrine and epinephrine in man. *Diabetes* **25:**1071, 1976.

197. Sarr N, Jackson R, Bachmann A, et al: Effect of sampling site and conditions on plasma levels of noradrenaline, adrenaline and dopamine. *Prog Biochem Pharmacol* **17:**90, 1980.

198. Brown MJ, Jenner DA, Allison DJ, et al: Increased sensitivity and accuracy of phaeochromocytoma diagnosis achieved by use of plasma-adrenaline estimations and a pentolinium-suppression test. *Lancet* **1:**174, 1981.

199. Melby JC: Identifying the adrenal lesion in primary aldosteronism (editorial). *Ann Intern Med* **76:**1039, 1972.

200. Kahn PC, Kelleher MD, Egdahl RH, et al: Adrenal arteriography and venography in primary aldosteronism. *Radiology* **101:**71, 1971.

201. Nicolis GL, Mitty HA, Modlinger RS, et al: Percutaneous adrenal venography. A clinical study of 50 patients. *Ann Intern Med* **76:**899, 1972.

202. Luetscher JA, Ganguly A, Melada GA, et al: Preoperative differentiation of adrenal adenoma from idiopathic adrenal hyperplasia in primary aldosteronism. *Circ Res* **34**(Suppl 1):175, 1974.

203. Hoefnagels WHL, Kloppenborg PWC: Hazards of long-term dexamethasone treatment in primary aldosteronism. *N Engl J Med* **306:**427, 1982.

204. Vetter H, Vetter W: Regulation of aldosterone secretion in primary aldosteronism. *Horm Metab Res* **7:**418, 1975.

205. Ganguly A, Chavarri M, Luetscher JA, et al: Transient fall and subsequent return of high aldosterone secretion by adrenal adenoma during continued dexamethasone administration. *J Clin Endocr* **44:**775, 1977.

206. Spark RF, Melby JC: Aldosterone in hypertension: The spironolactone response test. *Ann Intern Med* **69:**685, 1968.

207. Ross EJ: Conn's syndrome due to adrenal hyperplasia with hypertrophy of zona glomerulosa, relieved by unilateral adrenalectomy. *Am J Med* **39:**994, 1965.

208. Kawasaki T, Omae T, Tanaka K, et al: Remission of recurrent hyperaldosteronism resulting from subtotal adrenalectomy of adenomatous hyperplastic adrenal glands. *J Clin Endocrinol Metab* **33:**474, 1971.

209. Ganguly A, Zager PG, Luetscher JA: Primary aldosteronism due to unilateral adrenal hyperplasia. *J Clin Endocrinol Metab* **51:**1190, 1980.

210. Biglieri EG, Stockigt JR, Schambelan M: Adrenal mineralocorticoids causing hypertension. *Am J Med* **52:**623, 1972.

211. Collins RD, Weinberger MH, Dowdy AJ, et al: Abnormally sustained aldosterone secretion during salt loading in patients with various forms of benign hypertension: relation to plasma renin activity. *J Clin Invest* **45:**1415, 1970.

212. Salassa TM, Weeks RE, Northcutt RC, et al: Primary aldosteronism and malignant adrenocortical neoplasia. *Trans Am Clin Climatol Assoc* **86:**163, 1975.

213. Neville AM, Ohare MJ: The human adrenal cortex. In *Pathology and Biology, and Integrated Approach.* Springer-Verlag, Heidelberg, 1982, p. 215.

214. Arteaga E, Biglieri EG, Kater CE, et al: Aldosterone-producing adrenocortical carcinoma. Preoperative recognition and course in three cases. *Ann Intern Med* **101:**216, 1984.

215. Richie JP, Gittes RF: Carcinoma of the adrenal cortex. *Cancer* **45:**1957, 1980.

216. Alterman SL, Dominguez C, Lopez-Gomez A, et al: Primary adrenocortical carcinoma causing aldosteronism. *Cancer* **24:**602, 1969.

217. Foye LV Jr, Feichtmeir TV: Adrenal cortical carcinoma producing solely mineralocorticoid effect. *Am J Med* **19:**966, 1955.

218. Hutter AM, Kayhoe DE: Adrenal cortical carcinoma. Clinical features of 138 patients. *Am J Med* **41:**572, 1966.

219. Powell-Jackson JD, Calin A, Fraser R, et al: Excess deoxycorticosterone secretion from adrenocortical carcinoma. *Br Med J* **2:**32, 1974.

220. Kahn M, Melby JC, Jacobs DR: Isolated DOC excess: A unique syndrome simulating hyperaldosteronism with marked fluid retention. *Clin Res* **14:**282, 1966.

221. Kelly WF, O'Hare MJ, Loizou S, et al: Hypermineralocorticism without excessive aldosterone secretion: An adrenal carcinoma producing deoxycorticosterone. *Clin Endocrinol* **17:**353, 1982.

222. Brown JJ, Strott CA: Plasma deoxycorticosterone in man. *J Clin Endocrinol* **36:**44S, 1973.

223. Ehrlich EN, Dominguez OV, Samuel LT, et al: Aldosteronism and precocious puberty due to an ovarian androblastoma (Sertoli cell tumor). *J Clin Endocrinol Metab* **23:**358, 1963.

224. Todesco S, Terribile V, Borsatti A, et al: Primary aldosteronism due to a malignant ovarian tumor. *J Clin Endocrinol Metab* **41:**809, 1975.

225. Molhuysen JA, Gerbrandy J, de Vries LA, et al: A liquorice extract with desoxycortone-like action. *Lancet* **2:**381, 1950.

226. Louis LH, Conn JW: Preparation of glycyrrhizinic acid, the electrolyte-active principle of

licorice: Its effect upon metabolism and upon pituitary–adrenal function in man. *J Lab Clin Med* **47**:20, 1956.

227. Cotterill JA, Cunliffe WJ: Self-medication with liquorice in a patient with Addison's disease. *Lancet* **1**:294, 1973.

228. Conn JW, Rovner DR, Cohen EL: Licorice-induced pseudoaldosteronism. *JAMA* **205**:492, 1968.

229. Liddle GW, Melmon KL: The adrenals. In Williams RH (ed): *Textbook of Endocrinology*, 5th ed. Saunders, Philadelphia, 1974, p. 274.

230. Epstein MT, Espiner EA, Donald RA, et al: Effect of eating liquorice on the renin–angiotensin aldosterone axis in normal subjects. *Br Med J* **1**:488, 1977.

231. Blachley JD, Knochel JP: Tobacco chewer's hypokalemia: Licorice revisited. *N Engl J Med* **302**:784, 1980.

232. Turpie AG, Thompson TJ: Carbenexolone sodium in the treatment of gastric ulcer with special reference to side effects. *Gut* **6**:591, 1956.

233. Mantero F, Armanini D, Opocher G, et al: Mineralocorticoid-hypertension due to a nasal spray containing 9α-fluoroprednisolone. *Am J Med* **71**:352, 1981.

234. Montoliu J, Botey A, Trilla A, et al: Pseudoprimary aldostetonism from the topical application of 9-α-fluorprednisolone to the skin. *Clin Nephrol* **22**:262, 1984.

235. Baulieu EE, Phillon F, Migeon CJ. In Einsestein AB (ed): *The Adrenal Cortex*. J. & A. Churchill, London, 1967, p. 553.

236. Biglieri EG, Herron MA, Brust M: 17-Hydroxylation deficiency in man. *J Clin Invest* **45**:1945, 1966.

237. New MI, Bradlow L, Fishman J, et al: Deficiency of cortisol 11β-reduclase. A new metabolic defect. *Pediatr Res* **12**:416, 1976.

238. Brown JJ, Ferriss JB, Fraser R, et al: Apparently isolated excess deoxycorticosterone in hypertension: A variant of the mineralocorticoid-excess syndrome. *Lancet* **2**:243, 1972.

239. Fraser R, James WTH, Landon J, et al: Clinical and biochemical studies of a patient with a corticosterone-secreting tumor. *Lancet* **2**:1116, 1968.

240. Kondo K, Saruta T, Saito I, et al: Benign desoxycorticosterone-producing adrenal tumor. *JAMA* **236**:1042, 1976.

241. Sanders LL, Melby JC: Aldosterone and edema of congestive heart failure. *Arch Intern Med* **113**:331, 1964.

242. Bartter FC, Pronove P, Gill JR Jr, et al: Hyperplasia of the juxtaglomerular complex with hyperaldosteronism and hypokalemic alkalosis. *Am J Med* **33**:811, 1962.

243. White MG: Bartter's syndrome. *Arch Intern Med* **129**:41, 1972.

244. Goodman AD, Vagnucci AH, Hartroft PM: Pathogenesis of Bartter's syndrome. *N Engl J Med* **281**:1435, 1969.

245. Kurtzman NA, Gutierrez LF: The pathophysiology of Bartter syndrome. *JAMA* **234**:758, 1975.

246. White MG: Bartter's syndrome: A manifestation of renal tubular defects. *Arch Intern Med* **129**:41, 1972.

247. Brackett NC Jr, Koppel M, Randall RE Jr, et al: Hyperplasia of the juxtaglomerular complex with secondary aldosteronism without hypertension (Bartter's syndrome). *Am J Med* **44**:803, 1968.

248. Holley KE, Hunt JC, Brown AL, et al: Renal artery stenosis: A clinical pathologic study in normotensive and hypertensive patients. *Am J Med* **37**:14, 1964.

249. Eyler WR, Clark MD, Garman JE, et al: Angiography of the renal areas including a comparative study of renal arterial stenosis in patients with and without hypertension. *Radiology* **78**:879, 1962.

250. Dustan HP, Humphries AW, deWolfe VG, et al: Normal arterial pressure in patients with renal arterial stenosis. *JAMA* **187**:1028, 1964.

251. Bourgoignie J, Shieber W, Sunshine H, et al: Renovascular hypertension. *Arch Intern Med* **131**:596, 1973.

252. Sutton D, Brunton FJ, Starer F: Renal artery stenosis. *Clin Radiol* **12**:80, 1961.

253. Brown JJ, Davies DL, Lever AF, et al: Plasma renin concentration in human hypertension: II. Renin relation to etiology. *Br Med J* **2**:1215, 1965.

254. Bourgoignie J, Kurz S, Catanzaro FJ, et al: Renal venous renin in hypertension. *Am J Med* **48**:332, 1970.

255. Kaneko J, Ikeda T, Takeda T, et al: Renin release during acute reduction of arterial pressure in normotensive subjects and patients with renovascular hypertension. *J Clin Invest* **46:**705, 1967.
256. Ueda H, Yagi S, Kaneko Y: Hydralazine and plasma renin activity. *Arch Intern Med* **122:**387, 1968.
257. Hunt JC, Strong CG, Sheps SG, et al: Diagnosis and management of renovascular hypertension. *Am J Cardiol* **23:**434, 1969.
258. Robertson PW, Klidjian A, Harding DK, et al: Hypertension due to a renin-secreting renal tumour. *Am J Med* **43:**963, 1967.
259. Kihara I, Kitamura S, Hoshino T, et al: A hitherto unreported vascular tumor of the kidney: A proposal of "juxtaglomerular cell tumor." *Acta Pathol Jpn* **18:**197, 1968.
260. Eddy RL, Sanchez SA: Renin-secreting renal neoplasm and hypertension with hypokalemia. *Ann Intern Med* **75:**725, 1971.
261. Bonnin JM, Hodge RL, Lumbers ER: A renin-secreting renal tumor associated with hypertension. *Aust NZ J Med* **2:**178, 1972.
262. Conn JW, Cohen EL, Lucas CP, et al: Primary reninism: Hypertension, hyperreninemia, and secondary aldosteronism due to renin-producing juxtaglomerular cell tumors. *Arch Intern Med* **130:**682, 1972.
263. Schambelan M, Howes EL Jr, Stockigt JR, et al: Role of renin and aldosterone in hypertension due to a renin-secreting tumor. *Am J Med* **55:**86, 1973.
264. Lee MR: Renin-secreting kidney tumours. A rare but remediable cause of serious hypertension. *Lancet* **2:**254, 1971.
265. Brunner HR, Laragh JH, Baer L, et al: Essention hypertension: Renin and aldosterone, heart attack and stroke. *N Engl J Med* **286:**441, 1972.
266. Spark RF: Low renin hypertension and the adrenal cortex. *N Engl J Med* **287:**343, 1972.
267. Mroczek WJ, Finnerty FA, Catt, KJ: Lack of association between plasma–renin and history of heart-attack or stroke in patients with essential hypertension. *Lancet* **2:**464, 1973.
268. Christlieb AR, Gleason RE, Hickler RB, et al: Renin: A risk factor for cardiovascular disease? *Ann Intern Med* **81:**7, 1974.
269. Amery A, Stroobandt R, Fagard R: Prognosis in low renin hypertension. *N Engl J Med* **288:**267, 1973.
270. Kuchel O, Fishman LN, Liddle GW, et al: Effect of diazoxide on plasma renin activity in hypertensive patients. *Ann Intern Med* **67:**791, 1967.
271. Helmer OM, Judson WE: Metabolic studies on hypertension patients with suppressed plasma renin activity not due to hyperaldosteronism. *Circulation* **38:**965, 1968.
272. Woods JW, Liddle GW, Michelakis AM, et al: Effect of an adrenal inhibitor in hypertensive patients with suppressed renin. *Arch Intern Med* **123:**366, 1969.
273. Melby JC, Dole SL, Wilson TE: 18-Hydroxydeoxycorticosterone in human hypertension. *Circ Res* **28**(Suppl 2):143, 1971.
274. Genest J, Nowaczynski W. Kuchel O, et al: Plasma progesterone levels and 18-OH-DOC secretion rate in benign essential hypertension in humans. In Genest J, Koiw E (eds), *Hypertension.* Springer-Verlag, New York, 1972, pp. 293–298.
275. Carey RM, Douglas JG, Schweikert JR, et al: The syndrome of essential hypertension and suppressed plasma renin activity. *Arch Intern Med* **130:**849, 1972.
276. Grim CE, Keitzer WF, Esterly JA, et al: Adrenalectomy in low-renin "essential" hypertension. *Clin Res* **22:**340A, 1974.
277. Sadée W, Reigelman S, Jones SC: Plasma levels of spironolactone in dogs. *J Pharm Sci* **61;**1129, 1972.
278. Sadée W, Abshagen U, Finn C, et al: Conversion of spironolactone to canrenone and disposition kinetics of spironolactone and canrenoate potassium in rats. *Arch Pharmacol* **283:**303, 1974.
279. Mantero F, Armanini D, Urbani S: Antihypertensive effect of spironolactone in essential, renal and mineralocorticoid hypertension. *Clin Sci Mol Med* **45**(Suppl 1):219S, 1973.
280. Brown JJ, Davies DL, Ferriss JB, et al: Comparison of surgery and prolonged spironolactone therapy in patients with hypertension, aldosterone excess, and low plasma renin. *Br Med J* **2:**729, 1972.
281. Greenblatt DJ, Koch-Weser J: Adverse reactions to spironolactone: A report from the Boston Collaborative Drug Surveillance Program. *JAMA* **225:**40, 1973.

282. Clark E: Spironolactone therapy and gynecomastia. *JAMA* **193:**163, 1965.

283. Levitt JI: Spironolactone therapy and amenorrhea. *JAMA* **211:**2014, 1970.

284. Stripp B, Taylor AA, Bartter FC, et al: Effect of spironolactone on sex hormones in man. *J Clin Endocrinol Metab* **41:**777, 1975.

285. Taylor AA, Mitchell JR, Rollins DE, et al: Effect of spironolactone (S) on adrenal and gonadal steroids in man (Abstract). *Clin Res* **24:**279A, 1976.

286. Biffignandi P, Massuchetti C, Molinatti GM: Free estradiol increase with concomitant decrease of free testosterone in plasma from male subjects after incubation with spironolactone. *Horm Metab Res* **15:**55, 1983.

287. Evron S, Shapiro G, Diamant YZ: Induction of ovulation with spironolactone (Aldactone) in anovulatory oligomenorrhoic and hyperandrogenic women. *Fertil Steril* **36:**468, 1981.

288. Evron S: Induction of ovulation with spironolactone. *Fertil Steril* **38:**391, 1982.

289. Loriaux DL, Menard R, Taylor A, et al: Spironolactone and endocrine dysfunction. *Ann Intern Med* **85:**630, 1976.

290. Steelman SL, Brooks JR, Morgan ER, et al: Anti-androgenic activity of spironolactone. *Steroids* **14:**449, 1969.

291. Santen RJ, Leonard JM, Sherins RJ, et al: Short- and long-term effects of clomiphene citrate on the pituitary–testicular axis. *J Clin Endocrinol Metab* **33:**970, 1971.

292. Kulin HE, Grumbach MM, Kaplan SL: Gonadal–hypothalamic interaction in prepubertal and pubertal man: Effect of clomiphene citrate on urinary follicle-stimulating hormone and luteinizing hormone and plasma testosterone. *Pediatr Res* **6:**162, 1972.

Bartter's Syndrome

Introduction

In the early 1960s Bartter et al.[1,2] described a new syndrome characterized by hypokalemic alkalosis, hyperaldosteronism, hyperreninemia, hyperplasia of the juxtaglomerular apparatus, normal blood pressure, and decreased pressor responsiveness to intravenous infusion of angiotensin II. Despite its rarity, Bartter's syndrome has generated considerable interest owing to the fact that much has been learned by studying this disorder, regarding the role of the renal tubules in the transport of several ions. While considerable controversy exists regarding the nature of the fundamental defect in Bartter's syndrome, several recent discoveries have narrowed the search for the missing piece in the elegant puzzle that is Bartter's syndrome. The heterogeneity in the clinical as well as the biochemical expression of this syndrome has led to a heightened awareness of the disorder.[3]

Pathophysiology

Several theories, past and present, have tried unsuccessfully to explain all the features of Bartter's syndrome. Indeed, one of the major hurdles in accepting any hypothesis for the pathogenesis of the syndrome resides in the fact that no single theory explains *all* the features of the syndrome. It is possible that multiple mechanisms may underlie the development of Bartter's syndrome. Any theory that attempts to explain the pathogenesis of the syndrome must attempt to explain the following 12 metabolic alterations seen in Bartter's syndrome.

1. K^+ wasting and hypokalemia
2. Cl^- wasting
3. Na^+ losses in the urine
4. Metabolic alkalosis
5. Hyperreninemia
6. Hyperaldosteronism
7. Normal blood pressure
8. Resistance to pressor effects of angiotensin II infusion
9. Defect in urinary concentration, even when K is normal
10. Hyperplasia of the JG apparatus
11. Elevated excretion of prostaglandins (PG)
12. Elevated excretion of kallikrein

None of the proposed theories provide a completely satisfactory and unifying hypothesis that explains all the facets of Bartter's syndrome. In attempting to explain the pathogenesis of the syndrome, first the various theories will be examined. This will be followed by individually analyzing each facet of Bartter's syndrome with explanations for each facet.

The five theories that have enjoyed varying degrees of acceptance in explaining Bartter's syndrome are

1. Insensitivity of the arterial tree to angiotensin as the primary event
2. Overproduction of PG as the primary event.
3. Increased production of bradykinin as an important mediator that sets the stage for subsequent events

4. Renal potassium wastage as the primary event
5. Failure of chloride reabsorption in the thick ascending limb of the loop of Henle as the primary event

Each of these theories has played a major role in our understanding of the intricate and intriguing mechanisms that underlie the development of Bartter's syndrome. Some of these theories, with the passage of time, have clearly fallen by the wayside. Others still continue to remain plausible explanations for at least some of the phenomena seen in Bartter's syndrome. The consensus of opinions favors the notion that defective chloride reabsorption at the thick ascending limb is the proximate cause of Bartter's syndrome.

Angiotensin Insensitivity

When Bartter and colleagues originally described the syndrome, they proposed that inability of the arterial vasculature to respond to angiotensin was the primary mechanism in the pathogenesis of the syndrome. As a consequence, slackening of the arterial circulation develops, which is sensed by the kidney as a contraction of the arterial volume. This results in increased secretion of renin by the juxtaglomerular (JG) apparatus, leading to increased concentrations of renin and angiotensin in the circulation. While the increased angiotensin has no effect on the intrinsically resistant vasculature, the adrenal cortices respond by secreting aldosterone, which causes kaliuresis, hypokalemia, and metabolic alkalosis. The original hypothesis attached importance to angiotensin insensitivity and placed Bartter's syndrome in the realm of endocrine disorders that belong to the class of target organ insensitivity. The most important piece of evidence that invalidates this theory is the observation that angiotensin insensitivity in Bartter's syndrome is reversible. Volume expansion has been observed to restore the pressor effect to angiotensin infusion. White[4] described two brothers with Bartter's syndrome in whom extracellular volume expansion by rapid infusion of saline restored angiotensin sensitivity. Similar observations have been made by others.[5,6] Reversal of angiotensin resistance has also been demonstrated by the administration of PG-synthetase–inhibiting drugs to patients with Bartter's syndrome.[7] These data indicate that the basic defect in Bartter's syndrome does not reside in the blood vessels. It is believed that the angiotensin insensitivity seen in Bartter's syndrome is not a primary event, but represents a secondary phenomenon: secondary to tachyphylaxis to angiotensin II owing to decreased number of receptors; or secondary to increase in circulating vadodilators such as PGE_2, prostacyclin, or bradykinin; or secondary to potassium depletion per se. Although angiotensin insensitivity is no longer a prime mechanism in the causation of the syndrome, it continues to represent a major element in this multifaceted syndrome. The vascular resistance in Bartter's syndrome is not merely limited to angiotensin II, since it also extends to other pressor agents such as noreprinephrine.

Overproduction of Prostaglandins

That increased production of PG may play an important role in the pathogenesis of Bartter's syndrome is supported by several lines of evidences. First, elevated levels of serum[8,9] and urinary prostaglandin E_2[8,10–15] have been reported in patients with Bartter's syndrome. Second, and of considerable interest, is the observation that hyperplasia of renomedullary interstitial cells is encountered in patients with Bartter's syndrome.[7,15] These cells are the major sites for production of renal prostaglandin.[16,17] Third, and more important, the use of PG synthetase inhibitors (such as aspirin, indomethacin, ibuprofen, and naprosyn) is associated with reversal of several abnormalities seen in conjunction with Bartter's syndrome.[7–11,18–20] Gill et al.[10] postulated that the hyperreninemia of Bartter's syndrome was dependent on PG synthesis. Indeed, by this time, such an important role in renin production had been ascribed to PG in particular, and arachidonic acid in general, that it seemed appropriate to implicate hyperproduction of PG in the development of Bartter's syndrome. Several actions of PG appear relevant in light of the observed phenomena in Bartter's syndrome. Prostaglandins reduce vascular tone,[21] modulate vascular responsiveness to several pressor agents including angiotensin II,[22] stimulate renin secretion,[23] and cause natriuresis.[24] These properties of PG, coupled with the fact that lowering of the elevated levels of PGE_2 in Bartter's syndrome by certain drugs reversed some of the features of the syndrome, strongly supported a proximate role for PG in the development of Bartter's syndrome.

However, subsequent data have shown that PG excess in Bartter's syndrome does not account for all the features of the syndrome. Administration of large doses of PG inhibiting agents, while restoring the pressor response to angiotensin II, does not fully correct the hypokalemia.[9,10] Gill and Bartter in 1978[25] proposed the existence of a PG-independent mechanism in the development of Bartter's syndrome. It was noted that the hypokalemia of Bartter's syndrome per se might be responsible for increased PG production. Hypokalemia stimulates PGE_2 production by the renal tissue, a consistent hemodynamic observation in dogs.[26] Zusman and Keiser[27] have shown that a decrease in the potassium concentration in the fluid surrounding the PG-producing cells can stimulate PGE_2 production. Gill et al.[28] have reported that urinary PGE_2 excretion is increased in habitual vomiters with hypokalemia. Thus, increased production of PG may be a phenomenon secondary to hypokalemia, rather than a primary cause of Bartter's syndrome. Elucidating the role of potassium and PG in Bartter's syndrome, Ferris[29] has indicated that a decrease in intracellular potassium concentration might increase PGE synthesis in both the collecting tubule and renal medullary interstitial cells, both of which can synthesize PGE_2. This hypothesis is supported by the observation that in animals treated with lithium, the drug can cause lowering of intracellular potassium, along with a significantly increased urinary PGE excretion.[30]

If the increased PGE excretion of Bartter's syndrome is due to hypokalemia, one would anticipate reduction in PGE excretion upon correction of the hypokalemia. Zipser et al.[31] measured urinary PG excretion during oral administration of potassium in two patients with Bartter's syndrome. Their study showed that the oral administration of potassium chloride and potassium citrate, bicarbonate, and acetate to patients with Bartter's syndrome did not diminish the excessive PGE_2 excretion or the other manifestations of the syndrome. Paradoxically, urinary PGE increased several times its basal levels following potassium administration. Failure of potassium loading, in terms of restoring angiotensin responsiveness in Bartter's syndrome, has previously been reported.[32] Another difficulty in ascribing the reason for the increased PG secretion to hypokalemia stems from the fact that PGE elevation is not encountered in other situations characterized by hypokalemia. For instance, PGE levels are normal and pressor sensitivity is actually increased in patients with primary aldosteronism[33] and following desoxycorticosterone administration.[34] Thus, the elevated PG levels seen in patients with Bartter's syndrome cannot be explained on the basis of K^+ depletion alone.

As the primary role of PGE in the causation of Bartter's syndrome was being hotly debated, attention shifted to measuring other PG, particularly metabolites of prostacyclin. Gullner et al.[35] measured urinary excretion of 6-keto-prostaglandin-$F_{1\alpha}$ and thromboxane B_2 in patients with Bartter's syndrome and found that urinary excretion of 6-keto-PG-$F_{1\alpha}$ was four times greater in patients with Bartter's syndrome when compared to normal controls, while the excretion of thromboxane B_2 was no different from that of controls. Prostacyclin is normally generated in the endothelium of all arterial and venous walls including those of the renal vessels.[36-38] Prostacyclin is not metabolized by passage through the lung and is at least 10 times as potent as PGE_2 in terms of its ability to stimulate renin release[39] and to cause vasodilation.[40] Since prostacyclin is also a potent natural inhibitor of platelet aggregation, reports that the plasma of patients with Bartter's syndrome contains a substance that blocks platelet aggregation[41] assume importance in ascribing a role to prostacyclin overproduction in this syndrome. Thus, while hyperproduction of PG is clearly a phenomenon associated with Bartter's syndrome, this phenomenon does not explain all the endocrine and vascular derangements seen in that syndrome.

Overproduction of Bradykinin

Vinci et al.[42] have noted elevated plasma levels of bradykinin in patients with Bartter's syndrome. Bradykinin is a potent vasodilator, and this could cause angiotensin resistance. Bradykinin also stimulates renal PG production and is also a natriuretic agent. The effect of PG synthetase inhibition on the kallikrein–kinin system in patients with Bartter's syndrome has been evaluated by Vinci et al.[42] They noted that with PG synthetase inhibition, values for urinary kallikrein, kinin, and plasma bradykinin returned to normal, with simultaneous reductions in plasma renin activity (PRA), aldosterone, and PGE levels. Similar results have been described by Halushka et al.[12] In contrast, Delaney et al.[43] were unable to demonstrate improvement in the metabolic parameters of Bartter's syndrome by the use of trasylol, a kallikrein inhibitor. It is unclear whether the elevated bradykinin is a reflection of an epiphenomenon secondary to the other features of Bartter's syndrome, or whether it represents a major mechanism responsible for the electrolyte and endocrine abnormalities of the syndrome.

Renal Potassium Wastage

Bardgette and Stein[44] have proposed that primary renal potassium wastage is the basic defect responsible for the development of Bartter's syndrome. The hypokalemia resulting from potassium wastage can explain the hyperreninemia, either as a consequence of increased production of renin indirectly by stimulating PG or directly by stimulating renin secretion by the JG cells. K^+ depletion can also cause angiotensin insensitivity. Persistent excretion of K^+ can result in obligatory anion (chloride) wastage and chloriuresis. K^+ wastage, however, does not adequately explain the lowering of PRA with volume expansion or restoration of angiotensin sensitivity with that maneuver. K^+ wastage also fails to explain the vasopressin-resistant urinary-concentrating defect that persists even when K^+ is restored to normal. Most important, K^+ wastage does not explain the reason for persistence of the elevated PGE_2 excretion and the other abnormalities of Bartter's syndrome upon K^+ repletion.

Failure of Chloride Reabsorption

Kurtzman and Gutierrez[45] have proposed that the basic defect in Bartter's syndrome is the inability to reabsorb chloride in the ascending limb of the loop of Henle. To understand this mechanism it is essential to review some basic concepts in the transport of NaCl by the thick ascending limb. These concepts are derived from the postulates proposed by Rocha and Kokko[46] and Burg and Green[47] that chloride is actively transported in the thick ascending limb. These workers postulated that the active transport of the negatively charged chloride ion provided the generation of positive potential difference across the thick ascending limb. Other workers[48,49] favor the view that chloride transport in this segment is "passive" while sodium transport here is "active."

Westenfelder and Kurtzman[50] have further elucidated the mechanisms of Na^+ and Cl^- transport in the thick ascending limb. They postulate that the only active transport mechanism in this segment is the "asymmetrical exchange" of sodium for potassium at the basolateral surface of the ascending limb. The active removal of sodium is accompanied by the passive obligatory removal of chloride. It has been postulated that back-diffusion of sodium from the paracellular space across the tight junction of the cell into the lumen accounts for the lumen-positive potential difference (PD). The movement of sodium into the lumen results in a high chloride concentration at the paracellular site and serves as the driving force that moves chloride across the contralateral membrane.

If the basic defect in Bartter's syndrome is a defect in transport of NaCl across the thick ascending limb, this would result in increased delivery of sodium, potassium, and chloride to the distal segments of the nephron. The reasons for the impairment in NaCl transport could be manifold. Westenfelder and Kurtzman[50] have proposed five possible mechanisms for defective chloride reabsorption:

1. Altered permeability for Cl^- in the entry step (from lumen into cell)
2. Altered permeability for Cl^- in the exit step (from epithelium into blood)
3. Increased back-diffusion of Cl^-
4. Defective Na^+, K^+-ATPase pump
5. PG-mediated impairment of Cl^- transport[51]

Regardless of the mechanism, the event that sets the stage for subsequent phenomena is the defective transport of Na^+, K^+, and Cl^- across the thick ascending limb of the loop of Henle. As a result, the distal segments receive increased amounts of Na^+, K^+, and Cl^-. Consequently, several sequelae evolve.[50]

1. Increased natriuresis, chloriuresis, and kaliuresis.
2. Loss of salt leads to contraction of effective arterial volume.
3. The contracted arterial volume signals renin secretion by the JG apparatus, resulting in hyperreninemia. The macula densa also senses the defective salt transport locally, thus enhancing renin secretion.
4. Secondary hyperaldosteronism develops in response to the hyperreninemia. The dual effects of hyperaldosteronism are conservation of sodium at the cortical collecting duct, in exchange for K^+, and further kaliuresis.
5. The intrinsic failure to absorb potassium by the thick ascending limb (owing to defective NaCl transport) coupled with the effect of secondary hyperaldosteronism results in hypokalemia. The increased proton secretion causes hypokalemic metabolic alkalosis.
6. The inability to reabsorb salt effectively across the thick ascending limb results in a state where the urine can neither be diluted nor concentrated. This causes the vasopressin-resistant state seen in Bartter's syndrome.
7. Increased PG secretion, increased angiotensin II tachyphylaxis, and the potassium-depleted state result in the decreased pressor responsiveness to angiotensin II.

While initially all three ions—Na^+, K^+, and Cl^-—are inadequately reabsorbed at the thick ascending limb and, therefore, delivered to the distal segments in large quantities, Na^+ is reabsorbed at the collecting ducts due to the effect of aldosterone. Chloride, on the other hand, cannot be reabsorbed at this site, and therefore, continued chloriuresis occurs. Similarly, K^+ loss continues to occur due to K^+ secretion by the tubules in exchange for Na^+ effected by the secondary hyperaldosteronism.

Thus, it appears that defective NaCl and K^+ transport across the ascending limb of the loop of Henle explains several features seen in Bartter's syndrome. This hypothesis is currently favored by most workers. Gill and Bartter,[25] in 1978, proposed that a PG-independent defect in chloride reabsorption in the loop of Henle may be a proximal cause of Bartter's syndrome. They noted abnormally low free-water clearance in association with a high clearance of chloride in all patients with Bartter's syndrome. The increased fractional clearance of chloride in patients with Bartter's syndrome was not corrected by indomethacin, which led the authors to conclude that the defect was PG independent. Early observations on the ineffectiveness of salt depletion in correcting the hypokalemia are also in keeping with the assumption that tubular reabsorption of NaCl is impaired in patients with Bartter's syndrome. Numerous workers have subsequently demonstrated that an abnormally large fraction of the filtered load is delivered to the distal tubules. Baehler et al.[52] have noted that the abnormally low free-water clearance and distal fractional chloride reabsorption seen in Bartter's syndrome were not corrected by potassium repletion, indomethacin therapy, or magnesium repletion. Their observations also suggest that the defect in the reabsorption of sodium chloride is a primary event, unrelated to potassium depletion, magnesium depletion, or PG overproduction. Thus, it appears that the most plausible explanation for the defect in Bartter's syndrome resides in the transport of NaCl in the thick ascending limb of the loop of Henle.

The 12 Facets of Bartter's Syndrome

The several theories proposed have been instrumental in attempting to explain the metabolic and endocrine alterations associated with Bartter's syndrome. In this section, the 12 facets of Bartter's syndrome are individually analyzed, in terms of the mechanism of causation.

Potassium Wasting and Hypokalemia

This, of course, is the biochemical marker of Bartter's syndrome. While originally it was proposed that the hypokalemia of Bartter's syndrome was mediated by secondary hyperaldosteronism,[1] this factor cannot be solely responsible. It is well recognized that the hypokalemia of Bartter's syndrome responds only partially to aminoglutethimide or spironolactone[5] or even to bilateral adrenalectomy.[53,54] Besides, there is poor correlation between the degree of hypokalemia and the severity of aldosterone excess. In fact, plasma aldosterone levels in Bartter's syndrome may be in the normal range since profound hypokalemia blunts the response of the zona glomerulosa to angiotensin II. A combined defect in sodium and potassium transport, or in chloride transport,[25,45] coupled with the effects of secondary hyperaldosteronism on the collecting tubules, could provide an alternate explanation. A defect in chloride reabsorption may result in renal loss of potassium by (1) decreased absorption in the loop of Henle, if it is assumed that K^+ reabsorption at this site is passive to chloride reabsorption[55] and (2) enhancing potassium secretion at the distal sites by increasing the tubular flow rate.[56] Thus, decreased reabsorption and increased secretion of potassium provide a mechanism for the hypokalemia. An additional mechanism proposed is the notion that hypomagnesemia may contribute to or aggravate the hypokalemia. Both hypomagnesmia and low intracellular magnesium levels in the muscle have been recognized in Bartter's syndrome.[57,58] Since magnesium depletion can be associated with renal potassium wasting, it is possible that hypomagnesemia can contribute to the potassium-depleted state. The impact of this phenomenon on the potassium wasting in Bartter's syndrome has not been assessed. Baehler et al.[52] described a patient with Bartter's syndrome with hypomagnesemia and noted that magnesium infusion eliminated renal potassium wasting, while having no effect on the decreased free-water clearance or the distal fractional chloride reabsorption.

Regardless of the mechanism, hypokalemia has several important ramifications upon other phenomena associated with Bartter's syndrome. For instance, hypokalemia has been implicated in the PG overproduction seen in patients with Bartter's syndrome. Hypokalemia may also contribute to the decreased pressor response to angiotensin II, a consistent finding in Bartter's syndrome. Interestingly enough, potassium depletion depresses the chloride reabsorption in the loop of Henle in rats.[59] Finally, hypokalemia plays a role in the development of the vasopressin-resistant concentrating defect in Bartter's syndrome.

Chloride Wasting

This feature is also a hallmark of Bartter's syndrome. When contrasted with habitual vomiting, a disorder that shares several similarities with Bartter's syndrome, the presence of chloriuresis is striking in Bartter's. Both disorders are characterized by urinary potassium losses, hypokalemia, profound potassium depletion, elevated PRA, and increased PGE_2 synthesis. The unique feature of decreased distal fractional reabsorption of chloride is restricted to Bartter's syndrome.

The decreased chloride reabsorption seen in Bartter's syndrome is believed to represent an inherent abnormality in the renal tubule. Although severe potassium deficiency can produce a transient defect in chloride reabsorption,[60] this does not appear to be the prime mechanism for the chloriuresis of Bartter's syndrome. It is also highly improbable that increased PG secretion seen in Bartter's syndrome causes the defect in chloride reabsorption; this is because patients with psychogenic vomiting do not show a decrease in the distal fractional reabsorption of chloride despite an increase in the synthesis of PG.

The important phenomenon of decreased chloride reabsorption probably functions as the initiating event that sets the stage for the metabolic and endocrine events that characterize Bartter's syndrome. The renal loss of Na and K as well as the increase in renin angiotensin and aldosterone can be viewed as a sequential response to the inability to reabsorb sodium chloride at the thick ascending limb. While sodium loss can be ultimately minimized by the action of aldosterone on the cortical collecting tubules by exchange with K^+, no such exchange mechanism exists for chloride conservation. Therefore, continued chloriuresis is a diagnostic feature consistently observed in Bartter's syndrome.

Na^+ Losses in Urine

As already indicated, defective reabsorption of sodium and chloride at the thick ascending limb of the loop of Henle is an inherent feature of Bartter's syndrome. Renal sodium loss in Bartter's syndrome occurs with variable frequency. The natriuresis in some patients can be so profound as to result in hyponatremia, extracellular fluid volume contraction, and even dehydration, while in others the negative sodium balance is barely noticeable. The differences in the clinical and biochemical expression of Bartter's syndrome were noted by Robson et al.,[61] who postulated the existence of two groups of patients with Bartter's syndrome. The noticeable difference between the two, besides the age of onset, was the tendency for salt loss in group 1, often with volume depletion requiring hospitalization, while sodium balance was near normal in group 2.

Several noteworthy points bear emphasis with regard to sodium loss in Bartter's syndrome. First, sodium loss in the urine is somewhat curtailed owing to the effect of secondary aldosteronism at the cortical collecting tubules. When patients with Bartter's syndrome are placed on a very low intake of salt, most patients demonstrate the ability to conserve sodium. In this respect, patients with Bartter's syndrome resemble those with primary hyperaldosteronism. However, in contrast to the latter, patients with Bartter's syndrome will not normalize the serum K^+ level by this maneuver. While obligatory sodium loss has been observed in a few patients with classic features of Bartter's syndrome,[62] by and large, most patients are able to conserve sodium when dietary sodium is restricted. Second, studies that evaluate tubular reabsorption of sodium following the infusion of normal or hypotonic saline have demonstrated subnormal distal tubular reabsorption of sodium in some but not all patients with Bartter's syndrome.[4,63–65] Third, the delivery of an abnormally large percentage of the filtered load to the distal tubules also plays a role in the suboptimal distal tubular reabsorption of sodium.[66] Fourth, patients with Bartter's syndrome excrete salt loads more rapidly than normal. This is perceived to be an effect of the impairment in the salt transport by the ascending limb.[50] Finally, the natriuresis that results is largely responsible for activating the renin–angiotensin system in patients with Bartter's syndrome.

Metabolic Alkalosis

The metabolic alkalosis of Bartter's syndrome is a consequence of the effects of secondary hyperaldosteronism. The increased distal delivery of sodium and the aldosterone-mediated exchange of H^+ for Na^+ results in loss of H^+ ions with consequent metabolic alkalosis ("increased proton secretion").

Hyperreninemia

Chronic stimulation of the JG apparatus is the mechanism for the hyperreninemia encountered in Bartter's syndrome. The stimuli for increased renin production can be multifactorial, the most obvious one being the decreased effective arterial blood volume that results from chronic loss of sodium chloride in the urine. Two other mechanisms that play an equally important role in causing hyperreninemia are defective salt transport at the macula densa (brought about by defective sodium chloride transport at the thick ascending limb) and increased PG synthesis by the renomedullary interstitial cells. The effects of PG in stimulating renin synthesis are well established.[23,67] The primary versus secondary nature of the phenomenon, however, is controversial, since it is believed that the PG excess of Bartter's syndrome may be a reflection of the effect of hypokalemia. In addition to decreased effective arterial blood volume, defective sodium transport to the macula densa, and PG mediation, a direct effect of hypokalemia in stimulating the JG cells may also play a minimal role in the causation of the hyperreninemia of Bartter's syndrome.

Regardless of the mechanism, the excessive renin production in Bartter's syndrome is clearly not autonomous. Thus, the elevated PRA in patients with Bartter's syndrome can be suppressed by volume expansion[4] and can be lowered by drugs that inhibit the enzyme PG synthetase.[68] The increased production of renin is responsible for the characteristic histological appearance of JG hyperplasia seen in the biopsied kidneys of patients with Bartter's syndrome. The physiological consequences of hyperreninemia are increased levels of angiotensin II and aldosterone in the circulation.

Increased Aldosterone

The aldosterone excess in Bartter's syndrome is clearly secondary to the hyperreninemia. The aldosterone excess in patients with Bartter's syndrome is disproportionately low, when viewed in light of the markedly elevated PRA levels in Bartter's syndrome. This discrepancy is explained by the fact that the response of the zona glomerulosa to angiotensin II is blunted in the presence of hypokalemia. Dillon et al.[68] have reinforced the well-recognized observation that the plasma aldosterone levels in Bartter's syndrome rise following potassium repletion, indicating restoration of full sensitivity of the zona glomerulosa to angiotensin II. Occasionally the plasma aldosterone levels can be normal in patients with Bartter's syndrome. The combination of a normal aldosterone in presence of extraordinarily elevated PRA might suggest, at first glance, some form of hypoaldosteronism. The serum K^+ levels, however, would clarify the issue, since patients with hypoaldosteronism are hyperkalemic, while patients with Bartter's syndrome are hypokalemic. Further, restoration of normokalemia results in a rise in plasma aldosterone in Bartter's syndrome.

The effects of the aldosterone excess on the electrolyte milieu are twofold; clearly, the secondary aldosterone excess promotes reabsorption of sodium at the collecting tubules and helps conserve sodium. In the process of doing so, K^+ is exchanged, and thus, secondary hyperaldosteronism promotes loss of urinary K^+. Nonetheless, it is believed that aldosterone excess does not account for the severity of potassium losses, since aminoglutethimide, spironolactone, and even bilateral adrenalectomy do not prevent the renal potassium wastage that is innate to Bartter's syndrome.

Normal Blood Pressure

The normal blood pressure in patients with Bartter's syndrome is due to two factors—the insensitivity of the vascular smooth muscle to angiotensin II, and the presence of vasodilators such as PG, bradykinin, and prostacyclin in the circulation. The decreased pressor response to angiotensin II is not permanent and can be reversed with volume expansion. It is concievable, therefore, that the blood pressure in patients with Bartter's syndrome can be slightly elevated in the presence of chronic extracellular fluid volume-expanded states such as pregnancy.

Angiotensin Insensitivity

Decreased or absent pressor resonse to infused angiotensin II is a characteristic feature of Bartter's syndrome. The decreased pressor response seen in these patients extends to other agents such as noreprinephrine.[69] Three mechanisms underlie this phenomenon: the development of tachyphylaxis to chronically elevated angiotensin II in the circulation (receptor down-regulation), the presence of vasodilator substances such as prostaglandin E_2 and bradykinin in the circulation, and chronic potassium depletion. As indicated earlier, angiotensin responsiveness can be restored by volume expansion, as well as by drugs that decrease PG synthesis. The fact that angiotensin antagonists can lower the blood pressure in patients with Bartter's syndrome suggests that the maintenance of blood pressure is, in part, mediated by angiotensin. Angiotensin resistance is not unique to Bartter's syndrome and is encountered in patients with chronic diuretic abuse, as well as with psychogenic vomiting. In both situations the metabolic mimicry simulating Bartter's is carried to the extreme ("pseudo-Bartter's).

Defective Urinary Concentration

Inability to adequately concentrate urine is a frequent accompaniment of Bartter's syndrome. While hypokalemia clearly can induce a vasopressin-resistant defect in urinary concentration, this phenomenon may persist even after potassium repletion. In such a setting the vasopressin-resistant concentrating defect is believed to be a reflection of decreased NaCl reabsorption in the thick ascending limb; if salt cannot be reabsorbed effectively across this segment, the urine can neither be concentrated nor diluted. The role of PGE_2 in causing the vasopressin-resistant concentrating defect is debatable.

Hyperplasia of JG Apparatus

The histological hallmark of Bartter's syndrome is the presence of hyperplasia of the JG apparatus. The appearance on renal biopsy is quite characteristic. Increased cellularity is evident in the afferent arterioles, surrounding mesangium, and macula densa. The increased population of cells is structurally normal. These hyperplastic cells can be shown to contain granules (renin granules) with the use of Bowie's stain. The demonstration of Bowie positive granules is not universal. In instances where granules are not seen, electron microscopy may reveal electrodense protogranular forms in the hyperplastic cells. These may represent "immature" granules. While JG hyperplasia is a pathognomonic finding for Bartter's syndrome, there has been a case report of Bartter's syndrome without JG hyperplasia.[70] Further, JG hyperplasia can be encountered in other states characterized by long-standing hypersecretion of renin, such as familial chloride diarrhea,[71] laxative-induced hyperaldosteronism,[72] and even Addison's disease.

Increased Excretion of Prostaglandin

Increased excretion of prostaglandin E is now a well-established feature of Bartter's symdrome.[73,74] The increase in urinary excretion of PGE_2 can be impressive,[10] minimal,[18] or occasionally absent.[75] Increased production of PG may be a response to potassium depletion[76] or alternately could be due to the effect of angiotensin II on PG synthesis. The finding of hyperplasia of the renomedullary interstitial cells reflects the increased secretory rate of PG by these cells.[7]

Increased Kallikrein Excretion

The reasons for increased excretion of kallikreins are believed to be the same as those for PG excretion.

Histopathology

Histopathologically the hallmark of Bartter's syndrome is hyperplasia of the JG apparatus. Increased cellularity of the afferent arterioles, the surrounding mesangium, and the macula densa is quite characteristic. The hyperplastic cells can be shown to contain renin granules by the Bowie stain. In addition, several histological findings that resemble glomerulonephritis can be seen. Thus, membranous thickening of glomerular capillary walls, crescent formation, concentric periglomerular fibrosis, and glomerulosclerosis have been associated with Bartter's syndromre.[3] These findings suggest that ongoing renal injury may be involved in the mediation of some of these changes. This is supported by observations that changes of proliferative glomerulonephritis have been described at autopsy in a patient with Bartter's syndrome[77] and deposits of immune complexes have been observed in some patients with Bartter's syndrome.[62] Juxtaglomerular hyperplasia merely reflects hyperactivity of the renin-producing cells and can be encountered in other situations such as chloride diarrhea, laxative abuse, and surrpetitious vomiting. In addition to juxtamedullary hyperplasia, patients with Bartter's syndrome may also demonstrate hyperplasia of the interstitial renomedullary cells—a major site of PG synthesis.[7] It is unclear whether this finding is specific for Bartter's or might be encountered in other situations such as habitual vomiting, which is also associated with hyperexcretion of PGE_2 and hyperreninemia.

Clinical Features

Bartter's syndrome is a rare disorder, with approximately 100 cases having been reported since 1962. Although earlier reports indicated a preponderance of Bartter's syndrome in blacks,[78] the disorder is known to occur in all races with the same frequency. Bartter's syndrome affects both sexes, with a slight predominance in women.[62]

Although patients of any age can be affected, a large majority of patients present with symptoms during childhood or young adulthood. A familial tendency is suggested by the occurrence of the disorder in siblings.[4,53,79] Sutherland et al.[58] have described the occurrence of Bartter's syndrome in three siblings, and Delaney et al.[43] have described the syndrome in six siblings. While the inheritance pattern of Bartter's syndrome has not been established, an autosomal recessive pattern of inheritance has been implied.

The severity of the clinical expression in Bartter's syndrome is quite variable. This heterogeneity in expression has led Robson et al.[61] to postulate the existence of two clinical groups of patients with Bartter's syndrome. Group 1 included patients who were very young, often under 3 years of age, with marked hypokalemia and a pronounced tendency for salt loss, hyponatremia, extracellular volume contraction, and dehydration. In contrast, group 2 consisted of patients with onset of disease at a later age, and in whom renal sodium balance was normal and the hypokalemia milder. The authors also postulated that group 1 patients evolved into group 2 with passage of time.

The clinical features of Bartter's syndrome can be viewed in terms of those due to hypokalemia, those due to tubular disease, those due to suboptimal somatic and intellectual growth, and those due to generalized metabolic phenomena.

Hypokalemic Features

The most frequent symptoms experienced by patients with Bartter's syndrome are a result of chronic hypokalemia. The serum potassium levels in Bartter's syndrome can be dangerously low, often even below 2.5 meq/liter. The most common symptoms experienced are muscle cramps and muscle weakness. Proximal muscle weakness can sometimes be objectively demonstrable. With severe hypokalemia, the spectrum of manifestations extends to exclude diverse features—neuromuscular irritability (twitching, tetany, positive Chvostek and Trousseau's signs, and even convulsions), truncal muscle weakness, paralytic episodes, constipation, and even ileus.

The degree of hypokalemia in patients with Bartter's syndrome shows considerable individual variation. The potassium deficit results from urinary losses of that ion and can exceed 300 meq/day. In general, dietary sodium intake has little or no influence on the urinary loss of potassium in Bartter's syndrome, a finding that contrasts sharply with primary hyperaldosteronism. The hypokalemia of Bartter's syndrome is particularly resistant to oral replacement of potassium.

Urinary Symptoms

Polyuria, nocturia, and enuresis are frequently encountered in Bartter's syndrome. The polyuria and polydipsia seen in some cases can be severe enough to cause dehydration, especially in younger children. While hypokalemia usually underlies the development of a vasopressin-resistant type of urinary concentrating defect, such a defect can be seen in the absence of hypokalemia. Hypercalciuria, nephrocalcinosis, and progressive renal impairment have been described in Bartter's syndrome.[80] Chronic polyuria can result in the development of hydroureter or hydronephrosis.

Somatic and Intellectual Development

Growth retardation and mental delay were noted in the first two patients with the syndrome described by Bartter. Failure to thrive is an important sign of Bartter's syndrome occurring during infancy. Mental delay and suboptimal intelligence have been reported in patients with Bartter's syndrome. It is not clear whether this feature is a sequel of long-standing alterations in potassium metabolism. Patients with Bartter's syndrome tend to be short, but well proportioned and symmetrical. James et al.[81] have described a characteristic facies in children with Bartter's syndrome, but this is not uniformly encountered. While the reason for short stature is unclear, the concomitant occurrence of rickets, at least in some cases, may contribute to delayed skeletal growth.[82,83] Simopoulos[84] has observed the growth characteristics in patients with Bartter's syndrome and has noted a delayed adolescent growth spurt with ultimate attainment of normal stature. Unfortunately, the outlook for mental retardation characteristics was noted to be less optimistic for children with Bartter's. Some degree of mental retardation was observed in nearly two-thirds of children with Bartter's.

Generalized Symptoms

Patients with Bartter's syndrome periodically suffer from anorexia, salt craving, vomiting, and a feeling of "ill health" and fatigue. It is difficult to separate several of these symptoms from the chronically potassium-depleted state. The occurrence of magnesium deficiency in Bartter's syndrome may play a role in the development of some of these symptoms.[57]

The physical examination of patients with Bartter's syndrome generally does not disclose any striking abnormalities. Short stature and mild mental delay may be seen in some children with the disorder. The characteristic feature in patients with Bartter's syndrome is the normal blood pressure. In several instances, orthostatic hypotension has been observed.[85,86]

Laboratory Features

The laboratory features of Bartter's syndrome can be viewed in terms of routine tests, endocrine studies, and histopathological studies.

Routine Tests

Patients with Bartter's syndrome are invariably hypokalemic, at times profoundly so. The low serum K^+ is associated with profuse kaliuresis, indicating the renal origin of the hypokalemia. The urine often shows a fixed specific gravity with inability to concentrate or maximally acidify the urine. In most, but not all patients these abnormalities are reversed upon restoration of normokalemia. The intravenous pyelogram is usually normal. Hydroureters and hydronephrosis may be seen in long-standing cases of polyuria. Rarely nephrocalcinosis may be seen. The cardiogram may demonstrate electrocardiographic changes of hypokalemia. The serum sodium is usually normal, while the chloride levels show mild to moderate lowering, and the bicarbonate levels are usually high. Thus, the biochemical triad of Bartter's syndrome is hypokalemia, hypochloremia, and metabolic alkalosis. Hypomagnesemia[58] and hyperuricemia[87] are variably encountered. Glucose intolerance is a frequent accompaniment of Bartter's syndrome and is probably a result of chronic hypokalemia. Polycythemia may be seen in Bartter's syndrome, but the true frequency of this feature is uncertain. The abnormalities encountered upon routine testing are outlined in Table 33.

Measurement of urine lytes confirms the pronounced increase in the 24-hr potassium excretion. The urinary sodium is highly variable in patients with Bartter's syndrome, since a steady state is attained, often rapidly. The most crucial piece of laboratory determination in the urine is measurement of chloride. Patients with Bartter's syndrome will invariably demonstrate chloriuresis. This is a reflection of the decreased reabsorption of chloride in the thick ascend-

TABLE 33.
Bartter's Syndrome: Abnormalities in
Routine Tests

Hypokalemia
Kaliuresis
Isosthenuria
Metabolic alkalosis
Hypochloremia
Increased chloride excretion
Hypomagnesemia
Hyperuricemia
Glucose intolerance
Polycythemia
Abnormal EKG
Abnormal IVP

ing limb of the loop of Henle. Consequently, the fractional clearance of chloride is increased. This is a valuable diagnostic aid in differentiating Bartter's syndrome from psychogenic or surreptitious vomiting, where the urinary chloride excretion is strikingly low.

Endocrine Studies

The endocrine triad that characterizes Bartter's syndrome is hyperreninemia, hyperaldosteronism, and angiotensin insensitivity.

Hyperreninemia

The hyperreninemia of Bartter's syndrome is striking. The elevated PRA responds to physiological cues appropriately. Thus, volume expansion lowers PRA, but does not completely suppress it.[88] The failure of PRA to completely normalize with volume expansion suggests that other factors besides extracellular fluid volume are involved in the regulation of renin secretion in Bartter's syndrome. The PRA response to salt depletion is also predictable in Bartter's syndrome, i.e., an exaggerated rise in renin following salt restriction. The incomplete suppression to saline or volume expansion and the exaggerated response to salt depletion may reflect the chronic hypersecretion of renin by the hyperplastic population of the JG apparatus. The PRA activity can be almost normalized by the use of indomethacin.

The combination in a normotensive patient with hypokalemic alkalosis of a high PRA and a low serum chloride does not always indicate Bartter's syndrome. An identical profile would be expected in patients with habitual vomiting and chronic diuretic abuse.

Hyperaldosteronism

The plasma aldosterone levels in patients with Bartter's syndrome show a mild to moderate increment. The inhibitory effect of hypokalemia on aldosterone secretion is conducive to the disparity between the magnitude of renin elevation and the hyperaldosteronism. Since hypokalemia stimulates renin and inhibits aldosterone secretion, the hypokalemic patient with Bartter's syndrome may show a profile of "hyperreninemic hypoaldosteronism." This, however, would revert to a pattern of "hyperreninemic hyperaldosteronism" when potassium is repleted. Plasma aldosterone levels wax and wane in patients with Bartter's syndrome depending on the magnitude of potassium depletion. The high aldosterone levels of Bartter's syndrome do respond to physiological cues, declining with salt loading and hypokalemia, while increasing with salt depletion.

Angiotensin Insensitivity

The endogenous angiotensin levels are elevated in patients with Bartter's syndrome. In keeping with receptor regulation concepts, exogenous administration of angiotensin II fails to induce a pressor response in patients with Bartter's syndrome. When angiotensin II levels are lowered by using indomethacin, there is a brisk restoration of normal pressor responsiveness to angiotensin infusion. While angiotensin responsiveness can also be restored by saline infusion, the effects are more dramatic and prompt with indomethacin.

Histopathological Studies

When performed, the renal biopsy in patients with Bartter's syndrome demonstrates the characteristic appearance of JG hyperplasia. When the biochemical and endocrine data have provided clear-cut and incontrovertible evidence for Bartter's syndrome, there is no need to resort to renal biopsy for confirmation. Proper evaluation of serum and urinary electrolytes, coupled with measurement of the renin–angiotensin–aldosterone activity, offers a safe, definitive, and noninvasive means for establishing the diagnosis of Bartter's syndrome.

Differential Diagnosis

The biochemical and hormonal features of Bartter's syndrome are characteristic enough to permit specific diagnosis. The constellation of hypokalemia, metabolic alkalosis, kaliuresis, normotensive hyperreninemia, hyperaldosteronism, and insensitivity to angiotensin II is restricted to Bartter's syndrome and to simulacra of Bartter's syndrome, such as surreptitious habitual vomiting and surreptitious diuretic abuse (Pseudo-Bartter's syndrome). It is unclear whether incomplete expressions of Bartter's syndrome exist. Gullner et al.[89] reported two sisters and their brother, with ages ranging from 10 to 14 years, who presented with hypokalemia, hyperreninemic hyperaldosteronism, and excessive PGE_2 excretion. These patients, however, differed from the classic Bartter's syndrome in three respects; first, the pressor response to angiotensin II was hyperresponsive in these patients; second, JG hyperplasia was absent; and third, the distal fractional chloride reabsorption was normal, in contrast to classic Bartter's syndrome, where decreased distal fractional chloride reabsorption is invariably decreased. The nature of this syndrome is unclear.

The differential diagnosis of Bartter's syndrome can be viewed from several vantage points and can be discussed in terms of hypokalemia, hyperreninemia, and hyperaldosteronism.

Based on Hypokalemia

Since hypokalemia is the predominant manifestation of Bartter's syndrome, the initial clinical spectrum of differential diagnosis revolves around several common and uncommon entities. These include hypokalemic familial periodic paralysis, mineralocorticoid or glucocorticoid excess, and renal tubular disease; chronic diarrhea, vomiting, villous adenomas, secretory (endocrine) diarrheas, and laxative abuse are important gastrointestinal causes of hypokalemia. All patients with hypokalemia should be closely questioned regarding the use or abuse of steroids, mineralocorticoids, licorice, diuretics, and laxatives. The absence of diarrhea or vomiting, the normal blood pressure, and the documentation of urinary loss of potassium despite hypokalemia quickly narrow the initial differential diagnostic spectrum to five entities—Bartter's syndrome, normotensive primary hyperaldosteronism, renal tubular disease (salt-wasting nephritis), surreptitious vomiting or diuretic abuse, and rarely hypomagnesemia.

Normotensive Primary Hyperaldosteronism

Normotensive primary hyperaldosteronism is rare. Snow et al.[90] reported a patient with an adlosterone-secreting adrenal adenoma and normotensive hyperaldosteronism. Zipser and Speckart[91] reported a 45-year-old woman with hypokalemia, normal blood pressure with an enlarged left adrenal gland by venography, and bilateral adrenal hyperplasia during exploration. Shiroto et al.[92] described a 25-year-old woman with hypokalemia, increased urinary K^+ clearance, high aldosterone, low PRA, and a typical unilateral adrenocortical adenoma. The possible mechanisms for normotensive hyperaldosteronism are several and include early phase of disease, essential hypotension associated with primary hyperaldosteronism, and the existence of mechanisms that counteract the increase in blood pressure in primary aldosteronism. The clinical picture of normotensive primary hyperaldosteronism mimics that of Bartter's syndrome. However, the diagnosis can be readily established when PRA is measured. The PRA in normotensive hyperaldosteronism is suppressed in contrast to Bartter's, where the PRA is invariably and impressively high.

Renal Tubular Disease

Intrinsic tubular disease, resulting in salt-losing nephropathies, can lead to a state of sodium depletion, setting up a cascade of events—decreased effective circulating extracellular fluid volume, hyperreninemia, secondary hyperaldosteronism, urinary potassium loss, hypokalemia, and JG hyperplasia. The pressor response to angiotensin can be diminished, bringing the resemblance to Bartter's syndrome closer. Several significant differences, however, can help make the distinction. First, the hypokalemia of salt-losing tubular disease is milder in comparison to Bartter's syndrome; second, patients with salt-losing nephritides demonstrate moderate to large amounts of sodium in the urine, in contrast to patients with Bartter's, who attain a steady state; third, the natriuresis in salt-losing nephropathy is resistant to correction by salt loading. While patients with Bartter's syndrome, when given a salt load, do excrete a greater quantity of sodium following salt load in comparison to normal, they are not "salt losers" in the true sense of the term, because they start to rapidly conserve sodium after attainment of a steady state. Finally, other abnormalities in tubular function can be demonstrated in patients with salt-losing nephropathy.

Psychogenic, Habitual, or Covert Vomiting

The phenomenon of covert vomiting can mimic Bartter's syndrome to a remarkable extent. Patients with habitual covert vomiting can present with moderate to severe hypokalemia often associated with muscle weakness, tetany, and other consequences of potassium depletion. The constellation of hypokalemic alkalosis, hypochloremia, normal or low blood pressure, high PRA, and high aldosterone in plasma resembles Bartter's syndrome. The metabolic mimicry is carried further in that patients with habitual vomiting also demonstrate moderate kaliuresis, hyperexcretion of PGE_2, and decreased vascular response to angiotensin II. The kaliuresis in habitual vomiting is intriguing since the classic teaching is that conditions characterized by extrarenal potassium losses are associated with potassium sparing, while primary renal disorders are associated with potassium wasting in the urine. Several workers have pointed out the occurrence of persistent but inappropriate urinary potassium loss with protracted nasogastric suction or in patients with habitual vomiting.[93-95] Two mechanisms have been proposed to explain the continued potassium loss in the urine in these conditions. First, an enhanced delivery of bicarbonate to the distal tubule may create luminal electronegativity which facilitates potassium excretion; and second, the PGE_2-mediated hyperaldosteronism results in avid Na–K exchange at the distal tubule. Regardless of the mechanism, moderate kaliuresis can be seen in patients with habitual vomiting. With habitual vomiting, less chloride is available for simultaneous reabsorption of sodium. Therefore, more sodium reaches the diatal tubule, where, under the influence of aldosterone, it is reabsorbed in exchange for potassium, which is excreted at an accelerated rate. The similarities between Bartter's syndrome and habitual vomiting are so striking (Table 34) that is not surprising that the diagnosis of Bartter's is mistakenly made when patients strongly deny vomiting.[96] The distinction between Bartter's syndrome and habitual vomiting can be made by the simple measurement of urine chloride. Patients with habitual vomiting consistently and reproducibly demonstrate chloride sparing (often below 6 meq chloride per 24 hr), even when maintained on high dietary chloride intake.[97] This contrasts with the significant chloriuresis consistently encountered in Bartter's syndrome.

Surreptitious Diuretic Abuse

The principal action of furosemide is to inhibit active chloride transport in the thick ascending limb of the loop of Henle.[98] This action is identical to the basic defect in Bartter's syndrome. Thus, a normal subject on large doses of

TABLE 34.
Simulacra of Bartter's Syndrome

	Bartter's	Habitual vomiting	Surreptitious diuretic abuse
Hypokalemia	+++	++	++
Kaliuresis	++	+	++
Metabolic alkalosis	+	+	+
High PRA	+	+	+
High aldosterone	+	+	+
Decreased pressor response to angiotensin	+	+	+
Increased PGE excretion	+	+	+
JG hyperplasia	+	+	+
Urinary chloride	↑	↓	↑
Diuretic screen in urine	—	—	+

furosemide will develop almost every facet of Bartter's syndrome. The chronic volume contraction leads to hyperreninemia, hyperplasia of JG apparatus, and refractoriness to the pressor effects of angiotensin. The intrinsic effect of the diuretic leads to natriuresis, kaliuresis (compounded by secondary aldosteronism), and chloriuresis mimicking Bartter's. When patients deny the use of the diuretic while surreptitiously ingesting furosemide, the differentiation from Bartter's syndrome rests on performing a drug screen for diuretics in urine or plasma.

Hypomagnesemia

Hypomagnesemia can rarely result in a syndrome resembling Bartter's syndrome. It is well recognized that magnesium deficiency may be associated with hypokalemia and inappropriate kaliuresis.[99] Gitelman et al.[100] described a familial disorder characterized by hypokalemia and hypomagnesemia and several metabolic features similar to those seen in Bartter's syndrome. The nature of this "familial hypomagnesemic disease" is unclear.

Based on Hyperreninemia

The elevated PRA of Bartter's syndrome is identical to that of several other hyperreninemic states such as primary hyperreninism (renin-secreting tumors), malignant hypertension, accelerated hypertension, and renovascular hypertension. All these conditions can be excluded by the mere fact that the blood pressure is elevated in them, while being normal (or even low) in Bartter's syndrome. The hyperreninemia of edematous states (e.g., nephrotic syndrome, cirrhosis with ascites) is easily differentiated from Bartter's syndrome, which is characterized by an absence of edema.

Based on Hyperaldosteronemia

Bartter's syndrome can easily be differentiated from hyperaldosteronism (the major cause of elevated aldosterone and hypokalemia) by the elevated PRA seen in Bartter's. The aldosterone elevation in Bartter's syndrome is mild to moderate, owing to the inhibitory effect of hypokalemia on the secretion of aldosterone.

Treatment

The triple lines of therapeutic approach for treatment of Bartter's syndrome are potassium replacement, aldosterone antagonism with spironolactone, and the use of drugs that inhibit PG synthetase.

Potassium Replacement

Repletion of potassium is the sheet anchor of therapy in Bartter's syndrome and can be lifesaving. Large doses of oral potassium chloride (in excess of 160 meq/day), while providing symptomatic relief, may not completely normalize potassium levels. Intravenous potassium supplementation coupled with potassium-sparing diuretics such as triamterene or spironolactone may be required to normalize the potassium when the hypokalemia is profound. Patients often require maintenance therapy with variable doses of oral potassium supplementation. The adverse effects of large doses of oral potassium supplementation include the development of abdominal cramps, gastric upset, or diarrhea. When hypomagnesemia is present, it should be treated with magnesium chloride.

Spironolactone

Since hyperaldosteronism plays at least a partial role in the development of hypokalemia in Bartter's syndrome, the use of this drug is advocated to correct hypokalemia. However, complete normalization of serum potassium levels is seldom achieved with spironolactone alone, since the mechanisms for hypokalemia in Bartter's syndrome are multifactorial. Nevertheless, spironolactone therapy is a useful adjunct in the management of Bartter's syndrome.

Prostaglandin Synthetase Inhibitors

The observation that hyperexcretion of PGE_2 occurred in a large number of patients with Bartter's syndrome inevitably led to a focus on the use of PG synthetase inhibitors in that syndrome. The use of PG-lowering drugs such as indomethacin, and aspirin in the treatment of Bartter's syndrome has been extensively studied and reported. The histological demonstration of hyperplastic renomedullary interstitial cells, a major site of PG production, led Verberckmoes et al.[7] to treat successfully a patient with the PG synthetase inhibitor indomethacin. Bowden et al.[18] treated seven patients with Bartter's syndrome, five of whom had measurably increased urinary excretion of PGE. These workers noted a close correlation between the suppression of PGE excretion and correction of the clinical abnormalities of Bartter's syndrome. In addition, greater suppression of PGE was encountered with indomethacin in comparison to aspirin or ibuprofen. Thus, greater sodium and potassium retention, greater increases in serum potassium, and greater reductions in PRA were observed with indomethacin than with other inhibitors of PG synthetase. Garin et al.[20] used indomethacin at a dose of 2 mg/kg per day in two patients with Bartter's syndrome and demonstrated impressive weight gain, with a decrease in the rate of urinary excretion of sodium, inorganic phosphate, and aldosterone. The urinary excretion of potassium also decreased, albeit transiently, with an increase in serum potassium levels. These observations are in keeping with the generally agreed notion that despite excellent lowering of PG levels, indomethacin treatment does not completely normalize serum potassium concentrations.

The literature is flooded with papers attesting to the correction of high PG levels in Bartter's syndrome with short-term administration of indomethacin. Fichman et al.[9] treated a patient with indomethacin and showed a dramatic decrease in the whole-blood PGE levels. Norby et al.,[8] using aspirin, demonstrated a significant decrease in levels of PGE and PGF in urine and plasma of a 22-month-old child with Bartter's syndrome following aspirin therapy. In a detailed study of several PG derivatives, Gill et al.[10] demonstrated that treatment of four patients with Bartter's syndrome with indomethacin or ibuprofen caused significant decline in the urinary excretion of PGE_2, $PGF_{2\alpha}$, PGE-like material (i PGE), and 7-α-hydroxy 5-11-diketotetranorprostane-1,16-dioic acid. Despite such glowing reports of the use of indomethacin, aspirin, and ibuprofen in lowering, even normalizing, the PG levels, evidence is mounting to cast doubt on the primary role of PG in causation of Bartter's syndrome. It has been pointed out that biochemical and clinical escape can occur in patients with Bartter's syndrome on indomethacin therapy.[13,68] Kornerup et al.[70] reported a patient in whom the beneficial effects of indomethacin were lost subsequently, with persistent potassium leakage and hypokalemia despite long-term (9 months) indomethacin therapy.

The effects of indomethacin therapy on the renin–angiotensin system have also been elucidated fiarly well. Vierhapper and Waldhausl[101] studied PRA, aldosterone, and the effects of a competitive analog of angiotensin II before and after treatment in three patients with Bartter's syndrome. The use of the drug was associated with lowering of PRA and aldosterone levels and restoration of a pressor response to angiotensin II. Similarly, Radfar et al.[76] also noted that administration of indomethacin to patients with Bartter's syndrome resulted in correction of hyperreninemia and restoration of the pressor response to angiotensin II to normal, while only partially correcting the hypokalemia. The reversal of the altered vascular responsiveness seen in Bartter's syndrome has also been studied by the use of synthetic angiotensin in patients on indomethacin. Shimoyama[102] demonstrated that as early as 5 days following indomethacin administration, restoration of a normal pressor response to synthetic angiotensin II was attained. In the same patient, restoration of a normal pressor response was associated with the development of a hypotensive response when angitensin antagonists were administered. It appears that the reversal of altered vascular responsiveness in Bartter's syndrome is due to lowering of PG, PRA, or both, brought about by indomethacin therapy. In addition, the possible lowering of kallikrein levels may also have an impact on restoring the pressor response to normal following indomethacin therapy in Bartter's syndrome.

The effect of indomethacin therapy on reversing the abnormalities in the kallikrein system has also been evaluated. Sasaki et al.[103] showed that indomethacin administration to patients with Bartter's syndrome results in a marked decline in urinary kallikrein excretion, along with reductions in PGE excretion, aldosterone excretion, and PRA. These workers also noted that despite lowering of the above parameters and restoration of vascular responsiveness to angiotensin II, hypokalemia persisted. Lowering in urinary kallikrein excretion following indomethacin has also been demonstrated by three other groups of investigators.[12,42,104] The significance of these observations in terms of restoring angiotensin sensitivity has been a hotly debated issue.

In summary, use of indomethacin reverses several, but not all, metabolic abnormalities of Bartter's syndrome. It has been repeatedly observed that despite correction of several humoral abnormalities of Bartter's syndrome, hypokalemia and urinary potassium loss continue to persist. The effect of indomethacin in correcting the impaired absorption of chloride in the thick ascending limb is probably minimal. If one assumes that such a defect is the proximal cause of Bartter's syndrome, it is not surprising that kaliuresis persists despite indomethacin therapy. Yet, indomethacin clearly corrects three facets of abnormalities in Bartter's syndrome—the hyperreninemia, the PG excess, and the vascular insensitivity to angiotensin II. Clearly, more than any other drug used, indomethacin permits a greater degree of correction of potassium deficiency as well as a lowering of PRA than has been heretofore possible. Long-term treatment of children with Bartter's syndrome with indomethacin has resulted in high favorable outcomes. Dillon et al.[68] treated 10 children with the syndrome with indomethacin for 6–24 months and reported remarkable clinical and biochemical improvement with catchup growth demonstrable in all cases. Similar experiences have been reported by Delaney et al.,[43] Littlewood et al.,[14] and Gill et al.[10] Thus, with potassium replacement, spironolactone, and indomethacin, patients with Bartter's syndrome can be managed reasonably well. It remains to be seen whether therapy with indomethacin would prevent the development of proliferative and sclerotic changes in the renal vessels that can be seen in some patients with Bartter's syndrome.

References

1. Bartter, FC, Pronove P, Gill JR Jr, et al: Huperplasia of the juxtaglomerular complex with hyperaldosteronism and hypokalemic alkalosis. A new syndrome. *Am J Med* **33:**811, 1962.
2. Pronove P, MacCardle RC, Bartter PC: Aldosteronism, hypokalemia and a unique renal lesion in a five year old boy. *Acta Endocr (Copenh)* (Suppl. 51): 167, 1960.
3. Gill JR Jr: Bartter's syndrome. *Annu Rev Med* **31:**405, 1980.
4. White MG: Bartter's syndrome. A manifestation of renal tubular defects. *Arch Intern Med* **129:**41, 1972.
5. Goodman AD, Vagnucci AH, Hartroft PM: Pathogenesis of Bartter's syndrome. *N Engl J Med* **281:**1435, 1969.
6. Beilin LJ, Schiffman N, Crane M, et al: Hypokalaemic alkalosis and hyperplasia of the juxtaglomerular apparatus without hypertension or oedema. *Br Med J* **4:**327, 1967.
7. Verberckmoes R, Vandamme B, Clement J, et al: Bartter's syndrome with hyperplasia of renal medullary cells. Successful treatment with indomethacin. *Kidney Int* **9:**302, 1976.
8. Norby L, Flamenbaum W, Letz R, et al: Prostaglandin and aspirin therapy in Bartter's syndrome. *Lancet* **2:**604, 1976.
9. Fichman MP, Telfer N, Zia P, et al: Role of prostaglandins in the pathogenesis of Bartter's syndrome. *Am J Med* **60:**785, 1976.

10. Gill JR Jr, Frölich JC, Bowden RE, et al: Bartter's syndrome: A disorder characterized by urinary prostaglandins and a dependence of hyperreninemia on prostaglandin synthesis. *Am J Med* **61**:43, 1976.

11. Donker HAM, DeJong PE, Statiu S, et al: Indomethacin in Bartter's syndrome: Does this symdrome represent a state of hyperprostaglandinism? *Nephron* **19**:200, 1977.

12. Halushka PV, Wohltmann H, Privitera PJ, et al: Bartter's syndrome: Urinary prostaglandin E-like material and Kallikrein; indomethacin effects. *Ann Intern Med* **87**:281, 1977.

13. Bourke E, Delaney V: Pathogenesis of Bartter's syndrome: A family study. *Kidney Int* **12**:447, 1977.

14. Littlewood JM, Lee MR, Meadow SR: Treatment of Bartter's syndrome in early childhood with prostaglandin synthetase inhibitors. *Arch Dis Child* **53**:43, 1978.

15. Bartter FC, Gill JR Jr, Frölich J, et al: Prostaglandins are overproduced by the kidneys and mediate hyperreninemia in Bartter's syndrome. *Trans Assoc Am Physicians* **89**:77, 1976.

16. Janszen FH, Nugteren DH: Histochemical localization of prostaglandins synthetase. *Histochem J* **27**:159, 1971.

17. Muirhead EE, Germain G, Leach BE, et al: Production of renomedullary prostaglandins by renomedullary interstitial cells grown in tissue culture. *Circ Res* **31**(Suppl. 2):161, 1972.

18. Bowden RE, Gill JR Jr, Radfar N, et al: Prostaglandin synthetase inhibitors in Bartter's syndrome: Effect on immunoreactive prostaglandin E excretion. *JAMA* **239**:117, 1978.

19. Katz FH, Bortz AI: Treatment of Bartter's syndrome with naproxen. *N Engl J Med* **299**:100, 1978.

20. Garin EH, Fennell RS III, Iravani A, et al: Treatment of Bartter's syndrome with indomethacin. *Am J Dis Child* **134**:258, 1980.

21. Vane JR, McGiff JC: Possible contributions of endogenous prostaglandins to the control of blood pressure. *Circ Res* **36**(Suppl. 1):68, 1975.

22. Lonigro AJ, Terragno NA, Malik KU, et al: Differential inhibition by prostaglandins of renal actions of pressor stimuli. *Prostaglandins* **3**:595, 1973.

23. Larsson C, Weber P, Anggard E: Arachidonic acid increases and indomethacin decreases plasma renin activity in the rabbit. *Eur J Pharmacol* **28**:391, 1974.

24. Krafkoff LR, De Guia D, Vlachakis N, et al: Effect of sodium balance on arterial blood pressure and renal responses to prostaglandins A_1 in man. *Circ Res* **33**:539, 1973.

25. Gill JR Jr, Bartter FC: Evidence for a prostaglandin-independent defect in chloride reabsorption in the loop of Henle as a proximal cause of Bartter's syndrome. *Am J Med* **65**:766, 1978.

26. Galvez OG, Bay WH, Roberts BW, et al: The hemodynamic effects of potassium deficiency in the dog. *Cir Res* **40** (Suppl. 1):1, 1977.

27. Zusman RM, Keiser HR: Prostaglandin biosynthesis by rabbit renomedullary interstitial cells in tissue culture. Stimulation by angiotensin II, bradykinin and arginine vasopressin. *J Clin Invest* **60**:215, 1977.

28. Gill JR, Bartter FC, Taylor AA, et al: Impaired tubular chloride reabsorption as a proximal cause of Bartter's syndrome. *Clin Res* **25**:526, 1977 (Abstract).

29. Ferris TF: Prostaglandins, potassium, and Bartter's syndrome. *J Lab Clin Med* **92**:663, 1978.

30. Rutecki GW, Nally JV, Bay WH, et al: The acute effects of lithium on renal function. In *Proceedings of the American Society of Nephrology*, Washington, D.C. 1977, p. 119 (Abstract).

31. Zipser RD, Rude RK, Zia PK, et al: Regulation of urinary prostaglandins in Bartter's syndrome. *Am J Med* **67**:263, 1979.

32. Solomon RJ, Brown RS: Bartter's syndrome: New insights into pathogenesis and treatment. *Am J Med* **59**:575, 1975.

33. Tan SY, Bravo E, Mulrow PJ: Impaired renal prostaglandin E_2 biosynthesis in human hypertensive states. *Prostaglandins Med* **1**:76, 1978.

34. Zipser R, Zia P, Stone R, et al: The prostaglandin and kallikrein–kinin systems in mineralocorticoid escape. *J Clin Endocrinol Metab* **47**:996, 1978.

35. Gullner H-G, Bartter FC, Cerletti C, et al: Prostacyclin overproduction in Bartter's syndrome. *Lancet* **2**:767, 1979.

36. Moncada S, Higgs EA, Vane JR: Human arterial and venous tissues generate prostacyclin (prostaglandin X), a potent inhibitor of platelet aggregation. *Lancet* **1**:18, 1977.

37. Weksler BB, Marcus AJ, Jaffe EA: Synthesis of prostaglandin I_2 (prostacyclin) by cultured human and bovine endothelial cells. *Proc Natl Acad Sci USA* **74**:3922, 1977.

38. Frolich JC, Wilson TW, Sweetman BJ, et al: Urinary prostaglandins. Identification and origin. *J Clin Invest* **55:**763, 1975.

39. Bolger PM, Eisner GM, Ramwell PW, et al: Renal actions of prostacyclin. *Nature* **271:**467, 1978.

40. Armstrong JM, Lattimer N, Moncada S, et al: Comparison of the vasodepressor effects of prostacyclin and 6-oxo-prostaglandin F_1, with those of prostaglandin E_2 in rats and rabbits. *Br J Pharmacol* **62:**125, 1978.

41. O'Regan S, Rivard GE, Robitaille PO: Aspirin improves platelet dysfunction in Bartter's syndrome. In *Proceedings of the International Congress of Nephrology*, Montreal, S. Karger, New York 1978; p. 1.

42. Vinci JM, Gill JR Jr, Bowden RE, et al: The kallikrein–kinin system in Bartter's syndrome and its response to prostaglandin synthetase inhibition. *J Clin Invest* **61:**1671, 1978.

43. Delaney VB, Oliver JF, Simms M, et al: Bartter's syndrome: Physiological and pharmacological studies. *Q J Med* **50:**213, 1981.

44. Bardgette JJ, Stein JH: Pathophysiology of Bartter's syndrome. In Brenners, Stein J (eds): *Contemporary Issues in Nephrology: Acid–Base and Potassium Homeostasis.* Churchill Livingstone, New York, 1978, pp. 269–288.

45. Kurtzman NA, Guiterrez LF: The pathophysiology of Bartter's syndrome. *JAMA* **234:**758, 1975.

46. Rocha AS, Kokko JP: Sodium chloride and water transport in the medullary thick ascending limb of Henle. *J Clin Invest* **52:**612, 1973.

47. Burg MB, Green N: Function of the thick ascending limb of Henle's loop. *Am J Physiol* **224:**659, 1973.

48. Kiil F: Renal energy metabolism and regulation of sodium reabsorption. *Kidney Int* **11:**153, 1977.

49. Jørgensen PL: The function of (Na^+, K^+)-ATPase in the thick ascending limb of Henle's loop. *Curr Prob Clin Biochem* **6:**190, 1976.

50. Westenfelder C, Kurtzman NA: Bartter's syndrome: A disorder of active sodium and/or passive chloride transport in the thick ascending limb of Henle's loop. *Mineral Electrolyte Metab* **5:**135, 1981.

51. Stokes JB: Effect of Prostaglandin E_2 on chloride transport across the rabbit thick ascending limb of Henle. *J Clin Invest* **64:**495, 1979.

52. Baehler RW, Work J, Kotchen TA, et al: Studies on the pathogenesis of Bartter's syndrome. *Am J Med* **69:**933, 1980.

53. Trygstad CW, Mangos JA, Bloodworth JMB Jr, et al: A sibship with Bartter's syndrome. Failure of total adrenalectomy to correct the potassium wasting. *Pediatrics* **44:**234, 1969.

54. Takayasu H, Aso Y, Lakanchi K: A case of Bartter's syndrome with surgical treatment followed for four years. *J Clin Endocrinol Metab* **32:**842, 1971.

55. Malnic G, Klose RM, Giebisch G: Micropuncture study of renal potassium excretion in the rat. *Am J Physiol* **206:**674, 1964.

56. Khuri RN, Wiederholt M, Strieder N, et al: Effects of flow rate and potassium intake on distal tubular potassium transfer. *Am J Physiol* **228:**1249, 1975.

57. Mace JW, Hambridge KM, Gotlin RW, et al: Magnesium supplementation in Bartter's syndrome. *Arch Dis Child* **48:**485, 1973.

58. Sutherland LE, Hartroft P, Balis JU, et al: Bartter's syndrome. A report of four cases, including three in one sibship, with comparative histologic evaluation of the juxtaglomerular apparatuses and glomeruli. *Acta Paediatr Scand* **201**(Suppl.):1, 1970.

59. Luke RG, Wright FS, Fowler N, et al: Effects of potassium depletion on renal tubular chloride transport in the rat. *Kidney Int* **14:**414, 1978.

60. Garella S, Chazan JA, Cohen JJ: Saline-resistant metabolic alkalosis or "chloride-wasting nephropathy." *Ann Intern Med* **73:**31, 1970.

61. Robson WL, Arbus GS, Balfe JW: Bartter's syndrome. Differentiation into two clinical groups. *Am J Dis Child* **133:**636, 1979.

62. Cannon PJ, Leeming JM, Sommers SC, et al: Juxtaglomerular cell hyperplasia and secondary hyperaldosteronism (Bartter's syndrome): A reevaluation of the pathophysiology. *Medicine (Baltimore)* **47:**107, 1968.

63. Chaimovitc CJ, Levi J, Better US, et al: Studies on the site of renal salt loss in a patient with Bartter's syndrome. *Pediatr Res* **7:**89, 1973.

64. Chan JCM, Malekzedeh MH, Anand SK: Defect in renal tubular sodium reabsorption in a patient with Bartter's syndrome. *Clin Proc Children's Hosp Natl Med Cent* **31**:67, 1975.

65. Fujita T, Sakuguchi H, Shibagaki M, et al: The pathogenesis of Bartter's syndrome. *Am J Med* **63**:467, 1977.

66. Bartter FC, Delea CS, Kawasaki T, et al: The adrenal cortex and the kidney. *Kidney Int* **6**:272, 1974.

67. Weber PC, Larson C, Anggard E, et al: Stimulation of renin release from rabbit renal cortex by arachidonic acid and prostaglandin endoperoxides. *Cir Res* **39**:868, 1976.

68. Dillon MJ, Shah V, Mitchell MD: Bartter's syndrome: 10 cases in childhood. Results of long-term indomethacin therapy. *Q J Med* **48**:429, 1979.

69. Silverberg AB, Mewnes PA, Cryer PE, et al: Resistance to endogenous norepinephrine in Bartter's syndrome: Reversion during indomethacin administration. *Am J Med* **64**:231, 1978.

70. Kornerup HJ, Pedersen EB, Petersen VP: Bartter's syndrome without hyperplasia of the juxtaglomerular apparatus, treated with indomethacin. *Acta Med Scand* **204**:235, 1978.

71. Pasternack A, Perheentupa J, Launiala K, et al: Kidney biopsy findings in familial chloride diarrhoea. *Acta Endocrinol* **55**:1, 1967.

72. Fleischer N, Brown H, Graham D, et al: Chronic laxative-induced hyperaldosteronism and hypokalemia simulating Bartter's syndrome. *Ann Intern Med* **70**:791, 1969.

73. McGiff JC: Bartter's syndrome results from an imbalance of vasoactive hormones. *Ann Intern Med* **87**:369, 1977.

74. Kennedy BJ: Bartter's syndrome—Limelight on prostaglandins. *JAMA* **239**:137, 1978.

75. Dray F: Bartter's syndrome: Contrasting patterns of prostaglandin excretion in children and adults. *Clin Soc Mol Med* **54**:115, 1978.

76. Radfar N, Gill JR Jr, Bartter FC, et al: Hypokalemia, in Bartter's syndrome and other disorders, produces resistance to vasopressors via prostaglandin overproduction. *Proc. Soc Exp Biol Med* **158**:502, 1978.

77. Bryan GT, MacCardle RC, Bartter FC: Hyperaldosteronism, hyperplasia of the juxtaglomerular complex, normal blood pressure and dwarfism: Report of a case. *Pediatrics* **37**:43, 1966.

78. Hall BD: Preponderance of Bartter's syndrome among blacks. *N Engl J Med* **285**:581, 1971.

79. Desmit EM, Cost WS, Brown JJ, et al: An unusual type of hypokalaemic alkalosis with disturbance of renin and aldosterone. *Acta Endocrinol (Copenh)* **64**:75, 1970.

80. McCredie DA, Rotenberg E, Williams AL: Hypercalciuria in potassium-losing neuropathy: A variant of Bartter's syndrome. *Aust Paediatr J* **10**:286, 1974.

81. James T, Holland NH, Preston D: Bartter's syndrome: Typical facies and normal plasma volume. *Am J Dis Child* **129**:1205, 1975.

82. Fricker H, Frey K, Vallotton MB, et al: Bartter's syndrome and tubular functional disturbances. *Helv Paediatr Acta* **30**:61, 1975.

83. Sann L, David L, Bernheim J, et al: Hypophosphatemia and hyperparathyroidism in a case of Bartter's syndrome. *Helv Paediatr Acta* **33**:299, 1978.

84. Simopoulos AP: Growth characteristics in patients with Bartter's syndrome. *Nephron* **23**:130, 1979.

85. Kelsch RC, Gulhoed GW, Vander AJ, et al: Plasma renin in Bartter's syndrome: Response of depression of stimuli to renin release. *J Pediatr* **74**:821, 1969.

86. Ertel NH: Idiopathic hypovolemia and orthostatic hypotension with secondary hyper-aldosteronism. *Clin Res* **16**:265, 1968.

87. Meyer WJ III, Gill JR Jr, Bartter FC: Gout as a complication of Bartter's syndrome. *Ann Intern Med* **83**:56, 1975.

88. Modlinger RS, Nicolis GL, Krakoff LR, et al: Some observations on the pathogenesis of Bartter's syndrome. *N Engl J Med* **289**:1022, 1973.

89. Gullner HG, Gill JR Jr, Bartter FC, et al: A familial disorder with hypokalemic alkalosis, hyperreninemia, aldosteronism and high prostaglandins that is not Bartter's syndrome. *Clin Res* **27**:520, 1979 (Abstract).

90. Snow MH, Nicol P, Wilkinson R, et al: Normotensive primary aldosteronism. *Br Med J* **1**:1125, 1976.

91. Zipser RD, Speckart PF: "Normotensive" primary aldosteronism. *Ann Intern Med* **88**:655, 1978.

92. Shiroto H, Ando H, Ebitani I, et al: Normotensive primary aldosteronism. *Am J Med* **69**:603, 1980.

93. Makoff DT: Acid–base metabolism. In Maxwell MA, Kleeman CR (eds): *Clinical Disorders of Fluid and Electrolyte Metabolism,* 2nd ed. McGraw-Hill, New York, 1972, p. 332.
94. Harrington JT, Cohen JJ: Measurement of urinary electrolytes. *N Engl J Med* **293:**1241, 1975.
95. Nivet H, Grenier B, Rolland JC, et al: Raised urinary prostaglandins in patient without Bartter's syndrome. *Lancet* **1:**333, 1978.
96. Ramos E, Hall-Craggs M, Demers LM: Surreptitious habitual vomiting simulating Bartter's syndrome. *JAMA* **243:**1070, 1980.
97. Veldhuis JD, Bardin CW, Demers LM: Metabolic mimicry of Bartter's syndrome by covert vomiting. Utility of urinary chloride detemrinations. *Am J Med* **66:**361, 1979.
98. Burg M, Stoner L, Cardinal J, et al: Furosemide effect on isolated perfused tubules. *Am J Physiol* **225:**119, 1973.
99. Shils ME: Experimental human magnesium depletion. *Medicine (Baltimore)* **84:**61, 1969.
100. Gitelman HJ, Graham JB, Welt LG: A new familial disorder characterized by hypokalemia and hypomagnesemia. *Clin Res* **14:**108, 1966.
101. Vierhapper H, Waldhausl W: Effect of indomethacin upon the renin–angiotensin system in patients with Bartter's syndrome. *Eur J Clin Invest* **10:**119, 1980.
102. Shimoyama R: Reversal of altered vascular responsiveness in Bartter's syndrome by indomethacin treatment. *J Clin Endocrinol Metab* **51:**908, 1980.
103. Sasaki H, Kawasaki T, Okumura M, et al: Bartter's syndrome: Effect of indomethacin on prostaglandins, urinary kallikrein, renin–angiotensin–aldosterone system, and the response to angiotensin II antagonist. *Endocrinol Jpn* **27:**417, 1980.
104. Lechi A, Mantero F, Opocher G, et al: Effect of indomethacin on urinary kallikrein excretion in Bartter's syndrome of the adult. *J Endocrinol Invest* **4:**17, 1981.

8

Selective Hypoaldosteronism

Selective hypoaldosteronism refers to the development of aldosterone failure in the presence of adequate glucocorticoid function. This can occur in one of many ways. The most common and important form of selective aldosterone deficiency is secondary to hyporeninemia and is hence referred to as hyporeninemic hypoaldosteronism. Much of this chapter is devoted to this form of selective hypoaldosteronism. Isolated mineralocorticoid failure can also result from congenital defects in the biosynthetic pathway for aldosterone synthesis (primary aldosterone deficiency) or from acquired defects in the zona glomerulosa. The third mechanism for the development of selective aldosterone deficiency involves tubular resistance to the action of aldosterone ("pseudohypoaldosteronism"). In contrast to the other two varieties, the circulating aldosterone levels are elevated in pseudohypoaldosteronism, since it represents a classic example of target organ resistance. Table 35 illustrates the various types of aldosterone deficiency.

TABLE 35.
Types of Hypoaldosteronism

I.	Hyporeninemic hypoaldosteronism
II.	Hyperreninemic hypoaldosteronism
	Addison's disease
	Enzymatic defects in aldosterone synthesis
	Acquired defects in glomerulosa
	Critical illness
III.	Pseudohypoaldosteronism
	Tubular resistance
IV.	Miscellaneous
	Heparin

Hyporeninemic Hypoaldosteronism

Historical Perspectives

In 1957, Hudson et al.[1] first described a patient with Stokes–Adams attacks, hyperkalemia, low aldosterone levels in the urine, and intact glucocorticoid function. The first series of patients manifesting with hyperkalemia and hyperchloremic acidosis was reported in 1964 by Carroll and Farber.[2] During the sixties, several isolated case reports of "analdosteronism" appeared in the literature, with speculations on the mechanisms that underlie the development of isolated hypoaldosteronism.[3–7] It was not until 1972 that the first breakthrough arrived in unraveling the mechanism for the syndrome. Schambelan et al.[8] studied six adults with isolated hypoaldosteronism and demonstrated that primary deficiency of renin was the underlying cause for the syndrome. Originally thought to be a rarity, the syndrome of hyporeninemic hypoaldosteronism is being recognized with increasing frequency. For instance, until 1975, only 30 cases had been documented in the literature, but by 1986, nearly 100 had been reported.

In addition to the increased awareness of the existence of the syndrome, the literature of the past decade has focused on numerous mechanisms that cause the hyporeninemia of hypoaldosteronism. Theories abound as to the nature of the lesion that impairs the reserve of the juxtaglomerular (JG) apparatus. Thus, intrinsic damage to the JG apparatus, impaired conversion of prorenin to renin, insufficient sympathetic regulation, and subclinical volume expansion have all been proposed as mechanisms to explain the hyporeninemia. The notion that prostaglandin deficiency in general, and prostacyclin deficiency in particular, plays a central role in the development of hyporeninemia is becoming increasingly accepted. The development of the syndrome with the use of indomethacin has generated considerable interest in iatrogenic, reversible etiologies for the syndrome. Recent literature has also focused on the concept that additional and independent defects in aldosterone synthesis may also coexist in the syndrome of hyporeninemic hypoaldosteronism. The possibility that increased amounts of atrial natriuretic peptide may contribute to the suppression of aldosterone adds another dimension to the pathogenesis of this disorder. The numberous facets that have evolved around this once rare disorder make it one of the most intriguing disorders of the adrenal cortex. The definition, incidence, pathophysiology, and recognition of this disorder form the basis of this chapter.

Definition

Since there are several components to selective hyporeninemic hypoaldosteronism, this disorder is conceptually viewed as a syndrome. These components include:

1. Hyperkalemia
2. A unique variety of metabolic acidosis ("type IV renal tubular acidosis (RTA")
3. Varying degrees of renal insufficiency
4. Hyporeninemia
5. Hypoaldosteronism
6. Intact glucocorticoid reserve

The usual diagnostic clue that leads to eventual diagnosis is the presence of unexplained persistent hyperkalemia out of proportion to the impairment in the renal function.

Etiology

The clinical background that sets the stage for the development of hyporeninemic hypoaldosteronism is worth examining. Three main underlying factors, when taken together, are present in nearly all patients with hyporeninemic hypoaldosteronism. These are chronic renal disease, diabetes mellitus, and advanced age. The one factor common to all three is the propensity toward decreased renin production. Of the 50 reported cases reviewed by Michelis and Murdaugh,[9] mild to moderate renal failure was present in 74% and diabetes mellitus was seen in 46%. In the 81 cases reviewed by DeFronzo,[10] chronic renal failure and diabetes were present in 57 and 40 patients, respectively. The mean age of patients with hyporeninemic hypoaldosteronism is 65 years. The association of chronic renal failure and diabetes with hyporeninemic hypoaldosteronism is next reviewed briefly.

Chronic Renal Failure

Chronic renal failure of varying degrees is usually found in hyporeninemic hypoaldosteronism. The blood urea nitrogen and creatinine levels in these patients fluctuate widely. The spectrum of renal diseases that cause hyporeninemic hypoaldosteronism is also impressively variable. Thus, while diabetic nephropathy (particularly glomerulosclerosis) leads the list, the syndrome has also been found to occur in association with a variety of diseases that cause tubulointerstitial damage. Gout, chronic pyelonephritis, metabolic stone disease, analgesic abuse, hypertensive nephrosclerosis, and cystic disease have all been reported to cause hyporeninemic hypoaldosteronism.[10,11] It appears that regardless of the etiology, the metabolic alterations that result from damage to renal interstitium can set in motion a trail of events that culminate in hyporeninemic hypoaldosteronism. The alterations in renal function are particularly relevant to phenomena that, in addition to hyporeninism, contribute to the hyperkalemia and the acidosis associated with the syndrome. Alterations in renal function play an important, and possibly independent, role in causing changes in renal sodium handling, in reclamation of filtered bicarbonate, in impairing ammoniagenesis, and in causing inadequate secretion of hydrogen ion. These changes are independent of the reduced glomerular filtration rate (GFR) and are thought to represent important mechanisms in the mediation of several expressions of the syndrome.

Diabetes Mellitus

The association of diabetes with hyporeninemic hypoaldosteronism is so strong that it has come to represent a model for explaining the abnormal electrolyte and acid–base state of that syndrome. The pervasive effects of diabetes mellitus in causing the syndrome of hyporeninemic hypoaldosteronism are illustrated by multifactorial mechanisms.

1. Diabetic glomerulosclerosis can result in damage to the JG apparatus, resulting in decreased renin production. Of particular interest is the histological observation reported by Schindler and Sommers.[12] These workers examined renal biopsies from 30 patients with diabetes and found fibrosis between the JG cells and the macula densa, with a subnormal content of renin granules per JG apparatus in diabetic kidneys when compared to normal. This may provide an anatomical explanation for the hyporeninemia seen in hyporeninemic hypoaldosteronism.

2. The development of autonomic neuropathy may decrease the sympathetic signals necessary for release of renin from the JG apparatus. Two independent workers have reported lower basal and posture-stimulable plasma renin activity (PRA) in diabetes with neuropathy when compared with controls.[13,14]

3. Since insulin is responsible for potassium uptake by several tissues, such as muscle and liver, it is reasonable to assume that a similar situation exists for the zona glomerulosa cells as well. The natural extension of such a concept is the assumption that intracellular K depletion may exist in the adrenal cortex of diabetic patients. It is tenable that in the presence of intracellular potassium depletion the cells of the zona glomerulosa generate aldosterone poorly.[15,16] Tait and Tait[16] evaluated the effect of potassium depletion on the maximal steroidogenic response of purified zone glomerulosa cells to angiotensin II and observed a diminutive response. Given the known effect of K^+ concentrations on the conversion of cholesterol to DOC and corticosterone to aldosterone, cellular potassium depletion can clearly lead to impaired production of aldosterone. The occurrence of acquired enzymatic defects in aldosterone biosynthesis in diabetic patients has been described.[17]

4. The decreased GFR caused by diabetic nephropathy may have a permissive role in the development of hyperkalemia. In this regard it is to be

noted that, in general, significant impairment in potassium excretion does not occur in renal failure unless the GFR declines to below 10–15 ml/min. The GFR in most diabetics with the syndrome of hyporeninemic hypoaldosteronism exceeds 20 ml/min. Therefore, the role of decreased GFR in causing potassium retention in the syndrome of hyporeninemic hypoaldosteronism is perfunctory.

5. The development of interstitial disease and tubular damage may impair the ability of diabetics to excrete a potassium load. DeFronzo[10] examined the ability of four normokalemic patients with long-standing diabetes to excrete a load of intravenous potassium chloride. A significant impairment in the excretion of KCl was noted, despite a normal or minimally decreased GFR and an intact renin–angiotensin–aldosterone axis. Thus, a primary defect in tubular secretion of potassium may underlie the hyperkalemia, in at least some diabetics.

6. Finally, the conversion of prorenin to renin may in fact be impaired in some diabetics. DeLeiva et al.[17] have postulated such a defect in the development of hyporeninemic hypoaldosteronism in some diabetics.

Thus, in terms of etiology, patients with underlying renal disease, particularly of the tubulointerstitial variety, are prime candidates for the development of hyporeninemic hypoaldosteronism. In addition to diabetes, renal disease, and the effects of advanced age on renin secretion, two other etiological factors of relevance merit mention. First, lead intoxication can result in lowering of renin activity.[18] Since plumbism can also cause renal dysfunction, the development of hyporeninemic hypoaldosteronism can occur in a patient with that disease. Second, chronic heparin administration can lower aldosterone levels and result in selective hypoaldosteronism.[19]

Pathogenesis

The pathogenesis of hyporeninemic hypoaldosternism can be viewed from three vantage points: the phenomena that underlie the development of hyporeninemia, the hypoaldosteronism, and the metabolic acidosis.

Hyporeninemia

Since the report by Schambelan et al.[8] that hyporeninemia constituted an important mechanism in the development of isolated hypoaldosteronism, the literature is replete with proposed mechanisms to explain the hyporeninemia. Currently, five major theories are in vogue to account for the hyporeninemia:

1. Renin suppression by volume expansion.
2. Renin suppression due to impaired sympathetic signals.
3. Impaired conversion of "prorenin" to renin.
4. Impaired renin secretion due to prostaglandin deficiency.
5. Anatomical destruction of the JG apparatus.

Renin Suppression by Volume Expansion. According to this postulate, hyporeninemia is presumed to be a secondary phenomenon, reflecting the physiological response of renin to volume expansion. It presupposes that the reason for the low PRA in hyporeninemic hypoaldosteronism stems from the fact that these patients are volume expanded. Several observations support this theory.

First, patients with hyporeninemic hypoaldosteronism, in contrast to patients with Addison's disease, are generally not hypotensive. Orthostatic hypotension is not a feature of hyporeninemic hypoaldosteronism, unless the disorder is complicated by autonomic dysfunction, a combination frequently encountered in diabetics. In fact, the blood pressure of patients with hyporeninemic hypoaldosteronism is often elevated. Phelps et al.[11] carefully reviewed the blood pressure measurements of 86 patients with isolated hypoaldosteronism reported in the literature and pointed out that, with the available data, hypertension was observed in approximately 50% of the cases. It is well known that the hypertension of chronic renal disease is, at least in part, mediated by volume expansion. Indeed, this mechanism is believed to be responsible for the low PRA encountered in patients with essential hypertension of the low-renin variety ("low-renin essential hypertension"). To prove that such a mechanism plays a role in patients with hyporeninemic hypoaldosteronism, it is essential to (1) demonstrate an increase in total exchangeable sodium and extracellular fluid volume, (2) demonstrate a rise in PRA with sodium restriction and/or diuretic use, and (3) demonstrate that negative sodium balance improves the blood pressure as well as raises PRA level to normal. As will be shown, each of these phenomena has been shown to occur in patients with hyporeninemic hypoaldosteronism.

Second, several workers have demonstrated expansion of extracellular fluid volume in patients with hyporeninemic hypoaldosteronism. Oh et al.[20] have demonstrated that the total exchangeable sodium and extracellular fluid volume were increased in four patients with hyporeninemic hypoaldosteronism, three of whom showed elevated blood pressure. De Chatel et al.[21] have also noted a 10% increase in the total body sodium in some patients with diabetes, a disorder that is particularly likely to be associated with hyporeninemic hypoaldosteronism. The observation by Oster et al.[22] that chronic expansion of extracellular volume by sodium bicarbonate abuse can result in a syndrome analogous to hyporeninemic hypoaldosteronism is also supportive of this theory.

Third, the strongest line of evidence that supports a role for volume expansion in the causation of hyporeninemic hypoaldosteronism comes from studies that evaluate the effect of salt depletion on renin levels in patients with this syndrome. Oh et al.[20] measured PRA before and after several weeks of furosemide therapy in four patients with hyporeninemic hypoaldosteronism and demonstrated a rise in PRA with furosemide. Perez et al.[23] have evaluated the effect of severe sodium depletion by prolonged dietary restriction (10 meq/day) and furosemide administration in five patients with the syndrome of hyporeninemic hypoaldosteronism. Their data reveal substantial increments in both PRA and plasma aldosterone levels following sodium depletion. In a study of a single patient with isolated hypoaldosteronism, Szylman et al.[24] noted that the use of diuretics increased the patient's PRA with a simultaneous decrease in the patient's weight and blood pressure. These studies not only point to the presence of volume expansion in the development of hyporeninemia, but also indicate a valuable role for diuretic therapy in the management of these patients.

Renin Suppression Due to Impaired Sympathetic Signals. The secretion of renin by the JG apparatus is controlled by the sympathetic nervous system.[25] The demonstration of sympathetic nerve terminals in the JG apparatus,[26] as well as the ability of epinephrine to stimulate renin secretion,[27] is usually cited as an example of sympathetic mediation of renin release. Since both hyporeninemic hypoaldosteronism and autonomic neuropathy are particularly prevalent in diabetics, the effect of sympathetic dysfunction on the release of renin in diabetics has been under scrutiny. Tuck et al.[28] studied five diabetic patients with hyporeninemic hypoaldosteronism and noted impairment in the renin, aldosterone, and epinephrine responses to posture as well an impaired renin response to isoproteronol infusion. In general, however, the PRA and aldosterone responsiveness in diabetics is impressively heterogenous. De Chatel et al.[21] evaluated the postural responses of PRA and aldosterone in 60 diabetics, many of whom had hypertension with orthostasis, and found subnormal responses of both PRA and plasma aldosterone in 70%. In 22% of patients, a subnormal PRA response was observed in conjunction with a normal plasma aldosterone response, while 8% of the diabetics studied showed well-preserved responses of both PRA and aldosterone to posture. The correlation of the impairment in the PRA response in diabetics with autonomic neuropathy, peripheral neuropathy, renal disease, and orthostatic hypotension has yielded interesting results. Christlieb et al.[13,29] compared the PRA response to upright posture in diabetics with and without orthostatic hypotension and found significant impairment in the group with orthostatic hypotension. This, of course, is not surprising since patients with idiopathic (or primary) orthostatic hypotension typically demonstrate a flat PRA response to upright posture.[14,30] More interesting is the observation by Christlieb et al.[31] that diabetics whose disease is complicated by hypertension, nephropathy, or autonomic dysfunction are more likely to demonstrate decreased PRA with poor responses to posture. While most workers have found lower PRA in diabetics with neuropathy than those without neuropathy,[32] it is becoming evident that abnormal PRA responses may also be seen in diabetics without overt neuropathy or nephropathy. For instance, Perez et al.[32] have demonstrated that renin and aldosterone unresponsiveness can be encountered even in diabetics without hyperkalemia vasculopathy or detectable autonomic dysfunction. These observations are in keeping with those of Christlieb et al.,[33] who found hyporeninemia and hypoaldosteronism in some diabetics without any clinical evidence of nephropathy or neuropathy.

From the foregoing studies several conclusions emerge. First, as a group, diabetics with nephropathy seem to be the ones most prone to develop impairment in posture-mediated renin responses. Second, the syndrome of hyporeninemic hypoaldosteronism is also more frequently encountered in diabetics with neuropathy. Third, abnormal PRA and aldosterone response to posture may be encountered in diabetics even in the absence of overt nephropathy or detectable neuropathy. It is not clear whether this is a harbinger for the subsequent development of the syndrome of hyporeninemic hypoaldosteronism. Finally, hyperkalemia in some diabetic patients correlates closely with hyperglycemia, indicating that hyperglycemia can directly cause hypoaldosteronism.[33] In the final analysis, sympathetic dysfunction, while clearly an important factor, cannot solely explain the development of hyporeninemic hypoaldosteronism. Other mechanisms must be operative in the pathogenesis of the syndrome.

Impairment in Conversion of Prorenin to Renin.   Recent evidence suggests that renin is synthesized as a 50,000-dalton single-chain polypeptide. This prorenin ("big renin," "inactive renin") has generated considerable controversy. According to some workers, the inactive renin may represent renin inhibitor protein complexes, which would reflect an in vivo storage form of the enzyme, or alternatively might represent an in vitro artifact.[19,20] Day et al.[34] and DeLeiva et al.[17] have detected a renin precursor of molecular weight 63,000 daltons in diabetics. Theoretically, decreased formation of biologically active renin might account for the decreased plasma renin activity seen in patients with hyporeninemic hypoaldosteronism. The major difficulties in accepting impaired conversion of prorenin to renin in the causation of hyporeninemic hypoaldosteronism are twofold. First, considerable controversy shrouds the methodology for measuring prorenin.[35–37] The presence of renin activators and renin inhibitors in the plasma has further confounded the interpretation of "prorenin activity" in plasma. Second, even if one accepts the methodology, the numbers of patients reported with hyporeninemic hypoaldosteronism and increased prorenin concentrations are too few to permit conclusive interpretation. The importance of conversion of inactive renin to active renin has been readdressed in the study reported by Sowers et al.[38] In 10 patients with hyporeninemic hypoaldosteronism, the inactive renin (prorenin) was normal, while the active renin was decreased, with a significantly increased ratio of inactive renin to active renin. Although impaired conversion of prorenin to renin has been previously reported in diabetics, 9 of 10 patients in the study by Sowers et al.[38] were nondiabetics; these findings are in agreement with two other reports[39,40] that indicated the presence of impaired conversion of inactive renin in nondiabetics as well. The theory of impaired conversion of prorenin to active renin as a mechanism to explain hyporeninemia continues to remain controversial.

The Role of Renal Prostaglandins in the Development of Hyporeninemic Hypoaldosteronism.   Renal prostaglandins play a crucial role in the modulation of the renin–angiotensin–aldosterone axis. Experiments in rabbits have indicated that arachidonic acid increases PRA, while inhibitors of prostaglandin synthetase decrease PRA.[41] In human volunteers, the renin response to furosemide and posture can be markedly attenuated by the prior administration of indomethacin.[42] Numerous reports have documented that prostaglandin synthetase inhibition is associated with significant attenuation of renin release.[43,44] It is believed that prostaglandins directly stimulate release of renin from the JG apparatus.[45–47] Of the renal prostaglandins, prostacylin (prostaglandin I_2) is particularly potent in mediating renin release. Patrano et al.[48] have shown in humans that prostaglandin I_2 can directly stimulate renin secretion in a dose-dependent manner, quite independent of other mechanisms such as beta-receptor activation. The relationship between renal prostaglandins, renin, and angiotensin II is "symbiotic"; the production of vasodilatory prostaglandins by the renal tissue is augmented by angiotensin II,[49] while the production of vasoconstrictive angiotensin II (via renin) is activated by renal prostaglandins.

In view of the important effect exerted by prostaglandins on renin secretion, it was inevitable that the syndrome of hyporeninemic hypoaldosteronism would be described in association with the use of indomethacin. Tan et al.[50] described the occurrence of hyporeninemic hypoaldosteronism in a young woman with glomerulonephritis who was receiving indomethacin. The constellation of hyperkalemia despite only mild renal failure, low PRA, low aldosterone levels, and low urinary PGE_2 excretion was normalized upon withdrawal of indomethacin. The PRA and plasma aldosterone responses, which had been blunted while the patient was on indomethacin, showed a supranormal rebound following withdrawal of indomethacin, along with significant kaliuresis. More important, the suppression of PRA, renin, and aldosterone, as well as the development of antikaliuresis, could be reproduced upon reinstitution of indomethacin therapy. Thus, indomethacin, in susceptible individuals, could result in the development of hyporeninemic hypoaldosteronism.

A new dimension to the "prostaglandin connection" was added when Nadler et al.[51] demonstrated markedly reduced levels of prostacyclin in seven patients with hyporeninemic hypoaldosteronism. The urinary excretion of 6-keto-prostaglandin $F_{1\alpha}$, a stable metabolite that reflects prostacyclin activity, was markedly reduced in the group with hyporeninemic hypoaldosteronism in comparison to normal subjects and matched controls with renal insufficiency. This contrasted with the renal PGE_2 excretion, which was normal in all three groups. Further, infusions of calcium or norepinephrine failed to increase renal prostacyclin in patients with hyporeninemic hypoaldosteronism, in contrast to the other two groups who responded. The low prostacyclin generation, in this study, was unrelated to decreases in GFR or the presence of underlying diabetes mellitus. While the study by Nadler et al.[51] clearly imparts a role for prostacyclin deficiency in the evolution of hypoaldosteronism, it is unclear whether this is due to an anatomical or functional defect. Nevertheless, the impact of these findings to the clinician is to exercise caution in administering nonsteroidal anti-inflammatory agents to patients with renal disease. Indomethacin-induced hyporeninemic hypoaldosteronism appears to be limited to patients with underlying renal disease.

Anatomical Destruction of JG Apparatus. Since hyporeninemia is an important component of isolated hypoaldosteronism, and since underlying renal disease is generally present in this syndrome, it is not unreasonable to invoke anatomical reasons to explain the pathogenesis of this syndrome. In a single case report, Brown et al.[52] directly examined the JG cells of a patient who presented with recurrent hyperkalemia caused by isolated aldosterone deficiency and reported normal numbers of cells but with decreased granularity. The significance of the latter finding is open to question, since reduced granularity can reflect both anatomical and functional suppression of activity. The difficulty in proposing an anatomical basis for the development of hyporeninemia arises from the observation that destruction or distortion of the JG apparatus is seen in diverse renal diseases, and even with advanced age.[53] For instance, hyalinization of the afferent arteriole is frequently seen in patients with essential hypertension,[54] but only seldom is accompanied by hyporeninemic hypoaldosteronism. The frequency with which destruction or distortion of the JG apparatus is seen in renal disease contrasts sharply with the infrequency of hyporeninemic hypoaldo-steronism. Thus, the mere presence of demonstrable histological lesions, a par-ticularly common finding in diabetic kidneys, does not confer causality. While anatomical lesions of the JG apparatus constitute important histological observa-tions, at best they probably represent an epiphenomenon.

In summary, the search for a mechanism to explain the hyporeninemia of isolated hypoaldosteronism has taken the reader through a sojourn of multifac-torial mechanisms that have been proposed. It is indeed possible that more than one mechanism is involved. The combination of volume expansion, autonomic dysfunction, and impaired prostacyclin production provides the best explana-tion for the phenomenon of hyporeninemia, although such a combination may not be evident in all patients. It should also be pointed out that hyporeninemia is not an absolute phenomenon, and many patients with the syndrome of isolated hypoaldosteronism may demonstrate defects in aldosterone release independent of hyporeninemia. The defective regulation of aldosterone release in isolated hypoaldosteronism is discussed in the following section.

Hypoaldosteronism

The second facet that needs to be explored in the pathogenesis of the syndrome is the hypoaldosteronism. While it is widely believed that failure of aldosterone release is secondary to failure of the renin–angiotensin axis, several pieces of data in the literature suggest an independent, primary problem with aldosterone secretion.

1. Some patients with isolated hypoaldosteronism have impairment in al-dosterone release in response to direct stimulation by angiotensin II. The cases reported by Posner and Jacobs,[5,6] as well as by Gerstein et al.[55] and Vagnucci,[7] had evaluation of their aldosterone responses to exogenously administered an-giotensin II. These patients demonstrated an impairment in aldosterone pro-duction and or excretion following an infusion of angiotensin, indicating a pri-mary defect in the zona glomerulosa. In some instances the PRA was normal.[7,55] McGiff et al.[56] also reported a patient with isolated hypoaldosteronism who had normal PRA responses to salt restriction, but a diminished aldosterone response to the same maneuver. By measuring of aldosterone precursors in the urine, a distal defect in aldosterone synthesis was postulated by these authors. More recently Lebel and Grose[57] studied a patient with hyporeninemic hypoaldo-steronism and noted that angiotensin II infusion resulted in a stepwise increase in plasma pregnenolone, with distinct decreases in plasma progesterone, desox-ycorticosterone (DOC), and corticosterone, and a moderate increase in al-dosterone. It has been proposed that enzymatic defects of 3-β-ol-dehydrogenase and $\Delta^{4,5}$-isomerase present only within the zona glomerulosa might cause al-dosterone deficiency.

Defects in aldosterone biosynthesis constitute a separate category of causes for isolated hypoaldosteronism and are characterized by an elevated PRA. Sever-al case reports have characterized this entity, which is believed to occur as a consequence of enzymatic defects.[58,59] The condition is most often due to an inborn error in terminal step of aldosterone biosynthesis and involves the en-zyme corticosterone methyloxidase.

2. Defective aldosterone secretion can also occur as a result of impairment in

Insulin deficiency
↓
Decreased K+ uptake by zona glomerulosa
↓
Decreased aldosterone production
↓
Hyperkalemia
↓
Suppressed PRA

FIGURE 14.   Insulin deficiency as a possible mechanism for hyporeninemic hypoaldosteronism.

the sensitivity of the zona glomerulosa to angiotensin II. Morimoto et al.[60] have described a 62-year-old diabetic woman with hyperkalemia, low plasma and urinary aldosterone level, a high PRA, with inadequate aldosterone responses to sodium restriction, angiotensin infusion, and ACTH administration. The authors noted that the plasma levels of DOC and corticosterone, but not aldosterone, responded to exogenous ACTH administration. It was proposed that adrenal insensitivity to angiotensin II or a defect in angiotensin II receptors could explain the hypoaldosteronism with hypereninemia seen in their patient.

3. An independent abnormality in the synthesis of aldosterone in patients with hyporeninemic hypoaldosteronism is suggested by the fact that in these patients hyperkalemia fails to stimulate aldosterone appropriately.[61] This phenomenon, however, should be viewed in light of the observation that endogeneous angiotensin exerts a strong influence on the response of zona glomerulosa to increases in serum potassium.[62] An intrinsic defect in the zona glomerulosa can also occur in relation to hyperglycemia and insulin deficiency. If one accepts the notion that insulin promotes intracellular uptake of K^+ by the adrenal glomerulosa cells, and that intracellular potassium concentration is a determinant for the biosynthesis of aldosterone, then clearly insulin deficiency can result in decreased aldosterone production by the zona glomerulosa. The hyperkalemia (despite intracellular K^+ depletion) can directly suppress PRA, accounting for the dual findings of hypoaldosteronism and hyporeninemia (Fig. 14).

4. The role of atrial natriuretic factor in causing aldosterone suppression is an emerging one. Kudo and Baird,[63] from the Salk Institute, have evaluated aldosterone response to the synthetic replicate of this factor (ANF 8-33) and found this peptide to be a potent and direct inhibitor of aldosterone secretion by rat zona glomerulosa. Atrial natriuretic factor (ANF) not only lowered basal aldosterone secretion, but was powerful enough to impair aldosterone response to angiotensin II and ACTH. The role of this peptide in causing the functional hypoaldosteronism in volume-expanded states is speculative. Further work defining the basic mechanisms that integrate renin, volume status, ANF release, and aldosterone secretion are required before assigning a possible role to ANF in suppressing aldosterone secretion.

Acidosis

Patients with hyporeninemic hypoaldosteronism usually demonstrate mild to moderate hyperchloremic metabolic acidosis. A unique constellation of features characterizes the metabolic acidosis of hyporeninemic hypoaldosteronism[64]:

1. The urine is acidic and bicarbonate-free during spontaneously occurring acidosis.
2. Reduction in the bicarbonate threshold results in subnormal reabsorption of filtered bicarbonate at normal plasma bicarbonate concentrations. The magnitude of this reduction, however, is not sufficiently great to implicate an impairment in hydrogen ion excretion in the proximal tubule.
3. The urinary excretion of ammonia is greatly reduced, even in the presence of a urine that is highly acidic.
4. Absence of proximal tubular dysfunction, such as glucosuria, aminoaciduria, or phosphaturia.
5. Reduced renal clearance of potassium.

These features are quite distinct from RTA type 1 and 2.[65,66] The combined defect in renal tubular secretion of hydrogen ion and potassium has led Sebastian et al.[67] to designate this disorder "type 4 RTA." (Type 3 RTA is a term used to describe a variant of type 1.)

The occurrence of metabolic acidosis in any disorder characterized by aldosterone deficiency is hardly surprising since aldosterone normally promotes secretion of both hydrogen ion and potassium. Administration of aldosterone to normal subjects results in an increase in net acid excretion. Despite the important role of aldosterone on H ion excretion, patients with Addison's disease seldom have significant metabolic acidosis and are able to lower the urinary pH normally when challenged with acid.[68] This is not to imply a diminished role for aldosterone in H ion secretion, because the capacity to excrete acid is reduced in adrenalectomized animals.[69,70] Kurtzman et al.[71] evaluated renal acidification in adrenalectomized dogs maintained on glucocorticoids and concluded that the resulting acidosis was due to impaired generation of H^+ gradient. It is evident from the review of patients with hyporeninemic hypoaldosteronism that metabolic acidosis is prevalent in more than 75% of patients. This high incidence contrasts with the relative infrequency of severe metabolic acidosis in patients with Addison's disease. The presence of underlying renal damage in patients with hyporeninemic hypoaldosteronism is clearly responsible for the high frequency of metabolic acidosis in this syndrome. In the past decade, considerable interest has been focused on the mechanisms that underlie the development of hyperkalemic hypercholomeric metabolic acidosis in hyporeninemic hypoaldosteronism. While several extraneous factors, such as decreased GFR, sodium depletion, and secondary hyperparathyroidism, have clouded the issue, recent literature suggests the following mechanisms:

1. Shift of hydrogen ions out of the cells in exchange for potassium ions
2. Reduction of bicarbonate threshold
3. Impairment of renal ammoniagenesis, as a consequence of hyperkalemia
4. Impairment of hydrogen ion secretion due to lack of aldosterone

Shift of Hydrogen Ions out of the Cells. Reciprocal changes in the cellular content of H and K ions may, in part, account for the metabolic acidosis of hyporeninemic hypoaldosteronism. This occurs as hyperkalemia evolves, resulting in a shift of H^+ out of the cells in exchange for K^+. Sebastian et al.[64] evaluated the effect of oral administration of the synthetic mineralocorticoid 9-α-fludrocortisone on the acid–base and electrolyte composition of serum and urine in four patients with hyporeninemic hypoaldosteronism. While significant excretion in net acid excretion was indeed noted, the increment in plasma bicarbonate exceeded the increment that the kidneys could account for. The authors postulated that, in part, the increase in the plasma bicarbonate may have reflected intracellular buffering of H^+, which resulted as a consequence of a reduction in cellular K^+ concentration brought about by the mineralocorticoid.

Reduction in the Bicarbonate Threshold.   Two independent studies[68,72] have documented a small reduction in the bicarbonate threshold of patients with hyporeninemic hypoaldosteronism. Perez et al.[68] studied acid–base handling in a patient with this syndrome and noted that the urine pH was appropriately low when an acid load was administered to the patient. This indicated that the hydrogen ion gradient–generating capacity of the distalmost nephron was intact. However, the net acid excretion was markedly depressed owing to the decrease in both NH_3 and titratable acid excretion. The authors also noted substantial bicarbonaturia at normal serum bicarbonate levels, and during lesser degrees of acidosis. It was concluded that the patient's acidosis was related to both limitation of net acid excretion and bicarbonaturia when plasma bicarbonate was normal.

It is generally believed that the small degree of bicarbonaturia seen in patients with hyporeninemic hypoaldosteronism is not of sufficient magnitude to account for the degree of metabolic acidosis seen in this syndrome. Further, the reduction in bicarbonate threshold is generally not ameliorated by the administration of mineralocorticoid. The possibility that other factors such as volume expansion and the secondary hyperparathyroidism of renal failure contribute to this phenomenon also remains unproven. Perez et al.[68] postulated that loss of bicarbonate could be a contributing factor in the development of moderately severe metabolic acidosis in the following hypothetical fashion. Substantial loss of bicarbonate elevated the urine pH and therefore decreased renal acid excretion in the form of titrable acid and ammonium. As the serum bicarbonate declined as a consequence of bicarbonaturia and decreased acid excretion, the filtered bicarbonate also decreased, to a point where the absolute amount of bicarbonate reaching the distal nephron also decreased. Eventually, this resulted in allowance of complete removal of and attainment of an appropriately low urine pH. Yet, the titrable acid and ammonium excretion could not rise to normal since these phenomena were limited by low phosphorus excretion and low renal mass, which were consequences of the underlying renal disease. Therefore, this set of events culminated in the development of a complex variety of RTA, different from the conventional proximal and distal types.

Decreased Ammoniagenesis.   A new facet was added to the multifactorial nature of the metabolic acidosis in hyporeninemic hypoaldosteronism by Szylman et al.[24] They described a 60-year-old man with proven isolated hyporeninemic hypoaldosteronism, who showed a urinary pH of 4.9 despite severe systemic metabolic acidosis. The urinary excretion of ammonium was distinctly blunted. When the hyperkalemia was corrected by the use of an exchange resin, the urinary ammonium excretion increased markedly by severalfold, with disappearance of the metabolic acidosis. Importantly, the improvement in ammonium excretion and metabolic acidosis correlated with reduction of the serum potassium levels with no changes in other variables such as renal function, the hypervolemia, or the degree of suppression in the renin–angiotensin–aldosterone axis. The concept that hyperkalemia per se impairs urinary acidification by interfering with diffusion of ammonia from the tubular cell into the urine is not new.[73,74] The observation that this mechanism plays such a dominant role in the metabolic acidosis of patients with hyporeninemic hypoaldosteronism was indeed a novel one. Further, the ability to completely ameliorate the metabolic acidosis by mere correction of potassium without administering mineralocorticoids supported a primary role for hyperkalemia rather than aldosterone deficiency in the causation of the metabolic acidosis of this syndrome. Thus, reduced ammoniagenesis was established as an important mechanism that played a pivotal role in causing metabolic acidosis. Sebastian et al.[64] extended these observations to four patients with hyporeninemic hypoaldosteronism treated with the mineralocorticoid 9-α-fludrocortisone. It was demonstrated that the initial effect in these patients was an abrupt drop in the urine pH following mineralocorticoid therapy. This is probably due to the increase in renal hydrogen ion secretion and the consequent increase in the net acid secretion. As the hyperkalemia was corrected by the mineralocorticoid, the urinary ammonium excretion and the urine pH rose steadily, with gradual amelioration of the metabolic acidosis. In one patient, following 4 weeks of mineralocorticoid therapy, the excreted ammonium contributed to 75% of the cumulative increase in net acid excretion. Thus, diminished ammonia production was deemed as a major factor in the pathogenesis of metabolic acidosis, and this factor was a direct effect of hyperkalemia.

The relative roles of mineralocorticoid deficiency and hyperkalemia in causing the acidification defect and metabolic acidosis of hyporeninemic hypoaldosteronism complement each other. Thus, with progressive hypoaldosteronism, the urinary excretion of H ions and K ions is decreased, resulting in retention of potassium. With progressive hyperkalemia, renal ammoniageneis is impaired, which perpetuates the metabolic acidosis. Correction of hyperkalemia by either exchange resin or mineralocorticoid improves ammonium excretion and ameliorates the metabolic acidosis.

Decreased Hydrogen Ion Secretion.   Aldosterone and related mineralocorticoids stimulate secretion of both hydrogen ion and potassium ion by the kidney.[75] Therefore, the anticipated sequelae of loss of aldosterone are retention of potassium and decrease in net acid excretion. Yet, the urine pH of patients with hyporeninemic hypoaldosteronism is typically between 5 and 6, with the ability to normally acidify the urine following an acid load. The observation by Perez et al.[76] that adrenalectomized subjects with normal renal function seldom develop acidosis or hyperchloremia upon withdrawal of mineralocorticoid therapy suggests that other factors, besides mere aldosterone deficiency, must play a role in the metabolic acidosis of hyporeninemic hypoaldosteronism. Presence of impaired renal function is clearly one factor to be considered. When normal renal function is present with mineralocorticoid deficiency, even mild reductions in serum bicarbonate will reduce the distal bicarbonate delivery per nephron. In normal kidneys, this would occur in a manner sufficient to permit preservation of normal net acid excretion by the mineralocorticoid-deficient distal nephron. Such protective safety valve mechanisms are lacking in patients with hyporeninemic hypoaldosteronism, most of whom are renally insufficient. The combination of decreased threshold for bicarbonate reabsorption and decreased renal ammoniagenesis compounds the effect of aldosterone deficiency at the distal tubule, result ting in the development of moderate to sometimes severe metabolic acidosis.

Clearly aldosterone deficiency alone is not the sole cause of metabolic acidosis in patients with hyporeninemic hypoaldosteronism, since administration of physiological amounts (0.05–0.15 mg/day) of fludrocortisone or large amounts DOC has little effect on ameliorating the metabolic acidosis of patients with hyporeninemic hypoaldosteronism.[7,68,72,77] The presence of chronic renal disease must be a limiting factor that renders the tubules resistant to the effects of physiological amounts of mineralocorticoids. The complex nature of the metabolic acidosis seen in hyporeninemic hypoaldosteronism has a multifactorial origin, underlying renal disease and setting the stage for several physiological dysfunctions in acid–base and electrolyte balance.

Incidence

Since its original description in 1957, hyporeninemic hypoaldosteronism has evolved from being a rarity to a frequently encountered cause of hyperkalemia. This is illustrated by the study reported by Tan and Burton.[78] In an attempt to establish the frequency of this syndrome, these workers reviewed 100 consecutive cases of hyperkalemia and found that 10% of patients had biochemical and hormonal data consistent with hyporeninemic hypoaldosteronism. While renal failure, potassium medication, spironolactone, acidotic states, and hemolysis represented common causes of hyperkalemia, no obvious cause was descernible in 24% of patients; of this group, 19 had studies performed to exclude hyporeninemic hypoaldosteronism, a diagnosis that was established in 10 of the 19 patients. Thus, when no obvious cause was evident for the hyperkalemia, hyporeninemic hypoaldosteronism was present in nearly 50% of instances. While this study was undertaken primarily to determine the frequency with which the syndrome is encountered as a cause of hyperkalemia, and is hence clearly a selective study, the message is obvious; hyporeninemic hypoaldosteronism is an overlooked cause of hyperkalemia. The frequency of the syndrome is directly related to the zeal with which the diagnosis is pursued in patients with unexplained hyperkalemia.

Clinical Features

There are no specific clinical features that are unique to patients with hyporeninemic hypoaldosteronism. The characteristic patient with hyporeninemic hypoaldosteronism is usually middle aged or elderly, generally asymptomatic, and almost always has underlying renal disease, particularly related to diabetes.

The blood pressure is usually elevated, and orthostatic changes may or may not be present.

Since hyperkalemia is the major feature of the syndrome, the symptoms experienced by patients with hyporeninemic hypoaldosteronism are related to this biochemical perturbation. Muscle weakness, cardiac arrythmias, and resultant neurological dysfunction may accompany the hyperkalemia.[1,4,52] In nearly all instances the abnormal serum potassium is the reason for considering the diagnosis of hyporeninemic hypoaldosteronism. Symptoms of glucocorticoid deficiency, such as myopathy, increased pigmentation, and weight loss, are conspicuously absent in these patients. Dizziness may occasionally be noted in some patients with hyporeninemic hypoaldosteronism due to the combination of sodium loss and mineralocorticoid deficiency. Symptoms related to renal failure are also uncommon, since the degree of renal failure is mild. While dizziness and postural syncope may be seen, these features are less commonly observed in comparison to Addison's disease.

Laboratory Diagnosis

The laboratory findings in hyporeninemic hypoaldosteronism can be viewed in terms of biochemical and hormonal abnormalities.

Biochemical Abnormalities

The triad of (1) hyperkalemia disproportionate to the degree of renal failure, (2) hyperchloremia, and (3) metabolic acidosis characterizes the biochemical abnormalities seen in hyporeninemic hypoaldosteronism.

Hyperkalemia. Hyperkalemia, defined as a serum potassium concentration greater than 5.3 meq/liter, is a universal feature of hyporeninemic hypoaldosteronism. Clearly, the hyperkalemia is out of proportion to the degree of renal failure, since potassium retention in renal disease occurs only when the GFR declines below 20 ml/min. The GFR in patients with hyporeninemic hypoaldosteronism is most often above 20 ml/min. While deficiency of aldosterone superficially appears to be the sole reason to explain the hyperkalemia, renal mechanisms do indeed contribute to the development of the hyperkalemia. Although occasionally normal GFR has been documented in patients with hyporeninemic hypoaldosteronism,[23,79] the decrease in GFR seen in the majority of patients with the syndrome may contribute to hyperkalemia. In addition, some patients with hyporeninemic hypoaldosteronism may have volume expansion, which results in activating renal mechanisms that cause eventual reduction in distal tubular reabsorption of sodium. As a result there would be a reduction in the luminal sodium reaching the distal tubule. This, coupled with aldosterone deficiency, magnifies the tendency for potassium retention.

Acid–Base Disturbances. The consistent acid–base abnormality seen in patients with hyporeninemic hypoaldosteronism is hyperchloremic metabolic acidosis. The urinary excretion of potassium is low, and some degree of salt wasting is usually observed. The urine pH is low, and most patients are able to lower urinary pH during acid loading. The mechanisms that underlie the disturbances in acid–base balance have been alluded to in a previous section.

Hormonal Abnormalities

PRA and Aldosterone Response to Posture. The classic hormonal feature of hyporeninemic hypoaldosteronism is the subnormal response of renin and aldosterone to furosemide and posture. While hyporeninemia constitutes an important diagnostic feature, a substantial minority of patients with isolated hypoaldosteronism may not demonstrate this facet. DeFronzo,[10] in a review of the subject, points out that the baseline or the stimulated PRA was found to be normal in 13 of the 74 patients with the syndrome. An occasional patient may even show hyperreninemia with selective hypoaldosteronism, implying the presence of intrinsic adrenal defects in aldosterone production.[60] Despite these discordant observations, by and large the vast majority of patients with hyporeninemic hypoaldosteronism demonstrate a subnormal renin–aldosterone response to salt depletion and ambulation.[8,79–83]

Plasma Aldosterone Response to Direct Stimulation. Theoretically, since the aldosterone deficiency in the syndrome of hyporeninemic hypoaldosteronism is secondary to renin deficiency, direct stimulation of the zona glomerulosa by angiotensin II or ACTH should provide normal responses. Unfortunately, this does not seem to be the case in all patients with the syndrome. In a careful review of the published cases, DeFronzo[10] notes that the aldosterone response to angiotensin II infusion was normal in only 4 of 27 patients tested, and the aldosterone response to ACTH was preserved in only 7 of 34 patients studied. Blunted aldosterone responses to direct stimulation have also been reported by others.[52,84–86] Schambelan et al.,[8] who originally demonstrated that hyporeninemia was the basic cause of the syndrome, also noted a failure of aldosterone responsiveness to both ACTH and angiotensin II. The interpretation of this finding revolves around the functional versus anatomical defect in the zona glomerulosa. The aldosterone unresponsiveness could have been acquired from dormancy of the zona glomerulosa due to long-standing deprivation of renin. Chronic lack of stimulation can result in suboptimal aldosterone responses to direct stimulation with angiotensin II and ACTH. The theory of atrophy of the zona glomerulosa from chronic understimulation is not supported by the observation that nephrectomized patients, with absolutely no renin, show normal aldosterone responses. Besides, the effects of prolonged adrenal stimulation with ACTH or angiotensin II have not been studied to test the validity of the hypothesis of atrophied zona glomerulosa. Clearly, some patients with hyporeninemic hypoaldosteronism are unable to generate aldosterone in response to direct stimulation. Whether this represents chronic deprivation of renin with resultant atrophy, receptor insensitivity to angiotensin II, or an intrinsic defect in adrenal biosynthesis of aldosterone remains unclarified. Recently, Sowers et al.[38] evaluated renin activation and regulation of aldosterone as well as 18-hydroxycorticosterone (18-OH) biosynthesis in 10 patients with hyporeninemic hypoaldosteronism and found that the primary etiological factor in the syndrome of hyporeninemic hypoaldosteronism was impairment in the renal activation of renin. The basal levels of aldosterone and 18-OH-B were low in all patients. While the 18-OH-B response to ACTH was normal, the maximal incremental aldosterone response after ACTH was considerably less in comparison to normal subjects. Thus, while suboptimal renin activation dominated the hormonal spectrum of abnormalities, subtle biosynthetic defects in the zona glomerulosa could not be excluded in patients with hyporeninemic hypoaldosteronism.

Glucocorticoid Reserve. The glucocorticoid reserve is normal in patients with hyporeninemic hypoaldosteronism, as indicated by the normal cortisol output in response to ACTH administration. This feature differentiates hyporeninemic hypoaldosteronism ("selective" hypoaldosteronism) from the conventional Addisonian type of adrenal failure.

Differential Diagnosis

The features that characterize hyporeninemic hypoaldosteronism are:

1. Hyperkalemia
2. Mild renal failure
3. Hyporeninemia and hypoaldosteronism unresponsive to posture and diuretic
4. Normal glucocorticoid reserve
5. Type IV RTA

The differential diagnosis of hyporeninemic hypoaldosteronism revolves around the numerous causes of hyperkalemia. The first step in evaluation of the patient with hyperkalemia is the exclusion of "pseudo" or factitious hyperkalemia caused by laboratory error, hemolysis, and increased numbers of platelets and white cells. In "pseudohyperkalemia" caused by thrombocytosis or leukocytosis, simultaneous measurement of plasma and serum potassium concentrations will reveal a gap in excess of 0.2 meq/liter.

The second step in evaluation of true hyperkalemia is to exclude exogenous sources of potassium. Potassium-containing salts and medications should be excluded by careful inquiry. The history should include a search for drugs that inhibit tubular secretion of potassium. In this category belong spironolactone, triamterene, amiloride, and rarely digoxin.

After iatrogenic causes have been excluded, the third step is to evaluate the pH. The development of acute acidosis (such as diabetic ketoacidosis (DKA) is a well-known cause of true hyperkalemia. Acute acidemia inhibits renal potassium secretion.[87] Further, as acidosis evolves, hydrogen ions move into cells and such a shift necessitates an efflux of potassium from the cells.[88]

The fourth step in the evaluation is to exclude renal disease. Acute and chronic renal failure are the most important and frequent causes of hyperkalemia. In chronic renal failure, hyperkalemia occurs only when the GFR declines to below 15–20 ml/min. If the GFR is >20 ml, the hyperkalemia has to be explained on some other basis.

The next step is the exclusion of certain systemic disorders that may rarely manifest with a primary defect in renal tubular secretion of potassium. The most widely studied of these disorders is, of course, sickle cell disease.[89] Less commonly a tubular defect may be encountered in systemic lupus erythematosus, amyloidosis, or in posttransplanted kidneys.[10] Rarely, tubular defects in potassium handling can result in a syndrome consisting of hyperkalemia, hypertension, and hyporeninemia. Such cases are usually caused by renal tubular defects in either sodium or potassium handling. Gordon et al.[90] have suggested that enhanced proximal tubular reabsorption of Na^+ and H_2O caused chronic extracellular volume expansion, which in turn suppressed both renin and aldosterone secretion. Brautbar et al.[91] described a 52-year-old man with hypertension, persistent hyperkalemia, and hyperchloremic acidosis with normal renal and adrenal function; the plasma renin was undetectable, and plasma aldosterone levels were within the normal range. The inability to increase potassium excretion by administration of mineralocorticoids suggested a distal tubular defect in potassium handling. The authors also demonstrated an identical syndrome in four other family members.

After systemic disease has been excluded, the combination of hyperkalemia in the presence of mild renal failure is a strong clue for the presence of hyporeninemic hypoaldosteronism, provided other disorders of adrenal cortex have been excluded. These include Addison's disease, enzymatic defects in aldosterone synthesis, and acquired defects in the zona glomerulosa. Addison's disease can be excluded by examining the glucocorticoid response to ACTH. Congenital defects in aldosterone biosynthesis, acquired autoimmune defects in zona glomerulosa function, and Addison's disease are all characterized by hyperreninemic hypoaldosteronism. Further, the metabolic acidosis and the unique type 4 RTA are characteristics of hyporeninemic hypoaldosteronism.

Treatment

The main indication for treatment in patients with hyporeninemic hypoaldosterism is the hyperkalemia. The main therapeutic modality is the use of synthetic mineralocorticoid such as 9-α-fludrocortisone. When indicated, adjunctive therapy with diuretics or ion exchange resins may be necessary.

Mineralocorticoid Therapy

Since the main reason for evolution of the hyperkalemia is the hypoaldosteronism, the physiological approach for treatment of hyporeninemic hypoaldosteronism is replacement with mineralocorticoid. The effectiveness of mineralocorticoid therapy in lowering hyperkalemia has been quite well established. However, much larger doses than those required for Addison's disease are necessary to maintain patients with hyporeninemic hypoaldosteronism in a normokalemic and nonacidotic state. Most patients require doses of fludrocortisone in excess of 0.2 mg/day. The most obvious reason for the relative "resistance" of patients with hyporeninemic hypoaldosteronism is the underlying renal disease. Renal disease, especially tubular disease, is a limiting factor that attenuates the renal response to aldosterone. It is a general observation that plasma aldosterone levels are elevated in chronic renal failure.[92] Even patients on replacement therapy for Addison's disease may require larger doses of mineralocorticoid replacement with the progressive development of renal failure. Sebastian et al.[64] demonstrated remarkable amelioration of metabolic acidosis and hyperkalemia within 4 weeks of therapy with mineralocorticoids in four patients with hyporeninemic aldosteronism. The initiation of 9-α-fludrocortisone was associated with a prompt increase in urinary excretion of potassium and hydrogen ion, with a remarkable increase in the net acid excretion. With continued therapy urinary ammonium excretion increased, indicating correction of defective ammoniagenesis associated with hyporeninemic hypoaldosteronism. Increases in extracellular fluid volume and sodium content, at least in part, underlie the development of the hyporeninemia.[20] In these patients, the use of diuretics, especially loop diuretics, improves the volume expansion and the hyperkalemia.

Diuretic Therapy

As indicated, some patients with hyporeninemic hypoaldosteronism are quite hypertensive with evidence of volume expansion. In such patients, decrease of the expanded extracellular fluid volume by the use of salt restriction and diuretic therapy increases renin and aldosterone secretion. Clearly, there is a subset of patients with hyporeninemic hypoaldosteronism in whom diuretic therapy is beneficial.

Potassium Exchange Resins

Use of potassium exchange resins, particularly sodium polystyrene sulfate, is an excellent method for reducing serum potassium levels. Szylman et al.[24] demonstrated that normalization of the hyperkalemia by the use of exchange resin alone was successful in reversing the metabolic acidosis as well, without having to use fludrocortisone. In addition to its potassium-lowering effect, polystyrene sulfate can promote exchange of sodium and hydrogen ions across the gastrointestinal tract in oliguric patients.[93] In selected patients who develop adverse effects to the powerful mineralocorticoid fludrocortisone, treatment with potassium exchange resin becomes the mainstay of therapy. In addition, a low-potassium diet and avoidance of all forms of exogenous potassium are mandatory.

Not all patients with hyporeninemic hypoaldosteronism require treatment with mineralocorticoids. In the presence of orthostatic hypotension, and of course hyperkalemia, mineralocorticoid therapy is indicated. In patients with hypertension, especially in the presence of latent volume expansion, mineralocorticoids can further expand the extracellular fluid volume owing to their sodium-retaining effect on the renal tubules. In this setting, use of mineralocorticoids can result in congestive cardiac failure, especially when salt intake is not curtailed.

Hyperreninemic Hypoaldosteronism

A primary disorder in mineralocorticoid secretion caused by an intrinsic abnormality in the zona glomerulosa characterizes this group of disorders. As a consequence, the dual features of hypoaldosteronism and hyperreninemia develop. Several disorders can result in hyperreninemic hypoaldosteronism.

1. The most common cause of hypoaldosteronism with hyperreninemia is Addison's disease. The combined glucocorticoid and mineralocorticoid deficiency with secondary increases in plasma ACTH and PRA characterize the hypoaldosteronism of Addison's disease, which is separately discussed in Chapter 2.
2. An enzymatic defect in mineralocorticoid biosynthesis is a rare but important cause of hyperreninemic hypoaldosteronism. Moderate to severe salt wasting is the hallmark of this syndrome, which can be seen in newborns, in children, and rarely in adults.
3. Acquired primary hypoaldosteronism due to an isolated zona glomerulosa defect is a new addition to the causes of isolated hypoaldosteronism with hyperreninemia.
4. Hyperreninemic hypoaldosteronism in the critically ill patient, especially in the intensive-care unit (ICU) setting, is being recognized with increasing frequency.

Enzymatic Defects

Several defects in aldosterone biosynthesis can result in the development of hypoaldosteronism. The defect in the biosynthetic pathways can involve early steps, intermediate steps, or terminal steps in the synthesis of aldosterone. Defectual pathways in the early steps are represented by desmolase deficiency. These are rare and often result in defective secretion of not only mineralocorticoids, but the glucocorticoids, as well as the sex steroids. Defectual pathways in the intermediate steps represent the most common forms of adrenogenital syndrome and are seen classically in complete 21-hydroxylase enzyme deficiency. Cortisol deficiency, ambiguous genitalia, and salt wasting, often to a life-threatening degree, is encountered in this entity.[94,95] Selective defects in the terminal steps of aldosterone biosynthesis involve the final steps in aldosterone biosynthesis. The hallmark here is the isolated involvement of only the mineralocorticoid pathway. Selective hypoaldosteronism caused by defective enzymes in the zona glomerulosa, with normal glucocorticoid and adrenal androgen production, characterizes defects in the final steps of aldosterone synthesis.

Figure 15 outlines the steps involved in the biosynthesis of aldosterone. The crucial enzymes involved in the final steps of aldosterone biosynthesis are the corticosterone methyloxidases I and II (CMO I, CMO II). These enzymes are unique to the zona glomerulosa. CMO I acts on corticosterone and converts it into a labile oxygenated steroid–metalloenzyme complex.[59] This enzyme complex can serve as a substrate for CMO II or alternately undergo decomposition to its 18-hydroxy derivative (18-hydroxycorticosterone). CMO II further hydroxylates the methyl group in the steroid–metalloenzyme complex; spontaneous dehydration of the methylated product yields the aldehyde aldosterone. Thus, a deficiency of CMO I would affect the penultimate step in aldosterone synthesis, while a deficiency of CMO II would affect the ultimate step in aldosterone biosynthesis. Since CMO I and CMO II are present only in the zona glomerulosa, the biosynthesis of glucocorticoids and androgens would proceed normally.

Isolated hypoaldosteronism caused by enzymatic defects in aldosterone biosynthesis has been well studied.[58,96–98] Since the original description of a famil-

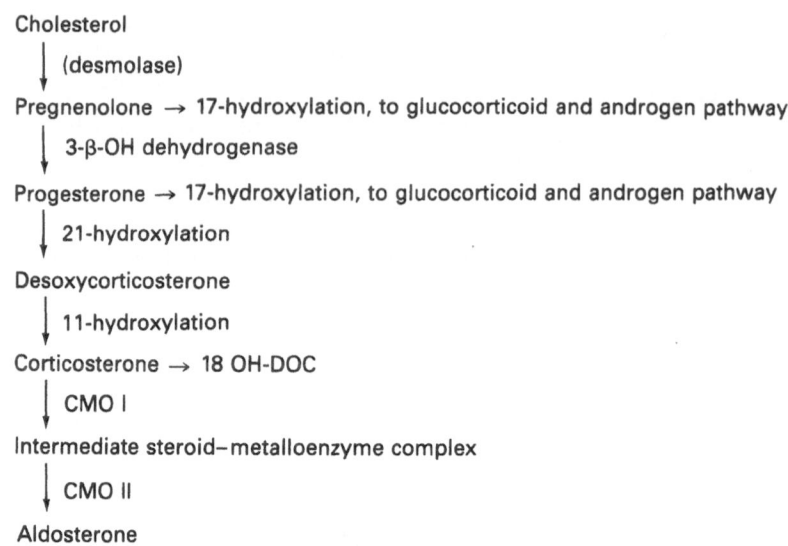

Cholesterol
↓ (desmolase)
Pregnenolone → 17-hydroxylation, to glucocorticoid and androgen pathway
↓ 3-β-OH dehydrogenase
Progesterone → 17-hydroxylation, to glucocorticoid and androgen pathway
↓ 21-hydroxylation
Desoxycorticosterone
↓ 11-hydroxylation
Corticosterone → 18 OH-DOC
↓ CMO I
Intermediate steroid–metalloenzyme complex
↓ CMO II
Aldosterone

FIGURE 15. Aldosterone biosynthesis. CMO = corticosterone methyloxidase.

ial defect in aldosterone biosynthesis by Royer et al.[99] from Paris, the condition and its familial nature have been recognized worldwide.[100–104] The clinical characteristics of children affected by defective synthesis of aldosterone vary with the degree of the defect. When severe, salt loss, dehydration, electrolyte imbalance, and the hyperkalemia can be life-threatening. Children with milder degrees of defect may manifest with failure to thrive, short stature, vomiting, or asymptomatic electrolyte abnormalities such as hyponatremia and hyperkalemia.

The hormonal characteristics of CMO II deficiency are as follows (Table 36).

1. Decreased or absent aldosterone in plasma and urine
2. Elevated plasma 18-OH corticosterone and its urinary metabolite tetrahydro-18-OH dehydrocorticosterone (18-OH-THA)
3. Elevated ratio of serum 18-OH corticosterone to aldosterone, as well as an elevated ratio of urinary 18-OH-THA to tetrahydroaldosterone

TABLE 36.
Characteristics of CMO II Deficiency

Hormone	Level
Plasma aldosterone, Urinary aldosterone, Urinary free aldosterone	Decreased
Plasma 18-OH DOC	Increased
Plasma 18-OH corticosterone	Increased
Ratio of plasma 18-OH corticosterone to plasma aldosterone	Increased
Ratio of urinary 18-OH-THA (18-hydroxy-11-dehydrotetrahydrocorticosterone) to urinary tetrahydroaldosterone	Increased
Plasma renin activity	Increased

In a detailed account of a pedigree affected by CMO type II deficiency, Veldhuis et al.[59] demonstrated that salt replacement but not hydrocortisone ameliroated the clinical and metabolic abnormalities of patients with this deficiency.

The inheritance pattern of CMO II deficiency is believed to be autosomal recessive. The convenient determination of urinary ratios of the corticosterone metabolite, 18-OH-THA, to tetrahydroaldosterone has been used as a marker that permits identification of CMO II deficiency in family members. Affected adults often demonstrate amelioration of the salt-losing tendency with increasing age. Corticosterone methyloxidase deficiency (type II) should be included as an important cause in the differential diagnosis of salt-losing congenital adrenogenital syndromes. The low aldosterone level, coupled with the increased precursor products (18-OH-DOC) and unresponsiveness to hydrocortisone while responding to salt repletion, should permit ready recognition of the disorder.

Corticosterone methyloxidase type I deficiency is rarer than CMO II deficiency. The clinical presentation is similar to type II deficiency and is characterized by aldosterone deficiency, salt loss, and hyperreninemia. The hormonal triad to detect CMO I deficiency is decreased aldosterone, decreased 18-OH corticosterone, but elevated corticosterone levels in the plasma.[105]

In contrast to the above disorder, which is congenital, decreased aldosterone biosynthesis can be acquired under certain circumstances. Heparin administration on a protracted basis, hypokalemia, and occasionally diabetes mellitus are acquired causes for defective aldosterone biosynthesis.

Acquired Defects

Williams et al.[106] described an asymptomatic 62-year-old diabetic man with hyperkalemia, moderate renal failure, and low aldosterone concentrations in the plasma that failed to respond to sodium depletion, upright posture, and ACTH administration. Contrary to the expected hyporeninemia that was anticipated in view of the diabetic background of the patient, the authors found the PRA levels to be high in this patient. Since the serum electrolytes had been normal on multiple previous occasions, an acquired defect in aldosterone secretion was postulated. Further studies revealed normal glucocorticoid responses to ACTH infusion, but defective release of aldosterone and its precursors. Careful evaluation of the precursor steroid responses to ACTH failed to reveal evidence of partial defects in desomolase, 21-hydroxylase, 3-β-hydroxysteroid dehydrogenase, 17- or 18-hydroxylase, or CMO I and II deficiencies.

The low levels of aldosterone and 18-OH corticosterone in the patient, together with normal zona fasciculata function, were considered evidence for isolated zona glomerulosa dysfunction. The elevated PRA clearly excluded the hyporeninemic variety of hypoaldosteronism frequently seen in diabetics. Among the postulated mechanisms, selective destruction of the zona glomerulosa by an autoimmune process was suggested. Selective testing for antibodies against the three layers of the adrenal cortex revealed a characteristic pattern of antiadrenal antibodies confined almost exclusively to the zona glomerulosa.[107]

Thus, it is becoming increasingly evident that rarely autoimmune adrenal destruction can be selective and limited to a single layer of the adrenal cortex. Marieb et al.,[108] in 1974, reported a 12-year-old girl with moniliasis, hypoparathyroidism, and isolated hypoaldosteronism. Although the clinical syndrome was consistent with Addison's disease, the serum cortisol and the urinary 17-hydroxycorticosteroids were normal and responded appropriately to corticotropin. Despite severe hyponatremia and excessive urinary sodium loss, the excretion of aldosterone and 18-OH corticosterone was markedly depressed. The authors postulated selective autoimmune destruction of the zona glomerulosa, but could not prove this by selective antibody studies. The availability of tests that detect organ-specific antibodies against each layer of zona glomerulosa[109] will undoubtedly result in detection and documentation of the entity of selective autoimmune adrenalitis exclusively involving the zona glomerulosa.

Critical Illness

Zipser et al.,[110] in 1981, identified a group of patients with critical illness, often in the ICU setting, who demonstrated hypoaldosteronism despite hyperreninemia. The prevalence of such a phenomenon in 100 consecutive patients evaluated by Davenport and Zipser[111] appears to be rather high; 22 of 100 patients demonstrated inappropriate plasma aldosterone levels in relation to the renin levels. The mechanism for this phenomenon is unclear. In the patients studied by Davenport and Zipser[111] there were no differences in electrolyte concentrations, nutrition, use of medications, or survival when compared to other hyperreninemic ICU patients. The one striking observation was the remarkably high incidence of persistent hypotension in the group with low plasma aldosterone concentrations. The significance of this finding in terms of the hypoaldosteronism seen in some patients with critical illness remains to be elucidated.

Pseudohypoaldosteronism

So far, isolated hypoaldosteronism has been viewed in terms of etiological factors related to primary renin deficiency and intrinsic aldosterone deficiency. A third category of disorders that can lead to clinical expression of hypoaldosteronism is the entity of pseudohypoaldosteronism. The term implies unresponsiveness of the renal tubule despite the presence of high levels of aldosterone in the circulation.

The classic syndrome of pseudohypoaldosteronism is that encountered in infancy. Since the first description of this syndrome in 1958 by Cheek and Perry[112] in an infant with severe salt-losing syndrome, approximately 40 cases have been described in the literature.[113–121] The hallmark of the syndrome is the occurrence of severe salt loss despite very high aldosterone concentrations in the plasma and urine. The condition is familial and has been reported to occur in twins[116] and triplets.[118]

The main clinical feature of pseudohypoaldosteronism is the development of severe renal salt wasting during infancy. Most patients are male, and the onset of the salt-losing syndrome is usually below age 1. The clinical presentation is dominated by dehydration, volume depletion, electrolyte imbalance, and failure to thrive. In this regard, pseudohypoaldosteronism resembles the various forms of salt-losing congenital adrenal hyperplasia caused by enzymatic defects in steroidogenesis.

The biochemical and hormonal features of pseudohypoaldosteronism show characteristic sequelae of aldosterone lack. Thus, hyponatremia, hyperkalemia (often severe), profound natriuresis, elevated PRA, and normal function of the zona fasciculata are evident upon testing. However, in contrast to true hypoaldosteronism, the aldosterone levels in plasma, as well as the excretion of aldosterone in the urine, are markedly elevated, indicating target organ resistance. The aldosterone secretory rate is increased. In addition to renal losses of sodium, these patients often demonstrate sodium wasting in the saliva, stool, and sweat.

The association of severe renal sodium loss in the presence of high aldosterone levels and normal renal and adrenal function places pseudohypoaldosteronism in the realm of disorders caused by unresponsiveness of target organs, in this instance, the distal tubule. Oberfield et al.[122] have demonstrated multiple target organ unresponsiveness to mineralocorticoid hormones. These workers studied a 7-month-old infant with pseudohypoaldosteronism and documented urinary, salivary, and sweat sodium wasting that was not corrected by administration of 9-α-fludrocortisol. Further, the colonic mucosal cells failed to respond to the exogenous administration of aldosterone. Thus, mineralocorticoid unresponsiveness was documented in all organs where sodium transport was mineralocorticoid dependent. While the exact mechanism for renal tubular responsiveness is unknown, Na,K-ATPase deficiency in the tubule has been implied as the cause of renal tubular salt loss.[123] The heterogeneity of pseudohypoaldosteronism is suggested by reports that have documented preservation of response to exogenous salt-retaining steroids. Rosler et al.[124] described a group of seven Jewish children from five Persian families, who demonstrated profound salt wastage in the presence of very high PRA with normal or high plasma aldosterone concentrations. These subjects differed from those with classical pseudohypoaldosteronism in that they responded to treatment with salt-retaining mineralocorticoids. The combination of very high PRA with only a slightly elevated plasma aldosterone could also be expalined by some degree of resistance of the zona glomerulosa to angiotensin II. Regardless of the mechanism, the severe salt loss associated with pseudohypoaldosteronism bears emphasis as an important etiology of salt wasting in childhood.

In addition to the congenital variety of pseudohypoaldosteronism, there are well-recognized clinical situations where tubular resistance to mineralocorticoids can be acquired. The usual disorders that result in such a phenomenon are renal in origin. Thus, chronic interstitial nephritis with salt wasting, renal tubular acidosis, and sickle cell nephropathy are associated with relative tubular resistance to aldosterone.

Heparin-Induced Hypoaldosteronism

Hypoaldosteronism caused by heparin administration has been recognized since 1964, when Wilson and Goetz[19] reported the development of selective hypoaldosteronism following chronic heparin administration. Reversible hypoaldosteronism following routine heparin therapy has been described to occur particularly in diabetics.[125-127] Several mechanisms have been proposed to explain the inhibitory effect of heparin on aldosterone production; these include (1) a direct effect of heparin on the aldosterone-secreting cells[128]; (2) inhibition of enzymes involved in biosynthesis[129]; and (3) accentuation of an underlying problem in the renin–angiotensin–aldosterone axis. Regardless of the mechanism involved, the deleterious effects of heparin must be borne in mind when this drug is administered, even in therapeutic dosage, to susceptible patients such as diabetics.

References

1. Hudson JB, Chaobanian AV, Relman AS: Hypoaldosteronism. *N Engl J Med* **257:**529, 1957.
2. Carroll HJ, Farber SJ: Hyperkalemia and hyperchloremic acidosis in chronic pyelonephritis. *Metabolism* **13:**808, 1964.
3. Hill SR, Nickerson JF, Chenault SB, et al: Studies in man on hyper- and hypoaldosteronism. *Arch Intern Med* **104:**982, 1959.
4. Lambrew CT, Carver ST, Peterson RE, et al: Hypoaldosteronism as a cause of hyperkalemia and syncopal attacks in a patient with complete heart block. *Am J Med* **31:**81, 1961.
5. Posner JB, Jacobs DR: Isolated analdosteronism. I. Clinical entity, with manifestations of persistent hyperkalemia, periodic paralysis, salt-losing tendency, and acidosis. *Metabolism* **13:**513, 1964.
6. Jacobs DR, Posner JB: Isolated analdosteronism. II. The nature of the adrenal cortical enzymatic defect, and the influence of diet and various agents on electrolyte balance. *Metabolism* **13:**522, 1964.
7. Vagnucci AH: Selective aldosterone deficiency. *J Clin Endocrinol Metab* **29:**279, 1969.
8. Schambelan M, Stockigt JR, Biglieri EG: Isolated hypoaldosteronism in adults. A renin-deficiency syndrome. *N Engl J Med* **287:**573, 1972.
9. Michelis MF, Murdaugh HV: Selective hypoaldosteronism. *Am J Med* **59:**1, 1975.
10. DeFronzo RA: Hyperkalemia and hyporeninemic hypoaldosteronism. *Kidney Int* **17:**118, 1980.
11. Phelps KR, Lieberman RL, Oh MS, et al: Pathophysiology of the syndrome of hyporeninemic hypoaldosteronism. *Metabolism* **29:**186, 1980.
12. Schindler AM, Sommers SC: Diabetic sclerosis of the renal juxtaglomerular apparatus. *Lab Invest* **15:**877, 1986.
13. Christlieb AR, Munichoodoppa C, Braaten JT: Decreased response of plasma renin activity to orthostasis in diabetic patients with orthostatic hypotension. *Diabetes* **23:**835, 1974.
14. Corder CN, Kanefsky TM, McDonald RH, et al: Postural hypotension: Adrenergic responsivity and levodopa therapy. *Neurology* **27:**921, 1977.
15. DeFronzo R, Sherwin R, Felig P, et al: Nonuremic diabetic hyperkalemia. Possible role of insulin deficiency. *Arch Intern Med* **137:**842, 1977.
16. Tait JF, Tait SAS: The effect of changes in potassium concentration on the maximal steroidogenic response of purified zona glomerulosa cells to angiotensin II. *J Steroid Biochem* **7:**687, 1976.
17. DeLeiva A, Christlieb R, Milby J, et al: Big renin and biosynthetic defect of aldosterone in diabetes mellitus. *N Engl J Med* **295:**639, 1976.
18. McAllister RG Jr, Michelakis AM, Sandstead HH: Plasma renin activity in chronic plumbism. *Arch Intern Med* **127:**919, 1971.
19. Wilson D, Goetz FC: Selective hypoaldosteronism after prolonged heparin administration. *Am J Med* **36:**635, 1964.
20. Oh MS, Carroll HJ, Clemmons JE, et al: A mechanism for hyporeninemic hypoaldosteronism in chronic renal disease. *Metabolism* **23:**1157, 1974.
21. De Chatel R, Weidmann P, Flammer J, et al: Sodium, renin, aldosterone, catecholamines and blood pressure in diabetes mellitus. *Kidney Int* **12:**412, 1977.

22. Oster JR, Perez GO, Rosen MS: Hyporeninemic hypoaldosteronism after chronic sodium bicarbonate abuse. *Arch Intern Med* **136**:1179, 1976.

23. Perez GO, Lespier LE, Oster JR, et al: Effect of alterations of sodium intake in patients with hyporeninemic hypoaldosteronism. *Nephron* **18**:259, 1977.

24. Szylman P, Better OS, Chaimowitz C, et al: Role of hyperkalemia in the metabolic acidosis of isolated hypoaldosteronism. *N Engl J Med* **294**:361, 1976.

25. Davis JO, Freeman RH: Mechanisms regulating renin release. *Physiol Rev* **56**:1, 1976.

26. Wagermark J, Ungersted U, Ljungquist A: Sympathetic innervation of the juxtaglomerular cells of the kidney. *Circ Res* **22**:149, 1968.

27. DeChamplain J, Genest J, Vegratt R: Factors controlling renin in man. *Arch Intern Med* **117**:355, 1966.

28. Tuck ML, Sambhi MP, Levin L: Hyporeninemic hypoaldosteronism in diabetes mellitus. Studies of the autonomic nervous system's control of renin release. *Diabetes* **28**:237, 1979.

29. Christlieb AR: Renin-angiotensin-aldosterone system in diabetes mellitus. *Diabetes* **25** (Suppl. 2):820, 1976.

30. Love DR, Brown JJ, Chinn RH, et al: Plasma renin in idiopathic orthostatic hypotension: Differential response in subjects with probable afferent and efferent autonomic failure. *Clin Sci* **41**:289, 1971.

31. Christlieb AR, Kaldany A, D'Elia JA: Plasma renin activity and hypertension in diabetes mellitus. *Diabetes* **25**:969, 1976.

32. Perez GO, Lespier L, Jacobi J, et al: Hyporeninemia and hypoaldosteronism in diabetes mellitus. *Arch Intern Med* **137**:852, 1977.

33. Christlieb AR, Kaldany A, D'Elia JA, et al: Aldosterone responsiveness in patients with diabetes mellitus. *Diabetes* **27**:732, 1978.

34. Day RP, Luetscher JA, Gonzales CM: Occurrence of big renin in human plasma, amniotic fluid and kidney extracts. *J Clin Endocrinol Metab* **40**:1078, 1975.

35. Boyd GW: An inactive higher-molecular-weight renin in normal subjects and hypertensive patients. *Lancet* **1**:215, 1977.

36. Weinberger M, Aoi W, Grim C: Dynamic response of active and inactive renin in normal and hypertensive humans. *Circ Res* **41**:21, 1977.

37. Derkx FHM, Wenting GJ, Man in't Veld AJ, et al: Inactive renin in human plasma. *Lancet* **2**:496, 1976.

38. Sowers JR, Beck FWJ, Waters BK, et al: Studies of renin activation and regulation of aldosterone and 18-hydroxycorticosterone biosynthesis in hyporeninemic hypoaldosteronism. *J Clin Endocrinol Metab* **61**:60, 1985.

39. Tan SY, Antonipilla I, Mulrow PJ: Inactive renin and prostaglandin E_2 production in hyporeninemic hypoaldosteronism. *Clin Endocrinol Metab* **51**:849, 1980.

40. Hahn JA, Zipser RD, Barg A, et al: Studies of the renal vasoactive systems in hyporeninemic hypoaldosteronism. *Prostaglandins Med* **6**:549, 1981.

41. Larsson C, Weber P, Anggard E: Arachidonic acid increases and indomethacin decreases plasma renin activity in rabbit. *Eur J Pharmacol* **28**:391, 1974.

42. Tan SY, Mulrow PJ: Inhibition of the renin–aldosterone response to furosemide by indomethacin. *J Clin Endocrinol Metab* **45**:174, 1977.

43. Romero JC, Dunlap CL, Strong CG: The effect of indomethacin and other anti-inflammatory drugs on the renin–angiotensin system. *J Clin Invest* **58**:282, 1976.

44. Speckart P, Zia P, Zipser R, et al: Effect of sodium restriction and prostaglandin inhibition on the renin-angiotensin system in man. *J Clin Endocrinol Metab* **44**:832, 1977.

45. Oates J, Whorton AR, Gerkens JF, et al: The participation of prostaglandins in the control of renin release. *Fed Proc* **38**:72, 1979.

46. Gerber JG, Olson RD, Nies AS: Interrelationship between prostaglandins and renin release. *Kidney Int* **19**:816, 1981.

47. Yun J, Kelly G, Bartter FC, et al: Role of prostaglandins in the control of renin secretion in the dog. *Circ Res* **40**:459, 1977.

48. Patrono C, Pugliese F, Ciabattoni G, et al: Evidence for a direct stimulatory effect of prostacyclin on renin release in man. *J Clin Invest* **69**:231, 1982.

49. Zusman RM: Renin- and non–renin-mediated antihypertensive actions of converting enzyme inhibitors. *Kidney Int* **25**:969, 1984.

50. Tan SY, Shapiro R, Franco R, et al: Indomethacin-induced prostaglandin inhibition with

hyperkalemia. A reversible cause of hyporeninemic hypoaldosteronism. *Ann Intern Med* **90:**783, 1979.

51. Nadler JL, Lee FO, Hsueh W, et al: Evidence of prostacyclin deficiency in the syndrome of hyporeninemic hypoaldosteronism. *N Engl J Med* **314:**1015, 1986.

52. Brown JJ, Chinn RH, Fraser R, et al: Recurrent hyperkalemia due to selective aldosterone deficiency: Correction by angiotensin infusion. *Br Med J* **1:**650, 1973.

53. Bell ET: Renal vascular disease in diabetes mellitus. *Diabetes* **2:**376, 1953.

54. Sommers SC: Hypertension and kidney disease. *Prog Cardiovasc Dis* **8:**210, 1965.

55. Gerstein AR, Kleeman CR, Gold EM, et al: Aldosterone deficiency in chronic renal failure. *Nephron* **5:**90, 1968.

56. McGiff JC, Muzzarelli RE, Duffy PA, et al: Interrelationships of renin and aldosterone in a patient with hypoaldosteronism. *Am J Med* **48:**247, 1970.

57. Lebel A, Grose JH: Selective hypoaldosteronism: Study of biosynthetic pathways under adrenocorticotrophin and angiotensin II infusion. *Clin Sci Mol Med* **51**(Suppl.):335, 1976.

58. Ulick S, Gautier E, Vetter KK, et al: An aldosterone biosynthetic defect in a salt-losing disorder. *J Clin Endocrinol Metab* **24:**669, 1964.

59. Veldhuis JD, Kulin HE, Santen RJ, et al: Inborn error in the terminal step of aldosterone biosynthesis: Cortitosterone methyl oxidase type II deficiency in a North American pedigree. *N Engl J Med* **303:**117, 1980.

60. Morimoto S, Kim KS, Yamamoto I, et al: Selective hypoaldosteronism with hyperreninemia in a diabetic patient. *J Clin Endocrinol Metab* **49:**742, 1979.

61. Knochel JP: The syndrome of hyporeninemic hypoaldosteronism. *Annu Rev Med* **30:**145, 1979.

62. Fredlund P, Saltman S, Kondo T, et al: Aldosterone production by isolated glomerulosa cells. Modulation of sensitivity to angiotensin II and ACTH by extracellular potassium concentration. *Endocrinology* **100:**481, 1977.

63. Kudo T, Baird A: Inhibition of aldosterone production in the adrenal glomerulosa by atrial natriuretic factor. *Nature* **312:**756, 1984.

64. Sebastian A, Schambelan M, Lindenfeld S, et al: Amelioration of metabolic acidosis with fludrocortisone therapy in hyporeninemic hypoaldosteronism. *N Engl J Med* **297:**576, 1977.

65. Sebastian A, McSherry E, Morris RC Jr: On the mechanism of renal potassium wasting in renal tubular acidosis associated with the Fanconi syndrome (type 2 RTA). *J Clin Invest* **50:**231, 1971.

66. Sebastian A, McSherry E, Morris RC Jr: Renal potassium wasting in renal tubular acidosis (RTA): Its occurrence in types 1 and 2 RTA despite sustained correction of systemic acidosis. *J Clin Invest* **50:**667, 1971.

67. Sebastian A, McSherry E, Morris RC Jr: Metabolic acidosis with special reference to the renal acidosis. In Brenner BM, Rector FC Jr (eds): *The Kindey.* Saunders, Philadelphia, 1976, pp. 615–660.

68. Perez GO, Oster JR, Vaamonde CA: Renal acidosis and renal potassium handling in selective hypoaldosteronism. *Am J Med* **57:**809, 1974.

69. Fukuda T, Koyama T: On the significance of acidosis in adrenal insufficiency. *Jpn J Physiol* **12:**176, 1962.

70. Luke RG, Levitin H: The renal and electrolyte response to respiratory acidosis in the adrenalectomized rat. *Yale J Biol Med* **39:**27, 1966.

71. Kurtzman NA, White MG, Rogers PW: Aldosterone deficiency and renal bicarbonate reabsorption. *J Lab Clin Med* **77:**931, 1971.

72. Sebastian A, McSherry E, Schambelan M, et al: Renal tubular acidosis in patients with hypoaldosteronism caused by renin deficiency. *Clin Res* **21:**706, 1973.

73. Tannen RL, Wedell E, Moore R: Renal adaptation to a high potassium intake: the role of hydrogen ion. *J Clin Invest* **52:**2089, 1973.

74. Kamm DE: Dissociation of urine pH and NH_3 excretion during KCl and NaCl loading. Presented at the 5th annual meeting of the American Society of Nephrology, Washington, DC, 1971, p. 36.

75. Lemann J Jr, Piering WF, Lennon EJ: Studies of the acute effects of aldosterone and cortisol on the interrelationship between renal sodium, calcium and magnesium excretion in normal man. *Nephron* **7:**117, 1970.

76. Perez GO, Oster JR, Vaamonde CA: Renal acidification in patients with mineralocorticoid deficiency. *Nephron* **17:**461, 1976.

77. Vagnucci AH: Selective aldosterone deficiency in chronic pyelonephritis. *Nephron* **7:**524, 1970.

78. Tan SY, Burton M: Hyporeninemic hypoaldosteronism. An overlooked cause of hyperkalemia. *Arch Intern Med* **141**:30, 1981.

79. Gossain VV, Ferrara EV, Werk EE, et al: Impaired renin responsiveness with secondary hypoaldosteronism. *Arch Intern Med* **132**:885, 1973.

80. Stockigt JR, Collins RD, Schambelan M, et al: Subordinate hypoaldosteronism. *Clin Res* **19**:174, 1971.

81. Brown JJ, Chinn RH, Fraser R, et al: Isolated analdosteronism with deficiency of renin and angiotensin. *Clin Sci* **42**:28, 1972.

82. Mellinger RC, Petermann FL, Jurgenson JC: Hyponatremia with low urinary aldosterone occurring in an old woman. *J Clin Endocrinol Metab* **34**:85, 1972.

83. Weidmann P, Reinhart R, Maxwell MH, et al: Syndrome of hyporeninemic hypoaldosteronism and hyperkalemia in renal disease. *J Clin Endocrinol Metab* **36**:965, 1973.

84. Schambelan M, Sebastian A, Biglieri EG: Prevelance, pathogenesis, and functional significance of aldosterone deficiency in hyperkalemic patients with chronic renal insufficiency. *Kidney Int* **17**:89, 1980.

85. Tuck ML, Mayes DM: Mineralocorticoid biosynthesis in patients with hypoaldosteronism. *J Clin Endocrinol Metab* **50**:341, 1980.

86. Bayard F, Cooke CR, Tiller DJ, et al: The regulation of aldosterone secretion in anephric man. *J Clin Invest* **50**:1585, 1970.

87. Wright FS, Giebisch G: Renal potassium transport: Contributions of individual nephron segments and populations. *Am J Physiol* **235**:F515, 1978.

88. Irvine ROH, Dow J: Muscle cell pH and potassium movement in metabolic acidosis. *Metabolism* **17**:563, 1968.

89. DeFronzo RA, Taufield PA, Black H, et al: Impaired renal tubular potassium secretion in sickle cell disease. *Ann Intern Med* **90**:310, 1979.

90. Gordon RD, Gedds AR, Pawsey CGK, et al: Hypertension and severe hyperkalemia associated with suppression of renin and aldosterone and completely reversed by dietary sodium restriction. *Aust Ann Med* **4**:287, 1970.

91. Brautbar N, Levi L, Rosler A, et al: Familial hyperkalemia, hypertension, and hyporeninemia with normal aldosterone levels. A tubular defect in potassium handling. *Arch Intern Med* **138**:607, 1978.

92. Gold EM, Kleeman CR, Ling S, et al: Sustained aldosterone secretion in chronic renal failure. *Clin Res* **13**:135, 1965.

93. Flinn RB, Merrill JP, Welzant WR: Treatment of the oliguric patient with a new sodium-exchange resin and sorbitol. *N Engl J Med* **264**:111, 1961.

94. Bongiovanni AM: The adrenogenital syndrome with deficiency of 3β-hydroxysteroid dehydrogenase. *J Clin Invest* **41**:2086, 1962.

95. Bryan GT, Kliman B, Bartter FC: Impaired aldosterone production in "salt-losing" congenital adrenal hyperplasia. *J Clin Invest* **44**:957, 1965.

96. Ulick S: Diagnosis and nomenclature of the disorders of the terminal portion of the aldosterone biosynthetic pathway. *J Clin Endocrinol Metab* **43**:92, 1976.

97. Neher R: Aldosterone: chemical aspects and related enzymology. *J Endocrinol* **81**:25, 1979.

98. Migeon CJ: Diagnosis and treatment of adrenogenital disorders. In DeGroot LJ, Cahill GF, Martini L, Nelson DH, ODell WD, Potts JT, Sternberger E, Winegrad AI (eds): *Endocrinology.* Grune & Stratton, New York, 1979, pp. 1203–1224.

99. Royer P, Lestradet H, de Menibus CH, et al: Hypoaldosteronisme familial chronique a debut neo-natal. *Ann Pediatr (Paris)* **8**:133, 1961.

100. Rappaport R, Dray F, Legrand JC, et al: Hypoaldosteronisme congenital familial par defaut de la 18-OH-dehydrogenase. *Pediatr Res* **2**:456, 1968.

101. David R, Golan S, Drucker W: Familial aldosterone deficiency: Enzyme defect, diagnosis, and clinical course. *Pediatrics* **41**:403, 1968.

102. Hamilton W, McCandless AE, Ireland JT, et al: Hypoaldosteronism in three sibs due to 18-dehydrogenase deficiency. *Arch Dis Child* **51**:576, 1976.

103. Rosler A, Rabinowitz D, Thedor R, et al: The nature of the defect in a salt-wasting disorder in Jews of Iran. *J Clin Endocrinol Metab* **44**:279, 1977.

104. David RR, Asnis M, Drucker WD: Disturbance of cortisol production in congenital aldosterone deficiency. *J Clin Endocrinol Metab* **35**:604, 1972.

105. Finkelstien M, Schaefer JM: Inborn errors of steroid biosynthesis. *Physiol Rev* **59**:353, 1979.

106. Williams FA Jr, Schambelan M, Biglieri EG, et al: Acquired primary hypoaldosteronism due to an isolated zona glomerulosa defect. *N Engl J Med* **309:**1623, 1983.

107. Carey RM, Schambelan M, Bigilieri EG: Primary hypoaldosteronism due to zona glomerulosa defect. *N Engl J Med* **310:**1394, 1984.

108. Marieb NJ, Melby JC, Lyall SS: Isolated hypoaldosteronism associated with idiopathic hypoparathyroidism. *Arch Intern Med* **134:**424, 1974.

109. Bright GM, Blizzard RM, Kaiser DL, et al: Organ-specific autoantibodies in children with common endocrine diseases. *J Pediatr* **100:**8, 1982.

110. Zipser RD, Davenport MW, Martin KL, et al: Hyperreninemic hypoaldosteronism in the critically ill: A new entity. *J Clin Endocrinol Metabol* **53:**867, 1981.

111. Davenport MW, Zipser RD: Association of hypotension with hyperreninemic hypoaldosteronism in the critically ill patient. *Arch Intern Med* **143:**735, 1983.

112. Cheek DB, Perry JW: A salt wasting syndrome in infancy. *Arch Dis Child* **33:**252, 1958.

113. Donnell GN, Litman N, Roland M: Pseudohypoadrenalocorticism. *Am J Dis Child* **97:**813, 1959.

114. Raine DN, Roy J: A salt-losing syndrome in infancy. *Arch Dis Child* **37:**548, 1962.

115. Proesmans W, Geussens H, Corbeel L, et al: Pseudohypoaldosteronism. *Am J Dis Child* **126:**510, 1973.

116. Alvarez M, Barnes N, Strickler G: Salt wasting nephropathy or "pseudohypoaldosteronism in twins." *Pediatr Res* **8:**453, 1974.

117. Postel-Vinay MC, Alberti GM, Ricour C, et al: Pseudohypoaldosteronism: Persistence of hyperaldosteronism and evidence for renal tubular and intestinal responsiveness to endogenous aldosterone. *J Clin Endocrinol Metab* **39:**1038, 1977.

118. Kaufman E, Hayek AI, Green R: Pseudohypoaldosteronism in triplets. *Pediatr Res* **11:**426, 1977.

119. Shackleton CHL, Snodgrass GHAI: Steroid excretion by an infant with an unusual salt-losing syndrome: A gas chromatographic-mass spectrometric study. *Ann Clin Biochem* **11:**91, 1974.

120. Roy C: Pseudohypoaldosteronisme familial. *Arch Fr Pediatr* **34:**37, 1977.

121. Limal JM, Rappaport R, Dechaux M, et al: Familial dominant pseudohypoaldosteronism. *Lancet* **1:**51, 1978.

122. Oberfield SE, Levine LS, Carey RM, et al: Pseudohypoaldosteronism: Multiple target organ unresponsiveness to mineralocorticoid hormones. *J Clin Endocrinol Metab* **48:**228, 1979.

123. Bierich JR, Schmidt U: Tubular Na, K-ATPase deficiency, the cause of the congenital renal salt-losing syndrome. *Eur J Pediatr* **121:**81, 1976.

124. Rosler A, Gazit E, Theodor R, et al: Salt wastage, raised plasma-renin activity, and normal or high plasma-aldosterone: A form of pseudohypoaldosteronism. *Lancet* **1:**959, 1973.

125. Leehey D, Gantt C, Lim V: Heparin-induced hypoaldosteronism: Report of a case. *JAMA* **246:**2189, 1981.

126. Phelps KR, Oh MS, Carroll HJ: Heparin-induced hyperkalemia: Report of a case. *Nephron* **25:**254, 1980.

127. O'Kelly R, Magee F, McKenna TJ: Routine heparin therapy inhibits adrenal aldosterone production. *J Clin Endocrinol Metab* **56:**108, 1983.

128. Conn JW, Rovner OE, Cohen EL, et al: Inhibition by heparinoid of aldosterone biosynthesis in man. *J Clin Endocrinol Metab* **26:**527, 1966.

129. Abbot EC, Gornall AG, Sutherland DJA, et al: The influences of a heparin-like compound on hypertension, electrolytes and aldosterone in man. *Can Med Assoc J* **94:**1155, 1966.

9

Hirsutism

Introduction

Female hirsutism is excessive growth of hair in sites where hair growth is considered a male secondary sexual characteristic. The phenomenon of male pattern hair growth in females is a complex one mediated by several factors, the two dominant ones being circulating androgen levels and sensitivity of the hair follicle to local androgen concentrations. Historically, hirsutism and virilization of females was recognized—and documented—centuries ago. Michael Kelly[1] of Melbourne, Australia, noted the antiquity of the hirsute syndrome in that clinical descriptions of two hirsute women have been found in translations of the writings of Hippocrates over 2000 years ago. The syndrome of male pattern hair growth in females continues to pose diagnostic and therapeutic challenges and sometimes represents a dark corridor between endocrinology and psychiatry. In addition to being cosmetically bothersome, hirsutism can take a severe toll on the psyche of the sensitive female, raising the specter of sexual ambivalence. At times frustrating, and at times menacing, even so-called "benign" idiopathic hirsutism can lead to significant cosmetic, financial, and psychological consequences. The clinical, laboratory, and therapeutic approach to female hirsutism constitutes the framework of this chapter.

Adrenal Androgens

The zona reticularis, the innermost portion of the adrenal cortex, synthesizes the sex steroids of the adrenals. The predominant secretory products by the zona reticularis are the adrenal androgens. Since, in both sexes, the gonads synthesize adequate amounts of androgens, it is not clear why the adrenal cortex must synthesize these androgens. Besides, in the functional sense of the term, these androgens are strictly only "preandrogens," since any androgenic activity of these steroids depends on conversion to testosterone, the most potent androgen. Thus, it is difficult to understand why the synthesis of these preandrogens by the adrenal cortex is necessary when testosterone synthesis is physiologically carried out by the gonads in both sexes. Yet the zona fasciculata–reticularis layer of human and animal adrenal cortices normally synthesizes and releases several C-19 androgenic steroids. This activity becomes enhanced in the presence of increased precursors within the zona fasciculata and when stimulated by trophic hormones such as ACTH. Regardless of the physiological significance, adrenal androgens are not mere by-products in adrenal steroidogenesis, but constitute significant synthetic products of the adrenal cortex. The importance of these "weak" androgens assumes significance when these products are synthesized in excess as a consequence of enzyme deletions in the pathways of steroidogenesis within the adrenal zona fasciculata.

Synthesis

The major adrenal androgen secreted by the zona reticularis is dehydroepiandrosterone (DHEA). The pathway for secretion of DHEA, a Δ^5 compound, and its Δ^4 product, androstenedione, begins at the zona fasciculata, with pregnenololone and progesterone serving as the substrates. Thus, it is customary to view the zona fascicula–reticularis complex as a single production unit from which both the glucocorticoids and the adrenal androgens are synthesized. Three pathways within the zona reticularis result in the synthesis of 17-ketosteroids, 17-β-hydroxysteroids, and estrogens.

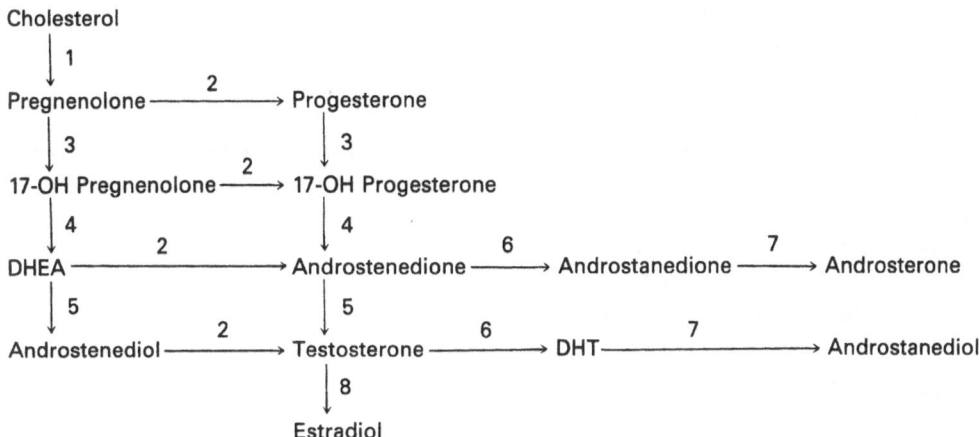

FIGURE 16. Androgen biosynthesis. The enzymes involved are: 1. cholesterol side-chain clearing enzyme, 2. 3-β-hydroxy steroid dehydrogenase, 3. 17-hydroxylase, 4. C-17, C-20 lyase, 5. 17-KS oxidoreduclase, 6. 5-α-reductase, 7. 3-KS oxidoreductase, and 8. aromatase.

17-Ketosteroid Pathway

The steroids formed in this pathway are DHEA, DHEA sulfate, and androstenedione. DHEA is formed from cholesterol, which is cleaved at its side chain to give pregnenolone, which is subsequently hxdroxylated at position 17 to form 17-α-hydroxypregnenolone. The conversion of the latter to DHEA is mediated by the enzyme C-17, C-20 lyase. The formation of androstenedione can be effected by two reactions. The first is conversion of DHEA, a Δ^5 compound, to androstenedione, a Δ^4 compound, by the action of the enzyme 3-β-hydroxysteroid dehydrogenase. The second pathway by which androstenedione can be derived is from conversion of 17-α-hydroxy progesterone by the action of the enzyme C-17, C-20 lyase. Androstenedione, thus formed, can be further reduced to androstanedione by the action of the enzyme 5-α-reductase. The final step in the 17-KS pathway is conversion of androstanedione to androsterone by the enzyme 3-ketosteroid oxidoreductase. Thus, the sequence of steroid products in the 17-KS pathway is DHEA, androstenedione, androstanedione, and androsterone (Fig. 16).

17-β-Hydroxy Pathway

This pathway leads to the formation of the more potent androgens from the preandrogens and consists of formation of androstenediol from DHEA and of testosterone from androstenedione. This single step is mediated by the enzyme 17-ketosteroid oxidoreductase.

Estrogen Pathway

The conversion of androstenedione to estrone and of testosterone to estradiol is mediated by the enzyme aromatase. The estrogen formation by the adrenal cortex is kept to a minimum, since availability of substrates and the necessary enzymes is considerably limited. According to the conventional schema of androgen synthesis by the adrenal cortex, the biosynthetic pathways leading to the synthesis of both cortisol and the C-19 androgens begin with 17-hydroxylation of pregnenolone or progesterone. Thus, the enzyme 17-hydroxylase plays a crucial role in the initial steps, with 17-α-hydroxy intermediates playing an important role as an intermediate precursors of androgen biosynthesis. These intermediates, which are 17-hydroxy-20-keto steroids, can either be 21-hydroxylated to form glucocorticoids or cleaved between C-17 and C-20 positions to form C-19 products.

Androgen Physiology

In both sexes, androgens are secreted by the gonads and by the adrenal cortex. The most potent androgens are testosterone and dihydrotestosterone, collectively referred to as 17-β-hydroxysteroid andorgens. The other group of androgens, the 17-ketosteroid androgens, are "weak" androgens and are represented by DHEA, DHEA sulfate, and androstenedione. In bioassay systems these androgens have negligible androgenic activity. However, these "weak" androgens have androgenic potential by virtue of their conversion to testosterone. Therefore, the term *preandrogens* is used to describe DHEA, DHEA sulfate, and androstenedione. Owing to the remarkable similarity in steroidogenic pathways in the adrenal cortex and in the gonads, both these endocrine glands can secrete both types of androgens. However, the adrenal cortex predominantly secretes DHEA, DHEA sulfate, and androstenedione. The enzyme that converts these 17-KS to 17-β-hydroxysteroids (testosterone and androstenediol) is 17-KS oxidoreductase. The activity of this enzyme in the adrenal cortex is several folds lower than in the gonads. Therefore, the contribution of the adrenal cortex to testosterone production is limited and rather modest. In contrast, the testis (and ovary) are rich in this enzyme, leading to almost total conversion of the preandrogens to testosterone and androstenediol. Therefore, DHEA, DHEA sulfate, and androstenedine are regarded as "adrenal androgens," while testosterone is regarded as the predominant gonadal androgen. Adrenal secretion normally accounts for the majority of 17-KS, and far less than 20% of testosterone production. Adrenal androgen production is under ACTH regulation but gonadal androgen production is not.

While conventionally ACTH is regarded as the major stimulus for adrenal androgen production, recent data suggest that the process may in fact be more complex. For instance, while DHEA and androstenedione clearly show concordant changes with a rise and fall of plasma ACTH concentrations,[2] DHEA sulfate seems to be dissociated from such changes.[3] It has been postulated that substances other than, or in addition to, ACTH must exist. One substance that does indeed qualify for such a role is the glycopeptide identified by Parker et al.[4] from the human pituitary gland. In addition, local alterations in steroid enzymology within the adrenal cortex can, per se, affect DHEA production.

The ovary secretes both testosterone and androstenedione. These steroids are regarded as obligate intermediates during the synthesis of estrogens by the granulosa cells of the ovary. The normal ovary produces twice as much androstenedione in comparison to testosterone. In addition to these two androgens, the normal ovary also secretes small amounts of DHEA and DHEA sulfate, possibly as metabolites of 17-α-OH progesterone secretion. In addition to being formed as intermediaries in the biosynthesis of estrogens, ovarian androgens can be synthesized de novo by the theca cells and by the interstitial compartment of the ovary (the hilar and stromal cells). This particular facet of androgen secretion by the ovary is controlled by pituitary gonadotropins, luteinizing hormone (LH) and follicle-stimulating hormone (FSH).[5,6] Androgen synthesis by the thecal cells is the major source of androgens during the follicular phase of the menstrual cycle.

In addition to direct synthesis by the ovary, androgens are produced in females by peripheral conversion of prehormones (17-KS) to testosterone by nonendocrine tissues such as liver, adipose tissue, and the skin.[7-9] The circulating testosterone in the plasma of normal females is derived from synthesis by the ovary, as well as from peripheral conversion of estrogens to testosterone. The range of testosterone in normal females is between 20 and 80 ng/dl. The testosterone in plasma circulates bound to a globulin called sex hormone–binding globulin (SHBG). Approximately 98–99% of testosterone is bound. This moiety is metabolically inert and unavailable to tissues. The free testosterone is the fraction that enters the cells and expresses androgen action. The importance of SHBG in terms of metabolic turnover and clearance of testosterone is discussed in the section dealing with etiology.

Clinical Considerations

In evaluation of female patients with hirsutism, the focus of attention revolves around the following aspects:

1. Degree of hirsutism
2. Presence or absence of virilization
3. Etiological clues in the history
4. Etiological clues in the physical examination

Each of these aspects requires brief mention.

Degree of Hirsutism

Generally "hirsutism" implies moderately excessive growth of terminal hair in androgen-sensitive areas. The term *terminal hair* refers to coarse, pigmented hair in contrast to "vellus hair," which is characterized by being fine, long, and unpigmented and occurs in both androgen-sensitive and androgen-insensitive parts of the body. Vellus-type hair is the type present before puberty, while terminal-type hair is that which develops after puberty. Thus, the conversion of vellus-type to terminal-type hair is a transitional phenomenon gradually mediated by the androgens in both sexes. Nine areas in the body contain androgen-sensitive hair follicles. In ascending order of severity, hirsutism can be semiquantified depending on the presence of terminal hair in these nine locations, which include the upper lip, chin, chest, upper abdomen, lower abdomen, upper arms, thighs, upper back, and lower back. In each of these areas, depending on the density of hair growth, the hirsutism can be graded from 1 through 4, grade 1 representing minimal terminal hair and grade 4 representing abundant terminal hair, which usually denotes virilism. Careful observation for the presence and severity of terminal hair growth in these nine sites has permitted the development of a scoring system.[10,11] A score greater than 8 indicates hirsutism.

The presence of "a few" hairs ("the argument of the beard") is not unusual in normal females and should not occasion any concern. While postmenopausal females may develop some hair groth in androgen-sensitive areas, only 5% of such women have scores exceeding 8.[12] Also, the presence and degree of hair growth in normal females is variable, depending on racial, ethnic, and genetic factors. Thus, Oriental females have comparatively less body hair, while women of Mediterranean descent have more.[13] In one study, conducted in the early sixties, McKnight[14] surveyed normal young English women and noted the presence of facial hair in 26%, hair around the areola in 17%, and lower abdominal hair in 35%. Thus, interpretation of body hair in females, and even the very definition of hirsutism in women, must take into account these factors. Further, a distinction must be made between hypertricosis and hirsutism. Hypertricosis refers to the presence of excessive hair growth in areas such as the forehead, forearms, and lower legs—areas that contain non–androgen-dependent hair growth. Hypertricosis is seldom caused by androgen imbalance, consists of predominantly vellous hair, and is usually related to hereditary factors and medication use, particularly glucocorticoids and anticonvulsants.

Normal hair growth physiologically occurs in cycles.[15] The active growth phase, called the anagen phase, is the single most important determinant of the length of hair. This phase alternates with the resting phase, or the telogen phase, of hair growth. Once the active phase of hair growth has started, treatment modalities do not appreciably alter it. Also, the anagen phase may be related to duration of exposure of the hair follicles to androgens, rather than to the circulating levels of androgens at a given time.

Virilization

The term *virilization* indicates the extreme effects of androgen excess. In addition to extreme degrees of hirsutism, virilization is often characterized by other signs, such as clitoromegaly, increased muscle mass, deepening of the voice, atrophy of the breast tissue, oligomenorrhea or even amenorrhea, and temporal recession of scalp hair (male pattern baldness). The presence of terminal hair growth on the shoulders, arms, upper back, and upper abdomen signifies an extreme degree of hirsutism and is usually associated with other virilizing features. Of these, clitoromegaly is an important one and can be objectively quantified. It has been suggested that calculation of the "clitoral index" provides a parameter to assess the degree of hyperandrogenicity. The "clitoral index," calculated as the product of the vertical and horizontal dimensions of the glans clitoris after retraction of the hood, normally measures between 9 and 35 mm^2.[16] The clitoral index can exceed 100 mm in cases of extreme hyperandrogenicity. The association of hirsutism with virilizing features denotes the presence of serious underlying diseases such as androgen-secreting tumors of the ovary or adrenal cortex, and less commonly adult-onset 21-hydroxylase deficiencies. Clitoromegaly in the absence of severe hirsutism is extremely unusual.

Etiological Clues in the History

The age of onset of hirsutism and the rate of progression provide important clues to the tiology of hirsutism. When the onset of hirsutism occurs in the peripubertal period and progresses gradually, without virilizing signs, benign disease underlies the hirsuitism. Such is the history in patients with idiopathic hirsutism. Significant menstrual abnormalities are less prevalent in patients with idiopathic hirsutism. The combination of oligomenorrhea, infertility, and hirsutism is consistent with both polycystic ovary syndrome and attenuated (late-onset) 21-hydroxylase deficiency. The onset and progression of hirsutism can be identical both in idiopathic hirsutism and in patients with attenuated 21-hydroxylase deficiency. When the hirsutism has its onset in adulthood, coupled with rapid progression and features of virilization, androgen-secreting tumors are an important and likely consideration. Many patients with hirsutism, regardless of etiology, may notice intensification of the hirsutism with rapid weight gain. Similarly, regardless of etiology, menstrual irregularities may coincide with intensification of the hirsutism.

A carefully obtained drug history is mandatory in all patients presenting with hirsutism of recent onset. This includes the use of diphenylhydantion; antihypertensive medications such as minoxidil or diazoxide; hormonal preparations such as testosterone, halotestone, anabolic steroids, and glucocorticoids; immunosuppressive therapy with cyclosporine; and very rarely the use of birth control pills. It should be emphasized that the true incidence of birth-control-pill–induced hirsutism is extremely low.

Etiological Clues in the Physical Examination

Several important clues can be found in the physical examination of hirsute women. The importance of virilizing signs has already been alluded to. The presence of pelvic masses in a hirsute female usually suggests polycystic ovary syndrome, or tumor of the ovary. However, a small but significant percentage of patients with congenital adrenal hyperplasia may also have cysts in the ovaries, detected by pelvic examination or ultrasonography. The combination of hirsutism, ovarian cyst, and acanthosis nigricans should immediaely suggest the possibility of the insulin-resistant syndrome associated with hirsutism. The presence of an abdominal mass in the hirsute patient points to an underlying adrenal tumor, usually an androgen-secreting carcinoma or, less commonly, an adenoma. The presence of Cushing's syndrome in a hirsute woman is consistent with both pituitary-dependent Cushing's disease and an adrenal carcinoma. However, the hirsutism is more pronounced in adrenal carcinoma. Further, nearly one-third to one-half of patients with adrenal carcinoma harbor an adrenal mass that is large enough to be palpable by abdominal examination.

With these clues in the history and physical examination of the hirsute female, the etiological perspective can be approached.

Etiology of Hirsutism

Hirsutism can be due to several etiological factors.

1. Excessive secretion of androgens from the adrenal cortex
2. Excessive secretion of androgens from the ovary
3. Excessive secretion of androgens from both the ovaries and the adrenals
4. Idiopathic factors
5. Drug-induced factors
6. Alterations in SHBG
7. Increased skin sensitivity to androgens

Excessive Secretion of Adrenal Androgens

The prototype of hirsutism resulting from excessive secretion of adrenal androgens is represented by the various forms of enzymatic defects in steroidogenesis. Thus, deficiencies of 21-hydroxylase, and less commonly of 11-hydroxylase and 3-β-hydroxysteroid dehydrogenase, can result in hirsutism secondary to adrenal androgen excess. The most common form of adult-onset disease is 21-hydroxylase deficiency.[17-23] The incidence, clinical features, and diagnostic aspects of the adult-onset (or attenuated) variety of 21-hydroxylase deficiency is discussed in Chapter 11.

Androgen-secreting adrenal carcinoma is a devastating disease associated with hirsutism, and often virilization. The clinical and diagnostic features of adrenal carcinoma are outlined in Chapter 10. Pure androgen-secreting adrenal adenomas are rare. In contrast to adrenal carcinomas, which are generally large, androgen-secreting adrenal adenomas tend to be small and can occasionally secrete testosterone instead of DHEA or androstenedione.

Another cause of hirsutism associated with adrenal androgen excess is pituitary-dependent Cushing's disease. Varying degrees of hirsutism occur with Cushing's disease. The consistency with which DHEA levels are elevated in this disorder has allowed the use of this parameter in differentiating hypercortisolism caused by pituitary-dependent disease from that caused by adrenal adenoma.

Ovarian Androgen Excess

The most common etiology of ovarian androgen excess is the polycystic ovarian syndrome (PCOS). The constellation of features in PCOS includes, in decreasing frequency, hirsutism (69%), dysfunctional uterine bleeding (29%), infertility or subfertility (74%), and obesity (41%). The anatomical hallmark of PCOS is an enlarged ovary with a pearly white surface, containing multiple small follicular cysts that are surrounded by a thickened luteinized theca. The luteinized theca cells (hyperthecosis) are the source of increased androgen production. When radioactive pregnenolone and progesterone are incubated with polycystic ovary tissue, excessive accumulation of DHEA (and to a lesser extent 19-hydroxyandrostenedione) has been described.[24,25] The increased accumulation of these C-19 androgenic steroids may be a consequence of deficiency in the aromatase enzyme in the ovaries, which normally converts androgenic steroids to estrogens. The exact mechanism for the increased androgen synthesis is less clear. The theories for the evolution of PCOS are several and include decreased activity of the aromatase enzyme, decreased ovulation due to hormonal imbalance (absolute or relative increase in luteinizing hormone levels), adverse effects of androgen on follicular growth and maturation, abnormally accelerated atresia of follicles, abnormally high androgen-to-estrogen ratios within the follicle, and a primary hypothalamic derangement that leads to tonic secretion of LH.[26–30] The multiplicity and diversity of these postulated mechanisms attest to the enigmatic nature of the etiology of PCOS. Probably several factors, in combination, are responsible for the clinical expressions of this common form of ovarian hyperandrogenism.

Androgen-producing ovarian tumors are a rare cause of hirsutism and virilization in postpubertal females. These virilizing tumors are well differentiated and usually benign, often attaining palpable size. Various histological types (arrhenoblastoma, hilus cell tumors, lipoid cell tumors, and Sertoli–Leydig cell tumors) have been described. These tumors, accounting for less than 0.2% of all ovarian tumors, are best regarded as sex cord–stromal tumors[31,32] that originate from androgen remnant tissue in the hilus portion of the ovary. The diagnostic triad of androgen-secreting ovarian neoplasms consists of moderate to marked elevation of plasma testosterone concentrations (>200 ng/dl), sudden onset and rapid progression of hirsutism and or virilization, and visualization by computed tomography or ultrasonography of the pelvic region.

Another condition that is probably associated with ovarian hyperandrogenism is so-called "idiopathic" hirsutism. This entity will be separately discussed, since ovarian hyperandrogenism is only one facet in the multifaceted etiology of idiopathic hirsutism.

Mixed Ovarian and Adrenal Hyperandrogenism

This category includes several conditions characterized by hirsutism associated with increased androgen secretion from both the adrenal and the ovary. Several entities belong in this group:

1. Enzymatic deficiencies in steroidogenesis that operate at dual levels—the adrenal as well as the ovaries. This group includes patients with 3-β-hydroxysteroid dehydrogenase deficiency.[33]
2. In some instances, excessive androgen production by the ovary can result in inhibition of adrenal steroidogenic enzymes. It has been shown, in vivo, that 11-β-hydroxylation can be inhibited in the dog by the administration of certain androgens.[34] It is possible that such a mechanism also exists in humans. Recently, iodocholesterol uptake by the adrenal cortex has been shown to be increased in some women with PCOS.[35]
3. Alternately, increased adrenal androgen secretion, as in some cases of 21-hydroxylase deficiency, can result in polycystic changes in the ovaries.[36,37] Such patients would present with hirsutism coupled with hormonal and radiological features consistent with both 21-hydroxylase deficiency and PCOS.

4. Finally, some patients with idiopathic hirsutism may demonstrate laboratory evidence of both ovarian and adrenal hyperandrogenism.[38,39] Such patients represent, at best, only a small subset of patients with idiopathic hirsutism.

Idiopathic Hirsutism

The term *idiopathic hirsutism* is used to denote patients with hirsutism in whom no clear-cut ovarian or adrenal pathology can be demonstrated. In general, the clinical presentation of these patients is characterized by hirsutism of varying severity, with normal or irregular menses. Virilization and amenorrhea are unusual, but can rarely develop. The important aspects to focus in this entity revolve around three basic questions: Is there a hormonal abnormality in patients with idiopathic hirsutism? If so, what is the nature of the androgen excess? What is the source of these androgens? Although the underlying defect in patients with idiopathic hirsutism remains to be clearly identified, some answers to these questions have been unraveled in the past decade. It appears that patients with idiopathic hirsutism do indeed have a hormonal abnormality. Nearly all women with idiopathic hirsutism have an abnormally elevated testosterone production rate.[40] The source of these androgens is most likely ovarian. An understanding of the pathogenesis of idiopathic hirsutism requires brief examination of the four cornerstones in the concept of this disorder—the nature of the androgen abnormality, the source of these androgens, the abnormalities in transport of these androgens, and, finally, the sensitivity of target tissue, the hair follicle, to androgens.

Abnormality in Androgen Levels

Several workers have demonstrated that idiopathic hirsutism is associated with excessive production rates and plasma levels of testosterone and androstenedione.[40–43] However, considerable debate exists as to the prevalence and degree of androgen hypersecretion in patients with idiopathic hirsutism. These controversies are a result of measurement of the total testosterone levels in idiopathic hirsutism. While the free (or unbound) testosterone levels in plasma are usually elevated in these women,[44,45] the total (or bound) fraction may be well within the normal range. The reason for this discrepancy is due to the fact that the SHBG is reduced in such patients, bringing down the proprotion of bound hormone. It has been estimated that the SHBG levels are decreased in the serum of approximately 50% of patients with idiopathic hirsutism. The reason for this phenomenon in hirsute women is unclear. Nevertheless, since the concentration of SHBG is a major determinant of the rate of turnover of plasma testosterone, the free testosterone concentrations in the plasma are usually elevated in idiopathic hirsutism. In addition to free testosterone elevation, patients with idiopathic hirsutism may demonstrate elevated plasma concentrations of other androgens (such as dihydrotestosterone) and preandrogens (such as androstenedione and DHEA sulfate). The incidence of detecting such abnormalities in women with idiopathic hirsutism has been variably reported.[46,47] The incidence of detecting abnormal elevations in one or more androgens and preandrogens is related to obtaining multiple samples, since these androgenic steroids demonstrate considerable pulsatility and fluctuation in the serum concentrations throughout the day.[48]

Source of Androgen Excess

After nearly two decades of conflicting opinions, there now seems to be some unanimity regarding the source of hyperandrogenism in idiopathic hirsutism. Various methods have been used to evaluate the source of hyperandrogenism in patients with idiopathic hirsutism. The two important diagnostic maneuvers to delineate an ovarian from an adrenal source of androgen excess are the use of dexamethasone suppression and selective catheterization of the adrenal and ovarian veins. The use of the dexamethasone suppression test in hirsute women to distinguish ovarian from adrenal androgen excess has suffered from lack of standardization of the measured parameters. The test has been fine-tuned for this purpose by Rosenfield et al.[49–51] and consists of comparing androgen profiles after the administration of 0.5 mg four times a day orally for 1 week. Patients with idiopathic hirsutism often demonstrate normalcy of ACTH-dependent androgen production. The criteria employed to determine suppression to dexamethasone by these workers include a decline in *free* testosterone concentrations below 8 pg/ml and a decrease in DHEA sulfate level by at least 50% of baseline (usually 75%), with both cortisol and DHEA sulfate falling below the normal adult control range. In normal women, dexamethasone has little or no effect on ovarian androgen production.[52,53] The difficulty in interpretation of androgen responses in idiopathic hirsutism lies in the fact that dexamethasone can decrease ovarian androgen production in hirsute women.[54,55] In addition, the heterogeneity in responses of patients with idiopathic hirsutism can further confound interpretation, even when strict criteria are employed. Moreover, it is now becoming increasingly accepted that DHEA sulfate, in hirsute women, is not exclusively due to adrenal hyperfunction and may reflect peripheral sulfation of DHEA. All these reservations have considerably shrouded the interpretation of the test. Owing to these reservations, it seemed desirable to quantitate the relative contribution of the ovaries and adrenals to the hyperandrogenicity of idiopathic hirsutism by direct measurement of androgen production rate obtained by selective catherization of the adrenal and ovarian veins. The largest body of data was collected by Kirschner and co-workers.[56] These workers calculated the production rates of testosterone and androstenedione in the ovaries and adrenals of 44 hirsute women and reported that the ovary is the primary site of hyperandrogenism in hirsute patients with idiopathic hirsutism and PCOS. The data, calculated from the arteriovenous differences in testosterone in samples collected from the ovarian and adrenal veins, rely on many assumptions in estimating the blood flow of the two organs as well as in calculation of the production rates. Yet, the results do show a clearly increased magnitude in the ovarian venous effluent. The findings of Kirscher et al.[56] are also supported by a different line of evidence that is derived from use of long-acting luteinizing hormone–releasing hormone (LHRH) analogs in patients with hirsutism. Chang et al.[57] employed analog therapy in five patients with hirsutism caused by PCOS and demonstrated significant reduction—even normalization—

of testosterone and androstenedione levels without significant changes in the level of DHEA sulfate. Thus, it appears that a primary ovarian abnormality underlies the hyperandrogenism of patients with PCOS as well as idiopathic hirsutism.

Androgen Transport

Sex hormone–binding globulin plays a crucial role in the transport and turnover of the 17-β-hydroxysteroids such as testosterone, dihydrotestosterone, and estradiol. Approximately 98–99% of testosterone circulates in the plasma bound to this globulin. An inverse correlation exists between the amount of SHBG and the free, unbound testosterone. Cell culture studies have revealed that SHBG retards cellular uptake of testosterone and its conversion to dihydro-testosterone.[58] Thus, a decrease in SHBG would result in the following se-quelae—reduction in the bound (total) testosterone, increase in the unbound (free) testosterone, increased turnover and kinetics of the hormone, increased uptake by target cells, and possibly an increased conversion of testosterone to dihyrotestosterone. While a significant percentage (as high as 50%) of hirsute women demonstrate varying degrees of lowering in the SHBG, the role of this phenomenon in the development of hirsutism is at best speculative.[59,60]

Tissue Metabolism of Androgens

The androgen effects exerted on the hair follicle are similar to the effects of testosterone on other target cells. Thus, testosterone and its reduced form di-hydrotestosterone (DHT) enter the cell, are transported to the cystosol, bind to cytoplasmic androgen receptor–binding protein, and eventually reach the nu-clear acceptor sites, where androgen effects are initiated by transcription.[61] It is possible that DHT is the crucial androgen for sexual hair growth since some patients with 5-α-reductase deficiency demonstrate a paucity of sexual hair.[62] However, this is not a universal finding, and many workers believe that testoster-one is the key hormone for sexual hair growth.[12,63,64] Studies that have evalu-ated 5-α-reductase activity in the pubic and labial skin of hirsute women suggest that such activity is indeed enhanced in hirsute women.[65–67] However, it is debatable whether a similar situation exists in the hair follicles.

Another parameter for androgen metabolism in the skin is measurement of 3-α-androstanediol in plasma and urine. 3-α-Androstanediol is a direct metabo-lite of DHT in the skin. Numerous studies have demonstrated increased con-centration of this metabolite in the plasma and urine of hirsute women.[68–71] Clearly, local metabolism of androgens may have a role in the development of idiopathic hirsutism.

Drug-Induced Hirsutism

Several drugs, used in pharmacological as well as physiological doses, can lead to hirsutism. While a well-obtained history would usually disclose the use or abuse of drugs that cause hirsutism, some transsexual patients who are on androgens and some women atheletes on anabolic steroids may not readily divulge the use of these drugs. In addition to hormones, the use of minoxidil, diazoxide, and diphenylhydantoin should be considered.

Decreased SHBG and Hirsutism

The important role played by SHBG in the transport and turnover of testosterone has been already alluded to. In addition to the observation that SHBG concentrations are found to be decreased in patients with hirsutism in general, and in the idiopathic variety in particular, two pertinent comments are noteworthy. First, the hirsutism associated with obesity may be related to changes in the concentrations of SHBG, and second, the prolactin-induced hyperandrogenism may be, at least in part, related to alterations in SHBG concentrations.

Obesity

The relationship between obesity and hirsutism is a well-known but ill-understood phenomenon. Several possible mechanisms have been suggested to explain this association:

Hyperinsulinemia. Obesity is nearly always associated with relative hyperinsulinemia, owing to the down-regulation of insulin receptors. It is well known that hyperinsulinemia enhances androgen secretion by the ovary. This is highlighted in the clinical situation characterized by acanthosis nigricans, insulin resistance, and hirsutism. In addition, a significant correlation has been shown to exist between hirsutism, insulin levels, and plasma testosterone concentrations in patients with PCOS.[72] In vitro studies using granulosa cells suggest that these cells secrete more androgens in the presence of insulin.[73] Further, weight loss is often associated with improvement in the hirsutism, with a simultaneous decline in insulin levels.

SHBG. It is well recognized that obesity results in lowering of the SHBG concentrations in the plasma.[74,75] This can result in decreased total and increased free testosterone concentrations.

Androgen–Estrogen Metabolism. Adipose tissue can convert preandrogens to testosterone.[76] Thus, it is conceivable that in obesity, greater amounts of androstenedione may be converted to testosterone. In a sense, the fat tissue becomes an extragonadal production site for production of testosterone. However, it is unclear whether such a mechanism plays any significant role in the development of hirsutism in the obese female. Paradoxically, it has been shown that adipocytes also contain appreciable amounts of aromatase (which converts testosterone into estradiol). It is believed that adipose tissue serves as an important source of estrogen in women.[77,78] These conflicting and opposing roles of adipose tissue in sex steroid synthesis make it difficult to assess the role of obesity in causing hirsutism.

Prolactin Excess

Hirsutism can be associated with hyperprolactinemic states. Several studies have reinforced the notion that DHEA and DHEA sulfate levels are elevated in some patients with hyperprolactinemia.[79-82] While it is conceivable that prolactin may exert a direct stimulatory effect on adrenal androgen secretion, more recent data suggest that hyperprolactinemia may increase free testosterone concentrations by decreasing SHBG levels.[83,84]

As indicated earlier, SHBG has been implicated in the pathogenesis of at least three conditions related to the development of hirsutism—idiopathic hirsutism, obesity, and hyperprolactinemia. While the role of SHBG in the pathogenesis of hirsutism has not been doubted, the relative magnitude of this factor is yet to be determined.

Increased Skin Sensitivity

While it is well known that endocrine syndromes can evolve from increased sensitivity of target tissues to normal levels of circulating hormone, proof for such a phenomenon in the causation of hirsutism is not convincing. The paucity of data on the subject is the main reason that precludes drawing firm conclusions in this regard. A single study, consisting of four hirsute women, has, if anything, demonstrated normal activity of 5-α-reductase in the hair follicle of these hirsute subjects.[85]

Laboratory Evaluation

The laboratory evaluation of patients with hirsutism evolves through three stages. The first phase consists of measuring androgen levels in the plasma. This phase involves obtaining a panel of androgens circulating in the plasma at the basal state. The second phase is initiated if hyperandrogenism is documented and consists of delineating the source of the androgen excess. Although this may not always be possible, the information obtained may have therapeutic importance. This phase involves performance of the dexamethasone suppression test, the utility of which has been disputed in the literature. The third phase of the workup includes tests based on whether the hyperandrogenism is thought to be of ovarian or adrenal origin. The tests during this phase of the workup are aimed at exclusion of adult-onset adrenogenital syndromes, adrenal tumors, PCOS, and ovarian tumors. The tests employed during this phase range from measuring steroid precursors (such as 17-α-OH progesterone) and evaluating the response to ACTH administration, to imaging procedures of the adrenal and ovary, and even laparoscopy. When appropriate, measurement of urinary free cortisol, prolactin, LH, and FSH may be indicated.

Phase 1. Androgen Levels

The first step in evaluation of the hirsute female is measuring the androgen levels in the circulation. This is attained by initially measuring levels of testosterone and DHEA in the plasma. Interpretation of plasma testosterone levels should take into consideration the following factors. First, the usual methods employed for measuring plasma testosterone concentrations determine the hormone bound to SHBG. The demonstration of a normal testosterone concentration does not conclusively exclude a testosterone-related problem. Second, testosterone levels in the circulation demonstrate considerable pulsatility; a single plasma determination can differ by as much as 38% from the 24-hr mean testosterone level. To obviate this problem, testosterone can be sampled hourly over a 3-hr period to obtain a mean level. If the testosterone level obtained by such a method is normal, the "free" testosterone level should be performed. As many as 60% of hirsute women with an initially normal plasma testosterone may show an abnormality in the free testosterone level. The free fraction can be directly measured by some laboratories or can be calculated by measuring fractional testosterone binding to plasma or by directly determining the SHBG level. Should the free and total testosterone remain normal despite multiple sampling, the patient's serum should be analyzed for the presence of other androgens such as androstenedione and DHEA sulfate.

Measurement of DHEA (and DHEA sulfate) in the plasma directly has obviated the need for measuring urinary 17-ketosteroids, a procedure that has proved to be of little value in the diagnosis or differential diagnosis of hirsutism. In fact, it is not routinely performed as part of the initial workup for hirsutism.[47,86]

Phase 2. Source of Androgen Excess

The second phase in the evaluation of hirsutism is establishing the source of hyperandrogenism. This may not always be easy. While a clearly elevated plasma testosterone level (>200 ng/dl) is strongly suggestive of ovarian neoplasm,[87] and a markedly elevated DHEA (or DHEA sulfate) suggests an adrenal etiology, mild to moderate elevation of either of these has only limited diagnostic specificity. For instance, mildly elevated testosterone levels can be seen in a variety of disorders that cause hirsutism, such as PCOS, hyperthecosis, idiopathic hirsutism, and even adrenogenital syndromes. Similarly, a mildly elevated DHEA level in the plasma can be encountered in all the aforementioned states.

Establishing the source of hyperandrogenism by selective venous catheterization and measuring androgen levels in the ovarian and adrenal effluents is an invasive procedure, which limits its application. Moreover, considerable controversy revolves around the ethical implications as well as the cost effectiveness of the routine application of this procedure in hirsute women. The use of the dexamethasone suppression test to determine the source of hyperandrogenism has also suffered the brunt of controversy. Opinions regarding use of dexamethasone suppression range from not recommending the test owing its nonspecificity[72] to routine application of the test with better-defined criteria.[12] In the experience of Hatch et al.,[12] subnormal suppression of plasma androgens to dexamethasone suggests four etiologies—PCOS (or its variants), tumor of the ovary, tumor of the adrenal, or Cushing's syndrome. Those patients with hyperandrogenemia who demonstrate preservation of dexamethasone suppressibility should be suspected of having some form of adrenogenital syndrome, obesity, a prolactin-related problem, or idiopathic disease. Clearly, the decision to perform the test depends on the epxerience of the individual physician with the predictive value of the test.

Phase 3. Radiological Studies

The studies performed in this phase depend on the degree and type of hyperandrogenemia encountered in the hirsute female. Thus, when the testosterone level is markedly elevated, evaluation for an ovarian neoplasm by ultrasonography and computerized tomography is indicated. In addition, when the hormonal and clinical features suggest PCOS, laparoscopy becomes an important diagnostic adjunct.

When the DHEA or its sulfate is elevated, the diagnostic workup should be directed to confirm or exclude various forms of adrenogenital syndrome. These include measurement of precursor steroids, particularly plasma levels of 17-α-OH progesterone. Additional tests may be required in some cases, such as evaluating the response of cortisol and 17-α-OH progesterone to ACTH and dexamethasone. Once adrenogenital syndrome has been excluded, the workup for elevated DHEA involves exclusion-of adrenal neoplasms by computed tomography of the adrenal.

Selective venous catheterization of the adrenal and ovarian veins for androgens, as well as evaluation of the androgen response to LHRH analog administration for 4 weeks, has not found wide application.

Treatment

The treatment of hirsutism depends on the underlying etiology of the condition. Even when a definite etiology has been identified, specific therapy may not completely reverse the hirsutism. However, specific therapy can arrest or decrease the progression of the problem. Patients must be informed that hormonal therapy will not affect actively growing terminal hairs, and hence therapeutic benefit may not be noticable for as long as 6–12 months. Many patients will already have tried mechanical modalities of therapy such as bleaching, mechanical hair removal, and even shaving.[88,89] Shaving on a regular basis can be terribly demoralizing to the self-image of women. Therefore, any medical treatment, sometimes even without a guaranteed success, is welcomed by hirsute women. Specific therapy for adrenogenital syndromes is discussed in Chapter 11.

The three modalities of drug therapy employed in the treatment of hirsutism are glucococorticoid therapy, estrogen–progestin therapy, and the use of antiandrogens. Glucocorticoid therapy is the primary mode of therapy for hirsutism caused by 21-hydroxylase deficiency. The use of glucocorticoid treatment for other etiologies of hirsutism, such as PCOS, or idiopathic disease is less well established. Some workers[72] do not recommend its use for forms of hirsutism other than that caused by adrenogenital syndrome, while others[12] recommend a trial of glucocorticoids in hyperandrogenic states suppressible with dexamethasone. The drug has to be given for protracted periods of time, and the potential for overtreatment is real. However, with doses that do not exceed a single dose of 0.5 mg of dexamethasone at bedtime, iatrogenic Cushing's syndrome and suppression of the hypothalamic pituitary adrenal axis (HPA) are less frequent. Measurement of androgen levels serves as a parameter to adjust the dose, which varies from patient to patient. Even when the androgen levels decline to normal levels, the treatment may not result in regression of hirsutism. Possibly, androgen levels must be suppressed to below normal ranges to be effective, but this may require dosages in excess of 0.5 mg of dexamethasone per day. In each patient, the risks of glucocorticoid therapy have to be weighed against its possible benefits. Patients with PCOS (or its variants) may resume their menses with glucocorticoid therapy; however, this may not necessarily be accompanied by improvement in the hirsutism.

Treatment with estrogens, in the form of oral contraceptive agents, continues to remain an enigmatic area. While the efficacy of low-dose oral estrogens is doubted,[90,91] high-dose estrogen therapy has been employed to retard excessive hair growth. This treatment may be associated with adverse side effects of estrogen therapy, especially in women over 35 years of age. Besides, such therapy is not an option in young women who desire conception. Like other forms of therapy for hirsutism, estrogen therapy (often combined with progesterone) is a protracted one, requiring careful and close follow-up. Patients who are young and have recent onset of hirsutism fare better than those who have been hirsute for long periods of time.

Finally, the use of antiandrogens—alone, or in combination with estrogens—may be helpful in some patients. Antiandrogens inhibit the effect of testosterone at the level of the receptor sites. The three prototypical drugs in this category are cimetidine, cyproterone acetate, and spironolactone. Considerable interest was originally generated when it was discovered that cimetidine possessed antiandrogenic properties.[92,93] However, cimetidine has enjoyed only a limited use for treatment of hirsutism.[94] As for cyproterone acetate, the drug is still considered investigational in the United States and therefore is not available for widespread use. Based on experience in Europe, when cyproterone acetate is used in doses exceeding 200 mg/day, diminished hair growth has been noted in 70–100% of hirsute women. The reported side effects are several and include nausea, weight gain, depression, decreased libido, and breast tenderness. These side effects are more pronounced with higher dosages of the drug.

Spironolactone is used for treatment of hirsutism based on its antiandrogenic effects. This drug has dual effects that are of benefit to patients with hirsutism. The drug is a strong inhibitor of DHT binding to androgen receptor,[95] and it also decreases the 5-α-reductase activity in the skin.[96] In addition, spironolactone also decreases biosynthesis of testosterone. In doses of 100–200 mg/day the drug can cause moderate to significant improvement in the amount and rate of hair growth.[95,97−100] In addition to adverse effects such as gastrointestinal disturbances, headaches, fatigue, and menstrual irregularities, spironolactone can result in potassium retention, especially in patients with suboptimal renal function.

The use of topical antiandrogens or LHRH analog therapy awaits elucidation. For the present, treatment of hirsutism revolves around mechanical means of hair removal, coupled with therapy directed against the underlying disorder.

When no underlying disorder can be identified, spironolactone therapy is the mainstay of medical therapy, coupled, when indicated, with estrogen progesterone combinations. As indicated earlier, success or failure of medical treatment can be assessed only with a protracted trial of drug therapy, often as long as a year.

References

1. Kelly, M: Hirsuitism. *Br Med J* **2**:1301, 1956.
2. James VHT, Turnbridge D, Wilson GA, et al: Central control of steroid hormones secretion. *J Steroid Biochem* **9**:429, 1978.
3. James VHT, Goodall A: Androgen production in women. In Jeffcoate SL (ed): *Androgens and Anti-androgen Therapy*. Wiley, New York, 1982, p. 23.
4. Parker LN, Lifrak ET, Odell WD: A 60,000 molecular weight human pituitary glycopeptide stimulates adrenal androgen secretion. *Endocrinology* **113**:2092, 1983.
5. McNatty, KP, Makris A, DeGrazia C, et al: The production of progesterone, androgens, and estrogens by granulsa cells, thecal tissue, and stromal tissue from human ovaries in vitro. *J Clin Endocrinol Metab* **49**:687, 1979.
6. Tsang BK, Armstrong DT, Whitfield JF: Steroid biosynthesis by isolated human ovarian follicular cells in vitro. *J Clin Endocrinol Metab* **51**:1407, 1980.
7. Rivarola MA, Singleton RT, Migeon CJ: Splanchnic extraction and interconversion of testosterone and androstenedione in man. *J Clin Invest* **46**:2095, 1967.
8. Nimrod A, Ryan KJ: Aromatization of androgens by human abdominal and breast fat tissue. *J Clin Endocrinol Metab* **40**:367, 1975.
9. Edman CD, MacDonald PC: Effect of obesity on conversion of plasma androstenedione to estrone in ovulatory and anovulatory young women. *Am J Obstet Gynecol* **130**:456, 1978.
10. Ferriman D, Gallwey JD: Clinical assessment of body hair growth in women. *J Clin Endocrinol Metab* **21**:1440, 1961.
11. Lorenzo EM: Familial study of hirsutism. *J Clin Endocrinol Metab* **31**:556, 1970.
12. Hatch R, Rosenfield RL, Kim MH, et al: Hirsutism: implications, etiology, and management. *Am J Obstet Gynecol* **140**:815, 1981.
13. Hamilton JB, Terada H: Interdependence of genetic, aging, and endocrine factors in hirsutism. In Greenblatt RB (ed): *The Hirsute Woman*. Charles C Thomas, Springfield, IL, 1962, p. 20.
14. McKnight E: The prevalence of "hirsutism" in young women. *Lancet* **1**:410, 1964.
15. Uno H: Biology of hair growth. *Semin Reprod Endocrinol* **4**:131, 1986.
16. Tagatz GE, Kopher RA, Nagel TC, et al: The clitoral index: a bioassay of androgenic stimulation. *Obstet Gynecol* **54**:562, 1979.
17. Rosenwaks Z, Lee PA, Jones GS, et al: An attenuated form of congenital virilizing adrenal hyperplasia. *J Clin Endocrinol Metab* **49**:335, 1979.
18. New MI, Lorenzen F, Pang S, et al: "Acquired" adrenal hyperplasia with 21-hydroxylase deficiency is not the same genetic disorder as congenital adrenal hyperplasia. *J Clin Endocrinol Metab* **48**:356, 1979.
19. Blankstein J, Faiman C, Reyes FI, et al: Adult-onset familial adrenal 21-hydroxylase deficiency. *Am J Med* **68**:441, 1980.
20. Kohn B, Levine LS, Pollack MS, et al: Late-onset steroid 21-hydroxylase deficiency: A variant of classical congenital adrenal hyperplasia. *J Clin Endocrinol Metab* **55**:817, 1982.
21. Chetkowski RJ, DeFazo J, Shamonki I, et al: The incidence of late-onset congenital adrenal hyperplasia due to 21-hydroxylase deficiency among hirsute women. *J Clin Endocrinol Metab* **58**:595, 1984.
22. Kuttenn F, Couillin P, Girard F, et al: Late-onset adrenal hyperplasia in hirsutism. *N Engl J Med* **313**:224, 1985.
23. Gabrilove JL, Sharma DC, Dorfman RI: Adrenocortical 11β-hydroxylase deficiency and virilism first manifest in the adult woman. *N Engl J Med* **272**:1189, 1965.

24. Axelrod LR, Goldzieher JW: The polycystic ovary. III. Steroid biosynthesis in normal and polycystic ovarian tissue. *J Clin Endocrinol Metab* **22**:431, 1962.
25. Mahesh VB, Greenblatt RB, Aydar CK, et al: Secretion of androgens by the polycystic ovary and its significance. *Fertil Steril* **13**:513, 1962.
26. Yen SS, Vela CP, Rankin J: Inappropriate secretion of follicle-stimulating hormone and luteinizing hormone in polycystic ovarian disease. *J Clin Endocrinol Metab* **30**:435, 1970.
27. Rebar R, Judd HL, Yen SS, et al: Characterization of the inappropriate gonadotropin secretion in polycystic ovary syndrome. *J Clin Invest* **57**:1320, 1976.
28. McNatty KP, Smith DM, Makris A, et al: The intraovarian sites of androgen and estrogen formation in normal and hyperandrogenic ovaries as judged by in vitro experiments. *J Clin Endocrinol Metab* **50**:755, 1980.
29. Louvet J-P, Harman SM, Schreiber JR, et al: Evidence for a role of androgens in follicular maturation. *Endocrinology* **97**:366, 1975.
30. Korth-Schutz S, Leving LS, Merkatz IR, et al: An unusual case of Cushing's syndrome, hilus cell tumor and polycystic ovaries. *J Clin Endocrinol Metab* **38**:794, 1974.
31. Young RH, Scully RE: Ovarian sex cord-stromal tumors: recent progress. *Int J Gynecol Pathol* **1**:101, 1982.
32. Fox H: Sex cord-stromal tumours of the ovary. *J Pathol* **145**:127, 1985.
33. Rosenfield RL, Rich BH, Wolfsdorf JI, et al: Pubertal presentation of congenital Δ5-3β hydroxysteroid dehydrogenase deficiency. *J Clin Endocrinol Metab* **51**:345, 1980.
34. Fragachan F, Nowaczynski W, Bertranu E, et al: Evidence of in vivo inhibition of 11β-hydroxylation of steroids by dehydroepiandrosterone in the dog. *J Endocrinol* **84**:98, 1969.
35. Gross MD, Wortsman J, Shapiro B, et al: Scintigraphic evidence of andrenal cortical dysfunction in the polycystic ovary syndrome. *J Clin Endocrinol Metab* **62**:197, 1986.
36. Bergman P, Sjogren B, Hakansson B: Hypertensive form of congenital adrenocortical hyperplasia. *Acta Endocrinol* **40**:555, 1962.
37. Kase N, Kowal J, Perolff W, et al: In vitro production of andorgens by a virilizing adrenal adenoma and associated polycystic ovaries. *Acta Endocrinol* **44**:15, 1963.
38. Steinberger E, Smith KD, Rodriguez-Rigau LJ: Testosterone, dehydroepiandrosterone, and dehydroepiandrosterone sulfate in hyperandrogenic women. *J Clin Endocrinol Metab* **59**:471, 1984.
39. Toscano V, Adamo MV, Caiola S, et al: Is hirsutism an evolving syndrome? *J Endocrinol* **97**:379, 1983.
40. Clark AF: Androgen biosynthesis, production, and transport in the normal and hyperandrogenic woman. *Semin Reprod Endocrinol* **4**:77, 1986.
41. Bardin CW, Lipsett MB: Testosterone and androstenedione blood production rates in normal women and women with idiopathic hirsutism or polycystic ovaries. *J Clin Invest* **46**:811, 1967.
42. Kirschner MA, Bardin CW: Androgen production and metabolism in normal and virilized women. *Metabolism* **21**:667, 1972.
43. Andre CM, James VHT: Plasma testosterone, androstenedione and urinary 17-oxosteroids in hirsutism. *J Endocrinol* **61**:31, 1974.
44. Rosenfield RL: Plasma testosterone binding globulin and indexes of the concentration of unbound androgens in normal and hirsute subjects. *J Clin Endocrinol Metab* **32**:717, 1971.
45. Vermeulen A, Stoica T, Verdonck F: The apparent free testosterone concentration, an index of androgenicity. *J Clin Endocrinol Metab* **33**:759, 1971.
46. Abraham GE, Maroulis GB, Buster JE, et al: Effect of dexamethasone on serum cortisol and androgen levels in hirsute patients. *Obstet Gynecol* **47**:395, 1975.
47. Maroulis GB, Manlimos FS, Abraham GE: Comparison between urinary 17-ketosteroids and serum androgens in hirsute patients. *Obstet Gynecol* **49**:454, 1977.
48. Rosenfield RL: Plasma free androgen patterns in hirsute women and their diagnostic implications. *Am J Med* **66**:417, 1979.
49. Rosenfield RL, Ehrlich EN, Cleary RE: Adrenal and ovarian contributions to the elevated free plasma androgen levels in hirsute women. *J Clin Endocrinol Metab* **34**:92, 1972.
50. Moll GW Jr, Rosenfield RL: Testosterone binding and free plasma androgen concentrations under physiological conditions: Characterization by flow dialysis technique. *J Clin Endocrinol Metab* **49**:730, 1979.

51. Kim MH, Rosenfield RL, Hosseinian AH, et al: Ovarian hyperandrogenism with normal and abnormal histologic findings of the ovaries. *Am J Obstet Gynecol* **134:**445, 1979.

52. Abrahm GE: Ovarian and adrenal contribution to peripheral androgens during the menstrual cycle. *J Clin Endocrinol Metab* **39:**340, 1974.

53. Kim MH, Hosseinian AH, Dupon C: Plasma levels of estrogens, androgens and progesterone during normal and dexamethasone-treated cycles. *J Clin Endocrinol Metab* **39:**706, 1974.

54. Yuen BH, Mincey EK: Role of androgens in menstrual disorders of nonhirsute and hirsute women, and the effect of glucocorticoid therapy on androgen levels in hirsute hyper-androgenic women. *Am J Obstet Gynecol* **145:**152, 1983.

55. Kirschner MA, Jacob JB: Combined ovarian and adrenal vein catheterization to determine the site(s) of androgen over-production in hirsute women. *J Clin Endocrinol Metab* **33:**199, 1971.

56. Kirschner MA, Zucker IR, Jespersen D. Idiopathic hirsutism-an ovarian abnormality. *N Engl J Med* **294:**637, 1976.

57. Chang RJ, Laufer LR, Meldrum DR, et al: Steroid secretion in polycystic ovarian disease after ovarian suppression by a long-acting gonadotropin-releasing hormone agonist. *J Clin Endocrinol Metab* **56:**897, 1983.

58. Lasnitzki I, Franklin HR: The influence of serum on uptake, conversion and action of testosterone in rat prostate glands in organ culture. *J Endocrinol* **54:**333, 1972.

59. Mathur RS, Moody LO, Landgrebe S, et al: Plasma androgens and sex hormone-binding globulin in the evaluation of hirsute females. *Fertil Steril* **35:**29, 1981.

60. Vermeulen A, Ando S: Metabolic clearance rate and interconversion of androgens and the influence of the free androgen fraction. *J Clin Endocrinol Metab* **48:**320, 1979.

61. Takayasu S: Metabolism and action of androgen in the skin. *Int J Dermatol* **18:**681, 1979.

62. Peterson RW, Imperato-McGinley J, Gautier T, et al: Male pseudohermaphroditism due to steroid 5α-reductase deficiency. *Am J Med* **62:**170, 1977.

63. Schweikert HU, Wilson JD: Regulation of human hair growth by steroid hormones. I. Testosterone metabolism in isolated hairs. *J Clin Endocrinol Metab* **38:**811, 1974.

64. Schweikert HU, Wilson JD: Regulation of human hair growth by steroid hormones. II. Androstenedione metabolism in isolated hairs. *J Clin Endocrinol Metab* **39:**1012, 1974.

65. Serafini P, Ablan F, Lobo RA: 5α-Reductase activity in the genital skin of hirsute women. *J Clin Endocrinol Metab* **60:**349, 1985.

66. Kuttenn F, Mowszowicz I, Schaison G, et al: Androgen production and skin metabolism in hirsutism. *J Endocrinol* **75:**83, 1977.

67. Mowszowicz I, Melanitou E, Doukani A, et al: Androgen binding capacity and 5α-reductase activity in pubic skin fibroblasts from hirsute patients. *J Clin Endocrinol Metab* **56:**1209, 1983.

68. Toscano V, Sciarra F, Adamo MV, et al: Is 3α-androstanediol a marker of peripheral hirsutism? *Acta Endocrinol* **99:**314, 1982.

69. Morimoto I, Edmiston A, Hawks D, et al: Studies on the origin of androstanediol and androstanediol glucuronide in young and elderly men. *J Clin Endocrinol Metab* **52:**772, 1981.

70. Horton R, Hawks D, Lobo R: 3α, 17β-androstanediol glucuronide in plasma: A marker of androgen action in idiopathic hirsutism. *J Clin Invest* **69:**1203, 1982.

71. Gompel A, Wright F, Kuttenn F, et al: Contribution of plasma androstenedione to 5α-androstanediol glucuronide in women with idiopathic hirsutism. *J Clin Endocrinol Metab* **62:**441, 1986.

72. Rittmaster RS, Loriaux DL: Hirsutism. *Ann Intern Med* **106:**95, 1987.

73. Barbieri RL, Makris A, Ryan KJ: Insulin stimulates androgen accumulation in incubations of human ovarian stroma and theca. *Obstet Gynecol* **64**(suppl 3):73S, 1984.

74. Rosenfield RL: Studies of the relation of plasma androgen levels to androgen action in women. *J Steroid Biochem* **6:**695, 1975.

75. Glass AR, Swerdloff RS, Bray GA, et al: Low serum testosterone and sex-hormone-binding-globulin in massively obsese men. *J Clin Endocrinol Metab* **45:**1211, 1977.

76. Longcope C, Pratt JH, Schneider SH, et al: The in vivo metabolism of androgens by muscle and adipose tissue of normal men. *Steroids* **28:**521, 1976.

77. Cleland WH, Mendelson CR, Simpson ER: Aromatase activity of membrane fractions of human adipose tissue stromal cells and adipocytes. *Endocrinology* **113:**2155, 1983.

78. Cleland WH, Mendelson CR, Simpson ER: Effects of aging and obesity on aromatase activity of human adipose cells. *J Clin Endocrinol Metab* **60:**174, 1985.

79. Carter JN, Tyson JE, Warne GL, et al: Adrenocortical function in hyperprolactinemic women. *J Clin Endocrinol Metab* **45:**973, 1977.

80. Vermeulen A, Ando S: Prolactin and adrenal androgen secretion. *Clin Endocrinol (Oxf)* **8:**295, 1978.

81. Evans WS, Schiebinger RJ, Kaiser DL, et al: Serum adrenal androgens in hyperprolactinaemic women prior to, during, and after chronic treatment with bromocritpine. *Acta Endocrinol (Copenh)* **101:**235, 1982.

82. Lobo RA, Kletzky OA, Kaptein EM, et al: Prolactin modulation of dehydroepiandrosterone sulfate secretion. *Am J Obstet Gynecol* **138:**632, 1980.

83. Glickman SP, Rosenfield RL, Bergenstal RM, et al: Multiple androgenic abnormalities including elevated free testosterone in hyperprolactinemic women. *J Clin Endocrinol Metab* **55:**251, 1982.

84. Vermeulen A, Ando S, Verdonck L: Prolactinomas, testosterone-binding globulin and androgen metabolism. *J Clin Endocrinol Metab* **54:**409, 1982.

85. Glickman SP, Rosenfield RL: Androgen metabolism by isolated hairs from women with idiopathic hirsuitism is usually normal. *J Invest Dermatol* **82:**62, 1984.

86. Judd HL, McPherson RA, Rekoff JS, et al: Correlation of the effects of dexamethasone administration on urinary 17-ketosteroid and serum androgen levels in patients with hirsutism. *Am J Obstet Gynecol* **128:**408, 1977.

87. Meldrum DR, Abraham GE: Peripheral and ovarian venous concentrations of various steroid hormones in virilizing ovarian tumors. *Obstet Gynecol* **53:**36, 1979.

88. Kirschner MA: Hirsutism and virilism in women. *Spec Top Endocrinol Metab* **6:**55, 1984.

89. Lynfield YL, MacWilliams P: Shaving and hair growth. *J Invest Dermatol* **55:**170, 1970.

90. Dewis P, Petsos P, Newman M, et al: The treatment of hirsutism with a combination of desogestrel and etinyl oestradiol. *Clin Endocrinol (Oxf)* **22:**29, 1985.

91. Aksel S: Therapeutic approaches to androgen excess. *Semin Reprod Endocrinol* **4:**211, 1986.

92. Funder JW, Mercer JE: Cimetidine, a histamine H2-receptor antagonist, occupies androgen receptors. *J Clin Endocrinol Metab* **48:**189, 1979.

93. Anand S, Van Thiel DH: Prenatal and neonatal exposure to cimetidine results in gonadal and sexual dysfunction in adult males. *Science* **218:**493, 1982.

94. Vigersky RA, Mehlamn I, Glass AR, et al: Treatment of hirsute women with cimetidine. A preliminary report. *N Engl J Med* **18:**1042, 1980.

95. Loriaux DL, Menard R, Taylor A, et al: Spironolactone and endocrine dysfunction. *Ann Intern Med* **85:**630, 1976.

96. Serafini PC, Catalino J, Lobo RA: The effect of spironolactone on genital skin 5α-reductase activity. *J Steroid Biochem* **23:**191, 1985.

97. Messina M, Manieri C, Biffignandi P, et al: Antiandrogenic properties of spironolactone: clinical trial in the management of female hirsutism. *J Endocrinol Invest* **6:**23, 1983.

98. Shapiro G, Evron S: A novel use of spironolactone: treatment of hirsutism. *J Clin Endocrinol Metab* **51:**429, 1980.

99. Cumming DC, Yang JC, Rebar RW, et al: Treatment of hirsutism with spironolactone. *JAMA* **247:**1295, 1982.

100. Gamborg Nielsen P: Treatment of idiopathic hirsutism with spironolactone. *Dermatologica* **165:**194, 1982.

Adrenocortical Carcinoma

Introduction

Adrenocortical carcinoma is a rare and devastating malignancy. The devastating nature of this tumor is due, in part, to the fact that in most instances the disease is quite advanced by the time it is diagnosed. Several general comments are worthy of mention regarding this endocrine malignancy. First, adrenocortical carcinoma is paralleled only by renal cell carcinoma in its tendency to invade vascular structures. Indeed, it is this nefarious propensity that is responsible for the extraordinary incidence of metastases at the time of diagnosis. Second, the adrenals are inaccessible to physical examination by virture of their deep, well-cushioned retroperitoneal position, protected by the rib cage posteriorly and laterally. Therefore, inevitably, the tumor is quite often advanced by the time it becomes a "palpable mass." Third, adrenocortical carcinoma is underdiagnosed because it is considered by most physicians only when hypersecretory syndromes are apparent. The fact that the proportion of nonfunctional adrenocortical carcinomas outnumbers those that are hyperfunctional is not widely appreciated. This is compounded by the fact that the most common expression of hypersecretion by adrenal cancer, i.e., androgen excess, goes unnoticed in males for obvious reasons, contributing to missed diagnosis. Fourth, the therapeutic nihilism that pervaded the treatment of this disease is not entirely justified since early and aggressive intervention does improve the outlook for these patients. Finally, although there are no parameters for early diagnosis, the increased use of computed tomography (CT) of the abdomen for unrelated causes has been conducive in the detection of asymptomatic adrenal masses. The emergence of magnetic resonance imaging may help in detecting carcinomatous lesions at an earlier stage than usual. The histological, clinical, and laboratory features of this malignancy, as well its treatment, form the focus of this chapter.

Incidence and Epidemiology

Adrenocortical carcinoma is a rare tumor. The most widely quoted figures are those provided by the Third National Cancer survey,[1] which estimates the incidence as approximately one case per 1,700,000 population; this tumor accounts for 0.02% of cancers. It occurs in both sexes, but with a slightly increased preponderance in females.[2–4] The median age of occurrence is in the mid- to late thirties, but it can occur as early as 6 months of age and as late as the seventh decade of life. Nonfunctioning carcinomas of the adrenal cortex occur most often in the age range 40–74 years and show a predominance in males.[5–7] Functioning adrenocortical carcinomas occur more frequently in females; in several series 75–78% of functioning adrenocortical carcinomas occurred in females.[3,4,8] The most common hypersecretory expressions of adrenocortical carcinoma in adult females are virilizing syndromes and hypercortisolism. The distribution of adrenocortical carcinoma appears to be equal in all races. The median time interval from the onset of symtpoms to the diagnosis of adrenal carcinoma can be as long as 6 months. Unfortunately, as many as 50–60% of patients found to harbor adrenal cancer demonstrate evidence of local or distal spread at the time of diagnosis. A higher prevalence of left-sided tumors is questionable and has been disputed.

Pathology

The pathology of adrenal carcinoma can be viewed in terms of its gross and microscopic appearances along with the unusual origin of such tumors from

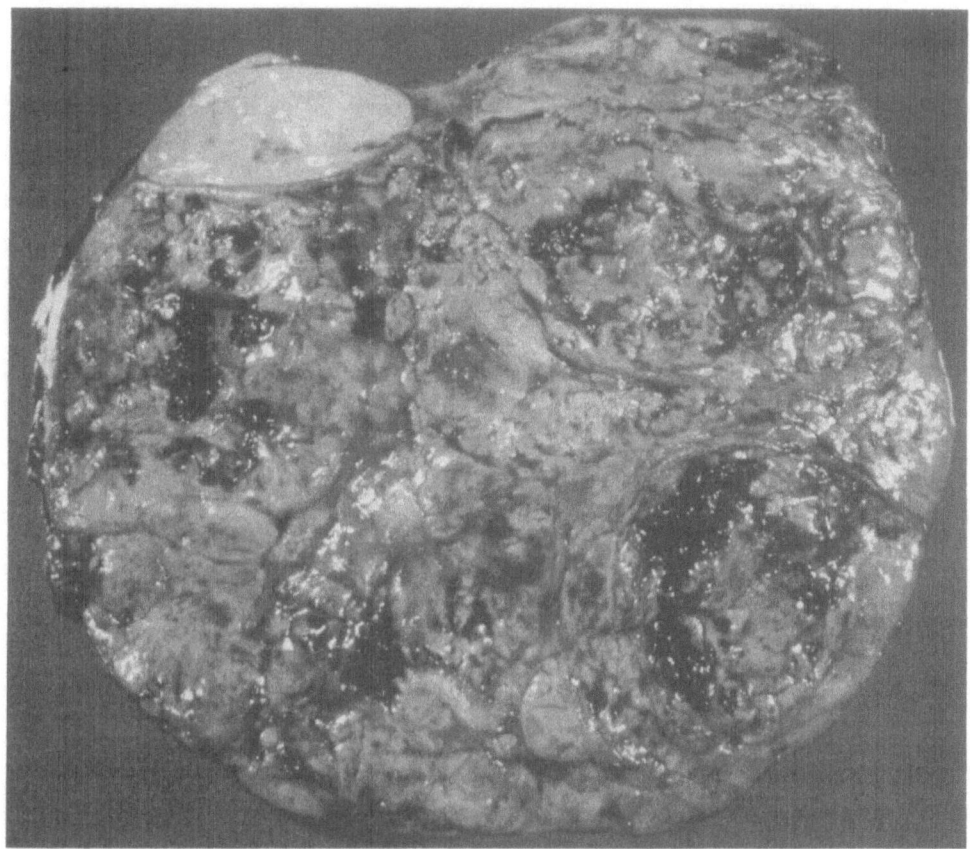

FIGURE 17. Gross appearance of adrenocortical carcinoma.

"adrenal rest cells." The relationship between the development of adrenocortical cancer and underlying chronic endogenous ACTH stimulation also requires brief comment.

Gross Apperance (Fig. 17)

Adrenal carcinoma is a highly vascular tumor. The cut surface is pinkish red, and the tumor demonstrates a characteristic softness, in contrast to the firm consistency of benign adenomas. Adrenal carcinoma demonstrates a proclivity to invade the adjacent vascular and retroperitoneal soft tissues.

Microscopic Appearance (Fig. 18)

Adrenal carcinomas can be divided into differentiated and undifferentiated (or pleomorphic) types. Differentiated adrenocortical carcinoma is composed of cells that resemble the normal adrenocortical cells. The tumor cells are polygonal, monotonous appearing, and are usually arranged in sheets, or in trabeculae. Less commonly, the neoplastic cells may demonstrate patterns that resemble nests or ribbons. The cells contain numerous mitotic figures and often

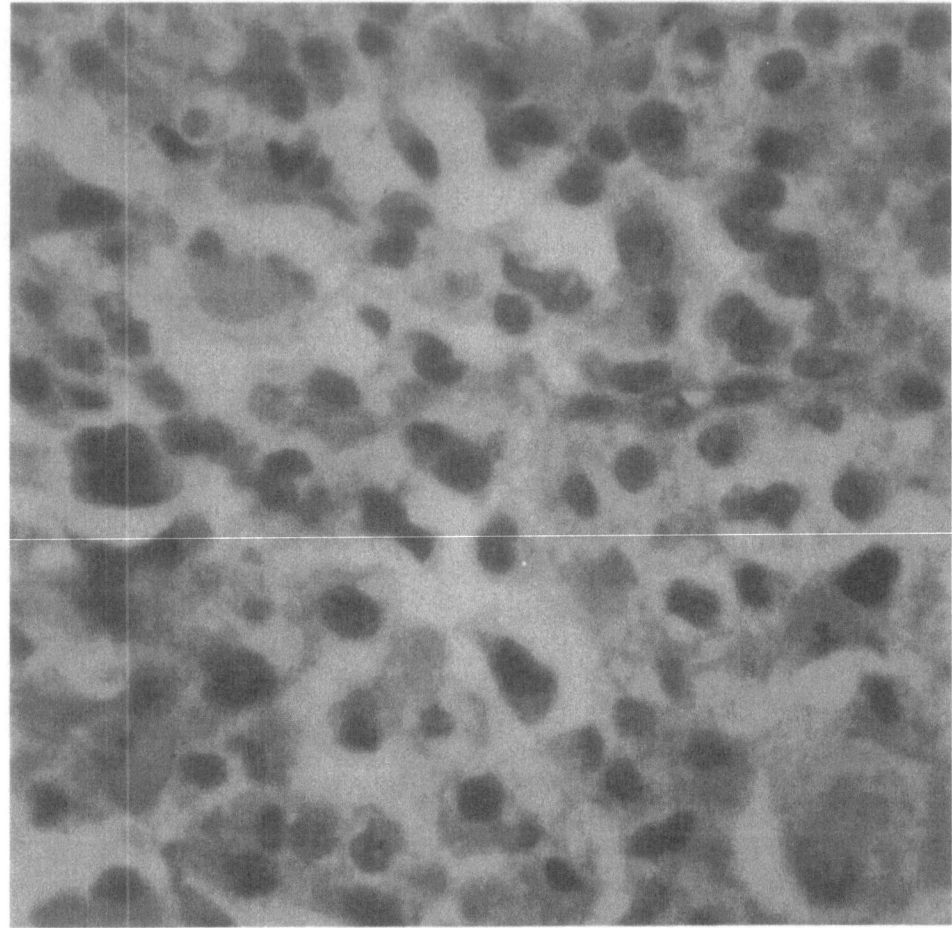

FIGURE 18. Microscopic appearance of adrenocortical carcinoma.

show signs of necrosis. The undifferentiated variety of adrenocortical carcinoma contains pleomorphic, bizarre nucleated cells, with a characteristic lack of cohesion. Some tumors contain both type of cells. Silva et al.[9] studied the ultrastructural features of 22 patients with adrenocortical carcinoma for the presence of ultrastructural characteristics of steroid-secreting cells. Smooth endoplasmic reticulum and tubular mitochondrial cristae were present in only some, underscoring the difficulties in delineating the tumor, particularly from clear cell hypernephroma.

Adrenal Carcinoma from "Adrenal Rest Cells"

Adrenocortical carcinoma can occasionally originate from "adrenal rest cells." Adrenal rests have been described in many locations. Accessory, or "aberrant" adrenocortical tissue has been most commonly found in the broad ligament,[10] in the testes, particularly in children,[11] and in the upper pole of the kidney.[12] Rarely, adrenocortical tissue has been described in distant locations such as intracranial sites,[13] the liver,[14] and the lung.[15] While description of adrenal rest cells in these locations has been appreciated by pathologists, the development of adrenocortical carcinoma from these unusual sites is an extremely rare phenomenon. Wallace et al.[16] reported a case of a 23-year-old woman who presented with hepatomegaly with virilization and mild hypercortisolism, both of which failed to suppress with dexamethasone. Inferior vena caval sampling demonstrated markedly increased steroid hormone levels from the right hepatic vein. At autopsy the right lobe of the liver was replaced by a massive tumor consisting of polygonal and round cells arranged in cords, resembling adrenocortical cancer. The tumor contained testosterone and cortisol, while the adrenals and ovaries were atrophic. Conteras et al.[17] also described a 21-year-old woman with Cushing's syndrome caused by a functioning adrenal rest tumor, with a macroscopic and microscopic appearance highly reminiscent of adrenocortical carcinoma. Wilkins and Ravitch[18] described a 3-year-old boy with virilism and Cushing's syndrome caused by a functioning adrenocortical tumor from the liver, the hormonal features of which ameliorated following partial resection. While functional adrenocortical rest tumors are distinctly rare, these tumors should be kept in mind especially when they arise from the superior pole of the kidney,[19,20] a location close to that of the normal adrenal gland. As with adrenocortical cancer, the malignant tumors that originate from adrenal rest cells can be nonfunctional. Hamperl,[21] in a review of "adrenal rest tumors" of the liver, reported that only 2 of 26 were found to be associated with endocrine abnormalities.

Chronic ACTH Excess and Adrenal Cancer

In terms of predisposing factors, none have been identified as risk factors. The development of adrenocortical cancer in adrenal glands that have been chronically stimulated by ACTH continues to be intriguing. It has been noted that in certain strains of mice adrenal hyperplasia may eventually progress to adrenocortical carcinoma.[22] Several case reports have periodically appeared regarding the development of adrenocortical carcinoma in patients with untreated or undertreated adrenogenital syndrome. The question was originally raised by Hamwi et al.,[23] who described a woman with untreated 21-hydroxylase deficiency who eventually developed adrenal carcinoma. Other cases of a similar sequence have been described.[24,25] The development of adrenal carcinoma in patients with long-standing pituitary-dependent Cushing's disease and macronodular hyperplasia has also been noted in the literature.[26,27] The significance of these reports remains unclear in terms of conferring a causal role to chronic ACTH excess in predisposing patients to adrenocortical cancer.

Clinical Features

The clinical features of adrenocortical carcinoma can be divided into three broad categories: those due to the adrenal mass, those due to local or distant

TABLE 37.
Clinical Features of Adrenocortical Cancer

I.	Features due to the mass lesion
	Palpable abdominal mass
	Abdominal pain (flank pain)
	GI symptoms, (nausea, vomiting, etc.)
II.	Features due to local invasion
	Venacaval obstruction, peripheral edema
	Renal vein thrombosis, hematuria, proteinuria, nephrotic syndrome
	Hepatic vein obstruction, Budd–Chiari syndrome
III.	Features due to metastastic spread
	To liver, lung, bone
IV.	Features due to complications
	Adrenal hemorrhage
	Pulmonary embolization
V.	Features due to hormonal hypersecretion
VI.	Generalized
	Fatigue, fever, weight loss

spread, and those secondary to endocrine hypersecretion (Table 37). Since only one-third to one-fourth of adrenocortical carcinomas are clinically hyperfunctional, it is important to emphasize the local and systemic manifestations of this cancer. For a relatively rare tumor, it is interesting to note that several series, consisting of impressive numbers of patients, have appeared in the literature. The earliest, and probably the largest, is that of Hutter and Kayhoe,[28] who described the clinical features of 138 patients with adrenocortical cancer, in 1966. The more recent series by Hajjar et al.[29] with 32 patients, Nader et al.[30] with 77 cases, Didolkar et al.[7] with 42 patients, and Huvos et al.[8] with 34 patients have all enhanced and enriched our understanding of the clinicopathological aspects of the dreaded tumor. The availability of improved hormonal assays, the emergence of CT, and the trials with adrenolytic drugs, particularly o p' DDD, have all contributed in studying the natural history of adrenocortical cancer.

Due to the Adrenal Mass

The tumor can be palpable as an abdominal mass in 33–36% of patients with adrenal carcinoma.[7,30] The observation that such tumors grow large enough to become palpable at the time of diagnosis implies that the malignancy is slow growing. Depending on the size and location of the mass, a variety of gastrointestinal symptoms may develop, resulting in early satiety, a feeling of distention, and even vomiting. Weight loss and fatigue are commonly experienced by patients with adrenocortical cancer. Abdominal pain and a low-grade fever can occur in patients with this tumor. It is not clear whether the pain is due to pressure on the nerves adjacent to the tumor or to hermorrhage and necrosis within the tumor.

The locally invasive properties of adrenal cancer can result in a variety of bizarre presentations. The propensity to invade vascular structures can result in thrombosis and even total occlusion of the neighboring veins. Thus, peripheral edema from venacaval obstruction, hematuria and proteinuria from renal vein thrombosis, and hepatosplenomegaly with portal hypertension from hepatic (or inferior venacaval) obstruction can develop in patients with adrenal carcinoma. These manifestations can result in atypical presentations where adrenal carcinoma is often the last to be considered. In addition to causing venous occlusion locally, tumor embolism can result in the development of distant occlusive syndromes such as pulmonary infarction. A good clinical aphorism to keep in mind is that adrenal carcinoma can mimic all the manifestations caused by renal cell carcinoma. The similarity to renal cell carcinoma is brought even closer, because abnormalities in the intravenous pyelogram are seen in approximately two-thirds of patients.[7] This, coupled with the fact that even experienced radiologists can occasionally mistake an adrenal mass by CT for a renal mass at the upper pole of the kidney, explains the frequency with which nonfunctioning adrenal carcinoma can be mistaken for an invasive renal cell carcinoma.

Due to Local or Distant Spread

Both functioning as well as nonfunctioning adrenal carcinomas demonstrate an enormous propensity for local and distant spread. In the 42 patients with histologically proven adrenocortical carcinoma reported by Didolkar et al.[7] one-third of patients came to medical attention as a consequence of signs and symptoms directly related to the metastases. The incidence of the presence of metastatic disease at the time of diagnosis was 52% in this series. The infrequency of strictly localized disease is even more impressive in the series reported by Nader et al.[30]; of the 77 patients with adrenocortical cancer only 3.9% had localized disease; 72% eventually developed metastatic spread, nearly half of whom had evidence of distant metastases even at the time of diagnosis.

Locally, the adrenal cancer spreads to the paraaortic nodes frequently enough to make curative resection an impossibility. In addition, tumors on the left side involve the diaphragm and the pancreas, while tumors on the right side can involve the liver directly. Tumors from either side can involve the kidney and the inferior vana cava by contiguous spread. The overt or subtle involvement of this great vein is the reason for the alarmingly high incidence of finding dissemination of cancer at, or soon after, the time of diagnosis. In addition to invasion of the retroperitoneal space, patients with unilateral adrenocortical cancer can demonstrate synchronous or metachronous presence of the cancer in the opposite adrenal.

In terms of distant metastatic spread, the lungs are the most frequently involved metastatic site (60%), followed by the liver (50%), lymph nodes (48%), bone (24%), and pleura and heart (10%). The interpretation of the presence of metastatic disease should take into account the fact that patients with adrenocortical cancer demonstrate a tendency for developing a histologically different second primary. As has been pointed out by Didolkar et al.,[7] 10 of 42 patients in their series had a second cancer; carcinoma of the breast and lymphomas led the list, followed by carcinoma of thyroid uterus, colon, and skin. Similar observations have been noted by others.[30] While it is beguiling to speculate a role for hypercortisolism (from adrenal cancer) in perpetuating an underlying cancer diathesis, this has not been borne out.

The clinical connotation to the physician is that adrenal carcinoma should

TABLE 38.
Hypersecretory Syndromes
Associated with Adrenal Cancer

Cushing's syndrome
Hyperandrogenism
    Virilization
    Precocious puberty
    Hirsutism
    Anabolic phenomena
Feminizing syndromes
Mineralocorticoid excess
Rare
    Precursor steroid hypersecretion
    Erythropoietin
    SIADH
    Hypoglycemia
    ACTH hypersecretion

be considered in all cases of metastatic disease with "occult primary." Regardless of the presence or absence of endocrine hypersecretory syndromes, CT of the adrenals must be part of the diagnostic tests in the patient with metastatic disease caused by an unknown primary.

Due to Hormonal Hypersecretion (Table 38)

Endocrine syndromes of hypersecretion often herald the diagnosis of adrenocortical cancer. In contrast to earlier belief, it is now recognized that nonfunctional adrenocortical cancer occurs more frequently than the functional ones; thus in one series of 42 patients[7] the ratio of clinically functional to clinically nonfunctioning adrenocortical cancer was 1:4. In another larger series, consisting of 77 cases,[30] two-thirds of patients had clinically nonfunctioning adrenal cancer. Thus, the practice of considering the diagnosis of adrenal cancer only in the presence of endocrine syndromes is to be deplored. Several general remarks are appropriate regarding the functional variety of adrenocortical cancer.

1. Functioning adrenal carcinomas tend to occur more frequently in females; nearly 75% of hyperfunctional adrenocortical cancers occur in females. The reason for this prevalence is not clear. While it is true that androgen hypersecretion is more likely to be detected in females than in males, careful hormonal assays in both sexes have reaffirmed a much higher prevalence of hypersecretory syndromes in females with adrenocortical cancer.

2. In general, the presence of hormonal hypersecretion cannot be correlated with the age of the patient, size of the tumor, or the presence or absence of metastases.

3. While several hormonal syndromes may evolve, the four most typical endocrinopathies seen in association with adrenocortical carcinoma are virilization of females, hypercortisolism in both sexes, feminization of males, and precocious puberty. Less commonly, adrenocortical carcinomas may result in endocrine syndromes characterized by hypersecretion of mineralocorticoids, insulin, erthropoietin, vasopressin, and ACTH.

4. Adrenal carcinomas are unique adrenal tumors that result in "mixed" hormonal hypersecretory syndromes. Indeed, this tendency was recognized in the fifties by Heinbecker et al.,[2] who noted that adrenal tumors are less likely to cause "pure" syndromes of hypersecretion.

5. The malignant adrenocortical tumor cells are inefficient hormone producers. This is due to the fact that these cells do not possess the entire enzymatic machinery required for effective synthesis of intact steroid hormones. As a consequence of these enzyme deletions, the tumor cells are capable of synthesizing a higher proportion of hormone precursors. Thus, depending on the particular steroid hormone, there is a preponderance in the synthesis and release of biologically weak steroids. Biochemically this translates into increased secretion of 11-deoxycortisol (compound S), dehydroepiandrosterone, estrone, estradiol, and 11-deoxycorticosterone. The clinical implications of such a phenomenon are fourfold. First, the tumor has to make large quantities of these weak hormones in order to cause clinical expression. In fact, it has been well recognized that adrenocorticol carcinoma can be biochemically hyperfunctional in the absence of overt clinical expression of hormonal hypersecretion. Second, the tumor must be large to produce enough weak hormones that cause clinical syndromes. This is perhaps the reason that such tumors have attained considerable size by the time the endocrine syndromes have attracted attention. Third, the presence of excessive precursor steroids in the circulation can mimic the hormonal profile of the virilizing adrenogenital syndromes. Of course, the nonsuppressibility to dexamethasone administration and the abnormality in CT would readily permit identification of the adrenal carcinoma. Finally, the clinical expression that ensues would depend on the type, potency, and magnitude of the precursors synthesized as well as the sex and age of the patient.

With these generalizations in mind, each of the endocrine syndromes will be described in brief.

Cushing's Syndrome (Fig. 19)

Hypercortisolism is one of the most frequent endocrine syndromes seen in association with hyperfunctional adrenocortical carcinoma; it occurs predominantly in female patients. The hypersecretion of glucocorticoids is autonomous and is therefore associated with the triad of suppressed plasma ACTH levels, nonsuppressibility to high-dose dexamethasone, and nonresponsiveness to metyrapone administration. In this regard, the hypercortisolism resembles that of cortisol-secreting adrenal adenomas. There are, however, several clinical, hormonal, and radiological differences unique to the hypercortisolism that results from adrenal carcinoma. The clinical and laboratory differences between the two entities have been compared by Bertagna and Orth.[31] Regardless of the etiology, certain features occurred in both. Thus, moon facies, the presence of supraclavicular and dorsocervical fat pads, plethora, and ecchymosis occurred with equal frequency in both. Patients with cortisol hypersecretion caused by adenomas tended to have a higher frequency of thinning of skin, striae, and left ventricular hypertrophy (despite an equal incidence of hypertension and hypokalemic alkalosis). Importantly, patients with cortisol-secreting adrenal carcinoma tended to demonstrate more signs of andorgen excess—ranging from

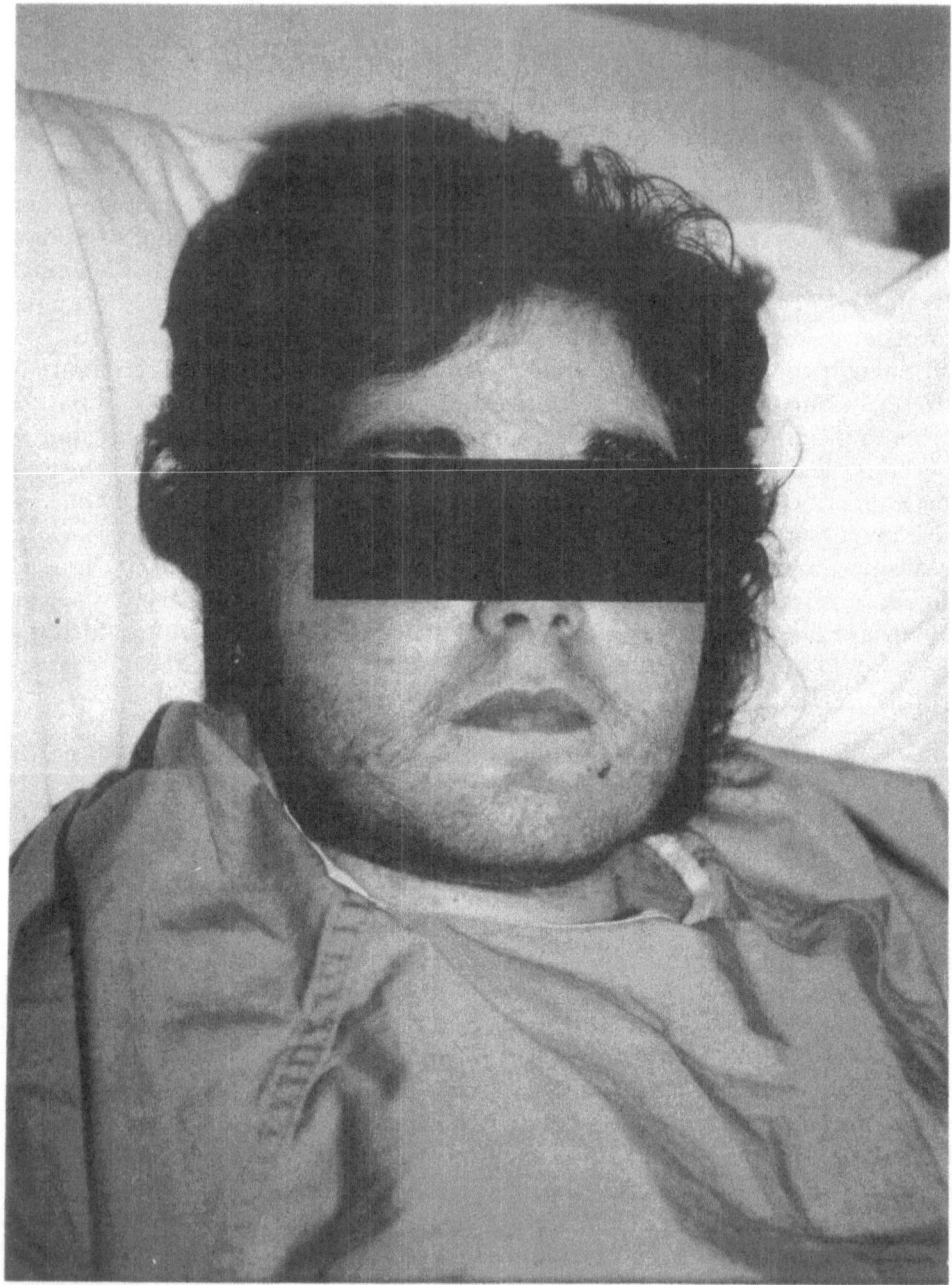

FIGURE 19. Marked hirsutism and clinical evidence of hypercortisolism in a 21-year-old woman.

thinning of hair, temporal hair loss, acne, and hirsutism to overt virilization. Indeed, the combination of hypercortisolism and hyperandrogenism is pathognomonic of adrenal carcinoma. Adenomas, in contrast, are characterized by defective secretion of androgens.[32] The increased secretion of androgens, which are anabolic, by the adrenal carcinoma may offset the catabolic effect of glucocorticoids on muscle. This may be the reason for the decreased incidence of objective myopathy in patients with Cushing's syndrome caused by adrenal carcinoma, in contrast to other forms of hypercortisolism.

Hormonally, the single most important difference between Cushing's syndrome caused by adenomas and that caused by carcinomas of the adrenal gland is seen in the urinary 17-ketosteroid levels. These levels are invariably greater than 15 mg/24 hr in patients with adrenocortical carcinoma, often averaging between 50 and 60 mg/24 hr. Elevated androgen levels in carcinoma may be seen even in the absence of overt symptoms and signs of virilization, because, as indicated earlier, precursor steroids are synthesized and excreted in excessive quantities. The importance of measuring 17-ketosteroids in patients with Cushing's syndrome to distinguish between hyperplasia, adenoma, and carcinoma was recognized as early as 1951, by Forbes and Albright,[33] and has been repeatedly confirmed by others since.[2-4,28,32] The radiological appearance of adrenal carcinoma and the features that distinguish it from adenoma are discussed in the section dealing with laboratory diagnosis.

Hyperandrogenic Syndromes

Androgen secretion is one of the most important and most frequently encountered phenomena seen in association with adrenal carcinoma. The effects of such androgen excess are expressed in many ways. In adult females, androgen excess can result in a spectrum of changes ranging from the development of hirsutism to complete virilization with amenorrhea, clitoromegaly, marked facial hair growth, and temporal recession and thinning of the scalp hair. In prepubertal boys, the cardinal manifestation of adrenal hyperandrogenism is the development of isosexual precocious puberty. In prepubertal girls, adrenal androgen excess can result in initiating early pubertal changes, but culminating in hirsutism and masculinization. In males, the hypersecretion of adrenal androgens is barely evident since these patients are already maximally virilized. The only symptom in these patients is weight gain, due to the anabolic effects of these steroids. The androgenic phenomena are mediated mainly through the action of the adrenal androgen dehydroepiandrosterone sulfate. Testosterone secretion by adrenocortical carcinoma is uncommon, owing to the presence of several enzyme deletions in the tumor. Therefore, androgen production by this carcinoma is limited to synthesis of the "weaker" androgens. It is necessary for the tumor to secrete large amounts of these androgens to effectively express their androgenic action. It is perhaps for this reason that these tumors have attained a considerable size by the time patients have manifested overt evidences of virilization.

As indicated earlier, "mixed" hormonal hypersecretion is frequently seen with adrenocortical carcinoma. Thus, in addition to "pure" androgen hypersecretion, these tumors demonstrate combined hypersecretion of cortisol and dehydroepiandrosterone sulfate. The clinical expressions, in such a setting, can be quite variable. In some, overt clinical evidence of Cushing's syndrome coexists with overt clinical evidence of hyperandrogenism. In others, clinical evidence of hypercortisolism may be lacking despite elevated cortisol levels, in addition to androgen excess. The reasons for the failure to develop clinical Cushing's syndrome are unclear, but possibly the anabolic effects of androgen excess may have masked the catabolic effects of glucocorticoids. A third variation consists of patients without overt clinical and hormonal evidence of virilization despite elevated 17-ketosteroid excretion in the urine. The reason for the lack of virilization in these patients may be a reflection of the diminished biological potency of these weak adrenal androgens. It has been repeatedly emphasized in the literature that the demonstration of 24-hr urinary 17-ketosteroid excretion in excess of 20 mg/gr of creatinine in a patient with Cushing's syndrome is strongly suspicious of adrenocortical carcinoma.[31] The final permutation of "mixed" hormonal syndromes is the occurrence of hyperandrogenism coupled with increased excretion of compound S (11-deoxycortisol). This phenomenon is a reflection of the fact that the enzyme deletions inherent to adrenal carcinoma (in this instance 11-β-hydroxylase) permit the increased production of the precursor compound S, but not the biologically active compound F. The clinical and hormonal profile of this group of patients resembles those with congenital adrenal hyperplasia caused by 11-β-hydroxylase deficiency. Regardless, the biochemical demonstration of increased dehydroepiandrone sulfate in plasma or 17-ketosteroid excretion in the urine should prompt the consideration of functioning adrenocortical carcinoma.

Feminizing Syndromes

A rare, but important endocrine expression of adrenocortical carcinoma is the hypersecretion of feminizing hormones. As early as 1948, Armstrong and Simpson[34] recognized "adrenal feminism" caused by carcinoma of the adrenal cortex. The feminizing changes are brought about by secretion of estrogens such as estradiol and estrone. The clinical expression, once again, depends on the age and sex of the patient as well as the magnitude of estrogen excess. When it occurs in prepubertal girls, the resultant endocrine syndrome is characterized by isosexual (pseudo) precocious puberty. When it occurs in adult males, the clinical expression is characterized by gynecomastia, loss of libido, and hypoandrogenism. In adult females, the effects of estrogen excess are often masked since these patients are already maximally feminized.

The syndrome of feminizing adrenocortical tumors, despite its rarity, has been the subject of several reports and reviews.[35–41] The most inclusive review is that by Gabrilove et al.,[36] who reviewed 52 cases of feminizing adrenocortical tumors, several of which were carcinomas. The pathophysiology of male hypogonadism associated with adrenal hyperestrogenism is probably a consequence of suppression of pituitary gonadotropins, particularly luteinizing hormone (LH). Veldhuis et al.[42] studied gonadotropin dynamics and Leydig cell function in a hypogonadal male with an adrenal adenoma. The data obtained were consistent with the existence of dual defects—profound suppression of circulating concentrations of biologically active LH, as well as a decrease in testicular Leydig cell reserve. Despite the severity in the hypogonadal state, the condition was reversible upon resection of the adenoma. It is unclear whether similar mechanisms underlie the hypogonadism in feminized males with adrenocortical carcinoma.

Mineralocorticoid Hypersecretion

Adrenocortical carcinoma is a very rare cause of hyperaldosteronism. Estimated to represent 1 to 2% of all cases of hyperaldosteronism, the mineralocorticoid excess that results from hypersecretion by adrenal carcinoma resembles that of adenoma. The rarity of aldosterone-secreting adrenal carcinoma has been emphasized in the literature.[43] As of 1980, only 29 cases of adrenal carcinoma with predominant aldosterone hypersecretion had been described.[44] There appear to be no differences in the age, sex, and clinical presentations between aldosterone-secreting adrenal adenoma and adrenal carcinoma. Thus, the initial presentation is characterized, like aldosterone-producing adenoma, by hypertension, hypokalemic alkalosis, and mineralocorticoid excess. The clue that adrenocortical carcinoma underlies the hyperaldosteronism is generally derived by one of four features—the appearance by CT, the concomitant hypersecretion of other hormones, the histopathological evidence of cancer in the excised tissue, and presence of dissemination to the liver and lung.

The hormonal features of aldosterone-secreting adrenal carcinoma are several and include the following:

1. The elevated plasma aldosterone levels, which show no suppression to salt loading, and the suppressed PRA activity in patients with aldosterone-secreting adrenal carcinoma resemble the hormonal milieu of aldosterone-producing adenoma. There is considerable overlap in the basal plasma aldosterone level in both groups, although generally plasma aldosterone tends to be higher and the hypokalemia more profound in carcinoma.

2. Patients with carcinoma often tend to show elevated basal levels of aldosterone precursors such as desoxycorticosterone (DOC) and 18-hydroxycorticosterone (18-OHB). While these steroids are also elevated in patients with adenoma, the degree of elevation is higher with carcinoma.

3. Adrenocortical carcinomas that secrete aldosterone fail to show ACTH dependency, a unique feature characteristic of adenoma. Thus, the plasma aldosterone levels in patients with aldosterone-secreting carcinoma often demonstrate lack of circadian periodicity and an absence of response to exogenous ACTH infusion. The ACTH independence of aldosterone secretion by adrenocortical carcinoma was impressively documented in the report of Arteaga et al.[45] These workers performed serial circadian studies in three patients with adrenocortical carcinoma and were able to demonstrate a correlation between progression of disease and loss of circadian variation in aldosterone levels. In one patient the loss in circadian periodicity of the elevated aldosterone and DOC was observed before surgery and when the disease recurred 8 months later. Cortisol and corticosterone, on the other hand, remained normal and maintained circadian rhythmicity. As the patient's disease progressed, there was impressive loss of circadian periodicity in aldosterone, DOC, cortisol, and cor-

ticosterone. In addition, the aldosterone secretion is unaffected by the administration of exogenous ACTH. This finding contrasts sharply with the aldosterone secretion by adenomas, where sensitivity to even small doses of ACTH is usually observed. Finally, patients with aldosterone-secreting carcinomas may fail to demonstrate an anomalous fall in aldosterone level with posture, in contrast to aldosterone-producing adenoma.

4. "Pure" mineralocorticoid hypersecretion by adrenocortical carcinoma has been described,[43,46-49] but it is exceedingly rare. More often there is concomitant hypersecretion of other adrenocortical steroids by the cancer. Even in the case of tumors with sole secretion of mineralocorticoids, autonomous secretion of cortisol and androgens by the tumor becomes noticeable as the cancer progresses. The hypersecretion of cortisol may be clinically silent or manifest as Cushing's syndrome. This mixed spectrum of hormone secretion is consistent with the behavior pattern of adrenal cancer in terms of indiscriminately secreting steroid hormones.[50]

5. Rarely, adrenal carcinomas can hypersecrete DOC without excessive aldosterone secretion. While such a phenomenon has been described,[51,52] it must be exceptional. Kelly et al.[53] described a 51-year-old woman with hypertension, hypokalemia, low PRA, normal aldosterone levels, and extremely high 11-DOC levels due to ineffective 11-β-steroid hydroxylation caused by a large adrenal carcinoma metastatic to the liver. In tissue culture, it was shown that the tumor predominantly produced DOC from tritiated pregnenolone without detectable aldosterone secretion, indicating absence of 11-β-steroid hydroxylation. Isolated DOC hypersecretion is a common observation in patients with ectopic ACTH syndrome.[54] However, isolated DOC hypersecretion is an exceptionally uncommon feature of adrenocortical carcinoma.

In summary, aldosterone hypersecretion by adrenocortical carcinoma is characterized by the following:

1. Its rarity.
2. A close resemblence to aldosterone-producing adenoma in its clinical presentation.
3. The underlying tumor is large, often with locally invasive propensities.
4. Subtle or overt abnormalities in glucocorticoid and/or androgen hypersecretion are usually encountered.
5. The aldosterone secretion by adrenocortical carcinoma is clearly independent of ACTH and therefore demonstrates no circadian periodicity or response to exogenous administration of ACTH.
6. Marked elevation in precursor steroids, such as 11-DOC, may be encountered in adrenal carcinoma. Isolated hypersecretion of 11-DOC (in the absence of aldosterone excess) is exceedingly rare.

Rare Endocrine Syndromes

Very rarely adrenocortical carcinoma can be associated with other "ectopic" hormonal syndromes. Polycythemia,[4] inappropriate ADH secretion,[55] and hypoglycemia[56,57] have been described in association with adrenocortical carcinoma. These, however, are extremely rare. Ectopic secretion of ACTH by

TABLE 39.
Laboratory Diagnosis of
Adrenal Cancer

I.   Hormonal
        Urine free cortisol
        Plasma cortisol
        Plasma ACTH
        DXM suppression test
        DHEA sulfate
        Testosterone
        Aldosterone dynamic studies
        17-β-estradiol
II. Radiological
        CT of adrenals
        MRI of adrenals
        Idiocholesterol imaging
III. For metastastic spread
        Liver scan
        Bone scan
        CT of chest
IV. Preoperative
        Angiography

adrenal carcinoma in an 11-year-old boy has been reported in the literature.[58] In the rare event of this combination, the contralateral adrenal would be expected to demonstrate hyperplastic changes. Cushing's syndrome resulting from ectopic secretion of ACTH by adrenocortical carcinoma must be extremely rare, since its description is limited to the single case report by Liddle et al.[58]

Laboratory Diagnosis (Table 39)

The confirmatory diagnosis of adrenocortical carcinoma ultimately rests on the histological demonstration of the malignancy. The laboratory approach to a patient suspected of harboring adrenal carcinoma consists of hormonal and radiological tests. The hormonal tests are primarily aimed at determining the presence of functioning adrenocortical carcinoma. The radiological tests are aimed at demonstrating the tumor, as well as identifying characteristics that indicate local invasion and delineating the anatomical extent. These can be attained by CT, isotopic adrenal scanning (where available), and invasive imaging procedures such as adrenal venography and arteriography. In addition, tests to determine metastatic spread to the lungs, liver, bone, and so forth are required to stage the adrenal carcinoma.

Hormonal Tests

Since the aim of hormonal tests is to document functional adreno-cortical carcinoma, the tests performed depend on the endocrine syndrome suspected. While several surgical series[7,30] have indicated that nonfunctional tumors ex-

TABLE 40.
Hypercortisolism from Adrenal Adenoma versus Carcinoma

Feature	Adenoma	Carcinoma
1. High-dose dexamethasone suppression	Nonsuppressible	Nonsuppressible
2. Metryapone administration	Nonresponsive	Nonresponsive
3. Plasma ACTH level	Suppressed	Suppressed
4. Urinary 17-KS excretion	Below 15 mg	Above 20 mg
5. Plasma DHEA-sulfate level	Low	Elevated
6. Circulating steroid precursors (Compound 5)	Normal	Often elevated
7. Computed tomography	Small, <6 cm	Larger than 6 cm, may show vascular invasion

ceed functional adrenocortical carcinoma, it is felt that, in general, when steroid production is subject to greater scrutiny, the proportion of "hormonally active" adrenocortical carcinoma increases. The hormonal strategy involves a search for hypersecretion of glucocorticoids, androgens, estrogens, mineralocorticoids, and precursor steroids.

Hypercortisolism

The first-line investigation consists of demonstrating nonsuppressible hypercortisolism. Patients with adrenocortical carcinoma resemble those with adrenal adenoma in that the resultant hypercortisolism is characterized by nonsuppressibility to high-dose dexamethasone, as well as nonresponsiveness to metyrapone, and is associated with a suppressed plasma ACTH level. The similarity, however, stops there. The three unique features that further characterize the hypercortisolism of adrenocortical cancer are consistent elevations in the 24-hr 17-ketosteroids (17-KS) (or plasma DHEA-S), increased circulating precursor steroids, and abnormalities in the imaging studies that suggest a malignant disease (Table 40).

Androgen Hypersecretion

Patients with suspected adrenocortical carcinoma quite often demonstrate isolated increases in plasma DHEA-S levels or the 24-hr excretion of urinary 17-KS. Plasma testosterone may occasionally be elevated in the presence of normal 17-KS. This phenomenon is more likely in women with virilizing adenomas.[59] In men with hyperandrogenism caused by adrenal tumors, testosterone elevation almost always occurs in association with elevation in 17-KS excretion.

Estrogen Secretion

Measurement of 17-β-estradiol, estriol, and estrone is necessary only when feminizing syndromes are encountered in males, and when isosexual precocious puberty is encountered in girls. In males who undergo feminization due to adrenocortical carcinoma, the 17-KS are also usually elevated. In an impressive review by Gabrilove et al.[36] of 52 cases of feminization of males due to adrenal tumors, the 17-KS were also found to be elevated when adrenal carcinoma was the etiology; thus, 19 of 24 patients with carcinoma demonstrated elevated 17-KS in urine.

Mineralocorticoid Excretion

While the clinical syndrome that results from mineralocorticoid secretion by carcinoma resembles that of an adenoma, several differences, when present, help to distinguish between the two. In general, patients with aldosterone-secreting carcinoma tend to have higher plasma aldosterone levels. In contrast to adenoma, these patients fail to show circadian periodicity in aldosterone levels; thus the anomalous decline in plasma aldosterone when patients stand between 8 A.M. and noon, characteristic of adenoma, may be absent in carcinoma. Furthermore, circulating concentrations of 18-OH and DOC are higher in carcinoma than in adenoma. Finally, the CT appearance of carcinomas is often characterized by larger tumors.

Precursor Steroid Secretion

Adrenal carcinoma cells are inefficient synthesizers of steroid hormone, owing to the presence of several enzyme deletions within the tumor cells. As a consequence, adrenocortical carcinoma displays a tendency for synthesizing and releasing several precursor steroids into the circulation. The most common enzyme deletion involves 11-β-hydroxylase. This results in accumulation of 11-deoxycortisol (or compound S), which, following its conversion into tetrahydro-11-deoxycortisol, is measured in the urine as 17-hydroxycorticosteroids. Indeed such an occurrence is the reason that some patients with adrenocortical carcinoma have elevated urinary 17-hydroxycorticosteroids in the absence of Cushing's syndrome.[4,60] Of course, measurement of urinary free cortisol or fractionation of 17-hydroxycorticosteroids would delineate the spurious nature of the elevated 17-hydroxycorticosteroids in the urine. In addition to 11-β-hydroxylase deletions, the malfunction in the enzymatic machinery may involve other enzymes, such as 3-β-hydroxysteroid dehydrogenase-isomerase, 21-hydroxylase, and 17,20-desmolase. The deletion of 3-β-hydroxysteroid dehydrogenase could result in elevation of pregnanetriol, the urinary metabolite of pregnenolone.[61,62] In such a setting, the urinary 17-KS would be normal, while the urinary 17-ketogenic steroids (which measure steroids with a 21-deoxy, 20-keto, 17-hydroxy side chain) demonstrate elevations. It is unclear whether adrenocortical carcinoma with increased pregnenolone metabolites should be called functional or nonfunctional.

Radiological Procedures

The initial radiological imaging procedure in the evaluation of patients suspected of harboring adrenocortical carcinoma is CT examination of the adrenals. The use of idocholesterol imaging provides additional information regarding the "functional nature" of the tumor seen by CT. Invasive procedures such as inferior venacavagraphy or arteriography may be necessary to define the extent of tumor when surgery is contemplated.

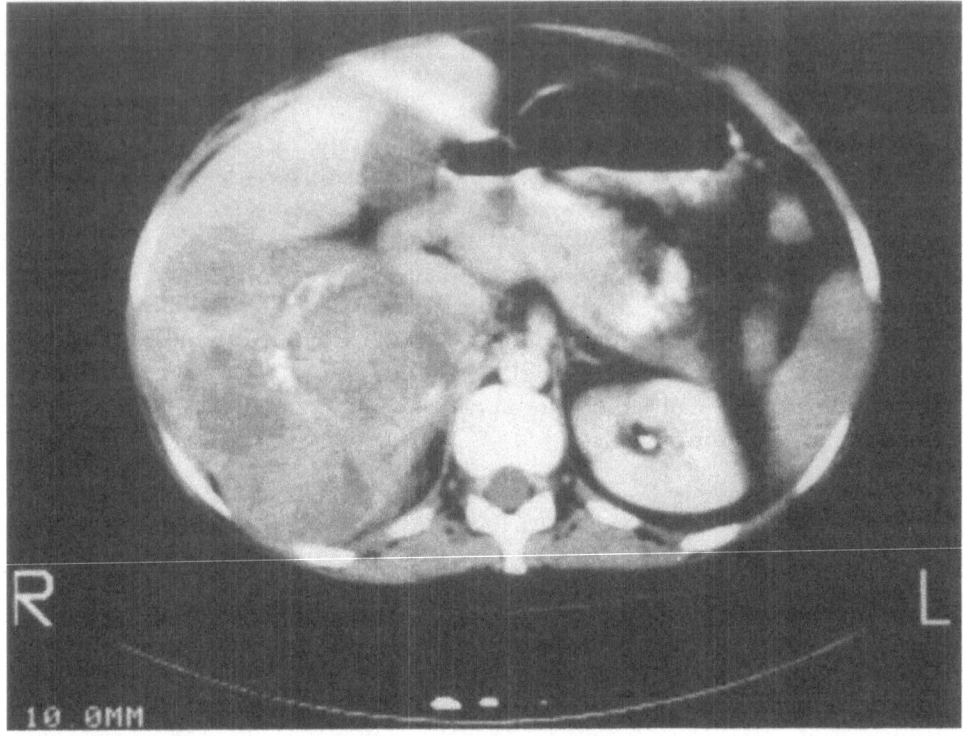

FIGURE 20. CT scan of the patient in Figure 19 demonstrates 9-cm mass in the right adrenal gland with metastatic lesions in the liver.

Computed Tomography (Fig. 20)

Adrenal carcinomas are almost always detected by CT of the adrenals. The most helpful parameter in differentiating adenoma from carcinoma of the adrenal is the size of the mass.[63] Adrenal adenomas seldom attain sizes greater than 6 cm in diameter. Copeland[63] has compiled the data from six reported series comprising 114 patients with adrenocortical carcinoma and noted that 105 of 114 carcinomas were greater than 6 cm in diameter. The large size of the adrenocortical carcinoma may, in part, be due to the long duration before the clinical features have attracted medical attention. The relationship between size of the adrenal mass and the possibility of malignancy has been emphasized by Hussain et al.[64] These workers also noted that adrenocortical carcinomas showed greater contrast enhancement and greater irregularity in the internal consistency of the tumor mass. Although size is an important discriminating parameter, occasionally benign adenomas may grow to dimensions greater than 6 cm, while a small percentage (<5%) of adrenocortical carcinomas can be smaller than 6 cm. Adrenal cysts can also attain large dimensions.[65] These cysts can be recognized by ultrasonography of the adrenals as well as by the low attenuation coefficient in CT.

The presence of calcification, although considered to be a suspicious sign of malignancy, may also occur in benign lesions, particularly cysts.[66,67] When the CT of adrenal demonstrates a large mass with evidence of vascular invasion or infiltration of the neighboring structures, the diagnosis of adrenocortical carcinoma is virtually certain. In summary, CT is enormously helpful in the diagnosis of adrenocortical carcinoma in a number of ways.

1. In patients with hormonal hypersecretion, the demonstration of a adrenal mass greater than 6 cm is almost diagnostic of adrenal carcinoma.[68]
2. In patients with no hormonal hypersecretion (nonfunctional), an adrenal mass greater than 6 cm is strongly suspicious, but not conclusive, of adrenocortical carcinoma. The differential diagnosis includes adrenal cysts, the rare "large adenoma," and other lesions such as lipoma or myelolipoma.
3. The demonstration of invasion of neighboring structures is absolutely diagnostic of adrenocortical carcinoma.
4. The differentiation between a small nonfunctioning carcinoma and an adenoma cannot be made on the grounds of CT appearance alone. It is this group that deserves accurate diagnosis because an inverse correlation has been recognized between size of the adrenocortical cancer and prognosis.[69–71] It is in this group that magnetic resonance imaging and CT-guided fine-needle aspiration may prove to be of assistance in the future. Reinig et al.[72] evaluated the ability of magnetic resonance to differentiate adenoma from carcinoma in intermediate-size adrenal masses. Their findings indicate that the ratio of the signal intensity of the mass to that of the liver is much greater in carcinoma than in adenoma. In terms of fine-needle aspiration, the distinction between benign and malignant lesions cannot be made with certainty unless extreme examples are encountered. The presence of a large number of mitoses is a specific, but not sensitive, indicator of malignancy. Despite reports of the successful use of fine-needle aspiration in the diagnosis of adrenocortical carcinoma in the literature,[73] the procedure has not found widespread application owing to the limitations in interpreting adrenal cytology.

Iodocholesterol Imaging

Nuclear imaging of the adrenals with iodocholesterol isotopes has proven to be helpful in distinguishing the various causes of Cushing's syndrome.[74] Adenomas characteristically concentrate the iodocholesterol, while generally adrenocortical carcinomas demonstrate no visualization bilaterally.[74,75] The reason for nonvisualization of even the "functional" adrenocortical carcinoma may be due to the fact that hormonal hypersecretion is carried out from cholesterol derived primarily from endogenous production, with very little serum cholesterol entering the metabolic pathway.[76] Schteingart et al.[77] estimated the concentration of [131I]iodocholesterol in neoplastic tissue obtained at surgery within 2–3 weeks following adrenal imaging and demonstrated a significantly lower concentration of the tracer in comparison to normal or adenomatous tissue. However, the typical appearance of bilateral nonvisualization may not be present when the adrenocortical carcinoma is well differentiated.[76]

The metastatic lesions from adrenocortical carcinoma are also not usually visualized by iodocholesterol imaging. However, functionless metastatic lesions have been reported to "pick up" iodocholesterol following removal of the primary tumor.[78-81] The situation is probably analogous to metastatic differentiated thyroid carcinoma, where the metastatic lesions often "light up" with radioactive iodine only after complete ablation of the thyroid gland has been carried out by surgery and/or radioactive iodine. Very rarely simultaneous visualization of adrenocortical carcinoma as well as its skeletal metastases has been reported.[76]

The role of adrenal imaging with iodocholesterol has been limited owing to the lack of easy availability of the isotope as well as the emergence of other equally noninvasive, but more convenient imaging techniques, such as CT and magnetic resonance imaging.

Invasive Procedures

The role of adrenal venography and arteriography in localizing adrenal masses has become limited, owing to excellent ability of CT to demonstrate adrenal masses. Inferior venacavagraphy (and even arteriography) may be resorted to for preoperative definition of the extent of the tumor as well as its vascular supply. Angiography may also play a diagnostic role in the case of adrenal cysts, which usually appear as vascular masses in contrast to carcinoma.

Differential Diagnosis

Adrenocortical carcinoma enters the differential diagnosis in the following clinical settings:

1. In the differential diagnoses of abdominal (retroperitoneal) masses. Adrenal cortical carcinoma should be considered while evaluating any mass located in the suprarenal location. This is especially important when the adrenocortical carcinoma is nonfunctional and hence lacks the typical endocrinopathies associated with this tumor. In children, nonfunctioning (or late-functioning) adrenocortical carcinoma can be mistaken for Wilms' tumor, neuroblastoma,[82] or nephroblastoma.[83] Adrenocortical carcinoma also enters the diagnosis of the "asymptomatic solid adrenal mass" seen by CT, especially those larger than 6 cm. The only definitive method to exclude malignancy in the serendipitously discovered large, solid adrenal cortical mass is by surgical exploration of the involved adrenal.

2. In the differential diagnosis of hypersecretory syndromes involving adrenocortical hormones. Since adrenocortical carcinoma can rarely underlie the etiology of Cushing's syndrome, virilization, and hypermineralocorticism, attempts should be made to exclude this cause. The simple measurement of adrenal androgen levels is often helpful in differentiating cortisol-secreting adenoma from cortisol-secreting carcinoma. The delineation of aldosterone-secreting adenoma from carcinoma can be facilitated by demonstrating loss of circadian periodicity, as well as the loss of ACTH responsiveness of aldosterone and marked elevation in precursor products of aldosterone in carcinomas. In addi-

tion, the CT appearance of carcinoma contrasts with the adenomas, which are generally much smaller and quite localized. The distinction between "pure" androgen-secreting adrenal carcinoma and the rare androgen-secreting adrenal adenoma may be impossible on hormonal grounds alone. It has been pointed out by Gabrilove et al.[59] that adrenal adenomas are frequently associated with elevated circulating testosterone levels in the plasma, in addition to elevation in the excretion of the neutral 17-KS. The "pure virilizing" syndromes associated with adrenocortical carcinoma are generally associated with predominant elevation of dehydroepiandrosterone sulfate. Significant elevations in testosterone levels are encountered with a distinctly lower frequency. However, these differences cannot be relied on consistently to distinguish "pure" virilizing adenoma from carcinoma.

3. In the evaluation of metastatic disease from an "unknown" or occult primary. While adrenal carcinoma metastasizes to the lungs and liver with an alarming frequency, in most instances the primary nature of the adrenocortical carcinoma is readily apparent. Occasionally, however, the metastatic manifestations of adrenocortical carcinoma antedate the symptoms and signs of the primary tumor. Marsden et al.[82] reported a unique case of a 5-year-old girl who presented with spinal cord compression due to a metastatic tumor, which histologically resembled the metastases from adrenal carcinoma. However, there was no hormonal or radiological evidence of an adrenal tumor, until 10 months later, when she developed Cushing's syndrome. Rare as it is, adrenocortical carcinoma must be kept in mind as a possible "occult" primary when evaluating patients with metastatic disease due to an unknown primary. As pointed out earlier, both functional as well as nonfunctional adrenocortical cancers metastasize with equal frequency. Metastases from the functional adrenal carcinoma usually manifest the same hormonal pattern as the primary tumor. However, there are cases of adrenocortical carcinoma where originally functioning tumors recurred as nonfunctioning types.[84] The ability of metastic adrenal tissue to function can be exploited in the hormonal evaluation of patients with reucrrent metastatic disease following therapy.

Complications

In addition to the enormous proclivity of adrenal carcinoma to metastasize to distant sites, the two complications encountered in this malignancy are thrombosis of the adrenal veins and hemorrhage into the adrenal gland harboring the tumor.

Thrombosis

Adrenocortical carcinoma can result in thrombosis of the adrenal veins, which can extend to eventually involve the renal veins, and even cause thrombo-occlusion of the inferior vana cava. Rarely, hepatic vein thrombosis may occur, resulting in the Budd–Chiari syndrome. In addition to the occlusive sequelae of these thrombotic phenomena (edema, proteinuria, hematuria, and portal hypertension), tumor embolism is a known complication of adrenocortical carcinoma.

Adrenal Hemorrhage

Adrenal hemorrhage is a potentially fatal complication in patients with adrenocortical carcinoma. The adrenal gland, the most vascular gland in the body on the basis of blood flow per weight, is fed by three main arteries, but is drained by a single vein. The inherent vascularity of the adrenal gland, coupled with the presence of a vascular tumor that often causes thrombosis of the single draining vein, is the reason for the predisposition to adrenal hemorrhage. The clinical picture of acute adrenal hemorrhage has been the subject of several reports.[85-89] The abrupt onset of upper abdominal pain, with or without radiation to the flank, in a patient harboring an adrenal tumor signals adrenal hemorrhage. A variety of neurological disturbances, ranging from confusion to seizures, may be associated with hemorrhage into the adrenal gland. The picture may be complicated by the rapid development of abdominal ileus, hypotension, and shcok, culminating in death if untreated. Adrenal insufficiency is likely to develop acutely since the contralateral gland has been suppressed and the cortisol-secreting tumor undergoes necrosis from the hemorrhage. Retroperitoneal hematoma formation and the development of peritonitis augur poorly for the prognosis.

Treatment

Adrenocortical carcinomas are highly malignant tumors. The short life expectancy in patients with adrenal carcinoma is due to the fact that the tumor is often advanced at the time of diagnosis, having metastasized in 50–60% of patients. The relative rarity of this malignancy has precluded well-designed therapeutic trials that involve large numbers of patients. The best prognoses are usually seen in patients whose tumors are less than 5 cm in diameter, and without local spread or distant metastasis. While there is no unanimity of opinion regarding the therapeutic options, the combination of adrenal resection (when feasible) and adjuvant therapy (o p' DDD) is generally accepted as the therapy for adrenocortical carcinoma. The prospects of a "surgical cure" appears to be dim in patients with adrenocortical carcinoma owing to the presence of advanced disease in most patients with this cancer. Yet, in a young patient, attempts of surgical debulking are often resorted to, since very little else can be offered. The adrenocortical carcinoma, at the time of presentation, is staged according to the SEER classification.[90] "Localized disease" refers to disease restricted to the adrenal confirmed at surgery, after the various studies have failed to disclose evidence of metastatic disease. Less than 5% of patients with adrenocortical carcinoma would belong in this category. "Regional disease" refers to disease that has extended beyond the adrenals to involve surrounding organs and tissues by contiguity, including the regional lymph nodes, but without evidence of distant metastases to the lungs, liver, bones, and so forth. As many as 20–30% of patients with adrenocortical carcinoma would fall in this category. "Distant disease" refers to patients with adrenocortical carcinoma that has disseminated to distant sites. Unfortunately, the bulk (40–60%) of patients with adrenocortical carcinoma fall in this dismal category. "Recurrent disease" refers to the development of local recurrence or the appearance of metastatic disease following the resection of the original adrenal lesion. Surgery, in the form of either "curative" or palliative resection, is often recommended in patients with localized or regional disease. The role of surgical debulking in patients with disseminated adrenocortical carcinoma has not been established and is indeed questionable. Patients with recurrent disease are often treated with adrenolytics and some form of chemotherapy, combined with surgical debulking for local recurrence.

Surgery

This is the primary mode of therapy in patients with localized or regional disease. The preoperative workup includes angiography (arteriography and inferior venacavagraphy) to delineate the blood supply of the tumor, and to evaluate for the presence of vena caval involvement. In addition, these studies may demonstrate the presence of otherwise undetected contralateral adrenal involvement—a sign that would necessitate bilateral adrenal surgery. Renal function should be scrutinized since nephrectomy is often performed during the en bloc dissection of the tumor. En bloc adrenal tumor removal is an operation of major proportions, and hence careful preoperative assessment of the cardiopulmonary status is mandatory. Glucocorticoid supplementation is initiated 2 days prior to surgery and continued for 2 weeks postoperatively. Great care should be exercised in the surgical handling of this tumor to avoid the risk of "tumor seeding" in the retroperitoneal bed. Most surgeons prefer a transabdominal or thoracoabdominal incision to obtain a good exposure to remove the often-large tumors, as well as for inspection of the contralateral adrenal.[91] Following the en block removal of the tumor, retroperitoneal lymph node dissection is carried out for better staging as well as therapeutic purposes. Attempts to resect the entire tumor mass should be undertaken[92] since this would diminish recurrence. However, this is not always possible owing to the presence of local extension into vital tissue. In such a circumstance, as much tumor as possible should be removed without otherwise endangering the patient's immediate survival. Unfortunately, even in instances where all visible tumor has been resected, small remnants and microscopic foci of the tumor may have been left behind, leading to local recurrence and/or metastatic spread.

Aggressive surgical therapy is considered the best primary mode of therapy for patients with localized and regional disease, and even for localized metastatic lesions. Repeated aggressive surgical resection of recurrent tumor has been reported to prolong survival in some patients.[71] In fact, it is believed that the improved survival of patients with adrenocortical carcinoma treated with surgery and adjuvant o p' DDD therapy is justification to excise and debulk recurrent tumors as and when they recur.[93] Postoperative abdominal radiation for adrenocortical carcinoma has been used on a limited basis, particularly in children.[94] The value of adjuvant radiotherapy has not been established. Postoperative adjuvant therapy with o p' DDD is believed to prolong and improve survival when instituted immediately following resection of primary tumor.[95]

The initial response to surgery is often characterized by a temporary remission.[70,92,96] Unfortunately, local recurrence and eventual metastases occur in a large number of patients who initially respond to surgery. It is not known whether these recurrences can be prevented by the early institution of o p' DDD therapy immediately following surgery.

o p' DDD (Mitotane)

o p' DDD or (1,1-dichloro-2[o-chlorophenyl]-2-[p-chlorophenyl]-ethane) is an adrenolytic drug. This drug is usually employed to decrease steroid hormone production (chapt. 3). The role of mitotane in the treatment of adrenocortical carcinoma can be viewed from several perspectives. First, the drug is commonly used to treat patients with overt metastatic disease resulting from the adrenocortical cancer. Second, the drug can be effectively used to lower hypercortisolism in patients with Cushing's syndrome caused by adrenocortical carcinoma. This action, in addition to providing obvious metabolic advantages to the patient, can be used to prepare the patient for surgery. Third, the drug does improve survival when given as adjuvant therapy immediately following surgical resection of the primary tumor.[95] Fourth, mitotane therapy is often the only palliative measure that can be offered to patients with nonresectable tumors and those with extensive recurrence.

Mitotane for Metastatic Disease

Since the original report by Bergenstal et al.[97] that o p' DDD treatment caused regression of adrenal carcinoma in humans, the drug has been extensively used. Two large series, one by Hutter and Kayhoe[98] and another by Lubitz et al.,[99] have reported measurable tumor response in 34% and 61% of patients, respectively, treated with this drug. In both series the survival was prolonged when compared to that of patients who had not received mitotane. These favorable reports have been clouded by other studies that have suggested that mitotane does not significantly affect the overall course of the disease.[29,100] While there is disagreement concerning the overall effectiveness of mitotane based on results from large series, several reports of individual patients who responded to the drug, sometimes dramatically, have appeared in the literature.[101–106] Prolonged remission, sometimes with exceptionally long survival, has been reported.[104] It is not known whether the wide variability in the individual responses is related to the dose employed or is due to the individual differences in biological behavior of the metastatic lesions. The type of tumor (functional or nonfunctional) has no bearing on the responsiveness to mitotane. The sustained effectiveness of the drug, seen in some patients, may be related to its deposition in subcutaneous fat, liver, brain, and other tissues. The drug can be detected in the blood several months after discontinuation.[107] This fact underscores the need to continue steroid replacement even following withdrawal of therapy. The tumorolytic effect has been well documented, often by histological demonstration of absence of disease.[105] Clearly, patients with metastatic adrenal carcinoma should be given the benefit of mitotane therapy.

The dosage of mitotane is also controversial. The optimal plasma level of mitotane required to induce an objective remission is not known. Hogan et al.[107] have reported objective responses using 5–6 g/day of mitotane, which provided plasma levels of the drug in the range of 15–20 µg/ml. This dosage is lower than the originally recommended dose of 8–10 g/day of mitotane to treat metastatic adrenocortical carcinoma.[98,99]

The toxicity of mitotane has been one of the major limitations in therapy with this drug. Nausea, vomiting, weakness, dermatitis, and neurotoxicity are frequent side effects of mitotane therapy. Hogan et al.[107] have noted a decrease in the incidence of side effects with lower (5–6 g) doses of the drug.

Mitotane for Lowering Hypercortisolemia

Mitotane, at doses of 0.5–3 g/day, effectively blocks adrenal 11-hydroxylation, resulting in a decrease in the concentrations of metabolically active cortisol.[108] Even at this lower dosage the drug can alter mitochondrial morphology of zona fasciculata cells. At higher doses, and with chronic administration, the drug causes adrenal atrophy, with combined gluco- and mineralocorticoid deficiency.

In terms of drug kinetics, approximately 40% of the orally administered drug is absorbed from the gastrointestinal tract. Being fat soluble, the drug deposits itself in several tissues, such as subcutaneous tissue, fat, liver, brain, and of course the normal as well as the malignant adrenal tissue.[109] This "depot effect" is the reason for the persistence of blood levels of mitotane even after discontinuation of the drug.

In addition to its adrenolytic effects, mitotane has important actions on the peripheral metabolism of both cortisol and androgens.[110–112] Mitotane causes increased excretion of unconjugated 6-β-OH derivatives. This may result in an impressive decline in the urinary 17-hydroxycorticosteroid levels during the first weeks of therapy at a time when the plasma cortisol level and the cortisol secretory rate remain unaltered. While originally doses of 8–10 g were used in the treatment, it is currently believed that 5–6 g of mitotane daily results in impressive objective improvement in the hypercortisolemia. Since the drug persists for months in the tissues and the plasma, the optimal level needed to achieve a hormonal and tumoral response is not clearly known. Because of this drug persistence, careful monitoring of adrenocortical function is necessary even following withdrawal of therapy. Of course, gluco- and mineralocorticoid replacement is necessary during mitotane therapy. Greater-than-normal replacement doses may be necessary in mitotane-treated patients owing to the alterations in peripheral steroid metabolism caused by mitotane. Recovery of adrenal function after treatment with mitotane has been described.[113]

Improved Survival with Mitotane

While this issue has been somewhat disputed, more recent trials suggest that mitotane therapy does indeed prolong survival of patients with adrenocortical cancer. The early studies had focused on mitotane therapy administered to patients with metastatic disease, a group with a dismal prognosis regardless of type of therapy given. Schteingart et al.,[95] on the other hand, demonstrated that response to mitotane therapy depended on the timing of institution of mitotane during the course of disease. The longest survivals were noted in patients who received mitotane as adjuvant therapy *before* clinical evidence of metatases had occurred, as well as in those who had received mitotane and underwent subsequent surgery for recurrent tumor. The survival rate in this group was 74 ± 33 months, the longest survival reported for patients with this cancer. In contrast, the mean survival rate was a dismal 10.3 ± 8.7 months for patients who simply had resection of primary tumor and/or local radiation with no mitotane therapy. The overall mean survival for patients who received mitotane was 46.6 ± 42.7 months, with a median of 24 months and a range of 1–120 months. The authors noted that, although survival for the mitotane-treated group was longer, the difference was not statistically significant. However, the group of patients who had received mitotane as adjuvant therapy after resection of primary tumor and before evidence of metastases had an impressively significant prolongation in the survival rate, averaging 74 ± 33 months. Thus, the best results from mitotane are to be expected when the drug is started before the appearance of metastases.

In summary, treatment of adrenocortical carcinoma continues to remain a frustrating problem. Aggressive surgery and early institution of adjuvant therapy hold the most promise. Nonspecific chemotherapy and radiation do not significantly alter the outcome of this disease. In the series reported by Didolkar et al.[7] statistically significant prolonged survivals were seen in women, those with localized tumors, and those who underwent extirpative procedures. Patients who showed a disease-free interval greater than 12 months also had a slightly prolonged survival. Although it is felt that size of the tumor is inversely related to prognosis, in the series reported by Nader et al.[30] 10 of the 18 patients who survived 5 years or longer had the largest tumors at the time of presentation. In general, it appears that the patients who respond with the best survivals are those with small, localized tumors that are completely resected, and who are placed immediately on adjuvant chermotherapy with low-dose, long-term mitotane therapy.

References

1. *Third National Cancer Survey: Incidence Data.* National Cancer Institute monograph 41. March 1975. DHEW publication no. (NIH) 75–787. US Department of Health, Education and Welfare, Public Health Service, National Institute of Health, National Cancer Institute, Bethesda, MD.
2. Heinbecker P, O'Neal LW, Ackerman LV: Functioning and nonfunctioning adrenal cortical tumors. *Surg Gynecol Obstet* **105:**21, 1957.
3. Macfarlane DA: Cancer of the adrenal cortex; the natural history prognosis and treatment in a study of fifty-five cases. *Ann R Coll Surg Engl* **23:**155, 1958.
4. Lipsett MB, Hertz R, Ross GT: Clinical and pathophysiologic aspects of adrenocortical carcinoma. *Am J Med* **35:**374, 1963.
5. Knight CD, Trichel BE, Mathews WR: Non-functioning carcinoma of the adrenal cortex. *Ann Surg* **151:**349, 1960.
6. Lewinsky BS, Grigor KM, Symington T, et al: The clinical and pathologic features of "nonhormonal" adrenocortical tumors. Report of 20 new cases and review of the literature. *Cancer* **33:**778, 1974.

7. Didolkar MS, Bescher RA, Elias EG, et al: Natural history of adrenal cortical carcinoma. A clinicopathologic study of 42 patients. *Cancer* **47:**2153, 1981.
8. Huvos AG, Hajdu SI, Brasfield RD, et al: Adrenal cortical carcinoma—Clinicopathologic study of 34 cases. *Cancer* **25:**354, 1970.
9. Silva EG, Mackay B, Samaan NA, et al: Adreno cortical carcinoma: An ultrastructional study of 22 cases. *Ultrastruct Pathol* **3:**1, 1982.
10. Falls JL: Accessory adrenal cortex in the broad ligament: Incidence and functional significance. *Cancer* **8:**143, 1955.
11. Dahl EV, Bahn RC: Aberrant adrenal cortical tissue near the testis in human infants. *Am J Pathol* **40:**587, 1962.
12. Nelson AA: Accessory adrenal cortical tissue. *Arch Pathol* **27:**955, 1939.
13. Wiener MF, Dallgaard SA: Intracranial adrenal gland: A case report. *Arch Pathol* **67:**228, 1959.
14. Vestfried MA: Ectopic adrenal cortex in neonatal liver. *Histopathology* **4:**669, 1980.
15. Armin A, Castelli M: Congenital adrenal tissue in the lung with adrenal cytomegaly: Case report and review of the literature. *Am J Clin Pathol* **82:**225, 1984.
16. Wallace EZ, Leonidas J-R, Stanek AE, et al: Endocrine studies in a patient with functioning adrenal rest tumor of the liver. *Am J Med* **70:**1122, 1981.
17. Contreras P, Altieri E, Liberman C, et al: Adrenal rest tumor of the liver causing Cushing's syndrome: Treatment with ketoconazole preceding an apparent surgical cure. *J Clin Endocrinol Metab* **60:**21, 1985.
18. Wilkins L, Ravitch MM: Adrenocortical tumor arising in the liver of a 3 year old boy, with signs of virilism and Cushing's syndrome: Report of a case with cure after partial resection of right lobe of liver. *Pediatrics* **9:**671, 1952.
19. Scully RE, Mark EJ, McNeely WF, et al: Case 50-1986. Case records of the Massachusetts General Hospital. *N Engl J Med* **315:**1595, 1986.
20. Leger L, Bouvresse M, Deslingneres S: Cortico-surénalome malin sur surrénale accessoire: trois surrénalectomies successives chez le même sujet. *J Chir (Paris)* **110:**7, 1975.
21. Hamperl H: On the "adrenal rest tumors" (hypernephromas) of the liver. *Z Krebsforsch* **74:**310, 1970.
22. Wooley GW: Experimental endocrine tumours with special reference to the adrenal cortex. *Recent Prog Horm Res* **5:**383, 1950.
23. Hamwi GJ, Serbin RA, Kruger FA: Does adrenal cortical hyperplasia result in adrenocortical carcinoma? *N Engl J Med* **257:**1153, 1957.
24. Dluhy RG, Barlow JJ, Mahoney EM, et al: Profile and possible origin of an adrenocortical carcinoma. *J Clin Endocrinol Metab* **33:**312, 1971.
25. Bauman A, Bauman CG: Virilizing adrenocortical carcinoma: Development in a patient with salt-losing congenital adrenal hyperplasia. *JAMA* **248:**3140, 1982.
26. Harrison RJ, Abelson D: Carcinoma of the adrenal cortex with endocrine manifestations. Report of a case. *Br Med J* **1:**303, 1952.
27. Anderson DC, Child DF, Sutcliffe CH, et al: Cushing's syndrome, nodular adrenal hyperplasia and virilizing carcinoma. *Clin Endocrinol* **9:**1, 1978.
28. Hutter AM, Kayhoe DE: Adrenal cortical carcinoma: Clinical features of 138 patients. *Am J Med* **41:**572, 1966.
29. Hajjar RA, Hickey RC, Samaan NA: Adrenal cortical carcinoma: A study of 32 patients. *Cancer* **35:**549, 1975.
30. Nader S, Hickey RC, Sellin RV, et al: Adrenal cortical carcinoma: A study of 77 cases. *Cancer* **52:**707, 1983.
31. Bertagna C, Orth DN: Clinical and laboratory findings and results of therapy in 58 patients with adrenocortical tumors admitted to a single medical center (1951 to 1978). *Am J Med* **71:**855, 1981.
32. Yamaji T, Ishibashi M, Sekihara H, et al: Serum dehydroepiandrosterone sulfate in Cushing's syndrome. *J Clin Endocrinol Metab* **59:**1164, 1984.
33. Forbes AP, Albright F: A comparison of the 17-ketosteroid excretion in Cushing's syndrome associated with adrenal tumor and with adrenal hyperplasia. *J Clin Endocrinol Metab* **11:**926, 1951.
34. Armstrong CN, Simpson J: Adrenal feminism due to carcinoma of adrenal cortex: A case report and review of the literature. *Br Med J* **1:**782, 1948.

35. Stewart WK, Fleming LW, Wotiz HH: The feminizing syndrome in male subjects with adrenocortical neoplasms. *Am J Med* **37**:455, 1964.

36. Gabrilove JL, Sharma DC, Wotiz HH, et al: Feminizing adrenocortical tumors in the male: A review of 52 cases. *Medicine* **44**:37, 1965.

37. Sohval AR, Gabrilove JL: Testicular histopathology in feminizing tumors of the adrenal cortex. *J Urol* **93**:711, 1965.

38. Case records of Massachusetts General Hospital (case 23-1979). *N Engl J Med* **300**:1322, 1979.

39. Wilkins L: A feminizing adrenal tumor causing gynecomastia in a boy of five years contrasted with virilizing tumor in a five-year-old girl: Classification of seventy cases of adrenal tumor in children according to their hormonal manifestations and a review of eleven cases of feminizing adrenal tumor in adults. *J Clin Endocrinol* **8**:111, 1948.

40. Bacon GE, Lowrey GH: Feminizing adrenal tumor in a six-year-old boy. *J Clin Endocrinol Metab* **25**:1403, 1965.

41. Smith AH: A case of feminizing adrenal tumor in a girl. *J Clin Endocrinol Metab* **18**:318, 1958.

42. Veldhuis JD, Sowers JR, Rogol AD, et al: Pathophysiology of male hypogonadism associated with endogenous hyperstrogenism: Evidence for dual defects in the gonadal axis. *N Engl J Med* **312**:1371, 1985.

43. Salassa TM, Weeks RE, Northcutt RC, et al: Primary aldosteronism and malignant adrenocortical neoplasia. *Trans Am Clin Climatol Assoc* **86**:163, 1975.

44. Neville AM, O'Hare MJ: The human adrenal cortex, in: *Pathology and Biology, an Integrated Approach*. Springer Verlag, Heidelberg, 1982, p. 215.

45. Arteaga E, Biglieri EG, Kater CE, et al: Aldosterone-producing adrenocortical carcinoma: Preoperative recognition and course in three cases. *Ann Intern Med* **101**:316, 1984.

46. Richie JP, Gittes RF: Carcinoma of the adrenal cortex. *Cancer* **45**:1957, 1980.

47. Alterman SL, Dominguez C, Lopez-Gomez A, et al: Primary adrenocortical carcinoma causing aldosteronism. *Cancer* **24**:602, 1969.

48. Foye LV Jr, Feichtmeir TV: Adrenal cortical carcinoma producing solely mineralocorticoid effect. *Am J Med* **19**:966, 1955.

49. Santander R, Gonzalez A, Suarez JA: Case of probable mineralocorticoid excess without hypercortisolism due to a carcinoma of the adrenal cortex. *J Clin Endocrinol Metab* **25**:1429, 1965.

50. Ross EJ: *Aldosterone and Aldosteronism*. Lloyd-Luke, London, 1975.

51. Powell-Jackson JD, Calin A, Fraser R, et al: Excess deoxycorticosterone secretion from adrenocortical carcinoma. *Br Med J* **2**:32, 1974.

52. Kahn M, Melby JC, Jacobs DR: Isolated DOC excess: A unique syndrome simulating hyperaldosteronism with marked fluid retention. *Clin Res* **14**:282, 1966.

53. Kelly WF, O'Hare MJ, Loizou S, et al: Hypermineralocorticism without excessive aldosterone secretion: An adrenal carcinoma producing deoxycorticosterone. *Clin Endocrinol* **17**:353, 1982.

54. Brown JJ, Strott CA: Plasma deoxycorticosterone in man. *J Clin Endocrinol* **36**:44S, 1973.

55. Falchuk KR: Case report: Inappropriate antidiuretic hormone-like syndrome associated with an adrenal carcinoma. *Am J Med Sci* **266**:393, 1973.

56. Williams R, Kellie AE, Wade AP, et al: Hypoglycaemia and abnormal steroid metabolism in adrenal tumours. *Q J Med* **30**:269, 1961.

57. Wallach S, Brown H, Englent E Jr, et al: Adrenocortical carcinoma with gynecomastia: A case report and review of the literature. *J Clin Endocrinol Metab* **17**:945, 1947.

58. Liddle GW, Nicholson WE, Island DP, et al: Clinical and laboratory studies of ectopic humoral syndromes. *Recent Prog Horm Res* **25**:283, 1969.

59. Gabrilove JL, Seman AT, Sabet R, et al: Virilizing adrenal adenoma with studies on the steroid content of the adrenal venous effluent and a review of the literature. *Endocrinol Rev* **2**:462, 1981.

60. Lipsett MB, Wilson H: Adrenocortical cancer: Steroid biosynthesis and metabolism evaluated by urinary metabolites. *J Clin Endocrinol Metab* **22**:906, 1962.

61. Fukushima DK, Gallagher TF: Steroid production in "nonfunctioning" adrenal cortical tumor. *J Clin Endocrinol Metab* **23**:923, 1963.

62. Fantl V, Booth M, Gray CH: Urinary pregn-5-ene, 3a, 16a, 20α-triol in adrenal dysfunction. *J Endocrinol* **57**:135, 1973.

63. Copeland PM: Diagnosis and treatment: The incidentally discovered adrenal mass. *Ann Intern Med* **98**:940, 1983.

64. Hussain S, Belldegrun A, Seltzer SE, et al: Differentiation of malignant from benign adrenal masses: Predictive indices on computed tomography. *Am J Roentgenol* **144:**61, 1985.
65. Scheible W, Coel M, Siemers PT, et al: Percutaneous aspiration of adrenal cysts. *Am J Roentgenol* **128:**1013, 1977.
66. Martin JF: Suprarenal calcification. *Radiol Clin North Am* **3:**129, 1965.
67. Kearny GP, Mahoney EM, Maher E, et al: Functioning and nonfunctioning cysts of the adrenal cortex and medulla. *Am J Surg* **134:**363, 1977.
68. Dunnick NR, Doppman JL, Gill JR Jr, et al: Localization of functional adrenal tumors by computed tomography and venous sampling. *Radiology* **142:**429, 1982.
69. Tang CK, Gray GF: Adrenocortical neoplasms: Prognosis and morphology. *Urology* **5:**691, 1975.
70. Sullivan M, Boileau M, Hodges CV: Adrenal cortical carcinoma. *J Urol* **120:**660, 1978.
71. Bradley EL III: Primary and adjunctive therapy in carcinoma of the adrenal cortex. *Surg Gynecol Obstet* **141:**507, 1975.
72. Reinig JW, Doppman JL, Dwyer AJ, et al: Adrenal masses differentiated by MR. *Radiology* **158:**81, 1986.
73. Levin NP: Fine needle aspiration and histology of adrenal cortical carcinoma: A case report. *Acta Cytol* **25:**421, 1981.
74. Thrall JH, Freitas JE, Beierwaltes WH: Adrenal scintigraphy. *Semin Nucl Med* **8:**23, 1978.
75. Moses DC, Schteingart DE, Sturman MF, et al: Efficacy of radiocholesterol imaging of the adrenal glands in Cushing's syndrome. *Surg Gynecol Obstet* **139:**201, 1974.
76. Drane WE, Graham MM, Nelp WB: Imaging of an adrenal cortical carcinoma and its skeletal metastasis. *J Nucl Med* **24:**710, 1983.
77. Schteingart DE, Seabold JE, Gross MD, et al: Iodocholesterol adrenal tissue uptake and imaging in adrenal neoplasms. *J Clin Endocrinol Metab* **52:**1156, 1981.
78. Seabold JE, Haynie TP, DeAsis DN: Clinical detection of metastatic adrenal carcinoma using ^{131}I-6-β-iodomethyl-19-norcholesterol total body scans. *J Clin Endocrinol Metab* **45:**788, 1977.
79. Chatel JF, Charbonnel B, Le Mevel BP, et al: Uptake of [131]I-19-iodocholesterol by an adrenal cortical carcinoma and its metastases. *J Clin Endocrinol Metab* **43:**248, 1976.
80. Watanabe K, Kamoi I, Nakayama C, et al: Scintigraphic detection of hepatic metastases with ^{131}I-labeled steroid in recurrent adrenal carcinoma: Case report. *J Nucl Med* **17:**904, 1976.
81. Forman BH, Antar MA, Touloukian RJ, et al: Localization of a metastatic adrenal carcinoma using ^{131}I-iodocholesterol. *J Nucl Med* **15:**332, 1974.
82. Marsden HB, Jones PM, Lees PD, et al: Late functioning adrenocortical carcinoma in a 5-year-old girl. *Arch Dis Chid* **53:**341, 1978.
83. Visconti EB, Peters RW, Cangir A, et al: Unusual case of adrenal cortical carcinoma in a female infant. *Arch Dis Child* **53:**343, 1978.
84. Rapaport E, Goldberg MB, Gordan GS, et al: Mortality in surgically treated adrenocortical tumors. Review of cases reported for 20 year period 1930–1949, inclusive. *Postgrad Med* **11:**325, 1952.
85. Berte SJ: Spontaneous adrenal hemorrhage in the adult: Literature review and report of two cases. *Ann Intern Med* **38:**28, 1953.
86. Clark OH, Hall AD, Schambelan M: Clinical manifestations of adrenal hemorrhage. *Am J Surg* **128:**219, 1974.
87. Lawson DW, Corry RJ, Patton AS, et al: Massive retropertioneal adrenal hemorrhage. *Surg Gynecol Obstet* **129:**989, 1969.
88. Greendyke RM: Adrenal hemorrhage. *Am J Clin Pathol* **43:**210, 1965.
89. Case Records of the Massachusetts General Hospital (case 46-1972). *N Engl J Med* **287:**1033, 1972.
90. *Summary Staging Guide for Cancer Surveillance, Epidemiology and End Result Reporting Program; April 1977.* US Department of Health, Education, and Welfare, Public Health Service, National Institute of Health, Bethesda, MD.
91. Flint LD: Surgical exposures for adrenal endocrinopathies. *Surg Clin North Am* **53:**445, 1973.
92. King DR, Lack EE: Adrenal cortical carcinoma: A clinical pathologic study of 49 cases. *Cancer* **44:**239, 1979.
93. Greenberg PH, Marks C: Adrenal cortical carcinoma: A presentation of 22 cases and a review of the literature. *Ann Surg* **44:**81, 1978.

94. Stewart DR, Jones PHM, Jolleys A: Carcinoma of the adrenal gland in children. *J Pediatr Surg* **9:**59, 1974.

95. Schteingart DE, Motazedi A, Noonan RA, et al: Treatment of adrenal carcinomas. *Arch Surg* **117:**1142, 1982.

96. Burger AR, Correa RJ: Experience with adrenal carcinoma. *Urology* **10:**12, 1977.

97. Bergenstal DM, Lipsett MB, Moy RH, et al: Regression of adrenal cancer and suppression of adrenal function in man by o p′ DDD. *Trans Assoc AM Physicians* **72:**341, 1959.

98. Hutter AM Jr, Kayhoe DE: Adrenal cortical carcinoma: Results of treatment with o p′ DDD in 138 patients. *Am J Med* **41:**581, 1966.

99. Lubitz JA, Freeman L, Okun R: Mitotane use in inoperable adrenal cortical carcinoma. *JAMA* **223:**1109, 1973.

100. Hoffman DL, Mattox VR: Treatment of adrenocortical carcinoma with o p′ DDD. *Med Clin North Am* **56:**999, 1972.

101. Becker D, Schumacher OP: o p′ DDD therapy in invasive adrenocortical carcinoma. *Ann Intern Med* **82:**677, 1975.

102. Ostuni JA, Roginsky MS: Metastatic adrenal cortical carcinoma. *Arch Intern Med* **135:**1257, 1975.

103. Downing V, Eule J, Huseby RA: Regression of an adrenal cortical carcinoma and its neovascular bed, following mitotane therapy: A case report. *Cancer* **34:**1882, 1974.

104. Jarabak J, Rice K: Metastatic adrenal cortical carcinoma: Prolonged regression with mitotane therapy. *JAMA* **246:**1706, 1981.

105. Boven E, Vermorken JB, van Slooten H, et al: Complete response of metastasized adrenal cortical carcinoma with o p′ DDD. Case report and literature review. *Cancer* **53:**26, 1984.

106. Helson L, Wollner N, Murphy MI, et al: Metastatic adrenal cortical carcinoma: Biochemical changes accompanying clinical regression during therapy with o p′ DDD. *Clin Chem* **17:**1191, 1971.

107. Hogan TF, Citrin DL, Johnson BM, et al: o p′ DDD (mitotane) therapy of adrenal cortical carcinoma. *Cancer* **42:**2177, 1978.

108. Brown RD, Nicholson WE, Chick WT, et al: Effect of o p′ DDD on human adrenal steroid 11-β-hydroxylation activity. *J Clin Endocrinol Metab* **36:**730, 1973.

109. Moy RH: Studies of the pharmacology of o p′ DDD in man. *J Lab Clin Med* **58:**296, 1961.

110. Bradlow HL, Zumoff B, Fukushima DK, et al: Drug induced alterations of steroid hormone metabolism in man. *Ann NY Acad Sci* **212:**148, 1973.

111. Hellman I, Bradlow HI, Zumoff B: Decreased conversion of androgens to normal 17-ketosteroid metabolites as a result of treatment with o p′ DDD. *J Clin Endocrinol Metab* **36:**801, 1973.

112. Schein PS: Chemotherapeutic management of the hormone-secreting endocrine malignancies. *Cancer* **30:**1616, 1972.

113. Greig F, Oberfield LS, Levine F, et al: Recovery of adrenal function after treatment of adrenocortical carcinoma with o p′ DDD. *Clin Endocrinol* **20:**389, 1984.

11

The Adrenogenital Syndromes

Introduction

The adrenogenital syndromes are inborn errors of metabolism that involve adrenal steroidogenesis and result in diverse hormonal, biochemical, and clinical effects. Recognition of these syndromes at birth in the child born with adrenogenital syndrome is crucial. In addition to facilitating assignment of sex of the child with ambiguous genitalia, recognition can be lifesaving, particularly in terms of prevention and treatment of electrolyte disorders. Several general statements are worthy of note regarding the adrenogenital syndromes.

1. The enzymatic blocks that cause adrenogenital syndrome can occur at several sites within the "steroidogenesis cascade." Some enzymatic blocks affect the patient more profoundly than others.
2. Regardless of the site of the block in steroidogenesis, the ultimate pathophysiological effect is failure to synthesize adequate amounts of cortisol—the metabolically active glucocorticoid. As a consequence of this phenomenon, release of ACTH is triggered.
3. The increased amounts of secreted ACTH stimulate the adrenal cortices on a chronic basis resulting in hyperplasia ("congenital adrenal hyperplasia"). Flagellating under the duress of chronic ACTH stimulation, the adrenal cortices put out excessive amounts of precursor steroids proximal to the enzymatic block. Active steroidogenesis proceeds in the open pathways. When the enzymatic block is complete and precludes formation of cortisol, the sustained secretion of endogenous ACTH results in profound activation of all the open pathways for adrenal steroidogenesis.
4. Adrenogenital syndromes are a heterogenous group of disorders with a spectrum characterized by varying degrees of severity. On one end are the patients born with sexual ambiguity where the block is severe, while the other end of the spectrum consists of patients who are completely feminized but infertile or oligomenorrheic.
5. The clinical expression of the adrenogenital syndromes depends on the site of the block (21, 11, etc.), type of the enzyme deficiency (whether the enzyme deficiency operates in the zona glomerulosa as well), the degree of enzyme deficiency (mild, moderate, or severe), the type of precursor steroids formed (androgens, neutral steroids, or mineralocorticoids) and the age of the patient (neonate, child, or adult). Thus, the clinical presentations of the adrenogenital syndrome exhibit considerable heterogeneity.
6. Since congenital adrenal hyperplasia results from inadequate production of cortisol which triggers ACTH release, the condition can be treated by the simple administration of glucocorticoid. For the most part, such a maneuver results in attenuation or abolition of the cycle of events that culminated in secretion of excessive precursor steroids by the adrenal cotices.
7. The genetics of 21-hydroxylase deficiency (21-OHD) has been well studied. It is clear that 21-OHD is transmitted as an autosomal recessive trait. The documentation that the 21-OH gene is linked to the HLA complex has led to elucidation of the human leukocyte antigen (HLA) markers associated with the nonclassical (or adult onset) form of 21-OHD.

8. Occasionally, the enzyme deficiency may involve the steroidogenic processes of both the adrenals as well as the gonads. This is most impressively illustrated by 17-α-hydroxylase deficiency (17-α-OHD).

9. Finally, sensitive immunoassay techniques have facilitated measurement of steroid precursors in the amniotic fluid. These techniques, coupled with the ability to study the amniotic fluid cell HLA haplotypes, have permitted identification of fetuses that have inherited the 21-OHD from the mother. In such cases of suspected congenital adrenal hyperplasia in the female fetus, abnormal masculinization of the external genitalia can perhaps be prevented by pharmacological suppression of the fetal adrenals in utero.

With this general preview, the various types of adrenogenital syndromes will be considered. The three most common forms—21-OHD, 11-β-hydroxylase deficiency (11-β-OHD), and 17-α-OHD—account for more than 90% of the adrenogenital syndromes. Before embarking on a discussion of the individual varieties of adrenogenital syndromes, it is crucial to review the basic aspects of adrenal steroidogenesis as well as the fundamentals of sexual differentiation.

Steroidogenesis

The framework for understanding adrenal steroidogenesis was detailed in Chapters 1 and 9. Therefore, the pathways essential for discussing the adrenogenital syndromes will be only briefly discussed in this chapter.

Formation of Pregnenolone

Figure 21 outlines the traditionally accepted schema for adrenal steroidogenesis. The conversion of cholesterol to pregnenolone is the crucial step that activates steroidogenesis. The enzyme that effects this conversion is the cholesterol side-chain–cleaving enzyme, or desmolase, present in the zona glomerulosa as well as in the zona fasciculata–reticularis. Pregnenolone is believed to be the principal substrate for all three pathways within the adrenal cortex (the mineralocorticoids, the glucocorticoids, and the sex steroids). The conversion of cholesterol to pregnenolone is the step activated by trophic hormones such as ACTH and angiotensin II. ACTH binds to receptors on the membrane of the adrenal cortical cells and stimulates production of cyclic AMP, which in turn activates intracellular phosphoprotein kinases. These enzymes catalyze the side-chain cleavage of cholesterol to pregnenolone. Cholesterol side-chain–cleaving enzyme is located in the mitochondria and requires the presence of cytochrome P-450.

Glucocorticoid Pathway

The production of glucocorticoids by the zona fasciculata involves the "17-hydroxy pathway." The substrate required for activation of this 17-hydroxy pathway is 17-α-hydroxyprogesterone (17-α-OHP). 17-α-OHP can be derived either from 17-OH pregnenolone (by action of 3-hydroxysteroid dehydrogen-

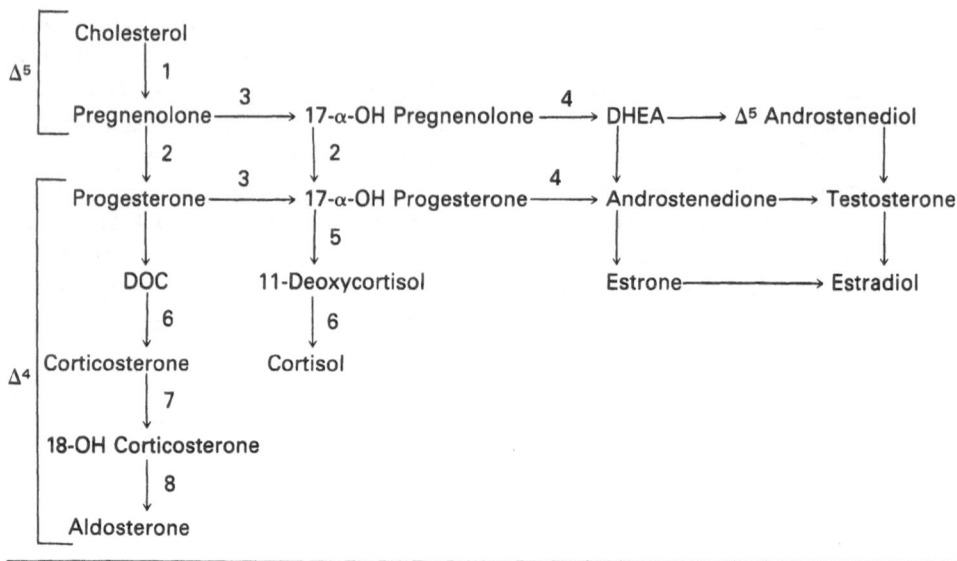

FIGURE 21. Adrenal Steroidogenesis. The enzymes involved are: 1. cholesterol side-chain cleaving enzyme, 2. 3-β-hydroxysteroid dehydrogenase, 3. 17-α-hydroxylase, 4. 17,20-lyase, 5. 21-hydroxylase, 6. 11-hydroxylase, 7. 18-hydroxylase, 8. hydroxydehydrogenase.

ase) or from progesterone (by action of 17-hydroxylase). 17-α-OHP, derived from either source, is 21-hydroxylated to form 11-deoxycortisol or compound S. This steroid, the immediate precursor of cortisol, undergoes hydroxylation at C-11 by 11-β-hydroxylase to form cortisol, or compound F.

Mineralocorticoid Pathway

The mineralocorticoid pathway takes place primarily in the zona glomerulosa. However, certain mineralocorticoid precursors are also synthesized in the zona fasciculata. Aldosterone and its immediate precursor 18-hydroxycorticosterone are synthesized only by the zona glomerulosa, under the regulatory influence of the renin–angiotensin–aldosterone system. The zona fasciculata, on the other hand, produces deoxycorticosterone (DOC), corticosterone, and 18-hydroxydeoxycorticosterone, under ACTH control. These precursors serve as substrates for the zona glomerulosa to synthesize 18-hydroxycorticosterone and aldosterone. The synthesis of mineralocorticoids is carried out by the 17-deoxy pathway, which operates in both the zona glomerulosa as well as in the zona fasciculata. The 17-deoxy pathway originates by the conversion of pregnenolone (a Δ^5 steroid) to progesterone (a Δ^4 steroid). This reaction is mediated by the enzyme 3-β-hydroxysteroid dehydrogenase (3-β-HSD). Conversion of progesterone to DOC requires 21 hydroxylase, while conversion of DOC to corticosterone (compound B) requires 11-hydroxylation. The emerging notion currently is that the 21- and 11-hydroxylases that govern the 17-deoxy pathway are different from the 21- and 11-hydroxylases that control the 17-hydroxy pathway. The end products of the 17-deoxy pathway in the zona fasciculata are DOC, corticosterone, and 18-OH DOC. In the zona glomerulosa, the pathway continues with conversion of the corticosterone to 18-OH corticosterone and eventually aldosterone.

Androgen Pathway

The sex steroid formation by the adrenal cortex is predominantly carried out by the fasciculata–reticularis zones. Dehydroepiandrosterone (DHEA), a Δ^5 steroid, is derived from 17-OH pregnenolone by the action of the enzyme 17,20-lyase (also known as 17,20-desmolase). DHEA, which has little androgenic activity, is converted to a Δ^4 compound, androstenedione, by the enzyme 3-β-HSD. Androstenedione, a moderately potent androgen, is converted to testosterone by a reduction reaction. Alternatively, testosterone can be derived from Δ^5 androstenediol, by the action of the enzyme 3-β-HSD. Androstenedione and testosterone are aromatized, respectively, to estrone and estradiol.

The Enzymes

The enzymes involved in the synthesis of adrenal steroids play a crucial role in steroidogenesis. Some of these enzymes are mitochondrial, while others are microsomal. The cholesterol side-chain–cleaving enzyme, C-11-hydroxylase, and C-18-hydroxylase are mitochondrial. C-21-hydroxylase, C-17-hydroxylase, and C-17,20-lyase and aromatases are examples of microsomal enzymes. Several important aspects of these enzymes bear emphasis.

1. The cholesterol side-chain–cleaving enzyme is critical for all steps of steroidogenesis. Complete lack of this enzyme is not compatible with synthesis of any of the adrenal hormones. The traditional concept that all pathways leading to all steroid hormones are initiated by this single enzyme is intriguing.
2. The Δ^5 steroids—pregnenolone, 17-OH pregnenolone, DHEA, and androstenediol—are biologically inactive and require the enzyme 3-β-HSD to convert these steroids to Δ^4 compounds, progesterone, 17-OH progesterone, androstenedione, and testosterone, respectively.
3. 17-Hydroxylase, the enzyme that effects hydroxylation at the C-17 position, is predominantly found in the zona fasciculata. Negligible 17-hydroxylase activity is found in the zona glomerulosa. In addition to 17-hydroxylating pregnenolone and progesterone to 17-OH pregnenolone and 17-OH progesterone, respectively, these enzymes may also possess 17, 20-lyase activity. This activity is responsible for shunting 17-OH progesterone and 17-OH pregnenolone into the androgen pathway. The possibility that different isoenzymes of 17-hydroxylase exist in different zones has not been excluded.
4. 21-Hydroxylation, the important step in conversion of progesterone to DOC and 17-OH progesterone to deoxycortisol, is mediated by the enzyme 21-hydroxylase. A great deal of evidence has been accumulated to indicate that more than one hydroxylating system may exist in the adrenal cortex. Thus, separate 21-hydroxylases may be involved in the 21-hydroxylation of progesterone to DOC and 17-OH progesterone to 11-

deoxycortisol. The presence of two distinct forms of 21-OHD (the simple virilizing and salt-losing forms), as well as studies that report adequate zona glomerulosa 21-hydroxylase activity in the simple virilizing form, supports the existence of two separate 21-hydroxylases.[1,2]

5. Finally, 11-hydroxylation may also be mediated by distinct isoenzymes in the zona fasciculata and zona glomerulosa. In addition, the 11-hydroxylase in the zona glomerulosa may also be intimately involved with 18-hydroxylation of corticosterone at the zona glomerulosa. The dependency of both these hydroxylations on P-450 enzymes may be one reason for such linkage.

The adrenogenital syndromes evolve due to inborn errors in cortisol synthesis. Depending on the site of the block, there is accumulation of precursors proximal ("upstream") to the block. The clinical consequences and hormonal features are due to dual effects—effects from the accumulated precursors proximal to the block and effects from the lack of synthesis of compounds distal to the block.

Normal Sexual Differentiation

The understanding of the consequences of enzymatic defects in adrenal steroidogenesis requires a basic overview of the process of sexual differentiation.[3]

Sexual differentiation is a coordinated process involving the sequential interaction of several hormonal and nonhormonal factors. Sexual differentiation can be perceived from three points of view; in order of appearance, these are the karyotypic sex, the gonadal sex, and the phenotypic sex. The term *karyotypic sex* refers to the chromosomal constitution conferred at the time of fertilization of the ovum by the sperm. The term *gonadal sex* refers to the development of a normal testis or ovary from the same primitive bipotential gonad. The term *phenotypic* sex refers to the development of male or female external genitalia with their appropriate accessory organs. Each of these processes is reviewed individually in terms of the factors involved in proper differentiation.

Karyotypic (Genetic) Sex

The karyotypic sex of the embryo is established at the time of fertilization. A heterogametic complement (XY) confers a male karyotype to the embryo, whereas a homogametic complement (XX) confers a female karyotype. The factors that determine the derivation of XY or an XX constitution are not clear.

Gonadal Sex

The primitive gonads, in embryos of both sexes, are bipotential. Around the 40th day, the bipotential gonad starts differentiating into a testis or an ovary, a process that is not hormone dependent. The crucial determinant of gonadal differentiation is the presence or absence of a Y chromosome. In the presence of a Y, the primitive bipotential gonad develops into a fetal testis, while in its absence it develops into a fetal ovary, provided a second X chromosome is present. Recent studies indicate that differentiation of the primitive bipotential gonad into a testis does not require the entire Y but requires an important portion of the Y called the H–Y antigen. This is a cell-surface antigen that probably binds to surface receptors on the primitive gonad to induce differentiation into a testis.

The development of the primitive gonad into the fetal testis is achieved by growth and proliferation of the cortical portion destined to become the testis and atrophy of the medullary portion, which in the absence of Y is destined to become an ovary. The exact mechanism that directs atrophy of the ovarian elements of the primitive gonads is unknown. By the 10th week of intrauterine life, the primitive bipotential gonad, in the male, develops into a normal fetal testes complete with the necessary secretory apparatus.

The development of the primitive gonad into the fetal ovary is achieved by growth and proliferation of the medullary portion of the gonad, destined to become the ovary. For proper differentiation, two factors are needed—the lack of Y and the presence of a normal second X. Thus, in an embryo of 46 XX, the primitive gonad progressively differentiates into a fetal ovary by the end of the 10th week.

Phenotypic Sex

The differentiation of the internal accessory organs and the external genitalia into their respective sex organs is referred to as the phenotypic sex. Fetuses of both sexes possess both wolffian (male) and mullerian (female) ducts. Therefore, for proper differentiation, if the wolffian ducts develop, then the mullerian ducts should involute, and vice versa. In the male fetus, the wolffian duct develops into epididymis, vas deferens, seminal vesicles, and ejaculatory duct, and the mullerian duct derivatives atrophy. These dual phenomena are secondary to two substances secreted by the fetal testes, testosterone and a peptide called mullerian involutional factor (MIF). The wolffian ducts, which have receptors for testosterone, differentiate under this hormonal influence into the male accessory organs, whereas the mullerian ducts, under the influence of MIF, atrophy.

In the female, the mullerian duct develops into the uterus, fallopian tube, and upper vagina, the only determinant required being the obligatory lack of MIF. The wolffian ducts will atrophy because of the lack of testosterone.

Subsequent to the differentiation of the wolffian and mullerian duct systems, the next important structure to undergo differentiation is the urogenital sinus. This common structure is the same in both sexes. In the male, the urogenital sinus will develop into the prostate and the male urethra, whereas in the female it becomes the lower vagina. The deciding factor here is the availability of dihydrotestosterone (DHT), a potent androgen derived from reduction of testosterone by the enzyme 5-α-reductase in target tissues. When this hormone is available, and the receptor action of the androgen is expressed, the urogenital sinus develops along male lines. The urogenital sinus of the female fetus, in

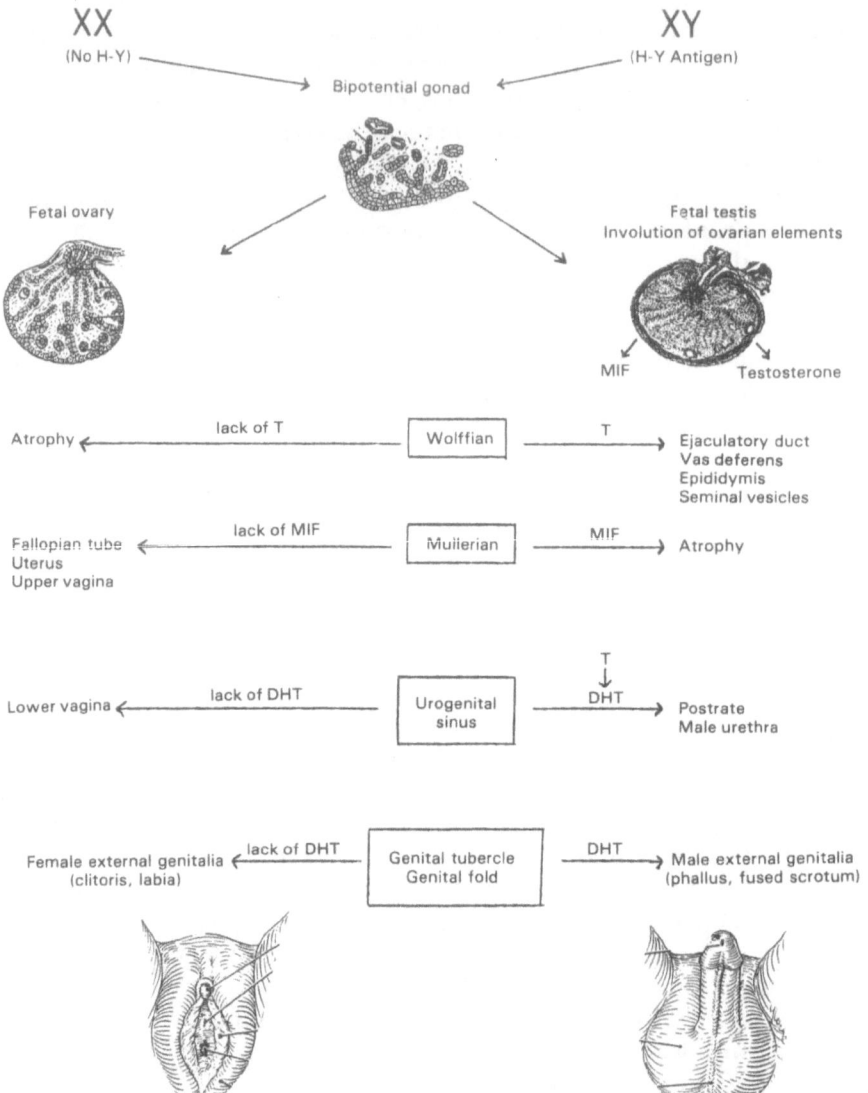

FIGURE 22. Normal sexual differentiation.

whom no testosterone is available for conversion into DHT, does not undergo such virilization.

Following differentiation of the urogenital sinus, the final sequential step is differentiation of genital tubercle and genital folds. These structures, in the presence of DHT, differentiate into the penis and scrotum, whereas in the absence of DHT, or when receptors for DHT are lacking, these structures fail to virilize and therefore develop along female lines. Figure 22 and Table 41 summarize the embryological events of sexual differentiation.

In genotypic females with congenital adrenal hyperplasia the marked in-

TABLE 41.
Normal Sexual Differentiation

Developing organ	Determining factor	Male embryo	Female embryo
Gonad bipotential up to 40 days	H-Y antigen	Testes	Ovary (requires XX)
Mullerian duct anlage	Mullerian duct involutional factor	Involutes	Uterus, fallopian tube, upper vagina
Wolffian duct anlage	Testosterone	Epidydimis, vas, seminal vesicles, ejaculatory duct	Regresses
Urogenital	DHT (not testosterone) Androgen receptors	Prostate and male urethra	Lower vagina
Genital tubercle	DHT Androgen receptors	Penis	Clitoris
Genital swelling	DHT Androgen receptors	Scrotum (fused)	Labia major

creases in androstenedione and DHEA result in virilization of the genital tubercle and the genital folds as well as the urogenital sinus. In extreme cases, complete fusion of the genital folds may result in the formation of a "scrotum," with a penile urethra and clitoromegaly resembling those of a male infant. In most cases, where the androgen excess is mild, the genitalia are ambiguous with varying degrees of virilization. The salt-losing form of 21-hydroxylase deficiency is associated with the highest prenatal androgen levels, and therefore with the most severe degrees of masculinization. In extreme cases the vagina can be posteriorly located and may open into the urethra near the neck of bladder. Such patients require extensive procedures for proper reconstruction of a vaginal introitus.

21-Hydroxylase Deficiency

21-Hydroxylase deficiency is the most common variety of adrenogenital syndrome encountered in clinical practice. Since its original description in 1865 by the neopolitan anatomist DeCrecchio,[4] 21-OHD has evolved from a rare mysterious disorder to a commonly encountered and extremely well-understood disorder. Sensitive immunoassays permit accurate diagnosis of this condition. Furthermore, these studies have facilitated recognition of heterozygotes with 21-OHD, a step in the right direction for genetic counseling. Even more astounding is the fact that 21-OHD can be diagnosed in utero. Measurement of precursor steroids in the amniotic fluid, as well as genetic studies of the cultured amniotic cells, provides answers as to the presence of 21-OHD in the fetus. Such information allows the patient, obstetrician, and endocrinologist to plan therapeutic maneuvers for the outcome of pregnancy. Thus, 21-OHD has become a classic example of an inborn error of metabolism where genetic counseling can decrease the future incidence of the "classic" (homozygous) form of the disease.

Incidence

The frequency of 21-OHD can be viewed from two perspectives—the incidence of classical 21-OHD in population surveys of newborns, and the incidence of nonclassical 21-OHD as the etiological condition in patients presenting with various clinical expressions in childhood and adulthood.

Several studies throughout the world have surveyed the incidence of 21-OHD in the newborn. The information in the literature started accumulating in an exponential fashion following the availability of a valid and reliable screening test for 21-OHD. Such a test was developed in 1977 in the laboratories of Pang et al.[5] and involved a microfilter paper method for 17-α-OH progesterone radioimmunoassay using a heel-prick capillary blood specimen. The incidence of classical 21-OHD in Yupik-speaking Eskimos of western Alaska appears to be the highest in the world, approaching 1/280 to 1/684.[6–8] The incidence of the classical 21-OHD has been estimated to be 1/15,000 in the United States,[9] 1/13,000 in Canada,[10] 1/15,472 in Switzerland,[11] 1/23,044 in metropolitan France,[12] and 1/7255 in Birmingham, England. With the exception of the Eskimos, there are no indications that classical 21-OHD occurs with greater frequency in any particular ethnic group. Since the late-onset nonclassical forms of 21-OHD cannot be screened with the microfilter paper technique, other methods have been used to define the incidence of this disorder. It has been estimated in one study[13] that 32% of patients referred for evaluation of premature adrenarche were found to have nonclassical 21-OHD. In another study[14] 14% of women referred for diagnostic evaluation of hirsutism had nonclassical 21-OHD. Although these studies provide a perspective in terms of understanding the magnitude of the incidence of nonclassical 21-OHD, they do not provide data on the incidence of the condition in the asymptomatic general population. This has been attempted by evaluating the "gene frequency" for nonclassical 21-OHD in various ethnic groups.

Speiser et al.[15] pointed out the high frequency with which nonclassical 21-OHD occurs in various subpopulations. Using genetic linkage markers for nonclassical 21-OHD in conjunction with ACTH testing, it has been noted that the gene frequency for nonclassical 21-OHD is highest in Ashkenazi Jews, Hispanics, Yugoslavians, and Italians. This high gene frequency, evident in obligate heterozygote parents, contributes to the high incidence of both classical and nonclassical forms in the ethnic groups mentioned.

The gene frequency, the HLA linkage, and the modes of inheritance of 21-OHD will be separately discussed under genetics of the disease. It should be pointed out that both the classic and the nonclassic forms of 21-OHD are inherited as autosomal recessive traits. For an infant to develop the classic form of the disease, both parents must be heterozygous carries of the severe deficiency gene for congenital adrenal hyperplasia. Patients with nonclassical symptomatic 21-OHD have both a severe 21-OHD (classical) gene and a mild deficiency (nonclassical) gene.

Types of 21-OHD

For epidemiological and genetic purposes 21-OHD has been classified into classical and nonclassical forms. The classical variety represents a severe expression and manifests at birth. The nonclassical variant is known by a diversity of terms such as adult-onset type, late-onset congenital adrenal hyperplasia, mild or attenuated adrenogenital syndrome, and acquired adrenal hyperplasia; in this form, the biochemical characteristics are less pronounced and the clinical manifestations appear during or after puberty.

Another classification of 21-OHD is based on the presence or absence of mineralocorticoid deficiency: thus, the conventional and often used classification of 21-OHD into "simple virilizing" and "salt-losing" forms. The simple virilizing form of adrenogenital syndrome is characterized by androgen excess only, without mineralocorticoid deficiency. In contrast, the salt-losing form is characterized by androgen excess and mineralocorticoid deficiency. It should be noted that "pure" salt-losing forms of 21-OHD do not occur. The salt-losing variety should be viewed as a global block in 21-hydroxylation involving both the zona fasciculata and zona glomerulosa. Salt-wasting phenomena occur in two-thirds of patients with classical 21-OHD.[16] The differences between the evolution of the simple virilizing form and the salt-wasting form have been based on whether one accepts the "one-enzyme theory" or the "two-enzyme theory." The one-enzyme theory proposes that the differences between the two forms simply lie in the degree of 21-OHD, i.e., the more severe the deficiency, the greater the likelihood of developing salt loss from mineralocorticoid deficiency. The two-enzyme theory postulates the existence of two different 21-hydroxylating enzymes, one that controls the 17-hydroxy pathway (that forms glucocorticoids), and another that controls the 17-deoxy pathway (that forms mineralocorticoids). According to the two-enzyme theory, the simple virilizers are deficient only in the 21-hydroxylase enzyme involved in the 17-hydroxy pathway, while the salt wasters have deficiency of both the enzymes, resulting in defective steroidogenesis involving both the 17-hydroxy and the 17-deoxy pathways. The latter theory is supported by two observations. First, enzymological data suggest that the kinetics of 21-hydroxylase differ for the 17-deoxy and 17-hydroxy pathways.[17] Second, New et al.[18] have shown normal (or even increased) aldosterone production in simple virilizers, while impaired aldosterone production is the rule in salt wasters. The absence of a defect in aldosterone biosynthesis in the zona glomerulosa of simple virilizers is also supported by the data of Kuhnle et al.[2] These workers showed that when the zona fasciculata was suppressed with dexamethasone, and the glomerulosa was stimulated by low sodium intake, normal subjects and simple virilizers increased their serum and urine aldosterone concentrations, while salt wasters did not. Thus, patients with the simple virilizing form are able to generate aldosterone secretion in response to renin stimulation, quite independent of the precursor steroids of the zona fasciculata.

A third theory, proposed by New et al.,[18] postulates the existence of an additional defect in the salt-losing form of 21-OHD. Accordingly, in both simple virilizers and salt wasters there exists a fasciculata defect involving hydroxylation in both the 17-hydroxy and the 17-deoxy pathways. In addition, salt wasters may have an enzyme defect at or distal to 21-hydroxylation in the glomerulosa, while patients with the simple virilizing form are spared this second defect. The intactness of the mineralocorticoid pathway in the simple virilizing form is supported by several lines of hormonal evidence. Kater and Biglieri[19] evaluated the mineralocorticoid profile of 12 patients with simple (non–salt-losing) form of 21-OHD and consistently found elevations in the serum levels of aldosterone and 18-

hydroxycorticosterone with normal levels of 18-hydroxy DOC. Further, Biglieri et al.[1] studied the effects of ACTH administration to patients with simple virilizing 21-OHD and demonstrated perfectly normal responses in aldosterone and 18-hydroxycorticosterone, while the zona fasciculata compounds showed clearly blunted responses. Thus, it has been proposed that the salt-losing form of 21-OHD represents more than merely a reflection of inhibition of 17-hydroxy and 17-deoxy pathways in the zona fasciculata. When this defect is compounded by another block in the enzymatic machinery distal to 21-hydroxylation in the zona glomerulosa, the salt-losing variety develops. Whether such a block in the zona glomerulosa involves the enzyme corticosterone methyloxidase is not clear. The corticosterone methyloxidase type II defect precludes aldosterone formation, resulting in profound salt wasting. Stoner et al.[20] have proposed that corticosterone methyloxidase type II defect represents the prototype of patients with isolated involvement of the zona glomerulosa while the zona fasciculata is intact. The simple virilizing form can be viewed as a reverse of this phenomenon, with isolated involvement of the zona fasciculata, while the zona glomerulosa is intact.

Hormonal Milieu (Fig. 23)

The hormonal characteristics of classical 21-OHD are fourfold:

1. Defective synthesis of cortisol and 11-deoxycortisol (S)—products distal to the block.
2. Increased accumulation of 17-hydroxyprogesterone, its urinary metabolite pregnanetriol, and to a lesser extent progesterone.
3. Increased secretion of adrenal androgens, DHEA, and androstenedione, which are eventually converted to testosterone.
4. Synthesis of aldosterone and 18-hydroxycorticosterone are normal, or even elevated in the simple virilizing form, but lowered in the salt-wasting form.

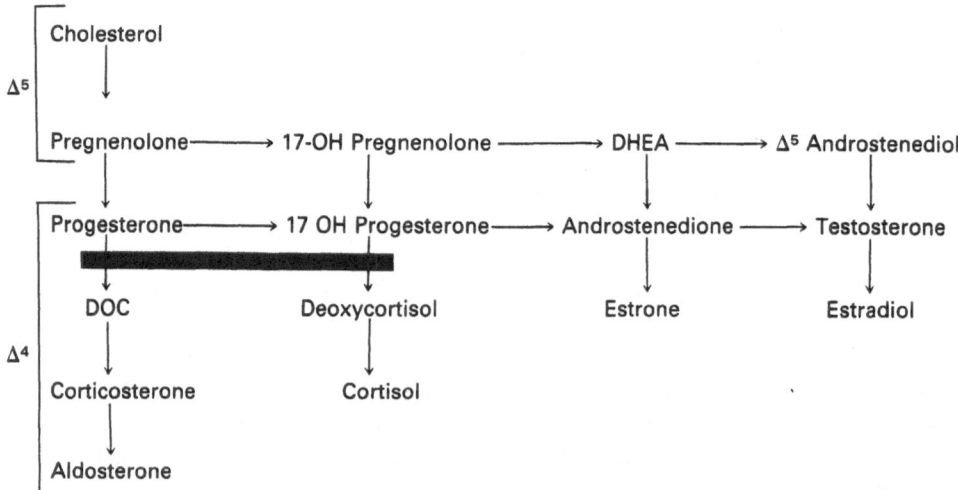

FIGURE 23.   21-Hydroxylase deficiency.

The three cardinal effects that arise as a consequence of these phenomena are a tendency for absolute or relative hypocortisolemia, hyperandrogenism, and a tendency to salt loss when aldosterone synthesis is impaired. Thus, in its complete form, the salt-losing 21-OHD resembles Addison's disease (primary adrenal failure) with the exceptional feature of excessive adrenal androgens in the circulation. It has been pointed out that the mineralocorticoid defect of the salt-losing variety of congenital adrenal hyperplasia may ameliorate with age.[20,21]

The hormonal milieu of nonclassical 21-OHD is highly variable. Thus, the androgen levels may be normal or mildly elevated. The basal serum 17-hydroxyprogesterone, the hormonal marker of complete 21-OHD, can also be minimally or mildly elevated. The characteristic hormonal abnormality in patients with nonclassical 21-OHD is the "explosive" response of 17-hydroxyprogesterone to ACTH administration, with a concomitant attenuation in the cortisol response to ACTH. The spectrum of hormonal abnormalities in the nonclassical form of 21-OHD is so varied that the diagnosis cannot be established by measurement of androgens, cortisol, or precursor steroids in the basal state, but requires challenge with ACTH administration.

Clinical Features

The clinical expressions of 21-OHD can manifest at birth, during puberty, or in adulthood (Table 42).

Neonatal Manifestations

Neonatal manifestations usually occur with only the classical 21-OHD, where the enzyme deficiency is complete. The effects of adrenal androgen excess on the genotypically female fetus are impressive. In extreme cases the external genitalia undergo complete masculinization, resulting in phallic growth, penile urethra, and even complete fusion of the labial folds ("scrotalization") resembling a male infant.[22] In less severe cases, the genitalia are ambiguous. Considerable heterogeneity exists in the virilization of the urogenital sinus, labial fusion, clitoromegaly, and scrotalization of the labia majora. The virilization of the genital folds and the genital tubercle are a result of the markedly elevated adrenal androgens levels, particularly androstenedione. The internal organs, however, do not show signs of masculinization. The gonads, in these patients, are normal fetal ovaries, and the mullerian duct derivatives are unaffected. The reason for the preservation of mullerian duct anlagen is because there are no fetal testes to secrete the mullerian involutional substance that causes regression of these structures in the normal testes-bearing male fetus. Interestingly, the wolffian duct derivatives do not undergo virilization. Possibly, the potency of the "weaker" adrenal androgens is inadequate—in comparison to testosterone—to effect virilization of the wolffian duct derivatives. The preservation of intact ovaries and intact mullerian duct derivatives (uterus, fallopian tube, etc.) is the reason for the capacity of these patients to subsequently become fertile. It is therefore crucial not to miss adrenogenital syndrome in the female neonate.

Males with the simple virilizing form of 21-OHD may not show any abnor-

TABLE 42.
Manifestations of 21-Hydroxylase Deficiency

I.	At birth (neonatal)
	Females
	Ambiguous genitalia
	Gross masculinization (female pseudohermaphroditism)
	Males
	Macrosomia genitalia precox
	In both sexes
	Salt wasting
	Hyponatremia, hyperkalemia, Elevated PRA
	Dehydration
	Inappropriate natriuresis
	Adrenal crisis during stress
II.	Prepubertal
	Female
	Virilization
	Premature adrenarche
	Males
	Isosexual pseudo–precocious puberty
	Both sexes
	Eventual short stature
III.	Postpubertal
	Females
	Hirsutism
	Virilization
	Oligomenorrhea, amenorrhea
	Infertility
	Males
	Oligospermia
	Infertility
	Both sexes
	Infertility
IV.	"Cryptic" 21-hydroxylase deficiency
	Both sexes
	Abnormal hormonal profile without obvious clinical manifestations

malities at birth. Occasionally, attention may be drawn to the slightly oversized external genitalia (macrosomia genitalia precox). These infants tend to show progressive virilization as they grow older.

In the salt-losing form, both males and females suffer the consequences of aldosterone deficiency. In the neonatal period this effect is manifested in a variety of ways—e.g., failure to thrive, electrolyte abnormalities, vomiting, dehydration. The condition is especially likely to be missed in the male infant who bears no stigmata of androgen excess at birth. In such instances the mistaken diagnosis of pyloric stenosis has often been erroneously made, with disastrous consequences. Equally catastrophic is the development of acute adrenal failure in infants of both sexes with classic 21-OHD, usually precipitated by stress. These patients are unable to generate glucocorticoids during stress and therefore succumb to adrenal crisis unless glucocorticoids are rapidly administered. Thus, classical 21-OHD is an important consideration not only in infants with ambiguous genitalia, but also in infants with electrolyte abnormalities, unexplained dehydration, and shock.

Prepubertal Manifestations

Prepubertal manifestations of 21-OHD are usually a consequence of the milder form of the disease (nonclassical) in girls or can be due to the classical form of the disease in boys. In girls, progressive prepubertal virilization and premature adrenarche represent the two important manifestations. In boys, precocious puberty is the most striking expression of classic 21-OHD. Characteristically, the striking phallic growth and adrenarche are contrasted by the small, prepubertal dimensions of the testes. The nonenlargement of the testes, in proportion to the virilizing changes, is due to suppression of pituitary gonadotropins ("pseudo" isosexual precocious puberty) caused by the increased adrenal androgens. The conventional concept that prolonged secretion of increased quantities of adrenal androgens results in suppression of pituitary gonadotropins has to be reevaluated in light of recent data. Boyar et al.[23] have reported that some children with untreated congenital adrenal hyperplasia with markedly elevated androgen levels showed "LH programming" similar to that of children with "true" precocious puberty, a condition where central mechanisms are believed to be activated prematurely. In this study[23] two boys with adrenal androgen excess demonstrated episodic secretion of luteinizing hormone (LH) and follicle-stimulating hormone (FSH), with augmentation of LH with sleep ("sleep synchrony"), a phenomenon consistent with early maturation of the pubertal "LH program." One of the two boys also had normal adult-sized testes. Thus, the role of excess of adrenal androgens in boys extends to more than simple and passive virilization of the external genitalia. These androgens may also possess an additional effect in premature initiation of central nervous system signals characteristic of central precocious puberty.

In addition to prepubertal virilization, a significant adverse effect of prolonged excess of adrenal androgens is the effect on growth. Premature bone maturation with an advanced bone age is conducive to epiphyseal fusion resulting in "stunting" of growth. Although the growth velocity may be accelerated in early childhood, eventually the net height attained is shorter than projected. This phenomenon is an important reason for early therapeutic intervention.

Postpubertal Manifestations

Postpubertal manifestations are usually seen in females with the nonclassical variants of 21-OHD. These patients are normal at birth without any ambiguity in the development of external genitalia. Premature adrenarche may be evident in some cases. The characteristic symptoms experienced include hirsutism (or occasionally virilization), oligomenorrhea, amenorrhea, and infertility. The late manifestation of a milder disease is explained on the basis that the nonclassical form of 21-OHD deficiency represents an allelic variant of the classical form of disease.[24-26] It has been estimated, by various investigators, that the nonclassical variant of 21-OHD accounts for 1.2–30% of patients referred for evaluation of hirsutism.[27-29] More recently, Kuttenn et al.[30] studied 400 women with hirsutism and found nonclassical 21-OHD in 24. The degree of hirsutism varies considerably in patients with nonclassical 21-OHD. One factor that could account for this wide variability is the role of skin sensitivity. The demonstration that 5-α-reductase activity is lowered in many patients with nonclassical 21-OHD[30,31] supports this notion. Reductions in 5-α-reductase are generally associated with varying degrees of resistance to androgen action. In males with the nonclassical 21-OHD, the clinical effects of androgen excess are barely noticeable. Oligospermia and subfertility have been reported in association with the nonclassical 21-OHD in males.[32] Presumably this occurs as a consequence of suppression of gonadotropins by chronic androgen excess.

The term *cryptic 21-OHD* has been used to denote patients with the hormonal abnormalities of the nonclassical form of 21-OHD without any symptoms. Since the hormonal abnormalities of both the symptomatic and asymptomatic patients with nonclassical 21-OHD are identical, it is not clear why patients with cryptic 21-OHD fail to develop clinical expressions of the disorder.

From the foregoing, it should be apparent that considerable variability exists in the clinical expression of 21-OHD. Even more interesting is the observation that the clinical severity of the disorder can wax and wane in the same patient over the years. The reasons for such fluctuations are far from clear. When the disorder occurs in several members of the same family, in general, the clinical and hormonal expressions tend to be identical. Thus, the simple virilizing, salt-wasting, and nonclassical types "breed true" within the family. However, exceptions to such concordance have been reported in the literature.[9,20]

Diagnostic Studies (Table 43)

As indicated earlier, the specific diagnosis of the type of enzymatic defect involved depends on measuring the precursor steroids that accumulate "upstream" of the enzyme block. The classical 21-OHD in the newborn is characterized by the following features:

1. Elevated 17-hydroxyprogesterone and progesterone in plasma. These are reflected, less reliably, by elevation in their urinary metabolites, pregnanetriol and pregnenediol, respectively.
2. Elevated serum levels of DHEA sulfate and androstenedione, reflected in the urine as elevated 17-ketosteroids.
3. Low (or low normal) cortisol levels in the plasma.
4. The levels of the mineralocorticoid hormones differ in the simple virilizing and salt-losing forms of 21-OHD. In the simple virilizing form, serum levels of aldosterone and 18-hydroxycorticosterone are usually normal or may even be elevated. In the salt-wasting form, these levels are decreased in the plasma.

The diagnostic studies in patients with nonclassical 21-OHD deficiency include measurement of basal 17-α-hydroxyprogesterone, DHEA, and androstenedione. However, the diagnostic maneuver to diagnose nonclassical 21-OHD involves evaluation of the 17-α-hydroxyprogesterone response to ACTH administration. Patients with this disorder demonstrate a markedly exaggerated 17-α-

TABLE 43.
21-Hydroxylase Deficiency

Diagnosis
    Accumulated products proximal to block
      ↑ 17-α-OH progesterone
      ↑ 17-α-OH pregnenolone
    Decrease in products distal to block
      ↓ 11-Deoxycortisol
      ↓ Cortisol
      ↓ 11-Deoxycorticosterone
      ↓ Corticosterone
      ↓ 18-OH corticosterone (in salt-losing form)
      ↓ Aldosterone (in salt-losing form)
    Increased ACTH drive
      ↑ Androgens, (androstenedione/DHEAS)
    Response to ACTH
      ↑ ↑ Response of 17-α-OHP to ACTH
      Attenuated response of cortisol to ACTH

---

hydroxyprogesterone response to ACTH. The cortisol response, on the contrary, is usually attenuated.

The direct measurement of adrenal androgens and precursor steroids in plasma has resulted in placing less reliance on measurements of metabolites in the 24-hr urine. However, measurements of 17-ketosteroids and pregnanetriol continue to be performed in several laboratories. The 17-ketosteroids in the 24-hr urine measure the unconjugated and conjugated neutral C-19 steroids with a 17-ketone ring (Zimmermann reaction). 17-Ketosteroid measurements reflect only the metabolites of adrenal androgens. Since normal infants, and to a lesser extent prepubertal children, excrete nonspecific pigments in their urine, it is necessary to read the Zimmermann color reaction at three different wavelengths to carry out corrections. Similarly, reliance on urinary pregnanetriol in newborns should take into account the fact that newborns have subnormal glucuronyltransferase activity, a fact that influences measurement of glucuronide conjugates of pregnanetriol. The elevations in plasma levels of 17-α-hydroxyprogesterone are far more striking than the elevations in 24-hr urinary pregnanetriol. Therefore, measurement of plasma 17-α-hydroxyprogesterone has become the hormonal marker for 21-OHD, especially in the classical form.

Differential Diagnosis (Table 44)

21-Hydroxylase deficiency has to be differentiated from several disorders characterized by androgen excess. In the female neonate, virilization caused by the classical simple virilizing form of 21-OHD needs to be differentiated from other forms of adrenogenital syndrome. The most important disorder that mimics 21-OHD is the congenital adrenal hyperplasia caused by 11-OHD. Both conditions are characterized by masculinzation of the female fetus. The elevation in circulating adrenal androgens and the increase in 17-α-hydroxyprogesterone are common to both entities. The distinction between the two conditions

TABLE 44.
Differential Diagnosis of 21-Hydroxylase Deficiency

Condition	Differentiating feature(s)
I. At birth	
1. 11-Hydroxylase deficiency	11-Deoxycortisol levels elevated
2. Nonadrenal female pseudohermaphroditism	Presence of virilization tumor in mother; history of drug intake (androgenic steroids) during
II. Prepubertal	
1. Adrenal tumors	Nonsuppression to DXM; CT study
2. Leydig cell tumors	Palpable testicular mass; testosterone markedly elevated
3. Ovarian tumors	Abnormal CT of ovaries; testosterone elevated
4. Ectopic HCG secretion	Usual in females; underlying malignancy obvious
III. Adulthood.	
1. Polycystic ovary syndrome	17-α-OH progesterone response to ACTH normal
2. Masculinizing ovarian tumor	Increased testosterone; abnormal CT, ultrasound
3. Adrenal tumors, usually carcinoma	Elevated adrenal androgens nonsuppressible to DXM; CT of adrenal

can be made by measurement of 11-deoxycortisol levels in the plasma. Elevated 11-deoxycortisol levels are seen in 11-OHD, while being low in 21-OHD.

Neonatal virilization of female infants can also be encountered when the fetus is exposed, through transplacental passage, to increased circulating maternal androgens. Two rare circumstances exemplify such an occurrence: the development of virilizing arrhenoblastomas of the ovary during pregnancy, and ingestion of androgenic steroids by the mother during pregnancy, especially during the critical months of sexual differentiation in the female fetus. History and physical examination of the mother as well as serial hormonal evaluation of the newborn assist in differentiating these two disorders from the classic form of 21-OHD. The hormonal profile of such infants is characterized by excessive testosterone, DHEA, or androstenedione without elevation in 17-hydroxyprogesterone. The elevated androgen titers, derived from maternal transmission, dissipate with passage of time, usually weeks. Prepubertal virilization that results from adrenal 21-OHD closely resembles androgen secretion from tumors of the adrenal or the gonads. Prepubertal virilization in both sexes can result from benign or malignant tumors of the adrenal cortex. These tumors can be differentiated from 21-OHD by hormonal tests, response to dexamethasone suppression, and appropriate imaging procedures. Adrenal carcinoma, in particular, can superficially resemble the hormonal features of adrenogenital syndrome. Since adrenal carcinomas are inefficient hormone synthesizers, several enzyme deletions may be present in these tumor cells. Therefore, the plasma of patients with adrenal carcinoma may demonstrate increased amounts of circulating precursor steroids, particularly 11-deoxycortisol, in addition to adrenal androgens, bringing the similarity to adrenogenital syndrome even closer. However, the 17-α-hydroxyprogesterone levels are usually not significantly elevated, and the androgen levels fail to suppress with dexamethasone suppression. Further, tumors of the adrenal can be readily recognized by computerized tomography of the adrenals.

In boys with isosexual precocious puberty, in addition to tumors of the adrenal, tumors of the Leydig cells constitute an important differential diagnosis. Leydig cell tumors are palpable and primarily secrete testosterone. It should be remembered that tumors of the testes can occur in boys with congenital adrenal hyperplasia. Such tumors originate from the adrenal rest cells located in the testes. The hormone features of adrenal rest cell tumors in the background of 21-OHD include strikingly elevated adrenal androgens in the circulation, coupled with adequate suppression to dexamethasone administration.

In girls, premature adrenarche and early changes of isosexual precocious puberty may result from 21-OHD. During this phase of presentation the clinical picture resembles isosexual precocious puberty resulting from other causes, such as idiopathic central precocious puberty, and ovarian tumors, such as granulosa cell tumors. However, when such children are followed, more virilization and less estrogenic effects develop due to the untreated 21-OHD.

In adult females with the nonclassical 21-OHD who present with hirsutism, oligomenorrhea or amenorrhea, and infertility, the differential diagnosis revolves around three entities—polycystic ovarian syndrome, benign or malignant androgen-secreting adrenal tumors, and androgen-secreting ovarian tumors. Since patients with tumors in the adrenals or ovaries can be readily identified by ultrasonography and/or computerized tomography, the main differential diagnosis revolves around distinction between late-onset 21-OHD and polycystic ovarian syndrome. The similarity is brought even closer due to the fact that a variable percentage of patients with 21-OHD may demonstrate "cysts" in the ovaries by ultrasound. The clinical, basal hormonal, and ultrasonographic studies therefore do not permit satisfactory distinction between the two entities. The diagnosis, however, can be established by demonstrating the characteristic rise in 17-hydroxyprogesterone levels in response to ACTH in patients with 21-OHD.

Genetics of 21-OHD

21-Hydroxylase deficiency is transmitted as an autosomal recessive disorder. Several surveys have convincingly established this nature of transmission.[22,33] Both males and females are at equal risk for development of the disorder. Genetic mapping studies performed by Carroll et al.[34] and White et al.[35] have established that the human genome includes two 21-hydroxylase genes that alternate with two genes for the fourth component of complement (C-4a and C-4b). These genes are located on the short arm of chromosome 6, between the loci of HLA-B and HLA-DR. It is believed that only one of the 21-hydroxylase genes is expressed, and the other one represents a pseudogene. The close genetic linkage between HLA and 21-OHD was first demonstrated by Dupont et al.[36] These workers performed HLA genotyping of the parents and children in six families with one or more children affected by 21-OHD. The crucial discovery that all affected children in a given family were HLA identical and different from their unaffected sibship had enormous impact in explaining the genetic transmission of 21-OHD. Thus, for the disorder to be transmitted to the child, it is essential that both parents be obligate heterozygote carriers, transmitting one HLA haplotype carrying the gene to the affected child. The similarity in HLA type shared by all affected members in a given family has been impressively outlined by several workers. Levine et al.[37] performed genetic mapping studies in 48 patients from 34 unrelated families and evaluated the HLA genotypes in parents, affected siblings, and the unaffected siblings. Their studies revealed that all affected siblings in a given pedigree were genotypically HLA identical. Thus, the siblings who inherited one haplotype linked to the 21-OH gene from *each* obligatory heterozygous parent developed the classic 21-OHD. The siblings who inherited only one haplotype linked to the 21-OH gene were presumed heterozygotes. The siblings who shared neither haplotype with the affected sib were presumed not to carry the gene for 21-OH at all. These studies have established that the HLA genotype is a marker for the congenital adrenal hyperplasia genotype.

In addition to the linkage between HLA genotype and 21-OHD, there is genetic linkage disequilibrium between 21-OHD and specific HLA alleles. It has been proposed that the most striking association for classical 21-OHD is that of HLA-Bw47, DR7. The HLA antigen Bw47 has been found to occur with remarkable frequency in certain surveys, approaching as high as 35%.[38] While HLA genotyping has enhanced the understanding of genetic transmission of 21-OHD, it must be realized that it is possible to have classical 21-OHD in the absence of the HLA markers linked with the disorder.

The inheritance pattern of the nonclassical form of 21-OHD has also been well studied. It is now believed that nonclassical 21-OHD is also genetically linked to HLA, and that the classical and nonclassical (late onset) types of the disease are allelic variants of the same disorder.[24,26,39,40] The existence of the classical and nonclassical forms of the disease in the same family is well known. The late-onset form of the disease can be inherited in one of several ways. First, it may result from inheriting two recessive genes, a severe 21-OHD gene ("21-OH-severe") and a mild nonclassical gene ("21-OH-mild"). These patients can be regarded as "genetic compounds." Alternatively, the nonclassical form can result from inheriting two mild 21-OHD genes, one from each heterozygous parent. These patients are considered homozygous for the 21-OH mild gene. Genetic analyses have established a close linkage between HLA genotype and the gene for the mild 21-OHD gene.[41] Thus, it appears that the nonclassical form of 21-OHD resembles its classical, congenital counterpart in its genetic linkage to a major histocompatibility complex. It is believed that different major histocompatibility antigens are associated with the classical and late-onset forms of the disease.[42] Kuttenn et al.[30] performed parallel hormonal and genetic studies in 24 patients with the late-onset form of 21-OHD and emphasized the close linkage of this disorder with the HLA locus. A high correlation between the disorder and the prevalence of HLA-B14 and Aw33 was noted. More important, an impressive similarity was noted in the biological profiles of patients and their HLA identical siblings. These data that link both forms of 21-OHD to major histocompatibility loci assume importance in genetic counseling. HLA markers also permit recognition of siblings who may be at high risk for developing the disorder. The development of gene-typing techniques also permits the detection of congenital adrenal hyperplasia in the fetus. This is done by HLA typing of the amniotic fluid cells, for closely linked complement loci (C4A, C4B, Bf, and C2).

Finally, genetic mapping studies permit identification of asymptomatic carriers, without a family history of congenital adrenal hyperplasia.[43]

11-β-Hydroxylase Deficiency

11-β-Hydroxylase deficiency is the second most frequently encountered form of congenital adrenal hyperplasia. This disorder differs from 21-OHD in its relative rarity and in the potential to cause mineralocorticoid excess. Unlike 21-OHD, adult onset of this disorder is uncommon.

Biochemical and Hormonal Milieu (Fig. 24)

The events that set the stage for the hormonal milieu of 11-β-OHD are as follows:

1. The occurrence of 11-β-hydroxylase deletions results in inadequate synthesis of cortisol from 11-deoxycortisol in the zona fasciculata. This results in stimulation of ACTH secretion by the pituitary gland.
2. As a consequence of ACTH drive, there is accumulation of precursor steroids "upstream" of the enzyme block. This leads to increased synthesis of 11-deoxycortisol, 17-hydroxyprogesterone, and 17-hydroxypregnenolone. The precursor steroids, under the influence of ACTH, are channeled into synthesis of increased amounts of adrenal androgens, particularly DHEA.
3. When the 11-β-OHD is severe, conversion of 11-deoxycorticosterone (DOC) to corticosterone is impaired, resulting in accumulation of DOC, which in large amounts leads to hypertension, hypokalemia, and alkalosis. This results in the characteristic combination of suppression of PRA due to DOC excess while aldosterone synthesis is impaired due to the lack of availability of corticosterone.

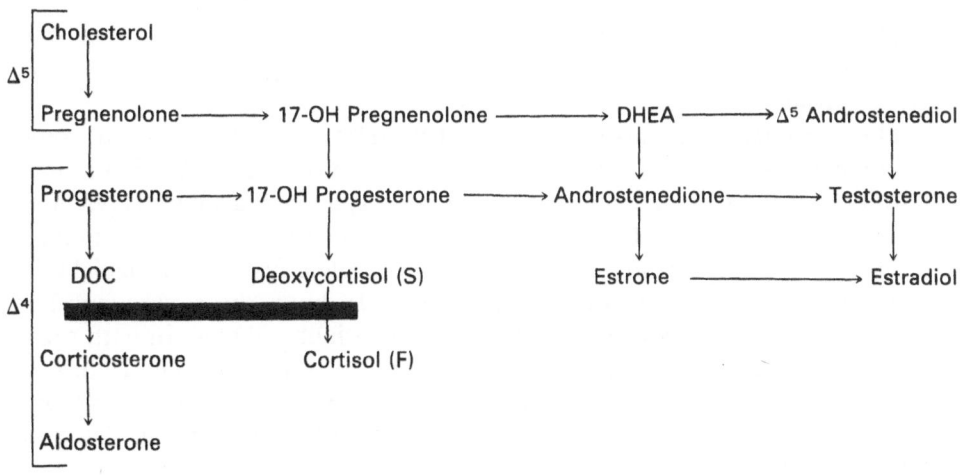

FIGURE 24. 11-β-Hydroxylase deficiency.

The deficient secretion of aldosterone in patients with 11-β-OHD has also been a matter of controversy. Simplistically, the decreased synthesis of aldosterone can be viewed as a result of decreased substrates (corticosterone and 18-hydroxycorticosterone) as a consequence of the 11-hydroxylation defect. However, emerging data emphasize the role of cytochrome P-450 in causing loss of both 11-hydroxylation as well as 18-hydroxylation of DOC. Ulick[44] has proposed that the same enzyme can catalyze both 11-hydroxylation as well as 18-hydroxylation of DOC. This concept—that the same enzyme mediates the hydroxylation of DOC at both the 11 and the 18 positions—might explain the inability to adequately synthesize aldosterone in presence of 11-hydroxylase deficiency. However, considerable controversy exists surrounding the issue of the presence of two different isoenzymes of 11-hydroxylase. In vitro studies using bovine adrenocortical tissue have suggested that the enzyme system that 11-β-hydroxylates DOC to corticosterone is different from the enzyme that hydroxylates 11-deoxycortisol to cortisol.[45] Experiments that have utilized heating the mitochondria have shown that the rates of hydroxylation of 11-deoxycortisol, DOC, and androstenedione are affected to different degrees by the procedure. Furthermore, competition experiments have indicated that the three substrates do not compete for the same site of 11-hydroxylation.[46] These studies contrast with reports by others[47,48] who have reported against the existence of different 11-hydroxylases within the various zones of the adrenal cortex. Amid this controversy, there have been reports of patients who have isolated deficiency in 11-hydroxylation of DOC but have adequate hydroxylation of deoxycortisol, and vice versa.[49,50] These reports have raised questions regarding the traditional concept that deficiency of a single 11-β-hydroxylase results in dual blockades in both systems. For practical purposes, it is convenient to visualize the hypertensive form of 11-β-OHD developing when the block is severe and complete. To this extent, the normotensive and hypertensive forms of 11-β-OHD may be viewed as partial and complete expressions, respectively, of the disorder.

Clinical Features

The clinical triad of 11-β-OHD consists of androgen excess, cortisol deficiency, particularly during stress, and mineralocorticoid (DOC) excess resulting in hypertension.

The androgen excess associated with 11-β-OHD results in virtually the same consequences as the androgen excess associated with 21-OHD. Thus, masculinization of the external genitalia in the genotypically female fetus results in female pseudohermaphroditism. The degree and severity of the virilization are variable and can range from mild ambiguity to complete fusion of the labial folds with even the development of a male urethra. In boys, the consequences of adrenal androgen excess result in isosexual precocious puberty. In both sexes, hypertension and hypokalemic alkolosis may evolve at any age. The laboratory features of 11-β-OHD are characterized by low levels of cortisol, corticosterone, 18-hydroxycorticosterone, 18-hydroxydeoxycorticosterone, and aldosterone, along with elevations in 11-deoxycorticosol, 17-hydroxyprogesterone, and DHEA.

17-α-Hydroxylase Deficiency[51,52]

17-α-Hydroxylase deficiency is a unique example of congenital adrenal hyperplasia in that it is not associated with virilization. The lack of 17-α-hydroxylase, when complete, precludes the formation of androgens, as well as glucocorticoids. In addition, the enzyme deficiency operates at the level of the adrenal glands as well as the gonads. When the deficiency in 17-hydroxylation is complete, the only pathway open for steroidogenesis is the mineralocorticoid pathway, which results in increased production of mineralocorticoids, particularly DOC.

Hormonal Milieu (Fig. 25)

The following sequence of events sets the stage for the effects that result from complete 17-OHD.

1. The presence of a defect in 17-hydroxylation of pregnenolone and progesterone results in limitation of synthesis of 17-OH progesterone, deoxycortisol, and cortisol.
2. The partial or complete inability to synthesize cortisol triggers release of ACTH.
3. The stimulatory effects of ACTH on the adrenal cortex result in increased conversion of cholesterol into pregnenolone and progesterone. As a consequence there is an accentuation of the mineralocorticoid synthetic pathway, with one notable feature—the mineralocorticoids synthesized in excess are DOC and corticosterone, but not aldosterone. Patients with 17-α-OHD paradoxically tend to have low levels of aldosterone, despite excessive DOC and corticosterone. Given the fact that the mineralocorticoid pathway is the only pathway open, one would expect these patients to show increased synthesis of all three steroids with mineralocorticoid activity (DOC, corticosterone, and aldosterone) to be synthe-

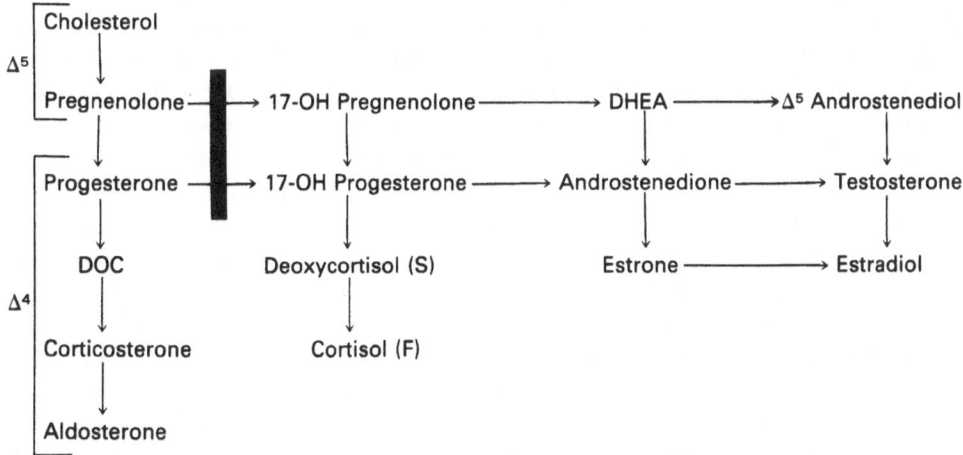

FIGURE 25. 17-α-Hydroxylase deficiency.

sized in excess. This, however, does not occur. It has been proposed that two separate pools for DOC secretion exist in the adrenal cortex—one controlled by ACTH (in the zona fasciculata–reticularis) and another controlled by angiotensin II in the zona glomerulosa. In patients with 17-α-OHD, the ACTH-dependent pool, operative in the fasciculata–reticularis, is markedly increased. This pool does not contribute much to aldosterone synthesis. The increased DOC and corticosterone formed in the fasciculata pool suppress the renin–angiotensin system, resulting in a decrease in the aldosterone by the zona glomerulosa. This might explain the combination of increased DOC and corticosterone, a reflection of ACTH drive, with a concomitant suppression of aldosterone, a reflection of the suppressed renin–angiotensin system due to DOC excess. Alternatively, patients with 17-α-OHD may have a second defect in 18-hydroxylation of corticosterone, precluding adequate aldosterone synthesis. This hypothesis, however, has not been conclusively proven. In fact, Kater and Biglieri[19] have shown that plasma level of 18-hydroxycorticosterone and 18-hydroxy DOC are elevated in patients with 17-α-OHD with a concurrent reduction in aldosterone.

4. The 17-α-OHD also results in an impairment in the synthesis of C-19 steroids, leading to decreased or absent synthesis of androgens and estrogens. Since 17-α-OHD occurs in both the adrenals and the gonads, these patients are partially or completely unable to secrete androgens or estrogens during fetal as well as adult life.

Clinical Features

The clinical features of 17-α-OHD are reflected in phenomena that involve all three pathways of steroidogenesis. The inability to synthesize androgens by the genotypically male fetus results in lack of virilization of the external genitalia of the male fetus. When the defect is complete, the male fetus is born with completely feminized external genitalia (male pseudohermaphroditism). The testes are undescended, the wolffian duct derivatives fail to develop, the urethra is female, and the external genitalia are those of a female infant. When the defect is only partial, the external genitalia are undervirilized, resulting in varying degrees of sexual ambiguity. In female infants with 17-α-OHD, the condition is difficult to diagnose at birth, owing to the presence of normal, nonvirilized, unambiguously female external genitalia. In both sexes, pubertal changes fail to occur, since neither androgens nor estrogens can be formed due to the 17-hydroxylase defect that operates at the level of the gonads and the adrenals. Lack of axillary hair is a striking feature in both sexes, coupled with varying degrees of sexual infantilism. The primary nature of the gonadal failure is evidenced by the combination of decreased sex steroids and elevated gonadotropins. The low level of sex steroids fails to respond to administration of human chorionic gonadotropin (HCG) or ACTH. The effects of glucocorticoid deficiency are seldom manifested overtly, since corticosterone in large amounts provides fairly adequate glucocorticoid support. However, during periods of stress adrenal decompensation may occur, resulting in the development of adrenocortical insufficiency. Excessive secretion of DOC and corticosterone leads to hypertension, hypokalemia, and alkalosis resembling primary hyperaldosteronism. However, the aldosterone levels are low, for the postulated reasons mentioned earlier.

Laboratory Features

The hormonal profile of 17-α-OHD is classic: low levels of DHEA, androstenedione, 17-β-estradiol, estrone, 17-α-hydroxyprogesterone, with high levels of DOC, corticosterone, and progesterone. The cortisol levels are usually subnormal, with a predictably attenuated response to ACTH administration.

3-β-Hydroxysteroid Dehydrogenase Deficiency (Fig. 26)

3-β-Hydroxysteroid deficiency is rare.[53-55] A complete defect in 3-β-hydroxysteroid dehydrogenase (3-β-HSD) deficiency results in failure to convert Δ^5 steroids to their respective Δ^4 steroids. The Δ^5 steroids are pregnenolone, 17-α-OH pregnenolone, DHEA, and androstenediol. These four steroids are converted by the enzyme 3-β-HSD into their respective Δ^4 compounds, which are progesterone, 17-α-OH progesterone, androstenedione, and testosterone. In the genotypically male fetus, complete 3-β-HSD deficiency precludes the formation of androstenedione and testosterone, resulting in a failure to virilize the external genitalia and producing male pseudohermaphroditism.[53,54] In the genotypically female fetus, complete deficiency of 3-β-HSD results in increased accumulation of DHEA and androstenediol, resulting in varying degrees of virilization of the external genitalia and causing female pseudohermaphroditism. Thus, complete 3-β-HSD deficiency is unique in that it can result in the development of both male and female pseudohermaphroditism. As boys with 3-β-HSD deficiency grow, they fail to show virilizing pubertal changes owing to inability to secrete testosterone and androstenedione.[56-58] It is assumed that

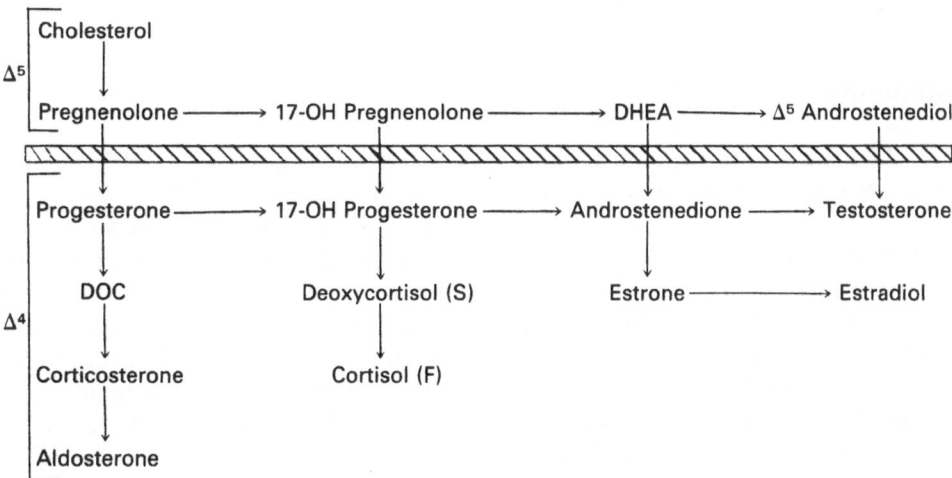

FIGURE 26.   3-β-Hydroxysteroid deficiency.

3-β-HSD deficiency, in a manner similar to 17-α-OHD, occurs in both the adrenals as well as in the gonads.

The hormonal profile of 17-α-HSD deficiency is characterized by increased accumulation of the steroids proximal to the block with decreased concentrations of compounds distal to the block. Thus, the major abnormality involves an elevation in the levels of 17-hydroxypregnenolone, DHEA, and androstenediol, with reduction in the levels of 17-hydroxyprogesterone, androstenedione, and testosterone. The effect of age on 3-β-HSD deficiency is intriguing. Kenny et al.[59] reported increasing levels of sex steroids with age and proposed that gonadal production of sex steroids improved with age. Bongiovanni et al.[60] reviewed the steroid data in nine patients with 3-β-HSD deficiency and noted increased levels of urinary pregnanetriol, a metabolite of 17-α-OH progesterone, (a Δ^4 compound). However, the predominant steroids in these patients were of Δ^5 configuration, suggesting that some conversion of compounds can occur in extraadrenal sites, particularly by hepatic 3-β-HSD activity. More recent work[55–57] has supported the hypothesis for the existence of extraadrenal 3-β-HSD activity in such patients, accounting for the presence of Δ^4 steroids in the circulation. The concept that the gonadal deficiency of 3-β-HSD coexists with the adrenal deficiency of the enzyme has also been supported by studies that evaluate ratios of Δ^5 to Δ^4 compounds in response to HCG administration.[55–57]

The difficulties in establishing a secure diagnosis of 3-β-HSD by measurement of Δ^5 and Δ^4 compounds have been highlighted by a recent report.[61] Cara et al.[61] described the case of a newborn genotypic male infant with grade IV (perineal) hypospadias due to complete 3-β-HSD deficiency. Serum concentrations of Δ^4 steroids (17-OH progesterone, androstenedionne, and testosterone) were found to be markedly elevated—a paradox in light of the complete 3-β-HSD deficiency, which theoretically should preclude the conversion of Δ^5 compounds to Δ^4 compounds. The authors proposed that in this case extraadrenal 3-β-HSD activity may have occurred in utero, but unfortunately came too late to produce normal virilization of the male fetus. The authors also cautioned that an elevated 17-hydroxyprogesterone level does not totally exclude the possibility of 3-β-HSD deficiency in the newborn or in the "adult."

Summary

The features of the four major enzyme deletions—21-OH, 11-OH, 17-OH, and 3-β-HSD—have been outlined in the preceding sections. A brief summary of the clinical and laboratory features of these four enzyme deficiencies is presented in Table 45. All four forms of adrenogenital syndrome can result in sexual ambiguity in the newborn. The importance of recognizing adrenogenital syndrome at birth cannot be overemphasized due to the profound psychosexual consequences caused by this disorder. In addition, the salt-losing form of 21-OHD can result in fatalities, if unrecognized. Screening for adrenogenital syndrome by measurement of plasma levels of 17-α-OH progesterone is an excellent method for 21- and 11-OH deficiencies, but not for 17-OH and 3-β-HSD deficiencies. The serum cortisol response to exogenous ACTH is impaired in all forms of adrenogenital syndrome. DHEA and androstenedione are elevated in

TABLE 45.
Enzymatic Blocks in Steroidogenesis—A Clinical Perspective

Clinical feature	21-OH deficiency	11-OH deficiency	17-OH deficiency	3-β-HSD deficiency
Neonatal virilization of the female fetus (female pseudohermaphroditism)	+	+	−	+
Suboptimal virilization or feminization of the male fetus (male pseudohermaphroditism)	−	−	+	+
Prepubertal virilization	+	+	−	−
Sexual infantilism at puberty	−	−	+	+
Salt loss (mineralocorticoid deficiency)	+	−	−	+
Hypertension	−	+	+	−
Growth retardation	+	+	−	−

21- and 11-OH deficiencies, while being low in 17-α-OH deficiency. When the defect is complete, aldosterone levels are low in all four varieties of adrenogenital syndrome (Table 46). While the general trends in precursor levels can be guessed with reasonably good predictability, these levels depend on the degree of block, degree of ACTH overdrive, and other factors such as extraadrenal conversion of precursors by the gonads and the liver.

Complications

The complications of untreated or undertreated congenital adrenal hyperplasia relate to androgen excess, sex steroid deficiency, and chronic stimulation by ACTH.

Androgen Excess

The effects of androgen excess in females with untreated adrenogenital syndrome are devastating; masculinization, failure to menstruate or ovulate,

TABLE 46.
Enzymatic Blocks in Steroidogenesis—Hormonal Profiles

Hormonal feature	21-OH	11-OH	17-OH	3-β-HSD
Cortisol	Subnormal	Subnormal	Subnormal	Subnormal
11-deoxycortisol	↓	↑	↓	↓
DHEA	↑	↑	↓	↑
Androstenedione	↑	↑	↓	↓
Testosterone	Mild ↑	Mild ↑	↓	↓
17-OH progesterone	↑ ↑	↑	↓	↓
Progesterone	↑	Mild ↑	↑	↓
DOC	↓	↑	Mild ↑	↓
Corticosterone	↓	↓	Mild ↑	↓
Aldosterone	↓	↓	↓	↓

infertility, and growth retardation result in tremendous psychosexual incapacitation. In males with untreated androgen excess, the result is premature epiphyseal closure resulting in short stature. Of course, depending on the type of underlying defect, salt loss and the tendency for adrenocortical insufficiency during periods of stress can develop in both sexes.

Sex Steroid Deficiency

The effects of sex steroid deficiency are most impressive in patients with complete 17-OH and 3-β-HSD deficiencies. Since the enzyme deficiencies occur in both the adrenals as well as the gonads, this results in the development of sexual infantilism and hypogonadism in both sexes.

Chronic ACTH Excess

Finally, the results of chronic ACTH stimulation of both adrenals can result in the formation of adrenal tumors. The development of adrenal tumors in patients with congenital adrenal hyperplasia has been reported in the literature.[62–64] More recently Van Seters et al.[65] presented data concerning a 60-year-old woman with untreated congenital adrenal hyperplasia caused by 21-OHD, who had developed a tumor of the left adrenal. The preoperative ACTH levels were high, with inadequate suppression to dexamethasone. After surgery, cortisol secretion decreased impressively, but ACTH dysregulation became even more prominent. The authors postulated that long-standing cortisol deficiency can result in the development of ACTH dysregulation, resulting in autonomous ACTH production and the eventual formation of adrenal tumor(s). The condition closely resembles Cushing's syndrome that results from bilateral macronodular hyperplasia. This phenomenon emphasizes the lifelong need for substitution glucocorticoid therapy in patients of both sexes with congenital adrenal hyperplasia.

In addition to stimulating adrenal tissue, chronic ACTH excess can result in stimulation of extraadrenal rest cells, particularly in the gonads. This is most impressively illustrated in boys with precocious puberty due to 21-OHD, who develop "tumors" in the testes.[66–69] These palpable tumors represent nodular hyperplasia of the adrenal (fetal) rest cells. Occasionally, these testicular tumors can synthesize cortisol.[70]

Almost all the complications of congenital adrenal hyperplasia can be prevented by the early and proper administration of glucocorticoids.

Prognosis and Natural History

The prognosis of congenital adrenal hyperplasia is generally good, if treatment is instituted early in the course of the disease. The prognosis of the various forms of adrenogenital syndrome should naturally focus on several important issues, such as

1. The ability of female infants with the virilizing forms of adrenogenital syndrome to lead a "hormonally normal" female life. This includes the potential for attaining pubertal feminization, menstruation, and fertility.

2. The effect of androgen excess on stature.
3. The effect of androgen excess on gender identification and in the assumption of a female role with comfort.
4. The ability to mount a cortisol response to stress.

Each of these facets deserves brief discussion.

Fertility

Genotypic females with the classical virilizing forms of adrenogenital syndrome (21-OHD and 11-OHD) have anatomically normal mullerian duct derivatives. Thus, theoretically if androgen production is kept to a minimum by glucocorticoid therapy, the occurrence of puberty, menarche, ovulation, and fertility should be no different than in normal females. The fertility rates in 80 female patients with congenital adrenal hyperplasia due to 21-OHD were evaluated by Mulaikal et al.[71] Their study highlights four important points. First, patients with the salt-losing form had a very low fertility rate compared to those with the simple virilizing form. The fertility rate was 60% for patients with the simple virilizing form, a rate much lower than in the general population. Second, the status of introitus seemed to have a more important role in sexual activity than the degree of prenatal exposure to androgens. Third, noncompliance with therapy played an important role, since approximately 25% of patients in the study were hirsute, with poor endocrine follow-up. Finally, the abnormalities in the introitus and the inability to conceive were more severe in patients with the salt-losing form, probably due to the severity of the enzyme block in this form of 21-OHD.

Several isolated reports in the literature have described the occurrence of successful pregnancies in patients with 21-OHD.[72-76] The study of Mulaikal et al.[71] is the first large study to document a lower fertility rate in patients with 21-OHD when compared to the general population. Besides, this study also clearly documents the extremely low potential for fertility in patients with the salt-losing form. Three factors influence the potential for fertility in patients with 21-OHD—type of defect (salt losing versus simple virilizing), compliance with therapy, and the status of the introitus. Since patients with the salt-losing form have the highest degree of masculinization,[77] the introitus is most severely malformed in this group, requiring major reconstruction, especially when the vagina is posterior and opens into the urethra, sometimes near the bladder neck. The surgical failure rate following reconstruction surgery in patients with the salt-losing form of 21-OHD approaches 54%.[78] Thus, limitation of coitus assumes an important role in the occurrence of pregnancy as well its outcome in patients with the salt-losing form of 21-OHD.

The fertility rate of patients with 11-OHD is probably the same as in patients with the simple virilizing form of 21-OHD. The potential for spontaneous menarche, ovulation, and fertility is negligible, almost zero, for patients with complete 17-OHD. The problem in adrenal steroidogenesis in these patients is compounded by a concurrent defect in gonadal steroidogenesis. Thus, such patients fail to develop pubertal feminization, menarche, or ovulation. Pregnancy is an unlikely event in these patients, who often require estrogen therapy to induce changes of feminization.

Adult Stature

When therapy is instituted early, it can be hoped that normal adult stature (and menarche) can be attained in females with congenital adrenal hyperplasia due to 21-OHD.[72,76,79,80] In a large series of 80 patients with 21-OHD the effect of treatment on adult height was evaluated by Mulaikal et al.[71] The mean height of the patients with the simple virilization approximated the 10th percentile for normal women, and the mean height of patients with the salt-losing form approximated the 25th percentile. Thus, prognostically, salt losers do poorly in comparison to the simple virilizers with 21-OHD.

Psychosexual Orientation

Adrenogenital syndromes caused by 21-OHD or 11-OHD represent a classical experiment of nature where all tissues of the genotypic female—including the brain hypothalamus—are exposed to exceedingly high androgen levels. The hypothalamus, which has receptors for both androgens and estrogens, can be viewed as a target organ for sex steroids, in a manner analogous to the developing external genitalia. Therefore, the issues posed by prenatal exposure to androgens in a female revolve around "masculinization" of the hypothalamus in addition to masculinization of the genitalia. Histological evidence in rats has clearly suggested that sexual behavior can be influenced by the developing sexually dimorphic nuclei, located in the preoptic area.[81] In finch, the three brain nuclei are larger in the male and are related to one particular sexual behavior in the male finch—its song.[82] Therefore, it is relevant to consider whether prenatal exposure to androgens in a female causes any significant changes in the sexual orientation (gender identity) as well as in sexual preference. Considerable controversy surrounds both issues.

As for gender identity, it is traditionally believed that gender identification is environmentally determined during the first 3–4 years of life. Clearly, an alternate view was proposed by Imperato-McGinley et al.,[83] when they showed reversal of gender role in patients with 5-α-reductase deficiency following pubertal virilization. Thus, awareness of "body image" based on genitalia may also have an important effect in gender identity. Most adult females with adrenogenital syndrome have female gender identity. It has been noted that the female gender role is reinforced in females with congenital adrenal hyperplasia after the appearance of female secondary sexual characteristics and periodic vaginal bleeding.[84] This observation confers importance to the effect of body image on gender identification. Patients with virilizing adrenal hyperplasia are often active and aggressive during childhood.[85] However, with the development of pubertal feminization, these children often have been known to change their gender role to females, perhaps as a consequence of the raising of their consciousness in terms of female gender identity.

The effect of androgen excess on sexual preference has also been a matter of dispute. In one recent study, Money et al.[86] observed a greater frequency of homosexuality and bisexuality in children and adolescents with congenital adrenal hyperplasia. The mating behavior and sexual preference of 80 women with 21-OHD were studied by Mulaikal et al.[71] Fifty percent of patients with the simple virilizing form had been married for over 4 years at the time of the study, and none were divorced. In contrast, only 13% of patients with the salt-losing form were married. In the same study, the authors contrasted the sexual experiences of patients with and without adequate vaginal introitus; in the 52 subjects with an adequate introitus, 75% had heterosexual experiences, 2% had homosexual or bisexual experiences, and 23% had no sexual activity. These data contrasted with the 28 patients with an inadequate introitus, where 25% had heterosexual activity, 11% had homosexual or bisexual activity, and a large 64% had no sexual activity. Thus, the incidence of asexual, homosexual, heterosexual, or bisexual behavior may be related more to anatomical than to hormonal reasons. However, the possibility that prenatal androgenization has an impact on adult psychosexual orientation has not been ruled out.

Adrenocortical Insufficiency

The fourth factor that influences the prognosis of patients with congenital adrenal hyperplasia is their ability to respond to stress. Since the underlying abnormality in all forms of congenital adrenal hyperplasia is suboptimal synthesis of cortisol, these patients do poorly during stress. It is important to recognize this during infections, surgery, anesthesia, etc., and provide adequate glucocorticoid coverage. In addition, some patients with adrenogenital syndrome may have been overtreated with glucocorticoid, adding an element of pituitary ACTH suppression to the already compromised cortisol status. Patients with adrenogenital syndrome must wear identification bracelets to highlight the need for intravenous hydrocortisone during stress.

Treatment

The treatment of adrenogenital syndromes rests on the proper application of endocrinological, surgical, and psychiatric principles. Several goals are formulated in the treatment of patients with virilizing adrenogenital syndromes, the prototype of which is 21-OHD:

1. The chain of events that led to androgen excess needs to be interrupted.
2. In females with virilizing forms of congenital adrenal hyperplasia, every attempt should be made to allow these patients to develop as normal females. Thus, with proper therapy, pubertal feminization, ovulation, and fertility are hopeful, but unguaranteed, goals.
3. Avoidance of electrolyte abnormalities (in the salt-losing forms) and correction of hypertension (in the hypertensive forms) are obvious therapeutic goals.
4. Attempt to avoid short adult stature is an important goal and is especially relevant since short stature can result from both undertreatment as well as overtreatment with glucocorticoids.
5. Every attempt should be made to provide these patients with external genitalia that match their genotypic sex. The role of the urologist and/or plastic surgeon in reconstructing a reasonably normal introitus has tre-

mendous bearing on the subsequent outcome and adjustments in these patients.

6. Intense psychological support is necessary to help these patients face life as normal females. The problems of prenatal masculinization notwithstanding, when properly handled, the majority of females with virilizing adrenogenital syndrome demonstrate excellent adaptability to the female gender.

With these general goals, therapy of adrenogenital syndromes can be viewed in terms of hormonal, surgical, and psychiatric support. Finally, the management of these patients during pregnancy also deserves brief comment.

Hormonal Therapy

The medical treatment for adrenogenital syndromes is replacement glucocorticoid therapy. The cascade of events in adrenogenital syndrome originates from the failure to adequately synthesize cortisol. Therefore, replacement glucocorticoid therapy would interrupt the chain of events that led to increased synthesis of precursors. Thus, in patients with 21-OHD and 11-OHD, cortisol replacement would result in decrease in the ACTH level, followed by a decrease in 17-OH progesterone, as well as decreased synthesis of adrenal androgens. When cortisone acetate became available in 1950, the therapeutic outlook for children with adrenogenital syndrome changed dramatically. 21-OHD deficiency is prototypical of the adrenogenital syndromes that respond to cortisone replacement. Treatment with cortisone acetate for patients with 21-OHD consists of an initial dose, a maintenance dose, and a "stress" dose during critical periods. The initial dose is generally a high dose of cortisone acetate. Wilkins et al., in 1951,[87] recommended initiating therapy with 25 mg/day i.m. for children below 2 years of age, 50 mg/day in children between 2 and 6 years of age, and 100 mg/day in children over 6 years of age. The high-dose regimen is continued until the 17-ketosteroids and pregnanetriol in the urine have declined to a normal range. When this has occurred, the patient may be switched to a maintenance regimen of cortisone or hydrocortisone. The route of administration depends on the age of child. If the child—or infant—has difficulty in ingesting or retaining the medication, obviously the intramuscular route is preferable.

The maintenance dose of cortisol roughly approximates 25 mg/M^2 per 24 hr; since the half-life of orally administered cortisol is short (60 min), it is necessary to administer the daily dose in three divided doses for maximal ACTH suppression. Hydrocortisone given in three divided doses achieves the desired effect i.e., normalization of 17-ketosteroid and pregnanetriol excretion. The availability of an excellent assay for 17-OH progesterone level in plasma has afforded a very sensitive and useful parameter to monitor the effectiveness of the therapy. When cortisone acetate is used, above 10% of the weight of cortisone acetate is contributed by the acetyl group, and for the most part, cortisone acetate has only 80% of the potency of cortisol (hydrocortisone). Hydrocortisone given in divided doses serves as an excellent oral therapy for suppressing androgen production in patients with congenital adrenal hyperplasia caused by 21-

OHD. The use of prednisone, methylprednisolone, or dexamethasone is seldom resorted to. Theoretically, dexamethasone, with its longer half-life, should be expected to exert more suppression on ACTH than hydroxortisone or cortisone. In addition, dexamethasone may be given as a single dose, a fact that is conducive for promoting compliance when lifelong therapy is contemplated. The unpredictability of the dosage of dexamethasone required, the higher incidence of side effects, and the lack of mineralocorticoid effects may render dexamethasone less reliable than hydrocortisone as replacement therapy for adrenogenital syndromes.

Treatment of patients with congenital adrenal hyperplasia during stress involves administration of additional glucocorticoid support to avoid the development of adrenal insufficiency. For minor infections such as sore throat or influenza, doubling the maintenance dose would suffice. When the infection is severe or is associated with symptoms of adrenal insufficiency, hydrocortisone in stress doses (100 mg parenterally t.i.d.) is administered. Preparation for surgery in this group of patients requires doubling the maintenance dose 2 days prior to surgery, while increasing the dosage to 37.5–50 mg/M² per day during the surgery and for 4–5 days thereafter.

Mineralocorticoid replacement, in the form of 9-α-fluorocortisol (fluorinef) or DOC, is recommended in patients with the salt-losing form of the entity. Unlike cortisol, fluorinef does not require adjustment of dosage with body size. The dosage of fluorinef ranges from 0.05 mg to 0.1 mg/day. During periods of stress or decompensation, intravenous hydration with normal saline often coupled with intramuscular administration of DOC acetate may be required to avert hypotension and severe hyponatremia. Additional hormonal therapy with estrogens or androgens is required in patients with 17-OHD. The presence of a concomitant defect in the gonads renders these patients profoundly hypogonadal. The hypertension associated with this form of congenital adrenal hyperplasia responds well to glucocorticoid replacement. However, the inability to synthesize sex steroids by the gonads requires lifelong replacement with gonadal steroids.

Medical treatment for congenital adrenal hyperplasia poses the inherent problems associated with any disorder that requires lifelong therapy. Therapy should be begun as early as possible and continued lifelong. Early and adequate therapy with hydrocortisone offers the best—and perhaps the only—chance for such patients to attain normal adult stature, as well as normal pubertal changes. When therapy is delayed, inadequate, or intermittent, girls with 21-OHD would be subjected to delayed puberty, diminished signs of feminization, short adult stature, and, worst of all, progressive signs of virilization. Thus, patient compliance becomes the cornerstone of successful therapy. In patients with the salt-losing form of 21-OHD, noncompliance can result in catastrophic sequelae such as electrolyte imbalance and dehydration. Every physician who has cared for patients with adrenogenital syndrome has been impressed with the compliance factor. Failure to achieve sustained ACTH suppression results in unsuccessful or only partially successful outcomes. Monitoring of patients on treatment includes hormonal measurements (urinary 17-ketosteroids, pregnanetriol, and/or plasma 17-OH progesterone levels), growth curves, and bone age. Attention should be

focused to detect side effects of glucocorticoid therapy, particularly growth retardation as well as the development of hypertension or glucose intolerance. The
need for lifelong compliance should be reinforced during each visit.

Surgical Treatment

In females with virilizing forms of adrenogenital syndrome, reconstruction
surgery is necessary. The type of surgery obviously depends on the degree of
virilization; particularly relevant are the degree of labial fusion, position of
urethra, degree of clitoromegaly, and position of the vagina. The type of surgery
can range from simple incision of the fused scrotal folds to major procedures
such as vaginal reconstruction. When virilization is severe, especially in female
children with genitalia that resemble those of a male, the reconstruction procedures can be extremely difficult, sometimes yielding poor results. Naturally, it
is advisable to perform corrective surgery as early as possible to avoid problems
with gender identity. In many patients two-stage operations may be necessary,
one done in the early years to "repair" the abnormalities, and one done at or
around puberty to provide a normal vaginal introitus by surgical reconstruction.
The importance of a normal introitus indeed has a major impact on the sexual
activity of these females, as well as on the successful outcome of pregnancy in the
future. Even in rare instances where the diagnosis was missed in early life and
the child was reared as a male, proper medical and surgical therapies can result
in excellent adjustment when handled properly. The role of psychological evaluation and support cannot be overemphasized in all patients with virilizing
adrenal hyperplasia.

Psychological Support

Psychiatric evaluation and psychological support play an extremely important role in the treatment of patients with adrenogenital syndromes. Although
many patients with 21-OHD adapt freely to female gender identity, professional
help is recommended for a healthy psychosexual adjustment. Even in patients
on adequate medical therapy, the very thought that virilization will occur in the
absence of medication can be quite unsettling. In addition, pubertal changes and
psychosexual changes can render adolescence a very trying period. In patients
with inadequate, ill-formed vaginal introituses, the tendency to withdraw altogether from sexual activity is understandable. Many patients are resigned to
the notion that nothing can be done and therefore tend to completely avoid the
issue of sexuality. The concerted efforts of the endocrinologist, psychologist,
and urologist given in a compassionate, caring setting makes a great deal of
difference to such patients. Psychological support is also crucial for patients who
have confusion in gender identity, as well as those who have to make a choice in
their sexual preference.

Pregnancy

As indicated earlier, patients with treated classical 21-OHD have lower rates
of pregnancy in comparison to the general population. Females with adult-onset

or nonclassical 21-OHD often seek attention for anovulation and infertility. The occurrence of pregnancy in patients with classical or nonclassical 21-OHD involves three major considerations in that setting:

1. Effect of pregnancy on the hormonal status of the mother
2. Effect of the treatment—or lack of it—on the fetus
3. Presence of disease in the infant, particularly the presence of a masculinized female fetus with homozygous classical 21-OHD

Effects of Pregnancy on the Mother

Since pregnancy imposes a measure of additional stress, it might be expected to accentuate the hormonal perturbations of 21-OHD. There is a paucity of information in the literature regarding this issue. Thus, very little information exists to support the notion that androgen excess becomes more severe with the onset of pregnancy. However, it is well recognized that some cases of nonclassical 21-OHD have had their onset during or immediately after pregnancy. It is unclear whether the metabolic demands of pregnancy unmask a latent 21-OH defect. Alternatively, it is unknown whether pregnancy is associated with a relative increase in ACTH secretion. For instance, it is well known that the intermediate lobe of the pituitary becomes active during pregnancy and can cause perturbations in ACTH dynamics. It is intriguing to speculate whether such a mechanism is conducive to magnification of the effects of a partial 21-OHD. Regardless, most patients with classical 21-OHD who become pregnant are maintained on replacement doses of hydrocortisone or cortisone acetate. Unless the patient shows renewed signs of androgen excess or marked increases in 17-OH progesterone or androgens, there is no indication to empirically increase the dose of glucocorticoids during pregnancy. During labor, however, these patients require coverage with "stress doses" of steroids, usually 100 mg hydrocortisone thrice a day intravenously.

Effects of Treatment on the Fetus

Theoretically, the developing fetus is at dual risks in the setting of therapy for maternal 21-OHD. On one hand, undertreatment of the mother, with the attendent maternal adrenal androgen excess, can result in transplacental passage of androgens with masculinization of the female fetus. Such a phenomenon is analogous to masculinization of the female fetus in the setting of virilizing tumors in the mother or intake of androgenic steroids by the mother. On the other hand, certain glucocorticoids administered to the mother, particularly dexamethasone, can cross the placenta and result in suppression of the fetal adrenals. Both these issues merit consideration.

Masculinization of the fetus from undertreated adrenogenital syndrome is very rare, mostly because the androgen levels in the mother are seldom high enough to cause fetal virilization. Thus, the androgen milieu of the exposed normal infant considerably differs from the degree of hyperandrogenism in infants with classical 21-OHD. Besides, the placenta may play a role in metabolizing the maternal androgens, "diluting" the effect of maternal androgens on the fetus. Thus, significant masculinization of a female fetus from maternal androgen excess secondary to undertreated 21-OHD is extremely unusual.

The effect of maternal steroid therapy on the fetus, especially when the steroid used is not dexamethasone, is also of minimal significance. Experience with the usage of cortisone, hydrocortisone, prednisone, or methylprednisolone in pregnant women for maternal diseases such as systemic lupus erythematosus and Crohn's disease has been reported by several workers.[88-91] The overall impression is that when used in pharmacological doses these glucocorticoids exert very little adverse effects on fetal development. The cortisol production rate of infants born to mothers treated with nondexamethasone type of glucocorticoids is usually normal.[90] Although animals treated with heavy doses of glucocorticoids during pregnancy may bear fetuses with congenital malformations such as cleft palate or growth retardation, this has not been the case in humans. The doses used in animals were extremely high, accounting for the high incidence of fetal deaths, congenital malformations, and placental degeneration.[92-95] Thus, the general consensus is that in humans, treatment of the mother with physiological doses of glucocorticoids, particularly cortisone, hydrocortisone, or prednisone, carries little or no risk to the fetus, in terms of congenital malformations or in terms of fetal adrenal suppression. The lack of significant adverse effects on the fetus may be related to binding of glucocorticoids to maternal globulins or due to metabolism of steroids by the placenta. In contrast to the use of cortisone, hydrocortisone, or prednisone, use of dexamethasone during pregnancy may result in suppression of fetal adrenals. Dexamethasone can be detected in amniotic fluid, a finding consistent with the concept of significant transplacental passage of the drug. This factor, coupled with its potency and lesser degree of binding to globulins, renders dexamethasone an effective agent in terms of its ability to suppress fetal adrenals. This effect can occur even when the drug is used in replacement doses. In fact, this effect of dexamethasone has been exploited by Evans et al.[96] in a novel manner; these workers attempted prevention of abnormal genital masculinization in a female fetus suspected to have congenital adrenal hyperplasia.[96] This practice, however, has not found widespread application yet.

Prenatal Diagnosis

Since 21-OHD is inherited as an autosomal recessive trait, the fetus can inherit the disorder only when a patient with classical or nonclassical forms of the disease mates with a spouse who is a heterozygous carrier for the 21-OH gene. Since the prevalence of heterozygous carrier state for 21-OHD is a relatively common occurrence, it is conceivable that patients with 21-OHD may have a child with classical 21-OHD, when the child has inherited two alleles for classical 21-OHD, one from each parent. Prenatal diagnosis of 21-OHD in the infant becomes important under several circumstances:

1. When the mother with classical or nonclassical 21-OHD has previously borne a child with classical or nonclassical 21-OHD and is pregnant again by the spouse who fathered the first child.
2. When the spouse of the patient with 21-OHD is known to have asymptomatic 21-OHD, documented by hormonal studies.

3. When the spouse of the patient with 21-OHD demonstrates HLA haplotypes that are known to be linked to the 21-OH gene. Genetic typing of closely linked complement loci, particularly (C4A, C4B, C2, and Bf) can reasonably predict carrier state for 21-OHD.

In these three settings it may be desirable to know whether the fetus, which is clearly at high risk for developing 21-OHD, has in fact inherited the disease. If the fetus is female and is affected by the classical form of the disorder, intra-uterine virilization is clearly a possible development. In such cases, documentation of a prenatally masculinized female fetus with classical 21-OHD may offer the parents an option. If the parents have already had a masculinized female child with the disease, the option to terminate the pregnancy is certainly the prerogative of the parents. Alternatively, the strategy to suppress the fetal adrenals by giving the mother dexamethasone is a newly emerging option. However, the bipotential external genital tubercle undergoes virilization between the 10th and 16th weeks of fetal life.[97,98] Therefore, to be of any value, dexamethasone suppression needs to be initiated at the end of the first trimester, at a time when neither the sex nor the presence of adrenogenital syndrome in the fetus can be determined with the currently available methods.

The available methods that permit the diagnosis of 21-OHD in the fetus are of two types—hormonal measurements (of 17-OH progesterone and adrenal androgens) in the amniotic fluid obtained by amniocentesis, and HLA typing of the cultured amniotic fluid cells.[99–102] Both these studies cannot be performed before 16–17 weeks of gestation, by which time the adverse effects of adrenal androgens in the developing female fetus may have already been exerted. Besides, the use of these methods is limited to specialized centers. Amniocentesis has not been routinely advocated in all patients with congenital adrenal hyperplasia who become pregnant.

References

1. Biglieri EG, Wajchenberg DA, Malerbi HO, et al: The zonal origins of the mineralocorticoid hormones in the 21-hydroxylation deficiency of congenital adrenal hyperplasia. *J Clin Endocrinol Metab* **53**:964, 1981.
2. Kuhnle U, Chow D, Rapaport R, et al: The 21-hydroxylase activity in the glomerulosa and the fasciculata of the adrenal cortex in congenital adrenal hyperplasia. *J Clin Endocrinol Metab* **52**:534, 1981.
3. Wilson JD: Sexual differentiation. *Annu Rev Physiol* **40**:279, 1978.
4. DeCrecchio L: Sopra un caso di apparenze virile in una donna. *Morgagni* **7**:1951, 1865.
5. Pang S, Hotchkiss J, Drash AL, et al: Microfilter paper method for 17α-progesterone radioimmunoassay: Its application for rapid screening for congenital adrenal hyperplasia. *J Clin Endocrinol Metab* **45**:1003, 1977.
6. Pang S, Spence DA, New MI: Newborn screening for congenital adrenal hyperplasia with special reference to screening in Alaska. *Ann NY Acad Sci* **458**:90, 1985.
7. Pang S, Murphey W, Levine LS, et al: A pilot newborn screening for congenital adrenal hyperplasia in Alaska. *J Clin Endocrinol Metab* **55**:413, 1982.
8. Hirschfeld AJ, Fleshman JK: An unusually high incidence of salt-losing congenital adrenal hyperplasia in the Alaskan Eskimo. *J Pediatr* **75**:492, 1969.
9. Rosenbloom AL, Smith DW: Varying expression for salt losing in related patients with congenital adrenal hyperplasia. *Pediatrics* **38**:215, 1966.
10. Qazi QH, Thompson MW: Incidence of salt-wasting form of congenital adrenal virilizing hyperplasia. *Arch Dis Child* **47**:302, 1972.

11. Werder EA, Siebenmann RE, Knorr-Murset G, et al: The incidence of congenital adrenal hyperplasia in Switzerland: A survey of patients born in 1960 to 1974. *Helv Paediatr Acta* **35:**5, 1980.

12. Bois E, Mornet E, Chompret A, et al: L'hyperplasie congénitale des surrénales (21-OH) en France: Génétique des populations. *Arch Fr Pediatr* **42:**175, 1985.

13. Temeck JW, Pang S, New MI: Premature adrenarche resulting from symptomatic nonclassical congenital adrenal hyperplasia. *Pediatr Res* **18:**178A/495, 1984 (Abstract).

14. Pang S, Lerner AJ, Stoner E, et al: Late-onset adrenal steroid 3βHSD deficiency. A cause of hirsutism in pubertal and postpubertal women. *J Clin Endocrinol Metab* **60:**428, 1985.

15. Speiser PW, Dupont B, Rubinstein P, et al: High frequency of nonclassical steroid 21-hydroxylase deficiency. *Am J Hum Genet* **37:**650, 1985.

16. Fife D, Rappaport EB: Prevalence of salt-losing among congenital adrenal hyperplasia patients. *Clin Endocrinol (Oxf)* **19:**259, 1983.

17. Nelson EA, Bryan GT: Steroid hydroxylations by human adrenal cortex microsomes. *J Clin Endocrinol Metab* **41:**7, 1975.

18. New MI, Dupont B, Pang S, et al: An update of congenital adrenal hyperplasia. *Recent Prog Horm Res* **37:**105, 1981.

19. Kater CE, Biglieri EG: Distinctive plasma aldosterone, 18-hydroxycorticosterone, and 18-hydroxydeoxycorticosterone profile in the 21-, 17α-, and 11β-hydroxylase deficiency types of congenital adrenal hyperplasia. *Am J Med* **75:**43, 1983.

20. Stoner E, DiMartino J, Kuhnle U, et al: Is salt-wasting in congenital adrenal hyperplasia genetic? *Clin Endocrinol* **24:**9, 1986.

21. Luetscher JA: Studies of aldosterone in relation to water and electrolyte balance in man. *Recent Prog Horm Res* **12:**175, 1956.

22. Wilkins L: Adrenal disorders. II. Congenital virilizing adrenal hyperplasia. *Arch Dis Child* **37:**231, 1962.

23. Boyar RM, Finkelstein JW, David R, et al: Twenty-four hour patterns of plasma luteinizing hormone and follicle-stimulating hormone in sexual precocity. *N Engl J Med* **289:**282, 1973.

24. Kohn B, Levine LS, Pollack MS, et al: Late-onset steroid 21-hydroxylase deficiency: A variant of classical congenital adrenal hyperplasia. *J Clin Endocrinol Metab* **55:**817, 1982.

25. Migeon CJ, Rosenwaks Z, Lee PA, et al: The attenuated form of congenital adrenal hyperplasia as an allelic form of 21-hydroxylase deficiency. *J Clin Endocrinol Metab* **51:**647, 1980.

26. Pollack MS, Levine LS, O'Neill GJ, et al: HLA linkage and B14, DR1, Bfs haplotype association with the genes for late onset and cryptic 21-hydroxylase deficiency. *Am J Hum Genet* **33:**540, 1981.

27. Lobo RA, Goebelsmann U: Adult manifestation of congenital adrenal hyperplasia due to incomplete 21-hydroxylase deficiency mimicking polycystic ovarian disease. *Am J Obstet Gynecol* **138:**720, 1980.

28. Chrousos GP, Loriaux DL, Mann DL, et al: Late onset 21-hydroxylase deficiency mimicking idiopathic hirsutism or polycystic ovarian disease: An allelic variant of congenital virilizing adrenal hyperplasia with a milder enzymatic defect. *Ann Intern Med* **96:**143, 1982.

29. Chetkowski RJ, DeFazio J, Shamonki I, et al: The incidence of late-onset congenital adrenal hyperplasia due to 21-hydroxylase deficiency among hirsute women. *J Clin Endocrinol Metab* **58:**595, 1984.

30. Kuttenn F, Couillin P, Girard F, et al: Late-onset adrenal hyperplasia in hirsutism. *N Engl J Med* **313:**224, 1985.

31. Bouchard P, Kuttenn F, Mowszowicz I, et al: Congenital adrenal hyperplasia due to partial 21-hydroxylase deficiency: A study of five cases. *Acta Endocrinol (Copenh)* **96:**107, 1981.

32. Wischusen J, Baker HWG, Hudson B: Reversible male infertility due to congenital adrenal hyperplasia. *Clin Endocrinol (Oxf)* **14:**571, 1981.

33. Childs B, Grumbach MM, van Wyk JJ: Virilizing adrenal hyperplasia: A genetic and hormonal study. *J Clin Invest* **35:**213, 1956.

34. Carroll MC, Campbell RD, Porter RR: Mapping of steroid 21-hydroxylase genes adjacent to complement component C4 genes in HLA, the major histocompatibility complex in man. *Proc Natl Acad Sci USA* **82:**521, 1985.

35. White PC, Grossberger D, Onulfer BJ, et al: Two genes encoding steroid 21-hydroxylase are located near the genes encoding the fourth component of complement in man. *Proc Natl Acad Sci USA* **82:**1089, 1985.

36. Dupont B, Oberfield SE, Smithwick EM, et al: Close genetic linkage between HLA and congenital adrenal hyperplasia (21-hydroxylase deficiency). *Lancet* **2:**1309, 1977.

37. Levine LS, Zachmann M, New MI, et al: Genetic mapping of the 21-hydroxylase deficiency gene within the HLA linkage group. *N Engl J Med* **299:**911, 1978.

38. Pucholt V, Fitzsimmons JS, Gelsthorpe K, et al: Location of the gene for 21 hydroxylase deficiency. *J Med Genet* **17:**447, 1980.

39. Laron Z, Pollack MS, Zamir R, et al: Late onset 21-hydroxylase deficiency and HLA in the Ashkenzai population: A new allele at the 21-hydroxylase locus. *Hum Immunol* **1:**55, 1980.

40. Blankstein J, Faiman C, Reyes FI, et al.: Adult-onset familial adrenal 21-hydroxylase deficiency. *Am J Med* **63:**441, 1980.

41. Levine LS, Dupont B, Lorenzen F, et al: Cryptic 21-hydroxylase deficiency in families of patients with classical congenital adrenal hyperplasia. *J Clin Endocrinol Metab* **51:**1316, 1980.

42. O'Neill GJ, Dupont B, Pollack MS, et al: Complement C4 allotypes in congenital adrenal hyperplasia due to 21-hydroxylase deficiency: Further evidence for different allelic variants at the 21-hydroxylase locus. *Clin Immunol Immunopathol* **23:**312, 1982.

43. McCluskey J, Kay PH, Stuckey M, et al.: MHC "Supratype" predicting heterozygous 21-hydroxylase deficiency. *Lancet* **1:**764, 1983.

44. Ulick S: Adrenocortical factors in hypertension. 1. Significance of 18-hydroxy 11-deoxycorticosterone. *Am J Cardiol* **38:**814, 1976.

45. Tompkins GM, Michael P, Curran JF: Studies on the nature of steroid 11β-hydroxylation. *Biochim Biophys Acta* **23:**655, 1957.

46. Hudson RW, Schachter H, Killinger DW: Studies of 11β-hydroxylations by beef adrenal mitochondria. *J Steroid Biochem* **7:**255, 1976.

47. Weiss M, Vardolov L: A study of steroid 11β-hydroxylation by adrenal mitochondria of marsupials. Part I. A comparison of 11β hydroxylase activity and specificity for different steroid substrates by possum (*Trechosurus vulpecula*), kangaroo (*Macropus major*) and beef. *J Steroid Biochem* **8:**1233, 1977.

48. Shibusawa H, Sano Y, Okinaga S, et al: Studies on 11β-hydroxylase of the human fetal adrenal gland. *J Steroid Biochem* **13:**881, 1980.

49. LAdodevoh BK, Engel LL, Shaw D, et al: Metabolism of progesterone-4-14C by adrenal tissue from a patient with Cushing's syndrome. *J Clin Endocrinol Metab* **25:**784, 1965.

50. Zachmann M, Vollmin JA, New MI, et al: Congenital adrenal hyperplasia due to deficiency of 11β-hydroxylation of 17α-hydroxylated steroids. *J Clin Endocrinol Metab* **33:**501, 1971.

51. Biglieri EG, Herron MA, Brust N: 17-hydroxylation deficiency in man. *J Clin Invest* **45:**1946, 1966.

52. New MI: Male pseudohermaphroditism due to 17α-hydroxylase deficiency. *J Clin Invest* **49:**1930, 1970.

53. Bongiovanni AM: The adrenogenital syndrome with deficiency of 3β-hydroxysteroid dehydrogenase. *J Clin Invest* **41:**2086, 1962.

54. Janne O, Perheentupa J, Viinikka L, et al: Testicular endocrine function in a pubertal boy with 3β-hydroxysteroid dehydrogenase deficiency. *J Clin Endocrinol Metab* **39:**206, 1974.

55. Pang S, Levine LS, Stoner E, et al: Nonsalt-losing congenital adrenal hyperplasia due to 3β-hydroxysteroid dehydrogenase deficiency with normal glomerulosa function. *J Clin Endocrinol Metab* **56:**808, 1983.

56. Rosenfield RL, Barmach de Niepomniszsze AA, Kenny FM, et al: The response to gonadotropin (HCG) administration in boys with and without Δ5-3β-hydroxysteroid dehydrogenase deficiency. *J Clin Endocrinol Metab* **39:**370, 1974.

57. Schneider G, Genel M, Bongiovanni AM, et al: Persistent testicular Δ5-isomerase-3β-hydroxysteroid dehydrogenase (Δ5-3-β-HSD) deficiency in the Δ5-3β-HSD form of congenital adrenal hyperplasia. *J Clin Invest* **55:**681, 1975.

58. Martin F, Perheentupa J, Adlercreutz H, et al: Plasma and urinary androgens and oestrogens in a pubertal boy with 3β-hydroxysteroid dehydrogenase deficiency. *J Steroid Biochem* **13:**197, 1979.

59. Kenny FM, Reynolds JW, Green OC: Partial 3β-hydroxysteroid dehydrogenase (3β-HSD) deficiency in a family with congenital adrenal hyperplasia: Evidence for increasing 3β-HSD activity with age. *Pediatrics* **48:**756, 1971.

60. Bongiovanni AM, Eberlein WR, Moshang T Jr: Urinary excretion of pregnanetriol and Δ5-pregnenetriol in two forms of congenital adrenal hyperplasia. *J Clin Invest* **50:**2751, 1971.

61. Cara JF, Moshang T Jr, Bongiovanni AM, et al: Elevated 17-hydroxyprogesterone and testosterone in a newborn with 3-beta-hydroxysteroid dehydrogenase deficiency. *N Engl J Med* **313**:618, 1985.

62. Wilkins L: *The Diagnosis and Treatment of Endocrine Disorders in Childhood and Adolescence.* Charles C. Thomas, Springfield, IL, 1950.

63. Pang S, Becker D, Foley T Jr, et al: *Adrenal Tissue Tumors in Patients with Congenital Adrenal Hyperplasia.* The Endocrine Society, San Francisco, No. 28, 1976.

64. Daeschner GL: Adrenal adenoma arising in a girl with congenital adrenogenital syndrome. *Pediatrics* **36**:140, 1965.

65. Van Seters AP, Van Aalderen W, Moolenaar AJ, et al: Adrenocortical tumour in untreated congenital adrenocortical hyperplasia associated with inadequate ACTH suppressibility. *Clin Endocrinol* **14**:325, 1981.

66. Landing BH, Gold E: The occurrence and significance of Leydig cell proliferation in familial adrenal cortical hyperplasia. *J Clin Endocr* **11**:1436, 1951.

67. Garvey FK, Daniel TB: Bilateral interstitial cell tumor of the testicle. *J Urol* **66**:713, 1951.

68. Miller EC Jr, Murray HL: Congenital adrenocortical hyperplasia: Case previously reported as "bilateral interstitial cell tumor of the testicle." *J Clin Endocrinol* **22**:655, 1962.

69. Cohen H: Hyperplasia of the adrenal cortex associated with bilateral testicular tumors. *Am J Pathol* **22**:157, 1946.

70. Fore, WW, Bledsoe T, Weber DM, et al: Cortisol production by testicular tumors in adrenogenital syndrome. *Arch Intern Med* **130**:59, 1972.

71. Mulaikal RM, Migeon CJ, Rock JA: Fertility rates in female patients with congenital adrenal hyperplasia due to 21-hydroxylase deficiency. *N Engl J Med* **316**:178, 1987.

72. Klingensmith GJ, Garcia SC, Jones HW Jr, et al: Glucocorticoid treatment of girls with congenital adrenal hyperplasia: Effects on height, sexual maturation, and fertility. *J Pediatr* **90**:996, 1977.

73. Swyer GIM, Bonham DG: Successful pregnancy in a female pseudohermaphrodite. *Br Med J* **1**:1005, 1961.

74. Avin J: Female pseudohermaphroditism with delivery of a normal child. *Pediatrics* **29**:828, 1962.

75. Eyton-Jones J. The adrenogenital syndrome and pregnancy. *J Obstet Gynaecol (Br Commonw)* **75**:1063, 1968.

76. Riddick DH, Hammond CB: Long-term steroid therapy in patients with adrenogenital syndrome. *Obstet Gynecol* **45**:15, 1975.

77. Weldon VV, Blizzard RM, Migeon CJ: Newborn girls misdiagnosed as bilaterally cryptorchid males. *N Engl J Med* **274**:829, 1966.

78. Aziz R, Mulaikal RM, Migeon CJ, et al: Congenital adrenal hyperplasia: Long term results following vaginal reconstruction. *Fertil Steril* **46**:1011, 1986.

79. Brook CGD, Zachmann M, Prader A, et al: Experience with long-term therapy in congenital adrenal hyperplasia. *J Pediatr* **85**:12, 1974.

80. Jones HW Jr, Verkauf BS: Congenital adrenal hyperplasia: Age at menarche and related events at puberty. *Am J Obstet Gynecol* **109**:292, 1971.

81. Gorski RA, Harlan RE, Jacobson CD, et al: Evidence for the existence of a sexually dimorphic nucleus in the preoptic area of the rat. *J Comp Neurol* **193**:529, 1980.

82. Nottebohm F, Arnold AP: Sexual dimorphism in vocal control areas of the songbird brain. *Science* **194**:211, 1976.

83. Imperato-McGinley J, Peterson RE, Gautier T, et al: Androgens and the evolution of male-gender identity among male pseudohermaphrodites with 5α-reductase deficiency. *N Engl J Med* **300**:1233, 1979.

84. Baker SW: Psychosexual differentiation in the human. *Biol Reprod* **22**:61, 1980.

85. Money J, Schwartz M: Dating, romantic and nonromantic friendships, and sexuality in 17 early-treated adreno genital females, aged 16–25, in Lee LP, Plotnick AA, Kowarski CJ, Migeon CJ (eds): *Congenital Adrenal Hyperplasia.* University Park, Baltimore, 1977, pp. 419–431.

86. Money J, Schwartz M, Lewis VG: Adult erotosexual status and fetal hormonal masculinization and demasculinization: 46, XX congenital virilizing adrenal hyperplasia and 46, XY androgen-insensitivity syndrome compared. *Psychoneuroendocrinology* **9**:405, 1984.

87. Wilkins L, Lewis RA, Klein R, et al: Treatment of congenital adrenal hyperplasia with cortisone. *J Clin Endocrinol Metab* **11**:1, 1951.
88. Bongiovanni AM, McFadden AJ: Steroids during pregrancy and possible fetal consequences. *Fertil Steril* **11**:181, 1960.
89. Warrell DW, Taylor. R: Outcome for the fetus of mothers receiving prednisolone during pregnancy. *Lancet* **1**:117, 1968.
90. Kenny FM, Preeyasombat C, Spaulding JS, et al: Cortisol production rate. IV. Infants born of steroid-treated mothers and of diabetic mothers: Infants with trisomy syndrome and with anencephaly. *Pediatrics* **37**:960, 1966.
91. Green OC: Steroid metabolism in the fetus and the newborn infant. *Pediatr Clin North Am* **12**:615, 1965.
92. Walsh SW, Norman RL, Novy MJ: In utero regulation of rhesus monkey fetal adrenals: Effects of dexamethasone, adrenocorticotropin, throtrophin releasing hormone, prolactin, human chorionic gonadotropin and alpha-melanocyte-stimulating hormone on fetal and maternal plasma steroids. *Endocrinology* **104**:1805, 1979.
93. Novy MJ, Walsh SW: Dexamethasone and estradiol treatment of pregnant rhesus macaques: Effects on gestational length, maternal plasma hormones, and fetal growth. *Am J Obstet Gynecol* **145**:920, 1983.
94. Wellman KF, Volk BW: Fine structure changes in the rabbit-placenta induced by cortisone. *Arch Pathol Lab Med* **94**:147, 1972.
95. Kotas RV, Mims LC, Hart LK: Reversible inhibition of lung cell number after glucocorticoid injection into fetal rabbits to enhance surfactant appearance. *Pediatrics* **53**:358, 1974.
96. Evans MI, Chrousos GP, Mann DW, et al: Pharmacologic suppression of the fetal adrenal gland in utero. Attempted prevention of abnormal external genital masculinization in suspected congenital adrenal hyperplasia. *JAMA* **253**:1015, 1985.
97. Villee DB: Development of endocrine function in the human placenta and fetus. I. *N Engl J Med* **281**:473, 1969.
98. Villee DB: Development of endocrine function in the human placenta and fetus. II. *N Engl J Med* **281**:533, 1969.
99. Pollack MS, Levine LS, Pang S, et al: Prenatal diagnosis of congenital adrenal hyperplasia (21-hydroxylase deficiency) by HLA typing. *Lancet* **1**:1107, 1979.
100. Pang S, Levine LS, Cederqvist LL, et al: Amniotic fluid concentrations of delta 5 and delta 4 steroids in fetuses with congenital adrenal hyperplasia due to 21-hydroxylase deficiency and anencephalic fetuses. *J Clin Endocrinol Metabol* **51**:223, 1980.
101. Warsof SL, Larsen JW, Kent SG, et al: Prenatal diagnosis of congenital adrenal hyperplasia. *Obstet Gynecol* **55**:751, 1980.
102. Forest MG, Betuel H, Gouillin P, et al: Prenatal diagnosis of congenital adrenal hyperplasia (CAH) due to 21-hydroxylase deficiency by steroid analysis in the amniotic fluid of mid-pregnancy: Comparison with HLA typing in 17 pregnancies at risk for CAH. *Prenat Diagn* **1**:197, 1981.

12

Pheochromocytoma

Historical Background

The original description of a pheochromocytoma dates back to 1886 and is credited to Frankel. The term "pheochromocytoma" (or black-celled tumor) was coined in 1912 by Pick. Dr. Mayo in 1926, when evaluating the first resectable pheochromocytoma at the Mayo Clinic, wrote that "toxins are evidently intermittently discharged affecting the sympathetic." In 1929 Rabin demonstrated that the tumor produced substances with properties similar to epinephrine. By 1949, it had become clear that in addition to epinephrine, these tumors secreted norepinephrine as well. Owing to its potential curability, pheochromocytoma should always be considered in the differential diagnosis of "secondary hypertension."

Incidence

Pheochromocytomas are rare tumors. The incidence varies among the reported series, ranging between 0.1% and 0.01% of the hypertensive population and approximately 2 in 100,000 of the adult population.[1] Its rarity notwithstanding, pheochromocytoma has earned its respect and reputation owing to the unique combination of being potentially lethal on one hand and yet benign and potentially curable on the other.[2,3] The true incidence of pheochromocytoma may in fact be underrepresented. This notion is fostered by the observation that the incidence of unexpectedly finding a pheochromocytoma at autopsy is impressive. Jones et al.[4] as well Modlin et al.[5] have observed that in one-third of more than 72 patients, seen over a period of 20 years, the tumor was found unexpectedly at autopsy. More important, in most cases the tumor was the immediate cause of death. In some instances, the first explosive manifestation of a pheochromocytoma may well be the last, costing the patient his/her life. Frequently, the diagnosis of pheochromocytoma is often ignored until a crisis develops, even though retrospectively several subtle clues may have been evident in the history. Considering the fatal consequences of a missed diagnosis, a high index of suspicion is essential to diagnose the tumor. It is unclear whether the incidence of pheochromocytoma would be higher if all hypertensive patients underwent routine screening for exclusion of pheochromocytoma.

Pheochromocytomas may occur at any age, but are encountered with the greatest frequency during the fourth and fifth decades of life. While in adults there is a slight predilection for women, in children boys are more commonly affected than girls.[6]

Histopathology

Site of Tumor

Pheochromocytomas originate from the chromaffin cells of the sympathoadrenal system. Nearly 85–90% of the time the tumor originates from the adrenal medulla. In 10–15% of cases, the tumor is extraadrenal. These sites include the paraganglia cells of the sympathetic chain (particularly from the lumbar and paraaortic sympathetic ganglia); the organ of Zuckerkandl, a structure composed of chromaffin-staining cells situated near the origin of the inferior mesenteric artery and often extending along the aorta as far down as its bifurcation[7-9]; the chest (<2%), particularly from the sympathetic chain in the posterior mediastinum; the neck (<0.1%); and the urinary bladder. Rarely, the pheochromocytoma can be intrapericardial.[10] The often quoted figures that approximately 10% of pheochromocytomas are extraadrenal, 10% are bilateral, and 10% are malignant appear to still hold good. For some inexplicable reason, right-sided pheochromocytomas occur twice as frequently as left-sided pheochromocytomas.[11] The familial variety of pheochromocytoma is regarded as a distinct entity, owing to the fact that tumors are commonly bilateral in this group of pheochromocytoma. Even when unilateral, the familial pheochromocytoma tends to demonstrate multicentricity within the gland. It is generally held that the incidence of extraadrenal pheochromocytoma is the same in familial as well as sporadic pheochromocytomas. However, recently Glowniak et al.[12] have described a kindred where extraadrenal pheochromocytomas developed in the same anatomical site (the right hilum) in three successive generations, suggesting that primary extraadrenal pheochromocytoma is a syndrome in which specific genetic abnormalities determine the sites of tumor development.

Macroscopic Appearance

Morphologically, adrenal pheochromocytomas average 100 g in weight and less than 10 cm in diameter. Extraadrenal pheochromocytomas weigh less and are usually less than 5 cm in diameter. Adrenal pheochromocytomas are highly vascular and often display a tendency to be supplied by all three suprarenal arteries—middle, superior, and inferior. Hemorrhage, necrosis, and cyst formation (from infarction and liquefaction) are often encountered in adrenal pheochromocytomas.

Microscopic Appearance

Microscopically, pheochromoctytomas are composed of large pleomorphic chromaffin cells. As with other endocrine tumors, it is difficult to determine whether the tumor is benign or malignant on the basis of the microscopic appearance alone. Less than 10% of adrenal pheochromocytomas are malignant, based on the presence of local invasion or distant spread. It is generally believed that the incidence of malignant pheochromocytomas is higher in three types of pheochromocytomas: the familial variant, the extraadrenal pheochromocytomas, and those in children. Ultrastructural studies of tumor tissue show that the tumor contains large amounts of chromaffin granules, which are subcellular particles that store catecholamines. These chromaffin granules, for the most part, appear no different than the granules seen within normal adrenal medullary cells. There are no consistent data to suggest differences in the mechanisms of intragranular binding, storage, and release of catecholamines between normal and tumorous adrenal medullary tissue. Considerable variability exists in the biosynthesis, storage, turnover, and release of catecholamine in pheochromocytomas. This precludes any assumptions in postulating differences in the manner in which tumor tissue synthesizes catecholamine in comparison to normal tissue. It is not clear whether a change in the rate-limiting step involving tyrosine hydroxylase leads to an uncontrolled rate of catecholamine synthesis by the tumor cells. There seems to be a correlation between the size of the tumor and the turnover rate of catecholamines within the tumor. Thus, small pheochromocytomas tend to show relatively lower concentrations of catecholamines within the tumor, with a high turnover rate of catecholamines and lower urinary ratios of vanillyl mandelic acid (VMA) to catecholamines. In contrast, the larger tumors contain higher concentrations of catecholamines within the tumor, with a lower turnover rate, resulting in a higher urinary ratio of VMA to catecholamine. It is possible that smaller tumors cause symptoms earlier, since more catecholamines are secreted into the system.

Associations of Pheochromocytoma

Although rare, several associations of pheochromocytomas need to be kept in mind and the disease sought in the presence of these "background disorders." The four most important associations are von Recklinghausen's disease (neurofibromatosis), von Hippel–Lindau disease, the multiple endocrine neoplasia syndromes type II and III (MEN II and III), and acromegaly. A fifth, cholelithiasis, is frequently mentioned as an association of pheochromocytoma.

von Recklinghausen's Disease

The classic concept has been that the prevalence of pheochromocytoma in patients with multiple cutaneous neurofibromatosis is approximately 1–2%,[13–15] a figure that suggests a 10-fold increase in the incidence of pheochromocytoma compared to the general population. Conversely, about 5% of patients with pheochromocytoma may have neurofibromatosis.[16] One of the clinical hallmarks of von Recklinghausen's disease is the presence of more than two café-au-lait spots greater than 2 cm in diameter with the presence of freckles in the axilla. Since von Recklinghausen's disease may be transmitted as an incomplete form, minor manifestations such as pigmentary or vertebral anomalies, may suggest the presence of the von Recklinghausen trait. The conventionally ascribed low rate of association (1–2%) of pheochromocytoma with von Recklinghausen's disease should be reevaluated in light of recent data. For instance, Kalff et al.[17] found an appreciable number of pheochromocytomas in 10 of 18 cases of neurofibromatosis with hypertension. The higher incidence noted by these workers was related to the high index of suspicion coupled with the intensity of screening, employing sophisticated diagnostic testing. The pheochromocytomas were detected within or adjacent to the adrenal gland and secreted both epinephrine as well as norepinephrine. The combination of neurofibromatosis, pheochromocytoma, and pancreatic somatostatinoma has also been described.[18]

von Hippel–Lindau Disease

von Hippel–Lindau disease refers to a familial syndrome characterized by the variable occurrence of retinal, cerebellar, and cerebral hemangioblastomas.

Pheochromocytoma is regarded as an unusual but notable association with von Hippel–Lindau disease, approximately 61 cases having been reported.[19] In a collected series of 50 patients with von Hippel–Lindau disease from nine families, the prevalence of pheochromocytoma was noted to be 10%.[20] In contrast to this low incidence, in another series, pheochromocytomas were found in 12 of 13 affected members with von Hippel–Lindau's disease.[21] Thus, it seems that the prevalence of pheochromocytomas in this syndrome is highly variable. The usual site of the tumor is adrenal, but rarely extraadrenal pheochromocytoma has been reported in patients with von Hippel–Lindau disease.[19]

MEN II and MEN III

The association of pheochromocytoma with MEN II and III is well recognized.[22,23] Type II MEN consists of medullary carcinoma of the thyroid, pheochromocytoma, and hyperparathyroidism. Type III MEN, which is believed to represent an entity distinct from type II MEN, consists of medullary thyroid carcinoma, pheochromocytoma, and a unique constellation of somatic stigmata.[24–27] These stigmata include a Marfanoid habitus, mucosal neuromas (particularly of the tongue, lips, and eyelids), thickening of corneal nerves, and ganglioneuromatosis of the alimentary tract. Type III MEN is also characterized by the lack of hyperparathyroidism, the heightened frequency of bilateral pheochromocytomas, the often aggressive nature of the medullary thyroid carcinoma, and the onset at a relatively younger age in comparison to MEN II. The pheochromocytomas that occur in association with familial syndromes, particularly the MEN types, are unique for the following reasons:

1. These tumors are often extremely small, increasing the likelihood of being missed by conventional localizational and imaging procedures.
2. Familial pheochromocytomas may remain clinically silent, with very few symptoms until a crisis supervenes. In some instances, the first manifestation of the pheochromocytoma, for instance during anesthesia, may well be the last.
3. The small size of these tumors, when combined with the rapid turnover within the tumor, may result in unimpressive biochemical data, which may result in missing the condition when conventional screening tests are employed.
4. The pheochromocytomas of familial variety are usually bilateral. Very often, even when the tumor is demonstrably unilateral, the contralateral adrenal medulla may show changes of "medullary hyperplasia." Whether the bilateral nature of such tumors is universal is rather controversial.
5. Familial pheochromocytomas tend to demonstrate a proclivity toward malignancy more often than the sporadic variants.
6. Familial pheochromocytomas often demonstrate a tendency to selectively or predominantly secrete epinephrine, a phenomenon that is relatively uncommon in the sporadic variety of pheochromocytoma.
7. Finally, the familial variety of pheochromocytoma does not often lend itself to successful diagnostic testing by employing provocative maneuvers such as intravenous glucagon.

For all these reasons, familial pheochromocytomas are intriguing and can pose considerable diagnostic as well as therapeutic challenges.

The frequent occurrence of familial pheochromocytoma with such diverse conditions as von Hippel–Lindau's disease, neurofibromatosis, the MEN syndromes, and carcinoid tumors has led to the concept that all these conditions may be ultimately related to an aberration in the migration, growth, and differentiation of the neural crest cells.[28] This concept is strengthened by the observation of a case of bilateral pheochromocytomas in combination with pancreatic islet cell tumors, which conventionally belong in the MEN I category.[29]

Acromegaly

The association between acromegaly and pheochromocytoma is less well recognized, although several case reports[30–34] had sporadically mentioned such an association. Anderson et al.[35] reported two cases of pheochromocytomas with acromegaly and cited eight, possibly 12, other instances in the literature, suggesting that the combination may represent a nonfamilial variety of multiple endocrine neoplasia syndrome. These patients tend to have sustained hypertension rather than the episodic hypertension characteristic of pheochromocytoma. In most of the patients reviewed by Anderson et al.,[35] symptoms of pheochromocytoma had persisted for an average of 7 years before the diagnosis was established. Even more important was the fact that 4 of the 10 patients died from causes directly related to pheochromocytoma. This emphasizes the importance of a careful search for pheochromocytoma in acromegalics with hypertension. Both patients reported by Anderson et al.[35] had malignant pheochromocytomas that had metastasized. Hyperparathyroidism has been variably associated with the syndrome. As with MEN I, aberrant neural crest development may underlie the development of both the pituitary tumor and the pheochromocytoma.

Cholelithiasis

The increased incidence (up to 30%) of cholelithiasis in association with pheochromocytoma is well recognized[2] but unexplained. The possibility that such a combination, at least in some instances, may represent gallstones caused by a somatostinoma is speculative.

Clinical Features

The clinical features of pheochromocytoma can be viewed from several perspectives: the acute paroxysms, which can be so characteristic as to permit instant recognition; the sustained manifestations, which may render the disease as unremarkable as any garden-variety essential hypertension; unusual manifestations, so kaleidoscopic in quality as to earn the deserved title of "the great masquerader"; and finally, the features that result from complications of chronic catecholamine excess, the brunt of which are shouldered by the cardiovascular system.

TABLE 47.
Precipitating Factors for a Paroxysm

1. Increase in intraabdominal pressure
   Massage, or steady pressure on the tumor
   Valsalva maneuver
   Straining at stool
   Coughing, sneezing, laughing, tight clothing
   Sexual intercourse
   Parturition
   Abdominal trauma
2. Anxiety
   Pain
   Nervousness
   Hyperventilation
3. Administration of certain drugs
   Histamine
   Glucagon
   ACTH
   Nicotine
   Phenothiazine
   Saralasin
   Epinephrine
   Tyramine
4. Procedures
   Anesthesia
   Surgery
   Endoscopy

Acute Paroxysms

Patients with pheochromocytoma often experience sudden "attacks," which are due to intermittent release of catecholamines from the tumor into the circulation. Paradoxically, smaller tumors are more often associated with more dramatic and explosive paroxysms than larger-sized tumors. These episodes of paroxysms are highlighted by the occurrence of moderate to severe hypertension. At times, the paroxysm can be so explosive as to result in stroke, myocardial infarction, or death.

The precipitating factors for the development of an attack can be several. Numerous "trigger factors" have been recognized (Table 47). Three factors seem to be the most commonly associated precipitating factors—increase in the intraabdominal pressure, anxiety, and use of certain drugs or procedures. Of particular interest is the precipitation of paroxysms from the use of drugs. There are primarily three mechanisms whereby drugs can precipitate a paroxysm, even a crisis.[36] First, certain drugs may result in direct release of catecholamines from the tumor. The precipitation of crisis by the use of histamine, glucagon, ACTH, saralasin, and opiates belongs in this category. More important, fentanyl, a potent opiate agonist, can provoke catecholamine release.[37] The second mechanism involves release of catecholamines stored in the sympathetic nerve endings. Methyldopa and decongestants containing sympathomimetic amines belong in this category. The third mechanism involves the use of drugs that inhibit reuptake of catecholamines by the sympathetic neurons; guanethidine and tricyclic antidepressants belong to this category.

The frequency of attacks can be highly variable in individual patients, ranging from occurring once in few months to occurring 25 times daily. The attacks, which evolve so dramatically and rapidly, resolve more slowly than they started. The symptoms that accompany the paroxysm of hypertension are headaches, excessive perspiration, and palpitations. In the series reported by Gifford et al.[38] of 76 patients with pheochromocytoma, all but three had one or more of these three symptoms, and 55 patients had at least two of these three symptoms. It is generally held that this triad of sweating, headache, and palpitations in a hypertensive patient has a specificity of 93.8% and a sensitivity of 90.9%.[39] In fact, absence of a pheochromocytoma can be presumed 99.9% of the time when all three symptoms are absent in a hypertensive patient.

Paroxysmal hypertension is encountered in more than 50% of patients with pheochromocytoma. The degree of hypertension during some of these attacks can be dangerously high, exceeding a diastolic greater than 130 mm Hg ("hypertensive crisis"). Headaches, which generally accompany the paroxysmal hypertension, are severe, are located in the frontal or occipital regions, and are described as throbbing or pounding. The increased diaphoresis can be severe enough to drench the patient in sweat. Palpitations, tachycardia, or both are present in more than 70% of patients who experience paroxysms of pheochromocytoma. In addition, patients may experience nervousness, panic, or a sense of impending doom, dizziness, weakness, and tremulousness during an attack. The classic attack of catecholamine excess is typical enough to be diagnosed by the history.

Sustained Hypertension

In approximately 49% of patients with pheochromocytoma the blood pressure elevation may be sustained without discernible evidence of the presence of paroxysms.[40] In such patients, the presentation and behavior of the hypertensive condition bear a close resemblance to those of patients with essential hypertension. Unremarkable as the hypertension may seem, an underlying pheochromocytoma should be suspected under the following conditions:

1. Fluctuation (lability) of the blood pressure, often spontaneously.
2. Little or no response to conventional antihypertensive medications.
3. The presence of orthostatic hypotension; spontaneous orthostatic hypotension can be a prominent feature of pheochromocytoma.[41,42] Such a finding in an untreated hypertensive patient should raise the suspicion of pheochromocytoma. The pathogenesis of orthostatic hypotension in pheochromocytoma is ill understood. The proposed mechanisms include increased venous tone with resultant decrease in plasma volume; an adverse effect of catecholamines on the sympathetic reflexes that normally maintain the cybernetic response; and tachyphylaxis to the catecholamines.

TABLE 48.
Unusual Manifestations of
Pheochromocytomas

1. Shock or circulatory collapse
2. "Normotensive" pheochromocytoma
3. Fever ("FUO")
4. Gastrointestinal manifestations
5. Vasomotor phenomena
6. Neuropsychiatric manifestations
7. Cardiovascular disease
8. "Paraneoplastic" disease
    Hypercalcemia
    Rarer syndromes
9. Hypermetabolic phenomena
10. Asymptomatic ("nonfunctional")

4. Adverse or paradoxical increases in blood pressure in response to beta-blockade or ganglion-blocking agents.
5. Presence of hypermetabolic features such as weight loss, heat intolerance, or glucose intolerance.
6. A marked pressor response during induction with any anesthetic agent.[2]

The question of screening all hypertensive patients for pheochromocytoma, in terms of cost-effectiveness, continues to remain controversial. While all hypertensive patients with the 4 Hs (headache, hyperhidrosis, hypermetabolism, and hyperglycemia) should be mandatorily screened for the presence of a pheochromocytoma, the reported rarity of the disease in the hypertensive population at large should be kept in mind. On the other hand, pheochromocytoma is a curable cause of hypertension, and therefore, the cost of prolonged antihypertensive therapy must be weighed against the cost of screening, as well as the price extracted by missed diagnosis. There is no unanimity of opinion in the matter of screening the entire hypertensive population for pheochromocytoma. Some degree of selectivity is of course desirable, and patients with even the most subtle clues in the history should be screened for the presence of this disease.

Unusual Manifestations (Table 48)

Pheochromocytoma, as indicated earlier, can be the proverbial "great masquerader." The diversity of manifestations is illustrated by the fact that no less than 80 manifestations have been reported.[43] Several unusual manifestations deserve brief comment.

Circulatory Collapse

Unexplained circulatory collapse following anesthesia, pregnancy, delivery or puerperium, surgery, or the administration of phenothiazine drugs may be related to predominant release of epinephrine, which has beta-adrenergic effects on the peripheral vasculature.

Normotensive Pheochromocytoma

While sustained or paroxysmal hypertension is the clinical marker of pheochromocytoma, very rarely pheochromocytomas have occurred in perfectly normotensive individuals.[44–47] Several mechanisms have been proposed to explain normotension in pheochromocytomas. These include predominant epinephrine secretion, tolerance of tissue receptors to catecholamines, and contemporaneous secretion of dopa or dopamine. Pheochromocytoma with predominant secretion of epinephrine is a well-documented entity.[47,48] In such patients, the clinical presentation is not only characterized by lack of hypertension, but may in fact be associated with hypotension, myocarditis, and cardiac arrythmias often mimicking "cardiomyopathy of unknown etiology." Tolerance of tissue receptors to catecholamines is also a well-documented phenomenon in laboratory animals.[49,50] In such a setting, even infusion of norepinephrine to patients with documented pheochromocytoma may elicit little or no response. The third mechanism, invoking predominant or coincident secretion of dopa or dopamine, may also modify the pressor response to catecholamines. Several workers have demonstrated dopa or dopamine secretion in pheochromocytoma.[51–53] These substances have vasoactive properties that can modify the pressor response of norepinephrine, in addition to their inherent vasodilatory effects on the mesenteric or renal vasculature. The importance of recognizing the existence of normotensive pheochromocytoma lies in the fact that the disorder should be considered even in patients with normal blood pressure who have additional symptomatology of catecholamine excess.

Fever

In 1950, Smithwick et al.[54] reviewed 118 patients with documented pheochromocytomas and noted an elevation in body temperature by at least 1°F in 70% of patients. This observation prompted the quite apt suggestion by these workers that pheochromocytoma should be considered in the differential diagnosis of any fever of unexplained origin. While hyperthermia is a frequent manifestation of "the pheochromocytoma in crisis," the incidence of pheochromocytoma masquerading as "FUO" is unknown. Several fascinating reports of pheochromocytoma (particularly the normotensive variety) masquerading as pyrexia of undetermined origin have illustrated this unique aspect.[44,55] The combination of weight loss, abdominal pain, nausea, sweating, and so forth in a patient with FUO often leads to the mistaken pursuit of an intraabdominal neoplasm or smouldering sepsis. In many instances the pheochromocytoma is discovered by the performance of gallium scan or arteriography studies in the pursuit of these diagnoses.

Gastrointestinal Manifestations

Gastrointestinal manifestations are rare in pheochromocytoma. The most common gastrointestinal symptom associated with pheochromocytoma is chronic constipation, which may be encountered in as many as 13% of patients with the tumor.[2] This symptom is a reflection of the effects of chronic hypersecretion of epinephrine and norepinephrine on the intestinal smooth muscle, which possesses both alpha$_2$ and beta$_2$ receptors that are responsive to catecholamines; chronic stimulation of these receptors results in relaxation and decrease of motility as well as tone. In addition, catecholamines can cause contraction of the pyloric and ileocecal sphincters and decrease in the splanchnic blood flow. These triple effects, i.e., decreased smooth muscle tone, increased sphincteric tone, and decreased splanchnic blood flow, are conducive to causing severe adynamic ileus, which in extreme cases can result in pseudoobstruction of the bowel.[56,57]

In addition to chronic constipation with pseudoobstruction at the extreme end of the spectrum, there are some very rarely reported gastrointestinal manifestations of pheochromocytoma. These include gastrointestinal bleeding, intestinal ischemia and necrosis, bowel perforation, and peritonitis.[58–60] In addition, spontaneous bleeding into a large pheochromocytoma can present as an acute abdomen with severe pain, hypotension, and local signs of peritoneal irritation. The dangers of laparotomy in the setting of a missed diagnosis of pheochromocytoma are catastrophic.

Vasomotor Phenomena

The most common vasomotor phenomenon associated with pheochromocytoma is pallor of the face and upper part of the body, which is particularly seen in patients with a paroxysm. As many as 60% of patients with paroxysmal hypertension and 28% of patients with sustained hypertension caused by a pheochromocytoma have been observed to demonstrate pallor of the face and upper part of the body.[2] Rarely flushing may be observed. The presence of these vasomotor phenomena may suggest the diagnoses of systemic mastocytosis or carcinoid syndrome, conditions that represent important differential diagnoses for pheochromocytoma.

Neuropsychiatric Manifestations

Anxiety, with or without a fear of impending death, and severe apprehension are frequently encountered in patients with pheochromocytoma. The similarity of pheochromocytoma to an anxiety disorder has been emphasized in the psychiatric literature.[61,62] Further, since acute anxiety states are generally associated with a heightened activity of the sympathoadrenal system, it is quite understandable that patients with anxiety often manifest "pheo-like" symptoms such as headaches, palpitations, and sweating. In this regard, the study by Starkman et al.[63] provides interesting insight into the prevalence of anxiety attacks in patients with pheochromocytoma. These workers studied 17 patients with active pheochromocytoma to determine whether they experienced anxiety that met the standardized criteria established to define it.[64] The interesting observation that emerged was that none of the 17 patients with pheochromocytoma described the severe apprehension or fear characteristic of panic attacks. Two patients experienced symptoms that met the criteria for generalized anxiety disorder, one received the diagnosis of "possible panic disorder," while two met criteria for a "major depressive episode." While these observations are of interest, it should be realized that several patients in this study were receiving alpha- and beta-adrenergic blocking agents, which could indeed alter or modify the psychiatric symptoms due to catecholamine excess.

The occurrence of depression in pheochromocytoma is of special interest since the use of certain tricyclic antidepressants can unmask the diagnosis of pheochromocytoma. Several case reports have described hemodynamic abnormalities triggered by the use of imipramine or desipramine therapy resulting in the recognition of pheochromocytoma.[65-68] This may be a reflection of the ability of tricyclic antidepressants to inhibit the reuptake of norepinephrine by the adrenergic nerve terminals.

Cardiovascular Disease

The cardiovascular disease caused by pheochromocytoma encompasses a spectrum of entities; one end is represented by hypertension with its sequelae on the heart, and the other end is represented by aortic dissection, a dreaded complication of pheochromocytoma. Somewhere in between lies the entity defined as "catecholamine cardiomyopathy." This entity has been the focus of intense recent interest. The fact that catecholamine administration to laboratory animals can result in myocarditis has been known since the late fifties.[69,70] Several clinical as well as autopsy studies have abundantly documented the existence of a similar phenomenon in humans with pheochromocytoma.[71-73] The histopathological characteristics of catecholamine-induced myocarditis include focal infiltration of leukocytes, small hemorrhages, and edema. These changes eventually lead to focal degeneration, myofibril necrosis, and fibrosis. Clinically, patients with pheochromocytoma may have abnormal electrocardiographic changes even in the absence of discrete cardiac symptoms. From a functional viewpoint, patients with pheochromocytoma may demonstrate a variety of echocardiographic abnormalities indicative of both dilated as well as obstructive cardiomyopathies.[74-76]

The concept of catecholamine-induced cardiomyopathy has evolved in the past decade. Several case reports have distinctly established pheochromocytoma as an etiology for cardiomyopathy.[77-81] The usual variety of cardiac involvement is the dilated form of cardiomyopathy. Evaluation of cardiac function before and after removal of pheochromocytoma has indicated partial to complete reversal of the cardiomyopathy after surgical removal of the tumor.[80,82] Thus, pheochromocytoma should be included in the conditions that cause reversible cardiomyopathy.

Paraneoplastic Syndromes

Several paraneoplastic syndromes can be associated with pheochromocytoma. The most common of these phehomena is hypercalcemia. Less commonly, pheochromocytoma can be associated with ectopic secretion of ACTH, erythropoietin, and several peptide hromones.

Hypercalcemia. The incidence of hypercalcemia associated with pheochromocytoma is not exactly known. Hypercalcemia can develop in patients with pheochromocytoma by a number of mechanisms.

1. Catecholamine induced parathyroid (PTH) secretion is an important mechanism for the development of hypercalcemia in patients with pheochromocytoma.[83] This is supported by the observation of normalization of hypercalcemia following extirpation of tumor.[83-85] Clearly, catecholamine mediation in the secretion and release of PTH may underlie this mechanism.
2. Ectopic secretion of PTH-like substances may also play a role in the genesis of hypercalcemia associated with pheochromocytoma.[86,87]
3. The association of pheochromocytoma with the MEN II syndrome (medullary thyroid carcinoma, hyperparathyroidism, and bilateral pheochromocytoma) is another mechanism for hypercalcemia in pheochromocytoma.
4. Catecholamines can directly induce osteoclastic bone resorption, resulting in hypercalcemia.[88,89]
5. Finally, Stewart et al.[90] have proposed that some patients with pheochromocytoma may secrete a bone-resorbing factor similar to that seen with humoral hypercalcemia of cancer. These workers described a child with pheochromocytoma and hypercalcemia with no evidence for excessive PTH secretion from the parathyroids or from the pheochromocytoma. Therapy with catecholamine synthesis inhibitors reversed the catecholamine excess, but had no effect on the hypercalcemia, which was reversed only by adrenalectomy. Tumor extracts possessed profound in vitro bone-resorbing activity as well as impressive adenylate cyclase–stimulating activity in renal cortical membranes.

Regardless of the mechanism of the hypercalcemia, pheochromocytoma should be considered in the differential diagnosis of unexplained hypercalcemia.

Rarer Paraneoplastic Syndromes. Rarely, pheochromocytomas have been reported to secrete ACTH.[91] The resulting clinical picture is highlighted by features of both Cushing's syndrome and pheochromocytoma. Erythropoietin secretion may result in polycythemia, an uncommon occurrence. Very rarely, immunocytochemical studies of excised pheochromocytomas have shown the presence of immunoreactive growth hormone–releasing factor within the tumor cells.[92] The secretory versatility of pheochromocytomas has been amply illustrated in a patient described by Viale et al.,[93] who described a 30-year-old man with the watery diarrhea syndrome (Verner–Morrison syndrome) due to excessive secretion of vasoactive intestinal polypeptide (VIP) by a pheochromocytoma. The patient had increased plasma levels of catecholamines, VIP, somatostatin calcitonin, and gastrin, all of which reverted to normal following excision of an adrenal pheochromocytoma. The excised neoplastic tissue was immunohistochemically shown to contain VIP, somatostatin, and calcitonin. The amine precursor uptake decarboxylation (APUD) origin of the adrenal medullary cells might explain the ability of the neoplastic cells to synthesize a variety of peptide hormones derived from embryologically related, but morphologically different, neuroendocrine cells.

Hypermetabolic Phenomena

Hypermetabolic phenomena may dominate the picture of pheochromo-cytoma in some patients. Thus, weight loss, diaphroresis, and heat intolerance can mimic the hyperthyroid state. The similarity to thyrotoxic disease is brought even closer by the observation of thyromegaly in some patients with phe-ochromocytoma. Even more interesting is the reported observation of acute thyroid swelling during paroxysms of hypertension due to pheochromo-cytoma.[94-96] The mechanism of acute thyroid swelling is unclear, but may be related to acute increases in the blood flow to the thyroid gland, consequent to "catecholamine surge." The role of adrenergic mediation of thyroid hormone synthesis is, at best, speculative.

Asymptomatic Pheochromocytoma

Occasionally, some pheochromocytomas may cause no signs or symptoms, being accidentally discovered during radilogical examination, surgery, or autop-sy.[97,98] The reason for the paucity of symptoms may relate to the "nonfunction-ing" nature of these tumors or be due to secretion of insignificant quantities of catecholamines into the circulation. The emergence of the increased use of computerized tomographic examination of the abdomen will undoubtedly result in detecting asymptomatic "nonfunctional" pheochromocytomas.

Complications

Occasionally patients with pheochromocytoma come to attention for the first time owing to manifestations from the complications of the disease. As can be seen from Table 49, cardiovascular complications lead the list.

TABLE 49.
Complications of
Pheochromocytoma

1. Cardiac
    Cardiomyopathy
    Congestive failure
    Myocarditis
    Myocardial infarction
    Dissecting aneurysm
2. Central nervous system
    Cerebrovascular accident
    Encephalopathy
3. Gastrointestinal
    Pseudoobstruction
    Ischemic enterocolitis
    Hemorrhage within tumor
4. Shock

Laboratory Diagnosis

The laboratory diagnosis of pheochromocytoma can be viewed from three perspectives: establishing the biochemical diagnosis, the use of dynamic studies, and localization of the tumor. It is unacceptable to "explore" a patient for the possible presence of a pheochromocytoma in the absence of documented biochemical or radiological proof of the tumor.

Biochemical Diagnosis

With the currently available hormonal tests to document catecholamine hypersecretion, the diagnosis of pheochromocytoma can be established with certainty in more than 90–95% of cases. When combined with dynamic testing (such as glucagon stimulation or clonidine suppression), the diagnostic accuracy approaches virtually 100%. The three major screening tests for pheochromocytoma involve determining the urinary excretion of catecholamine metabolites, measurement of free catecholamines in the urine, and determination of plasma catecholamines. Figure 27 outlines the synthesis and metabolism of catecholamines.

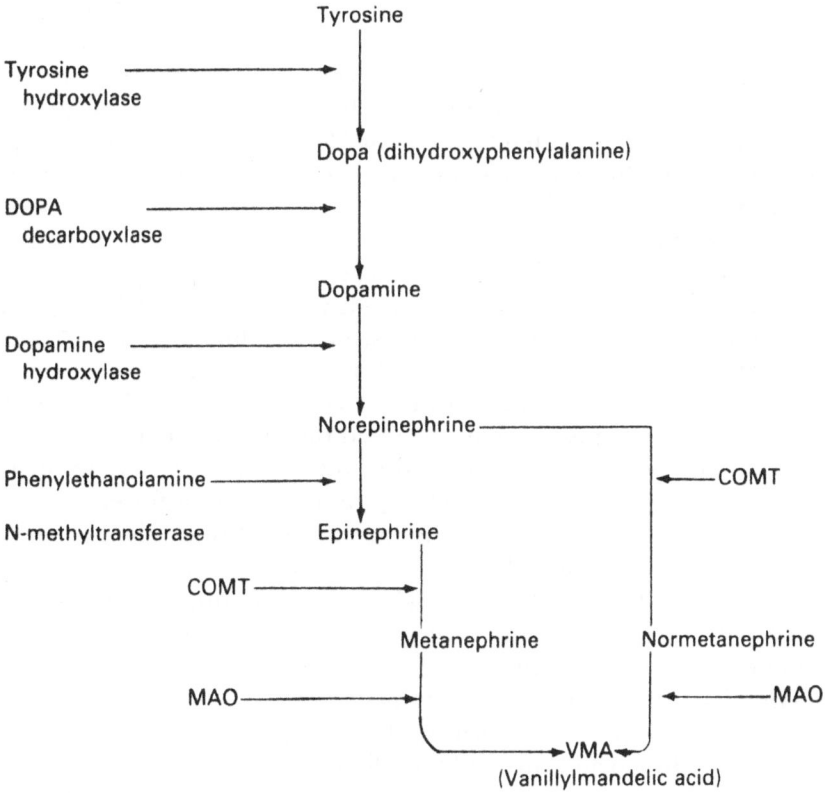

FIGURE 27. Synthesis and metabolism of catecholamines. COMT, catecholamine o-methyltransferase; MAO, monoamine oxidase.

Three important observations are clinically pertinent as they relate to catecholamine synthesis and metabolism. First, the enzyme phenylethanolamine N-methyltransferase is present only in the adrenal medulla and the organ of Zuckerkandl. Thus, elevation of epinephrine clearly indicates an origin from one of these structures. Second, both catecholamines are converted to their respective methylated compounds (normetanephrine and metanephrine) by the enzyme catecholamine-methyltransferase. These products are excreted in the urine. Third, these products are further inactivated by monoamine oxidase into VMA. With this background, the biochemical picture of pheochromocytoma can be reviewed.

Catecholamine Metabolites

Measurement of 24-hr urinary metanephrine, normetanephrine, and VMA have stood the test of time as simple, reasonably accurate, and inexpensive first-line screening tests for detection of a pheochromocytoma. The diagnostic yield of pheochromocytoma is highest when all three measurements are obtained by proper urine collection and when performed by a proficient laboratory.

Metanephrine, Normetanephrine. The metanephrine and normetanephrine levels in the 24-hr urine measure the methylated metabolites of epinephrine and norepinephrine; of the urinary indices of catecholamine production, measurement of metanephrine and normetanephrine provides the most reliable clues for the presence of an underlying pheochromocytoma. The normal range of these metabolites is from 0.02 to 1.7 mg/24 hr in a well-collected specimen. Collection of urine for catecholamines or their metabolites is done in an acidified bottle, preferably on ice. When the 24-hr meta- and normetanephrines exceed 1.8 mg/24 hr, the diagnosis of pheochromocytoma is highly suspect. It is essential to avoid the use of drugs that interfere with the assay; particular attention should be paid to the use of antihypertensive agents (methyldopa), catecholamine-containing drugs, benzodiazepines, MAO inhibitors, and alcohol, all of which can give an apparent increase in urinary metanephrine and normetanephrine. On the other end of the spectrum, spurious lowering of values might be seen in patients who have received radiographic dyes containing methyglucamine (renografin, renovist), and in those taking large doses of fenfluramine. Of course, in the hospital setting, the main reason for encountering elevated levels of urinary metanephrine, normetanephrine values is stress. Thus, patients with severe illness, burns, myocardial infarction, and stroke tend to show elevation of these metabolites. The impact of chronic renal failure on the urinary metabolites of catecholamines is also another factor to be kept in mind while interpreting these values.

The diagnostic value of urinary metanephrines and normetanephrines in the diagnosis of pheochromocytoma has been emphasized by several workers.[1,6,36,99,100] Bravo and Gifford[101] evaluated the utility of these measurements in 43 patients with surgically confirmed pheochromocytoma and found false negative results in only nine. This contrasted with VMA results, which were false negative in 25 of 43 patients. The referent values used by these authors were those obtained in a large population of essential hypertensives. In addition to serving as a reasonably good screening test, a positive correlation has been reported between urinary metanephrine and normetanephrine levels and tumor mass in pheochromocytoma.[102]

VMA. The 24-hr urinary VMA excretion is seldom recommended as the sole screening test for detection of pheochromocytoma. The normal range for VMA excretion in 24 hr in adults is 1–14 mg/24 hr. In children, the VMA excretion correlates with weight as well as creatinine excretion. VMA excretion is affected by diet (especially foods rich in vanillin) as well as the same drugs that affect metanephrine and normetanephrine values in the 24-hr urine. In addition, nalidexic acid, L-dopa, and rapid clonidine withdrawal may also spuriously increase VMA levels. In contrast, MAO inhibitors, clofibrate, disulfiram, and large doses of fenfluramine tend to decrease the apparent value of VMA measurements. Despite these disadvantages, when the VMA levels are greater than 13 mg/24 hr, the diagnosis of pheochromocytoma is strongly supported.

Urinary Catecholamines

Measurement of urinary excretion of free catecholamines, when collected and performed properly, provides 95% accuracy in detecting pheochromocytoma. However, free catecholamines are metabolized too rapidly to provide an accurate index of catecholamine production. It has been suggested that samples of urine obtained after sleep and analyzed for norepinephrine excretion, expressed as μg/hr, may provide accurate discrimination in detecting a pheochromocytoma.[103] The major limitations of using 24-hr free catecholamine levels to diagnose pheochromocytoma are the methodological problems associated with the assay, as well as the interference from physiological and chemical factors. Urinary catecholamines can increase with stress, as well as with the use of bronchodilators, sympathomimetic drugs, and in rare diseases such as certain posterior fossa tumors and acute intermittent porphyria. Despite these obvious limitations, several laboratories combine measurement of urinary total catecholamines [norepinephrine (NE) and epinephrine (E)] with metanephrines, normetanephrines, and VMA. When all three are completely normal, the diagnosis of pheochromocytoma can be excluded with 90% certainty. This is especially true when the urine collections have been performed during or shortly after a paroxysm.

Fractionation of free catecholamines and measuring E and NE in the urine may reveal important diagnostic information. If the patient secretes predominantly E or secretes both E and NE, the tumor is likely to be in adrenals. The conversion of NE to E takes place only in the adrenal medulla (and perhaps to a smaller extent in the organ of Zuckerkandl) and requires the permissive effect of adrenocortical hormones. However, rarely E-secreting tumors have been reported in the bladder, and even in the lungs.[104] In addition to providing insight into the possible site of the tumors, selective measurement of urinary E has been found to be extremely useful and sensitive in detection of the small-sized pheochromocytomas, particularly those associated with medullary thyroid carcinoma.[105]

Plasma Catecholamines

The measurement of basal plasma catecholamines for the diagnosis of pheochromocytoma has met with rave reviews as well as concerned criticism. The two most important factors that determine the usefulness of this test are the method of collection (and handling) of the sample, as well as the expertise in performing the assay. Meticulous attention should be paid to collection and handling of the blood. The patient should be supine, and rested for at least 20 to 30 min, and should have blood drawn through an indwelling needle to avoid the stress of venipuncture. The blood should be collected in special tubes, on ice, and immediately centrifuged to separate the red blood cells (RBC). This is because the RBCs quickly and avidly consume the catecholamines. It has been emphasized that approximately half the amount of catecholamines disappear from the plasma within 20 min, if the RBC are present in the sample. In addition, a highly specialized laboratory is needed to perform and interpret catecholamine levels in the plasma.

Bravo and Gifford[101] have reported their experience with measurement of basal plasma NE and E in 64 patients with pheochromocytoma and compared the results obtained in normals as well as 70 patients with essential hypertension. It was noted that none of the patients with pheochromocytoma had values that fell in the range for age- and sex-matched normotensive subjects. When compared to patients with essential hypertension, only four patients of the 64 with pheochromocytoma had basal plasma catecholamine concentrations that fell within the 95% confidence limits (i.e., 950 pg/ml) for values seen in essential hypertensives. The supine resting plasma catecholamines had a sensitivity of 94% in detecting pheochromocytoma, in this study. Other workers[106–108] have also reported excellent results with plasma catecholamine measurements for the diagnosis of pheochromocytoma.

The difficulties with methodology aside, elevation of plasma catecholamines can be encountered in several settings. Stress, hypoglycemia, acute myocardial infarction, and congestive failure are a few conditions where plasma catecholamines are elevated. More important, a small but significant number of patients with essential hypertension may show elevations in the supine resting plasma catecholamine concentrations.[4] While as a group the levels in patients with essential hypertension may not overlap the levels seen in association with pheochromocytoma, in individual patients the differentiation can be difficult. It is in such a setting that the clonidine suppression test can be employed to differentiate patients with essential hypertension and high basal plasma catecholamine levels from those with pheochromocytoma.

Nevertheless, the measurement of basal plasma concentrations is exceedingly valuable during a paroxysm. Impressive elevations in plasma levels of catecholamines (ranging from 2000 to 20,000 pg/ml) can be seen in patients with hypertensive crisis caused by pheochromocytoma. The plasma catecholamine levels are also invaluable while studying the response to glucagon or clonidine administration.

Arguments will continue to prevail as to the "best screening test" for diagnosis of pheochromocytoma. The combination of urine metabolites coupled with properly performed basal catecholamine levels in plasma would help in identifying the presence of pheochromocytoma in more than 95% of cases. Dynamic studies, using glucagon or clonidine, are undertaken when the clinical suspicion is strong but the first-line screening tests have not been diagnostic.

Dynamic Studies

The advent of sophisticated methodology to measure catecholamines and their metabolites has led to near abandonment of several provocative tests used in the not-so-distant past. Thus, the use of the tyramine test, histamine test, cold pressor tests, and phenotolamine test has become obsolete. Of these tests, only the phentolamine test may still have a diagnostic role in patients who present in hypertensive crisis. The failure to lower the diastolic pressure in a patient presenting in hypertensive crisis excludes pheochromocytoma as a cause of the hypertension. The presence of a response, however, does not establish the diagnosis of pheochromocytoma, since a small number of patients with severe essential hypertension my respond to phentolamine. Even in the setting of a hypertensive crisis, measurement of plasma catecholamines and initiation of urine collection (even a 4-hr collection) for metanephrines, normetanephrines, VMA, and free catecholamines has more diagnostic value than the response to phentolamine.

The two dynamic studies that still may have a role in the diagnosis of pheochromocytoma are the glucagon response study and the clonidine suppression test.

Clonidine Suppression Test

In 1981, Bravo et al.[109] introduced the clonidine suppression test as a useful aid in the diagnosis of pheochromocytoma. The clonidine suppression test is based on the principle that oral administration of this drug, which is a centrally acting alpha-adrenergic agonist, suppresses the release of neurogenically mediated catecholamine output, but not the release of catecholamines from a pheochromocytoma. Thus, it was anticipated that patients with essential hypertension and elevated plasma catecholamines could be distinguished from those with catecholamine excess from a pheochromocytoma by the plasma catecholamine response to oral administration of 0.3 mg of clonidine. Bravo and Gifford[101] reported their experience with the clonidine suppression test in 32 patients with documented pheochromocytoma and compared the response patterns in 70 sex- and age-matched patients with essential hypertension. The plasma catecholamine levels declined to below 500 pg/ml after clonidine in all but one patient with essential hypertension, while all but one of the patients with pheochromocytoma demonstrated postclonidine catecholamine levels in excess of 500 pg/ml. The authors felt that failure of plasma E and NE levels to suppress below 500 pg/ml 2 or 3 hr following the oral administration of 0.3 mg of clonidine is strong evidence for the presence of a pheochromocytoma. The authors[101] noted several important observations regarding the clonidine suppression test. First, the blood pressure response to clonidine does not differentiate patients with pheochromocytoma from those with essential hypertension. This is owing to the inhibition of central sympathetic outflow by clonidine in both groups of patients. Second, in their experience, significant or symptomatic hypotension was not a commonly encountered adverse side effect from cloni- dine. Such a complication is more likely to occur in patients with phe- ochromocytoma in whom clonidine is administered in the presence of other antihypertensive drugs. In particular, patients with pheochromocytoma receiv- ing beta blockers are more likely to develop hypotension with clonidine, which has potent vagotonic effects. It is advisable to avoid beta blockers and marked volume depletion for at least 48 hr prior to testing. Third, a false positive response, i.e., failure to suppress below 500 pg/ml in the absence of phe- ochromocytoma, is more likely to occur when radioimunoassays that measure both free and conjugated catecholamines are employed.[110] It is believed that measurement of free catecholamines by radioenzymatic assays that employ cate- cholamine-o-methyltransferase are superior and preferred for the proper in- terpretation of the clonidine suppression test. False positive responses to the clonidine suppression test can be also encountered when propranolol and other beta blockers are used. Beta blockers interefere with hepatic clearance of cate- cholamines, giving a falsely elevated level.[111]

Glucagon Provocative Test[112,113]

The administration of 1 mg of glucagon intravenously to patients harboring a pheochromocytoma leads to a prompt increase in the blood pressure, as well as in the circulating catecholamine levels. The test is considered positive if the pressor response to glucagon exceeded by 20/10 mm Hg the response caused by the nonspecific cold pressor stimulation.[113] The sensitivity of the glucagon stim- ulation test improves when plasma catecholamines are measured before and after glucagon administration.[114] The mechanisms that underlie the cate- cholamine release following glucagon in patients with pheochromocytoma are dual. Glucagon may directly enhance the release of catecholamines from the tumor, or alternatively the drug may cause vasodilatation with subsequent re- lease of catecholamines from extratumor storage sites. Administration of gluca- gon to an already hypertensive patient with pheochromocytoma may potentially result in severe hypertension. Therefore, parenteral phentolamine must be readily available in case of this adverse effect of the test.

The incidence of false negative responses to the glucagon stimulation test in patients with pheochromocytoma is probably quite low. However, one clinical situation has been emphasized in the literature to represent a classic setting where the test may be falsely negative. When pheochromocytoma occurs as part of the spectrum of MEN II, the glucagon challenge test may be normal. Siqueira- Filho et al.[115] performed glucagon challenge tests in six patients with phe- ochromocytoma associated with MEN II and encountered negative results in all six. This may represent low metabolic activity of the tumor, lack of abnormal catecholamine storage in the extra adrenal sites, or both. Regardless, it should be remembered that this group of pheochromocytomas often represent a challenge in diagnosis, no matter what test is employed. Therefore, it is not surprising that a high percentage of glucagon unresponsiveness can be encountered in this group of pheochromocytomas that are notorious for their "biochemical silence."

The availability of sensitive assays for plasma catecholamines, as well as the emergence of the clonidine suppression test, has, to some extent, dampened the enthusiasm for the glucagon challenge test. The test is probably most useful when quiescent tumor is suspected and provocative maneuvers are called upon to diagnose its presence. In comparison to the histamine or tyramine stimulation tests, the glucagon provocative test provides far superior diagnostic yield. Caution should be exercised in interpreting the catecholamine levels after glucagon injection, since these levels can rise in nonpheochromocytoma subjects as a nonspecific response to stress.[116]

Localization

Localization of a pheochromocytoma can be as simple as a single-step procedure utilizing computed tomography (CT) of the adrenals (and the abdomen) or can involve the use of arteriography, venography, or venous effluent studies. The development of [131I]meta-iodobenzylguanidine ([131I]MIBG) scintigraphy has added a new and exciting dimension in the localization of pheochromocytoma.

Computerized Tomography

Current experience suggests that a properly performed CT study of the adrenals and abdomen should virtually eliminate the need for invasive and potentially hazardous angiographic procedures. Several reports have repeatedly confirmed the high diagnostic yield by CT in locating the adrenal pheochromocytoma.[117–120] The procedure accurately localizes tumors larger than 1 cm, with a precision of approximately 90–95%. In one review of 60 histologically documented pheochromocytomas from a single institution over a 7-year period, CT correctly localized 51 of 57 tumors, in the group of 52 patients who presented for the first time with evidence of tumor.[121] In the eight patients who had presented with evidence of recurrence, CT accurately located the recurrent tumor in 73% of instances. The tumors that are most likely to be missed are those originating from the periaortic sympathetic chain and the organ of Zuckerkandl. If these areas are included in the scanning area, the likelihood of missing extraadrenal, intraabdominal pheochromocytomas would be minimized. Clearly, CT represents the first line of investigation for localization of pheochromocytoma. When the CT study provides no help, additional procedures such as arteriography or venous effluent studies may have to be performed. Of course, when [131I]MIBG photoscanning becomes widely available, the need for invasive studies should become obsolete.

[131I]Meta-Iodobenzylguanidine Scan

[131I]MIBG is a radiopharmaceutical with a striking affinity for the adrenal medulla. This analog of guanethedine resembles NE in its molecular structure and enters adrenergic tissues by a mechanism similar to the neurotransmitter. Thus, the radioisotope is concentrated in the catecholamine storage vesicles such as the adrenal medulla, or within adrenergic tumors such as pheochromocytoma or paraganglioma. A new approach to the localization of pheochromocytoma has been provided by investigators at the University of Michigan, utilizing [^{131}I]MIBG as an adrenergic tissue tracer.[122–127] The scintigraphic localization of pheochromocytoma is based on the principle that tissues that contain numerous adrenergic vesicles retain the radionuclide for a protracted period of time, usually days. The justifiable excitement generated by this new localizing agent was underscored by several observations. First, small tumors that are not visualized adequately by CT can be visualized by [^{131}I]MIBG; second, MIBG scintigraphy is probably unparalleled in its ability to detect extraadrenal and malignant pheochromocytomas; third, MIBG scintigraphy is uniquely applicable for anatomical localization of recurrent or residual disease; and finally, abnormal scintigraphy may provide a clue for the presence of medullary hyperplasia, a situation that is particularly likely to be encountered in familial pheochromocytomas, especially those in association with the spectrum of MEN II or MEN III.

The advantage of [^{131}I]MIBG scintigraphy is its ability to screen the whole body with remarkable specificity and precision for locating extraadrenal sites of the tumor, as well as metastatic lesions. The limitations of the procedure are the lack of widespread availability and the possibility of false positive results. The accuracy of [^{131}I]MIBG in localizing pheochromocytoma has been compared with that of CT by Chatal and Charbonnel.[128] These workers reported the results of [^{131}I]MIBG scintigraphy in a multicenter study involving 25 nuclear medicine departments and comprising 99 patients, 47 of whom had documented pheochromocytomas. They noted that in approximately 80% of patients with tumor, both methods gave concordant results as far as localizing the tumor. In the remaining 20%, the two procedures were complementary. While CT visualized nearly 95% of tumors, scintigraphy had the potential of false negative results in about 10% of patients. The experience by workers at the Mayo clinic,[129] however, appears to indicate that the overall specificity of the [^{131}I]MIBG scintigraphic study was 96%, with an accuracy of 85%. These workers pointed out the superiority of scintigraphy in (1) detecting lesions that were negative by CT, (2) detecting recurrent pheochromocytoma, and (3) detecting malignant, less differentiated catecholamine-secreting tumors such as paragangliomas. A similar high accuracy rate (>92%) of scintigraphy has been reported by Ackery, McEwan, et al.[130,131] from Southhampton, England.

Invasive Studies

The widespread availability of excellent CT evaluation and the limited availability of [^{131}I]MIBG scintigraphy have all but eliminated the need for invasive diagnostic studies for the localization of pheochromocytoma. When both CT and scintigraphy have failed, and the hormonal evidence of pheochromocytoma is indisputable, invasive studies are indicated for localization of the tumor. Arteriography, combined with and guided by the stepup in NE concentrations in the venous effluents, can be diagnostic for localization. Blood samples from sequential points in the vena cava demonstrate an abrupt rise in NE concentrations, permitting identification of the anatomical site of tumor.[132,133] Arteriography may also be helpful in identifying the presence of "cryptic" tumors.[134,135] Patients with pheochromocytoma undergoing invasive studies should be adequately alpha-blocked to prevent the occurrence of a crisis.

Treatment

The treatment of pheochromocytoma is surgical, after adequate preparation of the patient.

Medical Therapy

The preparation of the patient with pheochromocytoma consists of medical therapy with alpha-adrenergic blocking agents, the prototype of which is the alpha blocker phenoxybenzamine. The triple purposes that underlie such preparation are to control blood pressure, to prevent paroxysms, and to promote intravascular volume expansion.

Phenoxybenzamine

Phenoxybenzamine is a noncompetitive adrenergic blocking agent. In contrast to phenotolamine, which has equal affinity for both α_1 and α_2 receptors,[136] phenoxybenzamine is 100 times more potent in blocking α_1 than α_2 receptors.[137] The pharmacological importance of this property of phenoxybenzamine is readily apparent if one views the physiology of blocking both versus α_1 receptors alone: α_1 receptors are postsynaptic, while α_2 receptors are presynaptic. Stimulation of α_1 receptors leads to vasoconstriction and increase in peripheral resistance and in blood pressure. In contrast, the stimulation of α_2 receptors (which serve the function of providing "feedback" regulation for NE release from the nerve) leads to inhibition of NE release from the nerve terminal with an attendant diminution in constriction of the vascular bed.[138] Thus, if α_1 receptors are blocked, the blood pressure declines, while with α_2 blockade the blood pressure can rise owing to increased release of NE from the nerve terminal. When these phenomena are applied to α blocker therapy, important differences emerge in the effects exerted by phenotolamine and phenoxybenzamine. Phentolamine, because of its ability to block both α_1 and α_2 receptors to an equal degree, would result in dual effects—the α_1 blockade leads to a drop in the blood pressure immediately; however, with progressive α_2 blockade, more NE is released from the presynaptic terminal, leading to eventual displacement of phentolamine from the α_1 terminal. The net result would be an eventual rise in the blood pressure and loss of alpha$_1$-blocking effects of the drug. Phenoxybenzamine, on the other hand, has a much more potent blocking effect on the α_1 receptors in comparison to the α_2 receptors; as a result, there is negligible effect in causing NE release from the nerve terminal. Thus, chronic therapy is less likely to result in loss of α_1-blocking effect.

In addition to its predominant α_1-blocking effect, the advantages of phenoxybenzamine includes ease of administration (oral), few side effects, and a long half-life, which permits a twice-a-day regimen. The initial dose is 10 mg twice daily, with gradual increments to 20–40 mg twice (or even thrice) daily until the blood pressure is perfectly normalized. The track record of phenoxybenzamine for long-term control of pheochromocytic hypertension is excellent; only two cases of pheochromocytoma resistant to phenoxybenzamine had been reported as of 1983.[139,140]

Other Drugs

The role for other drugs in the medical management of pheochromocytoma is limited. Three, however, deserve comment—prazosin, beta blockers, and drugs that inhibit catecholamine synthesis.

Prazosin.   Prazosin hydrochloride has been used in isolated cases of pheochromocytoma for effective control of blood pressure, either alone[141] or in combination with beta blockers.[142] Prazosin is a specific α_1 postsynaptic antagonist and has seemingly impressive advantages over pheoxybenzamine therapy. First, the hypotensive action of prazosin is not associated with an increase in the heart rate, in contrast to nonselective alpha-adrenergic antagonists. Second, the adverse effects of complete alpha blockade are seldom seen with chronic prazosin therapy. Third, patients with pheochromocytoma are quite sensitive to the drug and respond with minimum dosage of the drug, a factor that may promote compliance. Finally, in at least one study[143] long-term dosage requirements could be predicted based on the blood pressure response to a single 1-mg oral dose. Despite these superficial advantages, prazosin has not been scrutinized in large numbers of patients with pheochromocytoma. Further, it has been noted that surgical management of the tumor, i.e., intra- and perioperative management, was less than optimal with prazosin alone, requiring the use of intravenous phentolamine to suppress the pressor surges during manipulation.[143]

Beta Blockers.   Treatment with propranolol is indicated only in the presence of persistent tachycardia, angina, or hazardous arrythmias. Propranolol should never be administered alone, owing to the unopposed alpha effects, which may result in severe hypertension. Even in the presence of alpha-blocker therapy, use of propranolol can elevate the blood pressure in patients with pheochromocytoma. If beta-blocker therapy is deemed necessary, it should be initiated cautiously by starting at a low dose (10 mg orally twice a day), with gradual increments of 10 mg every day or 2 until the tachycardia or arrythmia is brought under control. Labetalol, a drug that has both alpha- and beta-adrenergic blocking properties, has been effectively used in some patients with pheochromocytoma.[144]

Drugs That Inhibit Catecholamine Synthesis.   The drug alpha-methyl paratyrosine (metyrosine) blocks the synthesis of norepinephrine by inhibiting the enzyme tyrosine hydroxylase. In a dosage of 2–2.5 g/day, metyrosine therapy can result in acceptable reductions in catecholamine synthesis. The effectiveness of the drug can be monitored by following the reduction in the excretion of catecholamine metabolites in the urine. Treatment with metyrosine is often initiated with 250 mg, twice daily, and gradually increased to 2 g/day. The drug is relatively long-acting, and therefore, dosage increments should be made slowly, with the guide of urinary metabolite excretion. The side effects of metyrosine include drowsiness, diarrhea, and crystalluria. The latter can be minimized by liberalizing fluid intake. The majority of patients can be kept relatively asymptomatic with the use of metyrosine.

Other drugs that inhibit catecholamine synthesis are currently under investigation. One such drug, fusaric acid (a dopamine hydroxylase inhibitor), has been found to be useful by some investigators.[145] Finally, there is at least one report of pheochromocytoma associated with a profound decline in urinary NE levels following calcium channel blockade.[75] Radiopharmaceutical treatment of pheochromocytoma by the use of [^{131}I]MIBG is currently under study by investigators at the University of Michigan.[146]

Medical therapy with adrenergic blockade and or metyrosine is the mainstay of therapy for surgical preparation of the patient with pheochromocytoma. While there is no controversy as to the need for adrenergic blockade prior to surgical exploration of the tumor, some controversy exists as to the time frame of discontinuing adrenergic blockade before the surgery. There are two schools of thought in this regard. One subscribes to continuation of adrenergic blockade up to the time of surgery. The arguments in favor of this opinion include prevention of catecholamine surges during the operation, reversal of hypovolemia, promotion of smooth anesthetic induction, and prevention of severe postoperative manifestations of hypercatecholism. The other school of thought argues against continuation of alpha blockade up to the time of surgery based on the fact that complete alpha blockade removes the surgical advantage of locating the tumor by observing blood pressure responses to manipulation. The controversy remains unsettled, with individual preferences prevailing. One approach is to continue metyrosine until the day of surgery in the partially alpha-blocked patient, especially if tumor location is uncertain or if multiple tumors are anticipated. This approach, however, has not found universal application.

Surgical Treatment

Surgical excision of the tumor provides the best and possibly the only cure for unilateral benign pheochromocytoma. The mortality and morbidity from surgery, in experienced hands, is acceptably low, particularly when preoperative preparation is excellent. The incision usually employed is an anterior, transperitoneal approach that permits adequate exposure of both adrenals, the entire paraaortic sympathetic chain, and the urinary bladder. Even when one tumor is found, all areas of sympathetic tissue must be carefully examined to exclude the presence of a second tumor. Greater scrutiny is required in the familial pheochromocytomas. Should a hypertensive crisis occur during the period of anesthesia or tumor manipulation, this should be treated with intravenous phenotolamine or sodium nitroprusside. Cardiac arrythmias may be treated by the use of lidocaine. Postoperative hypotension can complicate the course of the patient recovering from surgery. Volume expansion, normalization of the blood pressure, and liberal replacement of blood lost during surgery would minimize the frequency and severity of postoperative hypotension.[147] Should postoperative hypotension develop, this is best managed by vigorous fluid expansion rather than by pressor agents, since considerable difficulty exists in withdrawal of these agents. The success rate of surgical excision of pheochromocytoma is excellent, approaching 95%. Failure to restore normotension by surgical excision indicates the presence of a second (or multiple) tumor(s) missed during exploration or the concomitant presence of nephrosclerosis as result of the chronic hypertensive process. Rarely the persistence of hypertension following surgery may be the clue to the presence of malignant pheochromocytoma with metastases.

References

1. van Heerden JA, Sheps SG, Hamberger B, et al: Pheochromocytoma: Current status and changing trends. *Surgery* **91**:367, 1982.
2. Manger WM, Gifford RW Jr: *Pheochromocytoma*. Springer-Verlag, New York, 1977.
3. Sheps SG: Pheochromocytoma. In Spittell JA Jr (ed): *Clinical Medicine*, vol. 7. Harper & Row, Philadelphia, 1981, pp. 1–22.
4. Jones DH, Reid JL, Hamilton CA, et al: The biochemical diagnosis, localization and follow up of phaeochromocytoma: The role of plasma and urinary catecholamine measurements. *Q J Med* **49**:341, 1980.
5. Modlin IM, Farndon JR, Shepherd A, et al: Phaeochromocytoma in 72 patients: Clinical and diagnostic features, treatment and long term results. *Br J Surg* **66**:456, 1979.
6. Manger WM, Gifford RW Jr: Hypertension secondary to pheochromocytoma. *Bull NY Acad Med* **58**:139, 1982.
7. Theron LL, Hendry DT: Functioning phaeochromocytoma of the organ of Zuckerkandl. *S Afr Med J* **49**:827, 1975.
8. Van Zyle JJW, Du Toit Van Zyle FD, Wicht CL: Phaeochromocytoma of the organ of Zuckerkandl. *S Afr J Surg* **4**:43, 1966.
9. Scharf Y, Nahir AM, Better OS, et al: Prolonged survival in malignant pheochromocytoma of the organ of Zuckerkandl with pharmacological treatment. *Cancer* **31**:746, 1973.
10. Saad MF, Frazier OH, Hickey RC, et al: Intrapericardial pheochromocytoma. *Am J Med* **75**:371, 1983.
11. Melicow MM: One hundred cases of pheochromocytoma (107 tumors) at the Columbia–Presbyterian Medical Center, 1926–1979. A clinicopathological analysis. *Cancer* **40**:1987, 1977.
12. Glowniak JV, Shapiro B, Sisson JC, et al: Familial extraadrenal pheochromocytoma: A new syndrome. *Arch Intern Med* **146**:257, 1985.
13. Glushian AS, Manrsuy MM, Littman DS: Pheochromocytoma. *Am J Med* **14**:318, 1953.
14. Veyre B, Saint Pierre Laffart G, Milon H, et al: Association phéochromoctome: Neurofibromatoma. *Nouv Presse Med* **4**:2873, 1975.
15. Lynch JD, Sheps SG, Bernatz PE, et al: Neurofibromatosis and hypertension. *Minn Med* **55**:25, 1972.
16. Manger WM, Gifford RW: Current concepts of pheochromocytoma. *Cardiovasc Med* **3**:289, 1978.
17. Kalff V, Shapiro B, Lloyd R, et al: The spectrum of pheochromocytoma in hypertensive patients with neurofibromatosis. *Arch Intern Med* **142**:2092, 1982.
18. Cantor AM, Rigby CC, Beck PR, et al: Neurofibromatosis, phaeochromocytoma, and somatostatinoma. *Br Med J* **285**:1619, 1982.
19. Hoffman RW, Gardner DW, Mitchell FL: Intrathoracic and multiple abdominal pheochromocytomas in von Hippel–Lindau disease. *Arch Intern Med* **142**:1962, 1982.
20. Horton WA, Wong V, Eldridge R: von Hippel–Lindau disease: Clinical and pathological manifestations in nine families with 50 affected members. *Arch Intern Med* **136**:76, 1976.
21. Atuk NO, McDonald T, Wood T, et al: Familial pheochromocytoma, hypercalcemia, and von Hippel–Lindau disease. *Medicine (Baltimore)* **58**:209, 1979.
22. Steiner AL, Goodman AD, Powers SR: Study of a kindred with pheochromocytoma, medullary thyroid carcinoma, hyperparathyroidism and Cushing's disease: Multiple endocrine neoplasia, type 2. *Medicine (Baltimore)* **47**:371, 1968.
23. Khairi MRA, Dexter RN, Burzynski NV, et al: Mucosal neuroma, pheochromocytoma and medullary thyroid carcinoma: Multiple endocrine neoplasia type 3. *Medicine (Baltimore)* **54**:89, 1975.
24. Carney JA, Go VL, Sizemore, et al: Alimentary-tract ganglioneuromatosis. A major component of the syndrome of multiple neoplasia, type 2b. *N Engl J Med* **295**:1287, 1976.

25. Carney JA, Sizemore GW, Lovestedt SA: Mucosal ganglioneuromatosis, medullary thyroid carcinoma, and pheochromocytoma: Multiple endocrine neoplasia, type 2b. *Oral Surg* **41**:739, 1976.

26. Chong GC, Beahrs OH, Sizemore GW, et al: Medullary carcinoma of the thyroid gland. *Cancer* **35**:695, 1975.

27. Robertson DM, Sizemore GW, Gordon H: Thickened corneal nerves as a manifestation of multiple endocrine neoplasia. *Trans Am Acad Opthalmol Otolaryngol* **79**:772, 1975.

28. Bolande RP: The neurocristopathies: A unifying concept of disease arising in neural crest maldevelopment. *Hum Pathol* **4**:409, 1974.

29. Carney JA, Go VLW, Gordon H, et al: Familial pheochromocytoma and islet cell tumor of the pancreas. *Am J Med* **68**:515, 1980.

30. Farhi F, Dikman SH, Lawson W: Paragangliomatosis associated with multiple endocrine adenomas. *Arch Pathol Lab Med* **100**:495, 1976.

31. German WJ, Flanigan S: Pituitary adenomas: A follow-up study of the Cushing series. *Clin Neurosurg* **10**:72, 1964.

32. Kadowaki S, Baba Y, Kakita T, et al: A case of acromegaly associated with pheochromocytoma. [In Japanese.] *Saishin-Igaku* **31**:1402, 1976.

33. Kahn MT, Mullon DA: Pheochromocytoma without hypertension: Report of a patient with acromegaly. *JAMA* **188**:74, 1964.

34. Miller GL, Wynn J: Acromegaly, pheochromocytoma, toxic goiter, diabetes mellitus, and endometriosis. *Arch Intern Med* **127**:299, 1971.

35. Anderson RJ, Lufkin EG, Sizemore GW, et al: Acromegaly and pituitary adenoma with phaeochromocytoma: A variant of multiple endocrine neoplasia. *Clin Endocrinol (Oxf)* **14**:605, 1981.

36. Landsberg L: Pheochromocytoma. *Med Grand Rounds* **2**:7, 1983.

37. Bittar DA: Innovar-induced hypertensive crises in patients with pheochromocytoma. *Anesthesiology* **50**:366, 1979.

38. Gifford RW Jr, Kvale WF, Maher FT, et al: Clinical features diagnosis and treatment of pheochromocytoma: A review of 76 cases. *Mayo Clin Proc* **39**:281, 1964.

39. Plouin P-F, Degoulet P, Tugaye A, et al: Le dépistage du phéochromocytome: chez quels hypertendus?: étude semiologique chez 2585 hypertendus dont 11 ayant un phéochromocytome. *Nouv Presse Med* **10**:869, 1981.

40. Remine WH, Chong GC, van Heerden JA, et al: Current management of pheochromocytoma. *Ann Surg* **179**:740, 1974.

41. Sjoerdsma A: Sympatho-adrenal system: Pheochromocytoma. In Beeson PB, McDermott W (eds): *Cecil Loeb Textbook of Medicine*. 13th ed. Saunders, Philadelphia, 1971, pp. 1832–1836.

42. Engleman K, Zelis R, Waldmann T, et al: Mechanism of orthostatic hypotension in pheochromocytoma. *Circulation* **38**(Suppl 6):72, 1968.

43. Hermann H, Mornex R: *Human Tumors Secreting Catecholamines: Clinical and Physiopathological Study*. Pergamon Press, Oxford, 1964.

44. Kirby BD, Ham J, Fairley HB, Benowitz, et al: Normotensive pheochromocytoma pharmacologic, paraneoplastic and anesthetic considerations. *West J Med* **139**:221, 1983.

45. Kahn MT, Millon DA: Pheochromocytoma without hypertension. *JAMA* **188**:74, 1964.

46. Ho AD, Feurle G, Gless KH, et al: Normotensive familial phaeochromocytoma with predominant noradrenaline secretion. *Br Med J* **1**:81, 1978.

47. Page LB, Raker JW, Berberich FR: Pheochromocytoma with predominant epinephrine secretion. *Am J Med* **47**:648, 1969.

48. Aronoff SL, Passamani E, Borowsky BA, et al: Norepinephrine and epinephrine secretion from a clinically epinephrine-secreting pheochromocytoma. *Am J Med* **69**:321, 1980.

49. Mukherjee C, Caron MG, Lefkowitz RJ: Regulation of adenylate cyclase coupled β-adrenergic receptors by β-adrenergic catecholamines. *Endocrinology* **99**:347, 1976.

50. Shenkman L, Saito M, Feit F, et al: Reduced number of cardiac β-adrenergic receptors in rat pheochromocytoma (abstr). *Clin Res* **27**:595A, 1979.

51. Louis WJ, Doyle AE, Heath WC, et al: Secretion of dopa in phaeochromocytoma. *Br Med J* **4**:325, 1972.

52. Kuchel O, Buu NT, Hamet P, et al: Free and conjugated dopamine in pheochromocytoma, primary aldosteronism and essential hypertension. *Hypertension* **1**:267, 1979.

53. Voorhess ML: Functioning neural tumors. *Pediatr Clin North Am* **13**:3, 1966.

54. Smithwick RH, Greer WER, Robertson CW, et al: Pheochromocytoma: A discussion of symptoms, signs, and procedure of diagnostic value. *N Engl J Med* **242:**252, 1950.

55. Dawson J, Harding LK: Phaeochromocytoma presenting as pyrexia of undetermined origin: Diagnosis using gallium-67. *Br Med J* **284:**1164, 1982.

56. Mullen JP, Cartwright RC, Tisherman SE, et al: Pathogenesis and pharmacologic management of pseudo-obstruction of the bowel in pheochromocytoma. *Am J Med Sci* **290**(4):155, 1985.

57. Turner CE: Gastrointestinal pseudo-obstruction due to pheochromocytoma. *Am J Gastroenterol* **78:**214, 1983.

58. Gendel BR, Ende M: Pheochromocytoma: Report of an unusual case with gastrointestinal bleeding. *Gastroenterology* **19:**344, 1951.

59. Fee HS, Fonkalsrud EW, Ament ME, et al: Enterocolitis with peritonitis in a child with phaeochromocytoma. *Ann Surg* **185:**448, 1977.

60. Rosati LA, Auger NA Jr: Ischemic enterocolitis in pheochromocytoma. *Gastroenterology* **60:**581, 1971.

61. Martin MJ: Psychiatry and medicine. In Freedman AM, Kaplan HI, Sadock BJ (eds): *Comprehensive Textbook of Psychiatry.* Williams & Wilkins, Baltimore, 1975, p. 1746.

62. Mackenzie TB, Popkin MK: Organic anxiety syndrome. *Am J Psychiatry* **140:**342, 1983.

63. Starkman MN, Zelnik TC, Nesse RM, et al: Anxiety in patients with pheochromocytomas. *Arch Intern Med* **145:**248, 1985.

64. American Psychiatric Association Committee on Nomenclature and Statistics: *Diagnostic and Statistical Manual of Mental Disorders,* 3rd ed. American Psychiatric Association, Washington, DC, 1980.

65. Kaufmann JS: Pheochromocytoma and tricyclic antidepressants. *JAMA* **229:**1282, 1974.

66. Mok J, Swann I: Diagnosis of pheochromocytoma after ingestion of imipramine. *Arch Dis Child* **53:**676, 1978.

67. Johnson ER, Jones MD, Stewart JN: Occurrence of a pheochromocytoma after ingestion of imipramine. *J Am Osteopathic Assoc* **78:**332, 1979.

68. Achong MR, Keane PM: Pheochromocytoma unmasked by desipramine therapy. *Ann Intern Med* **94:**358, 1981.

69. Sjoerdsma A, Engelman K, Waldmann TA, et al: Pheochromocytoma: Current concepts of diagnosis and treatment. *Ann Intern Med* **65:**1302, 1966.

70. Pearce RM: Experimental myocarditis: A study of the histological changes following intravenous injections of adrenalin. *J Exp Med* **8:**400, 1906.

71. Szakacs JE, Cannon A: Norepinephrine myocarditis. *Am J Clin Pathol* **30:**425, 1958.

72. Van Vliet PD, Burchell HB, Titus JL: Focal myocarditis associated with pheochromocytoma. *N Engl J Med* **274:**1102, 1966.

73. Kline IK: Myocardial alterations associated with pheochromocytomas. *Am J Pathol* **38:**539, 1961.

74. Mardini MK: Echocardiographic findings in pheochromocytoma. *Chest* **81:**394, 1982.

75. Serfas D, Shoback DM, Lorell BH: Phaeochromocytoma and hypertrophic cardiomyopathy: Apparent suppression of symptoms and noradrenaline secretion by calcium-channel blockade. *Lancet* **2:**711, 1983.

76. Shub C, Williamson MD, Tajik AJ, et al: Dynamic left ventricular outflow tract obstruction associated with phaeochromocytoma. *Am Heart J* **102:**286, 1981.

77. Velasquez G, D'Souza VJ, Hackshaw BT, et al: Phaeochromocytoma and cardiomyopathy. *Br J Radiol* **57:**89, 1984.

78. Porciani MC, Barletta G, Manneli M, et al: Pheochromocytoma presenting as possible cardiomyopathy. *G Ital Cardiol* **12:**826, 1982.

79. Schaffer MS, Zuberbuhler P, Wilson G, et al: Catecholamine cardiomyopathy: An unusual presentation of pheochromocytoma in children. *J Pediatr* **99:**276, 1981.

80. Lam JB, Shub C, Sheps SG: Reversible dilatation of hypertrophied left ventricle in pheochromocytoma: Serial two-dimensional echocardiographic observations. *Am Heart J* **109:**613, 1985.

81. Shapiro LM, Trethowan N, Singh SP: Normotensive cardiomyopathy and malignant hypertension in phaeochromocytoma. *Postgrad Med J* **58:**110, 1982.

82. Imperato-McGinley J, Gautier T, Ehlers K, et al: Reversibility of catecholamine-induced dilated cardiomyopathy in a child with a pheochromocytoma. *N Engl J Med* **316:**793, 1987.

83. Kukreja SC, Hargis GK, Rosenthal IM, et al: Pheochromocytoma causing excessive parathyroid hormone production and hypercalcemia. *Ann Intern Med* **79**:838, 1973.

84. Swinton NW Jr, Clerkin EP, Flint LD: Hypercalcemia and familial pheochromocytoma: Correction after adrenalectomy. *Ann Intern Med* **76**:455, 1972.

85. Finlayson JF, Casey JH: Hypercalcaemia and multiple pheochromocytomas. *Ann Intern Med* **82**:810, 1975.

86. Ghose RR, Winsey HS, Jemmett J, et al: Phaeochromocytoma and hypercalcaemia. *Postgrad Med J* **52**:593, 1976.

87. Fairhurst BJ, Shettar SP: Hypercalcaemia and phaeochromocytoma. *Postgrad Med J* **57**:459, 1981.

88. Kalager T, Gluck E, Heimann P, et al: Phaeochromocytoma with ectopic calcitonin production and parathyroid cyst. *Br Med J* **2**:21, 1977.

89. Heath H III, Edis AJ: Pheochromocytoma associated with hypercalcemia and ectopic secretion of calcitonin. *Ann Intern Med* **91**:208, 1979.

90. Stewart AF, Hoecker JL, Mallette LE, et al: Hypercalcemia in pheochromocytoma: Evidence for a novel mechanism. *Ann Intern Med* **102**:776, 1985.

91. Spark RF, Connolly PB, Gluckin DS, et al: ACTH secretion from a functioning pheochromocytoma. *N Engl J Med* **301**:416, 1979.

92. Sano T, Saito H, Yamasaki R, et al: Production and secretion of immunoreactive growth hormone-releasing factor by pheochromocytomas. *Cancer* **57**:1788, 1986.

93. Viale G, Dell'orto P, Moro E, et al: Vasoactive intestinal polypeptide-, somatostatin-, and calcitonin-producing adrenal pheochromocytoma associated with the watery diarrhea (WDHH) syndrome. *Cancer* **55**:1099, 1985.

94. Howard JE, Barker WH: Paroxysmal hypertension and other clinical manifestations associated with benign chromaffin cell tumours (phaeochromocytoma). *Bull Johns Hopkins Hosp* **61**:371, 1937.

95. Bauer J, Belt E: Paroxysmal hypertension with concomitant swelling of the thyroid due to pheochromocytoma of the right adrenal gland. Cure by surgical removal of the pheochromocytoma. *J Clin Endocrinol* **7**:130, 1947.

96. Buckels JAC, Webb AMC, Rhodes A: Is paroxysmal thyroid swelling due to phaeochromocytoma a forgotten physical sign? *Br Med J* **287**:1206, 1983.

97. Minno AM, Bennett WA, Kvale WF: Pheochromocytoma: A study of 15 cases diagnosed at autopsy. *N Engl J Med* **251**:959, 1954.

98. Taubman I, Pearson OH, Anton AH: An asymptomatic catecholamine-secreting pheochromocytoma. *Am J Med* **57**:953, 1974.

99. Engelman K, Portnoy B, Lovenberg W: Sensitive and specific double isotope derivative method for the determination of catecholamines in biologic specimens. *Am J Sci* **255**:259, 1968.

100. Pullerits J, Reynolds C: Pheochromocytoma: A clinical review with emphasis on pharmacologic aspects. *Clin Invest Med* **5**:259, 1982.

101. Bravo EL, Gifford RW Jr: Phaeochromocytoma: Diagnosis, localization and management. *N Engl J Med* **311**:1298, 1984.

102. Stenstrom G, Waldenstrom J: Positive correlation between urinary excretion of catecholamine metabolites and tumour mass in pheochromocytoma. *Acta Med Scand* **217**:73, 1985.

103. Ganguly A, Henry DP, Yune HY, et al: Diagnosis and localization of pheochromocytoma: Detection by measurement of urinary norepinephrine excretion during sleep, plasma norepinephrine concentration and computerized axial tomography (CT scan). *Am J Med* **67**:21, 1979.

104. Atuk NO: Pheochromocytoma: Diagnosis, localization, and treatment. *Hosp Pract* **18**:187, 1983.

105. Hamilton BP, Landsberg L, Levine RJ: Measurement of urinary epinephrine in screening for pheochromocytoma in multiple endocrine neoplasia type II. *Am J Med* **65**:1027, 1978.

106. Engelman K, Portnoy B, Sjoerdsma A: Plasma catecholamine concentrations in patients with hypertension. *Circ Res* **26, 27** (Suppl 1):141, 1970.

107. Geffen LB, Rush RA, Luis WJ, et al: Plasma dopamine β-hydroxylase and noradrenaline amounts in essential hypertension. *Clin Sci* **44**:617, 1973.

108. Cryer PE: Physiology and pathophysiology of the human sympathoadrenal neuroendocrine system. *N Engl J Med* **303**:436, 1980.

109. Bravo EL, Tarazi RC, Fouad FM, et al: Clonidine suppression test: A useful aid in the diagnosis of pheochromocytoma. *N Engl J Med* **305:**623, 1981.

110. Aron DC, Bravo EL, Kapcala LP: Erroneous plasma norepinephrine levels in radioimmunoassay. *Ann Intern Med* **98:**1023, 1983.

111. Esler J, Jackman G, Bobik A, et al: Norepinephrine kinetics is essential hypertension: Defective neuronal uptake of norepinephrine in some patients. *Hypertension* **3:**149, 1981.

112. Lawrence AM: Glucagon provocative test for pheochromocytoma. *Ann Intern Med* **66:**1091, 1967.

113. Sheps SG, Maher FT: Histamine and glucagon tests in diagnosis of pheochromocytoma. *JAMA* **205:**895, 1968.

114. White LW, Levy RP, Anton AH: Comparison of biochemical and pharmacological testing for pheochromocytoma. *Res Commun Chem Pathol Pharmacol* **5:**252, 1973.

115. Siqueira-Filho AG, Sheps SG, Maher FT, et al: Glucagon–blood catecholamine test: Use in isolated and familial pheochromocytoma. *Arch Intern Med* **135:**1227, 1975.

116. Levinson PD, Hamilton BP, Mersey JH, et al: Plasma norepinephrine and epinephrine responses to glucagon in patients with suspected pheochromocytomas. *Metabolism* **32:**998, 1983.

117. Dunnick NR, Doppman JL, Gill JR Jr, et al: Localization of functional adrenal tumors by computed tomography and venous sampling. *Radiology* **142:**429, 1982.

118. Stewart BH, Bravo EL, Haaga J, et al: Localization of pheochromocytoma by computed tomography. *N Engl J Med* **299:**460, 1978.

119. Eghrari M, McLoughlin MJ, Rosen IE, et al: The role of computed tomography in assessment of tumoral pathology of the adrenal glands. *J Comput Assist Tomogr* **4:**71, 1980.

120. Stewart BH, Straffon RA, Bravo EL, et al: A simplified, cost-effective approach to the diagnosis of pheochromocytoma. *Trans AM Assoc Genitourin Surg* **71:**101, 1979.

121. Welch TJ, Sheedy PF, van Heerden JA, et al: Pheochromocytoma: Value of computed tomography. *Radiology* **148:**501, 1983.

122. Sisson JC, Frager MS, Valk TW, et al: Scintigraphic localization of pheochromocytoma. *N Engl J Med* **305:**12, 1981.

123. Sisson JC, Shapiro B, Beierwaltes WH, et al: Locating pheochromocytomas by scintigraphy using 131I-metaiodobenzyl-guanidine. *Cancer J Clin* **34:**86, 1984.

124. Shapiro B, Sisson JC, Mangner T, et al: Experience with 131I-MIBG scintigraphy in the location of pheochromocytomas (abstr). *Eur J Nucl* **8:**A2, 1984.

125. Francis IR, Glazer GM, Shapiro B, et al: Complementary roles of CT and 131I-MIBG scintigraphy in diagnosing pheochromocytoma. *Am J Roentgenol Radium Nucl Med* **141:**719, 1983.

126. Nakajo M, Shapiro B, Copp J, et al: The normal and abnormal distribution of the adrenomedullary imaging agent 131I-MIBG in man: Evaluation by scintigraphy. *J Nucl Med* **24:**672, 1983.

127. Swanson DP, Carey JE, Brown LE, et al: Human absorbed dose calculations for iodine-131 and iodine-123 labeled meta-iodobenzylguanidine (MIBG): A potential myocardial and adrenal medulla imaging agent. In *Proceedings of the Third International Symposium on Radiopharmaceutical Dosimetry*, Oak Ridge, Tennessee. Oak Ridge Associated Universities, Oak Ridge, TN, 1981, pp. 213–224.

128. Chatal JF, Charbonnel B: Comparison of iodobenzylguanidine imaging with computed tomography in locating pheochromocytoma. *J Clin Endocrinol Metab* **61:**769, 1985.

129. Swensen SJ, Brown ML, Sheps SG, et al: Use of 131I-MIBG scintigraphy in the evaluation of suspected pheochromocytoma. *Mayo Clin Proc* **60:**299, 1985.

130. Ackery DM, Tippett PA, Condon BR, et al: New approach to the localisation of phaeochromocytoma: Imaging with iodine-131-meta-iodobenzylguanidine. *Br Med J* **288:**1587, 1984.

131. McEwan AJ, Shapiro B, Sisson JC, et al: Radio-iodobenzylguanidine for the scintigraphic location and therapy of adrenergic tumors. *Semin Nucl Med* **15:**132, 1985.

132. Harrison TS, Scaton JF, Cerny JC, et al: Localization of pheochromocytomata by caval catheterization. *Arch Surg* **95:**339, 1967.

133. Jones DH, Allison DJ, Hamilton CA, et al: Selective venous sampling in the diagnosis and localization of phaeochromocytoma. *Clin Endocrinol* **10:**179, 1979.

134. Kinkhabwala MN, Conradi H: Angiography of extra-adrenal pheochromocytomas. *J Urol* **108:**666, 1972.

135. Rossi P, Young IS, Panke WF: Techniques, usefulness, and hazards of arteriography of pheochromocytoma: Review of 99 cases. *JAMA* **205**:547, 1968.
136. Hoffman BB, Delean A, Wood CL, et al: Alpha-adrenergic receptor subtypes: Quantitative assessment by ligand binding. *Life Sci* **24**:1739, 1979.
137. Weiner N: Drugs that inhibit adrenergic nerves and block adrenergic receptors. In Gilman AG, Goodman LS, Gilman A (eds): *Basis of Therapeutics*. Macmillan, New York, 1980, pp. 176–210.
138. Graham RM, Kennedy P, Stephenson W, et al: In vivo evidence for the presynaptic alpha-adrenergic receptor controlling norepinephrine release. *Fed Proc* **37**:308, 1978.
139. Hauptman JB, Modlinger RS, Ertel N: Pheochromocytoma resistant to α-adrenergic blockade. *Arch Intern Med* **143**:2321, 1983.
140. Robinson RG, DeQuattro V, Grushkin CM, et al: Childhood pheochromocytoma: Treatment with alphamethyltyrosine for resistant hypertension. *J Pediatr* **91**:143, 1977.
141. Wallace JM, Gill DP: Prazosin in the diagnosis and treatment of pheochromocytoma. *JAMA* **240**:2752, 1978.
142. Cubeddu LX, Zarate NA, Rosales LB, et al: Prazosin and propranolol in preoperative management of pheochromocytoma. *Clin Pharmacol Ther* **32**:156, 1982.
143. Nicholson JP, Vaughn ED, Pickering TG, et al: Pheochromocytoma and prazosin. *Ann Intern Med* **99**:477, 1983.
144. Rosei EA, Brown JJ, Lever AF: Treatment of pheochromocytoma and of clonidine withdrawal hypertension with labetalol. *Br J Clin Pharmacol* **3**:809, 1976.
145. Nagasaka A, Hara I, Imai Y, et al: Effect of fusaric acid (a dopamine β-hydroxylase inhibitor) on phaeochromocytoma. *Clin Endocrinol* **22**:437, 1985.
146. Sisson JC, Shapiro B, Beierwaltes WH, et al: Radiopharmaceutical treatment of malignant pheochromocytoma. *J Nucl Med* **24**:197, 1984.
147. Dereo GA Jr, Stewart BH, Tarazi RW Jr: Preoperative blood transfusion in the safe surgical management of pheochromocytoma: A review of 46 cases. *J Urol* **111**:715, 1974.

Index